GAS TURBINE ANALYSIS AND PRACTICE

GAS TURBINE ANALYSIS AND PRACTICE

BURGESS H. JENNINGS
Associate Dean
Department of Mechanical Engineering
Northwestern University

WILLARD L. ROGERS
Professor of Aerospace and Mechanical Engineering
The University of Arizona

DOVER PUBLICATIONS, INC.
NEW YORK

Published in Canada by General Publishing Company, Ltd., 30 Lesmill Road, Don Mills, Toronto, Ontario.

Published in the United Kingdom by Constable and Company, Ltd., 10 Orange Street, London WC 2.

This Dover edition, first published in 1969, is an unabridged and revised republication of the work originally published in 1953 by the McGraw-Hill Book Company, Inc. The authors have added a Preface to the Dover Edition, and have updated some of the illustrative material.

Standard Book Number: 486-62022-0
Library of Congress Catalog Card Number: 74-85444

Manufactured in the United States of America
Dover Publications, Inc.
180 Varick Street
New York, N.Y. 10014

PREFACE TO THE DOVER EDITION

The growth of the gas turbine as a major prime mover in the last two decades has been little short of phenomenal. Units of much greater capacity and improved economy are now in daily production for many usages which were hardly envisioned when this book was written. Progress will continue, particularly in the area of materials and for additional applications, but there is evidence that standardization of design approaches and procedures is beginning to take place to the point that few fundamental changes can now be expected with the passage of time. For the Dover Edition an effort has been made to bring some of the illustrative material up to date and to give an indication of the sophisticated level to which the gas-turbine industry has risen.

B. H. Jennings
W. L. Rogers

January, 1969

PREFACE TO THE FIRST EDITION

This text on gas turbines was prepared by the authors to meet the need for a book presenting a broad, but not detailed, coverage of the operating characteristics, industrial scope, and position that the gas turbine now holds and will attain in coming years. When the authors started writing the manuscript some years ago, suitable text material was practically nonexistent. Since that time, several books have appeared on the market. However, the authors hope that this text will simplify the teaching of this important subject and believe some of the methods of presentation to be sufficiently different to merit an additional book in the field.

It is assumed that the readers have some knowledge of basic thermodynamics and mechanics. The material presented in this text deals specifically with the types of problems associated with gas compression and gas-turbine operations. For these applications the required basic considerations are given in some detail. The authors have prepared a set of tables of air properties, which are included for convenience in the text to simplify the work required in solving problems.

Up to the present time, by far the greatest use of the gas turbine has been in the field of aircraft propulsion, where it is supplanting in many instances the reciprocating engine. For this reason, it was felt desirable to give some coverage of the characteristics of the reciprocating internal-combustion engine as well as the gas turbine so that each type of unit could be placed in its proper perspective as a propulsion drive. The broad picture of the gas turbine as a prime mover in its own right has been adequately covered for potential uses in railway transportation and marine applications and for general power purposes.

Thermodynamic and flow-theory approaches to fundamental design and layout objectives are used. Although aerodynamic aspects of blading in compressors and turbines are discussed, they are not emphasized. The literature has been freely consulted, and a comprehensive bibliography of reference material is included in each chapter.

The authors appreciate the fine coöperation they received from a host of manufacturers, who furnished data, illustrative material, and basic information in regard to production and planned units. Credit lines for references and material used appear throughout the book.

This text is designed for a semester course meeting three times a week or for a briefer coverage in a similar class meeting for a quarter.

It is hoped that this book, in addition to meeting the need for a brief but comprehensive text, can furnish practicing engineers with reference data and an adequate background to enable them to understand the gas-turbine field.

B. H. Jennings
W. L. Rogers

Evanston, Ill.
January, 1953

CONTENTS

1. Development of the Internal-combustion Engine and Turbine . . . 1
2. Thermodynamics and Air Properties 22
3. Gas-turbine Cycles and Thermodynamic Analyses 92
4. Gas Turbine Types and Design Features 132
5. Axial-flow Compressors 184
6. Centrifugal and Displacement Compressors 223
7. Aircraft Propulsion and Jet Engines 253
8. Power Plant and Transportation Turbine Units 290
9. Fuels, Combustion and Combustion Chambers 328
10. Stresses in Turbine-plant Elements 381
11. Vibration and Balancing 415
12. Materials and Metallurgy 441

Appendix 477

Index 479

CHAPTER 1

DEVELOPMENT OF THE INTERNAL-COMBUSTION ENGINE AND TURBINE

1-1. General Historical Background. From the latter part of the eighteenth century and continuing into the twentieth century, the industrial world thrived and prospered in what might be called an *age of steam.* During this time the steam engine grew to full stature, reaching a peak with the enormous central-station engines which were in widespread use during the early part of this century. However, the importance of steam engines declined as steam turbines, which were first introduced about 1880, developed into the efficient and versatile units of the present day. The last stronghold of the steam engine was the railway locomotive, but more and more of these familiar "iron horses" are being converted into scrap iron, and their days are apparently numbered.

During the latter half of the nineteenth century, an additional competitor to the steam engine began to appear in various types of reciprocating internal-combustion engines. These early engines employed a combustible mixture of air and gaseous (or volatile-liquid) fuel which was ignited either by direct flame or by some sort of spark ignition. A still later development in this field (1892) was the diesel engine, which did not need a spark for ignition. In the diesel engine, fuel was sprayed into a chamber of air which, as a result of rapid compression, had reached a sufficiently high temperature to cause spontaneous ignition of the fuel.

In the present century the spark-ignition engine has reached a high degree of perfection in current automotive and airplane engines. The diesel engine has likewise been greatly improved and in one field, that of railroad propulsion, has reached a position of outstanding leadership.

Just as the steam turbine followed the reciprocating steam engine at a later date, a similar pattern held for the gas turbine. The gas turbine, as a satisfactory commercial device, did not appear until the internal-combustion engine had reached a very advanced stage of development and usage. The gas turbine and its air compressor are high-rotative-speed machines, and the passages for compressing and expanding the air must follow aerodynamic design principles and meet requirements for high thermodynamic effectiveness as well. The design experience of steam-turbine practice was available to gas-turbine builders. However, steam-turbine designers had not considered the reversed problem of compression, nor had operating practice with steam extended much above 1000 F, whereas gas-turbine temperatures of 1200 to 1500 F were needed. High gas-turbine temperatures required metals capable of operating at elevated temperatures under stressed conditions. Both design features and metallurgical developments have progressed to where the gas turbine is becoming an increasingly significant prime mover in many fields of power.

The first extensive use of gas turbines was in connection with the supercharging of reciprocating internal-combustion engines. The gas turbine is also a type of internal-combustion engine, and to place the turbine in its proper perspective, a brief background of the reciprocating engine will be presented before formally developing the turbine.

1-2. The Reciprocating Engine. What might be called the forerunners of modern reciprocating gasoline engines were the gas engines developed to working form by Lenoir, Otto, and Langen in the period between 1860 and 1880. A Frenchman, Beau de Rochas, in 1864 proposed a cycle which closely resembles the modern cycle. In his proposal, he explained the advantages of compression before ignition and effective use of a large ratio of expansion. Otto, using the ideas of de Rochas, developed a four-stroke cycle which was so outstanding and successful that the name of de Rochas almost lapsed into obscurity and the general four-stroke pattern and its prototype cycle came to be associated with the name of Otto.

From these early engine attempts, the reciprocating gasoline engine has been developed to reach the high degree of perfection which it holds in present-day automotive and aircraft engines. Most modern engines use gasoline as fuel with high-tension spark ignition. They range from single-cylinder to multicylinder units which in some cases have employed 28 cylinders. Automotive units are usually of *in-line* or *V-type* designs, using liquid cooling for the six or eight cylinders which usually constitute the engine. Aircraft engines are most frequently of the radial air-cooled type, although liquid-cooled V-type units have also been employed.

Most spark-ignition engines operate on the four-stroke principle, which is shown diagrammatically in Fig. 1-1. The suction stroke first occurs with the inlet valve open and the piston drawing in a combustible charge of fuel and air. On the second stroke, with both valves closed, the piston compresses the charge. Near the end of the compression stroke, ignition is started from the electric-spark device (appearing between the valves in the figure), and the rapid rise in temperature and pressure of the gases accelerates the piston downward on its power stroke with the valves continuing in the closed position. Near the end of this stroke, the exhaust valve opens, and on the following return stroke the

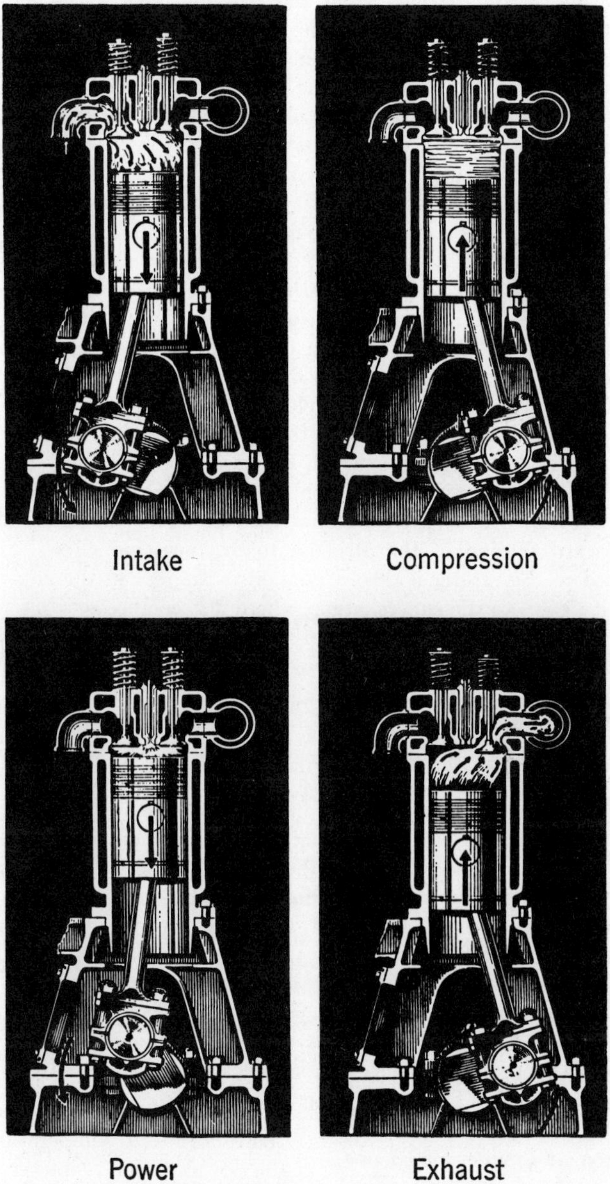

FIG. 1-1. Four-stroke-cycle events shown in connection with a stationary-type engine.

exhaust gases are expelled from the cylinder. Thus the cylinder is ready to receive a fresh charge on the next downstroke. It can be shown how the economy of such an engine is influenced by the *ratio of compression* of the gases (*viz.*, the volume occupied by the gases at the start of compression compared with the volume occupied after compression). In an engine of this type, the combustible mixture may consist of gaseous fuel and air or of liquid fuel delivered by a carburetor into the air stream entering the engine in such quantity that after atomization and subsequent vaporization a combustible mixture exists in the engine cylinder.

1-3. Reciprocating Aircraft Engines. In reciprocating spark-ignition engines a zenith was probably reached in the multirow radial aircraft engine. The engine shared with the gas turbine an important position in the field of aircraft propulsion. In particular for commercial and large transport airplanes at moderate speeds, the reciprocating engine is still employed, but infrequently. One development has been a combination of such an engine with integrated gas turbines for utilizing a portion of the energy in the exhaust gases which otherwise would largely be wasted.

Figure 1-2 is a three-quarter front view of a Wright Cyclone engine, model C18BD. Air enters the engine through a scoop so placed as to use the forward motion of the airplane to produce ram effect. This air is ducted to the master control chamber at the rear of the engine, from which it passes to the supercharger, where it is compressed to a pressure above the atmospheric pressure existing at the flight altitude of the plane. This compressed air then flows through tubing to each of the 18 cylinders. With the intake valve of a particular cylinder open, the piston moves toward the center of the engine, drawing in air while the fuel pump injects gasoline directly into the cylinder. The piston then returns on its compression stroke, and, with both the intake and the exhaust valve closed, the mixture is compressed into the combustion chamber at the top of the cylinder. Near the end of this compression stroke, the charge is ignited by dual spark plugs, and the power stroke then takes place with the intake and exhaust valves remaining closed. On the next stroke, the exhaust valve opens, and the burned gases are pushed into the exhaust pipe.

Observation of Fig. 1-2 will show that the engine has two parallel rows of cylinders. Each of these rows consists of nine cylinders with the firing so arranged that every other cylinder in a row fires in sequence, *viz.*, 1-3-5-7-9-2-4-6-8-1. The nose part of the engine contains planetary reduction gearing which reduces the speed of the propeller hub to less than half engine speed. This is necessary because propeller-tip speeds should be kept below the speed of sound to prevent undue noise, excessive vibration, and reduced efficiency. With lower engine speed, the propeller blades can be made longer, and if, in addition, they are given a deep pitch, a large amount of engine power is utilized per revolution of the propeller.

This particular engine has a cylinder bore of 6.125 in. and a stroke of 6.312 in. which, for the 18 cylinders, gives a piston displacement of 3347 in.^{3} The volume compression ratio is 6.5:1. This engine is provided with a centrifugal-type supercharger of which the 13-in.-diameter impeller is driven either at 6.46 or 8.67 times crankshaft speed. The supercharger compresses the air supplied to the engine. The engine develops 2500 hp at 2800 rpm during take-off, with cruising horsepower 1470 at 2300 rpm. The engine weighs 2884 lb without starter and has an over-all diameter of 55.6 in. and length of 78.5 in.

FIG. 1-2. Eighteen-cylinder reciprocating aircraft engine, Wright Cyclone C18BD. (*Wright Aeronautical Corp.*)

1-4. Diesel Engines. Early internal-combustion engines required either a spark or a flame to ignite the mixture charged into the cylinder. This charge consisted either of a combustible mixture of gas and air or of a volatile-liquid fuel with the air. Toward the end of the nineteenth century, increasing interest appeared in an engine which would compress air and then burn fuel injected into compressed air. One of the first successful engines of this type was the Hornsby-Ackroyd engine. This was a conventionally built engine except for an uncooled bulbous portion built into the cylinder head. Into this bulb, liquid fuel could be sprayed at timed intervals and ignited by the hot metal. The Hornsby-Ackroyd engine was the forerunner of a group of so-called "hot-bulb," "hot-spot," or "semidiesel" engines.

Several years after the development of hot-spot engines, Rudolph Diesel in 1892 obtained a patent on an engine which could compress the air by such an amount that the resulting temperature rise of the air would be sufficiently high to ignite any fuel which might be injected into the air. Diesel's first engine envisaged the use of pulverized coal injected into the compressed air, but this idea was never successful, and the greater part of Diesel's development was concerned with the use of a fuel oil sprayed into the hot compressed air. Progress on the diesel engine was not very rapid until after the original Diesel patents expired in 1913. However, before that time this type of engine had clearly demonstrated itself as an efficient and effective prime mover.

To inject the fuel oil into the cylinder of the engine, the early diesel engines used a separate high-pressure air compressor operating at a pressure higher than that which existed in the engine cylinder. Later and present-day engines have largely given up this so-called "air" injection and use suitable mechanically operated force-feed pumps which inject finely atomized fuel oil at high pressures into the engine cylinder or into a precombustion chamber.

Figure 1-1 shows the essential arrangement of a four-stroke-cycle engine. The same outline applies either to a diesel engine or to a spark-ignition engine. For the diesel engine, no spark is required, and a fuel-injection or spray valve appears instead of a spark device near the center of the cylinder head. In the diesel engine, the charge of air in the cylinder must be compressed through a much greater range than in the spark-ignition engine. In spark-ignition engines, volume-compression ratios from 6:1 to 8:1 are employed, whereas the diesel engine, in order to attain high temperatures from compression alone, requires ratios ranging from 14:1 up to 20:1. For example, if ordinary air is compressed through a volume ratio of 16:1, the resulting temperature of the air rises to about 1100 F. This air is sufficiently hot so that when atomized fuel oil is injected into it combustion starts and no additional form of ignition is required. Compression ratio is defined as the volume occupied by the gas at the beginning of the compression stroke divided by the volume occupied by this gas at the end of the compression stroke.

Many engines work on the four-stroke cycle as previously described, but a two-stroke cycle, particularly in diesel engines, has also been found to be effective. As the two-stroke cycle uses only two passages of the piston to complete a working cycle, it can sometimes result in a lighter weight unit for a given horsepower output. In such an engine, exhaust and charging take place while the piston is moving near the bottom of its stroke and the crank is turning through 60° or less. Most engines of this type employ a separate scavenging air compressor which not only can supply the air required but can give it a slight supercharge as well, thereby increasing engine capacity. A greater weight of air delivered into the cylinder makes it possible to burn more fuel per stroke and develop more power.

Supercharging, or the supplying of air under some pressure at the beginning of the compression stroke, is suitable not only with two-stroke-cycle engines but also with four-stroke-cycle engines, and Fig. 1-3 shows a supercharger mounted on a multicylinder diesel engine. This particular supercharger utilizes energy in the exhaust gases from the engine to drive a small turbine which in turn operates the centrifugal compressor of the supercharger.

The valve timing, or the period in a given stroke at which a respective inlet or exhaust valve opens, varies considerably from ideal conditions, as the movement of the gases into and from the cylinder requires a finite time interval and in a high-speed engine the time interval is necessarily short. In a four-stroke-cycle engine the exhaust valve usually opens when the power stroke is approximately 87 per cent completed. In a two-stroke-cycle engine the exhaust ports are uncovered even earlier when about 70 per cent of the stroke has been completed. Thus, there is usually slightly less work developed in the two-stroke-cycle engine per power stroke than in a four-stroke-cycle engine. Because there are two less strokes per cycle, there is less rubbing friction on the cylinder walls. However, an additional demand on the two-stroke cycle engine is the power required to drive the scavenging air compressor which recharges the cylinders of the engine. This takes an appreciable percentage of the gross power developed by the engine, in the neighborhood of 8 to 12 per cent.

Diesel engines are extremely efficient in converting the thermal energy of fuel into useful shaft power, probably the most efficient of all commercial prime movers. Efficiencies in excess of 35 per cent are obtained with many low-speed diesel engines. Although diesel engines use a refined fuel which is not inexpensive, they are competitive to most other types of prime movers. In particular, the American railroads are rapidly adopting them almost to the exclusion of steam for new locomotives. For many industrial jobs, small isolated power plants, pumping lines, and marine propulsion, the diesel engine is also important. It is built in commercial sizes, which range from less than 5 to more than 10,000 hp per unit. The four-stroke-cycle diesel engine of Fig. 1-3 is employed for power generation. Because of the higher capacity per unit weight that can be built into spark-ignition and gas-turbine units, the diesel engine has not been used to any extent for aircraft propulsion.

In addition to the conventional diesel-oil engines, a large number of dual-fuel engines are in operation. These engines are arranged to operate on natural (or process) gas and fuel oil or on fuel oil alone, depending upon which fuel is most available and economical. As natural gas is a cheaper fuel it is the most common basic fuel for these engines. In operation, the engine draws in a charge of natural gas and air which is compressed in a conventional manner in the cylinder. In the absence of spark ignition, it has been found that by injecting a small amount of diesel fuel, ignition is readily maintained and the combination of diesel fuel and gas in the same cylinder burns very satisfactorily. In accordance with load conditions and under governor action, the ratio of the amount of gas and diesel fuel can be varied to suit desired ends. The fuel oil acts merely as a pilot fuel. Many engines of this type are in use, particularly in the oil-field regions.

FIG. 1-3. Eight-cylinder 1370-hp supercharged diesel engine and generator. *(Nordberg Mfg. Co.)*

1-5. The Turbosupercharger. One of the earliest usages of a gas turbine was in connection with the exhaust-gas turbine-driven supercharger. The exhaust gases from a typical reciprocating internal-combustion engine are at high temperature, 600 to 1300 F, and if a back pressure above atmospheric is permissible on the reciprocating engine, it is possible then to expand these high-temperature moderate-pressure gases through a turbine and develop useful work. In an airplane engine, a useful way to employ these gases is for driving a supercharger. A supercharger operates so as to take the air at any altitude and compress it to a sufficiently high pressure so that the weight of air supplied to the engine cylinders does not seriously diminish at design altitudes. Thus a serious reduction in engine power is avoided. At the expense of a moderate back-pressure build-up, appreciable supercharge power can be obtained.

Figure 1-4 is a diagrammatic arrangement of a turbosupercharged power plant for an aircraft engine designed for operation at high altitudes. The figure shows the exhaust line from the engine leading to the nozzle box of the turbosupercharger. The supercharger has a single turbine wheel directly connected to the compressor impeller. Ram air slightly compressed by the forward progress of the airplane enters the impeller, where the pressure of this air is increased by some 10 to 20 in. Hg. Typical operating data on this unit would show that at 25,000 ft altitude gases would enter the turbine nozzles at some 1500 F and at a pressure of 25 in. Hg and expand to the 10 in. Hg pressure of the atmosphere after passing through the nozzles and turbine. The air enters the compressor at

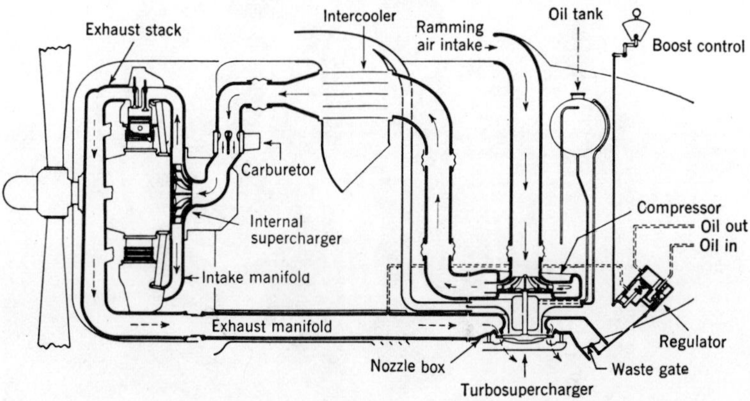

FIG. 1-4. Schematic diagram of turbosupercharged-aircraft power plant. (*General Electric Co.*)

about −30 F at a pressure of some 12 in. Hg; this higher pressure results from the ram effect of the forward-moving airplane. The gases in the compressor, as a result of compression, rise in temperature to about 150 F.

During sea-level operation, the hot gases from the engine are delivered at approximately 38 in. Hg and expand to 30 in. in passing through the turbine. The turbosupercharger, at starting, may raise the delivered air pressure to 35 in. Hg, giving extra power for take-off. Turbosupercharger units of this type rotate at speeds up to 25,000 rpm. In some cases, the turbosupercharger alone might be sufficient, but in many cases a second supercharger driven directly by the engine through gearing is supplied. This is illustrated in Fig. 1-4.

The benefits of supercharging are not limited to airplane engines alone, as it holds for any internal-combustion engine that the capacity of the engine is directly associated with the amount of oxygen per unit of time that can be made available in the engine cylinder or combustion chamber for combination with fuel. Consequently, it is possible to increase the capacity (power) of a diesel engine also through supercharging by providing a greater weight of air per unit volume in the engine cylinder. Figure 1-5 shows the details of an Elliott-Buchi turbosupercharger designed for use on a diesel engine. Here the hot gases from the diesel engine at a moderate back pressure (2 to 8 in. Hg gage) expand through

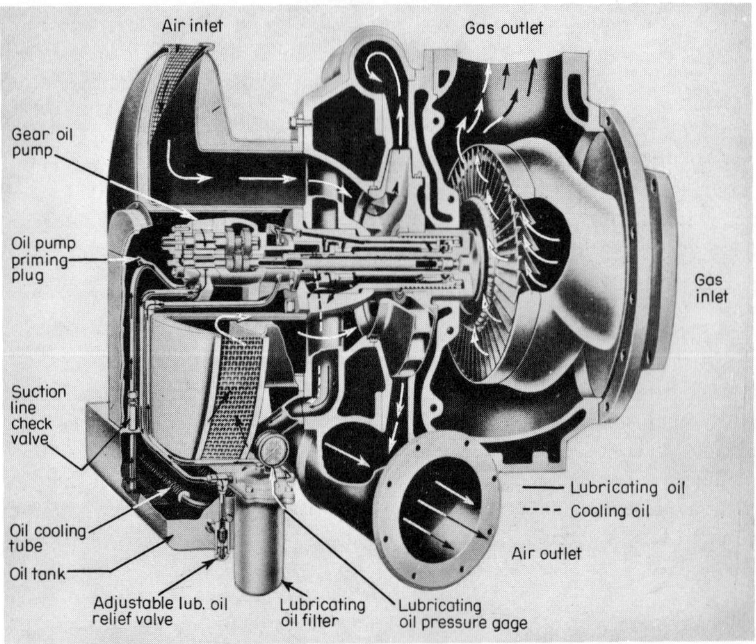

FIG. 1-5. Elliott-Buchi turbocharger. (*Elliott Co.*)

the nozzles and blading of a single-row turbine, giving up some energy as useful work before passing into the gas outlet. The power developed by the turbine wheel is used to drive a single-stage centrifugal-type compressor which takes atmospheric air and compresses it to a higher pressure before delivering it into the passage which supplies the various cylinders of the diesel engine. The exhaust gases are so hot (up to 1000 F) that it is desirable to interpose water-cooled shielding between the turbine exhaust section and the compressor section of the unit. The normal pressure ratio during compression is purposely not made high on these units, but runs about 1.35:1. Typical units of one manufacturer (the Elliott Company) are listed as follows to give some idea of sizes and performance.

Impeller and turbine wheel diameters, in.	Rpm	Capacity, cfm
10¼	21,000	2000
13¼	16,000	3900
17	12,500	6100

Thus, by utilization of some of the energy in the exhaust, it is possible to increase the power output of a diesel engine and also its economy without the expense of additional fuel. Although this, per se, is not a complete gas-turbine power plant, it nevertheless represents a basic element of such a plant, *viz.*, a turbine supplied with high-temperature gases under pressure for development of power. In this case the power is used to produce a supply of compressed air. The complete power-plant cycle operates in similar manner, with the compressed air supply being used in the cycle and with surplus energy delivered by the turbine being made available for other useful power outlets.

Figure 1-3 is a photograph of a diesel-engine installation, and in the upper right-hand corner of the engine can be seen an exhaust-gas supercharger of the type just described.

1-6. The Gas-turbine Plant. The idea of the gas turbine is not new, but its growth and development in recent years have been brought about most significantly because of two factors: (1) metallurgical advances have made it possible to employ high temperatures in the operating parts of the turbine and (2) the accumulated background of aerodynamic and thermodynamic knowledge has made it possible to design compressors and turbines of high efficiency. John Parsons of England probably crystallized present-day thinking on what a gas-turbine system might be in a patent application made in 1884. In his patent application, he brought forth the basic idea of a compressor, the burning of fuel in the compressed gas, and the utilization of this hot gas in a turbine for the purpose of generating power. In this country, Charles G. Curtis obtained a United States patent on a gas turbine in 1895, and in 1904 the General Electric Company operated gas turbines at Lynn, Mass., and Schenectady, N.Y. In 1907, there is a record of a gas turbine having been in successful operation in Paris. Most of these early designs were not particularly successful largely because of low compressor efficiencies and metallurgical limitations.

It is well known that a machine designed along the principles of a reaction-type turbine can be used as a compressor. This machine, when operated as a turbine, has an efficiency approaching 85 per cent; the same type of machine, as a compressor, would have an efficiency in the range of 60 to 65 per cent or less. There are two essential factors which account for this condition. In a multistage turbine, the reheat effect resulting from losses in the early stages aids performance in the later stages, whereas in the compressor this effect is reversed. The other important fundamental characteristic is that in a compressor the process of diffusion of a moving stream is much more difficult of accomplishment than is the corresponding process of accelerating a moving stream such as occurs in a turbine. Expressed in another way, flow in the turbine is in the direction of decreasing pressure, whereas in the compressor the flow occurs in the direction of pressure increase. This characteristic means that extreme

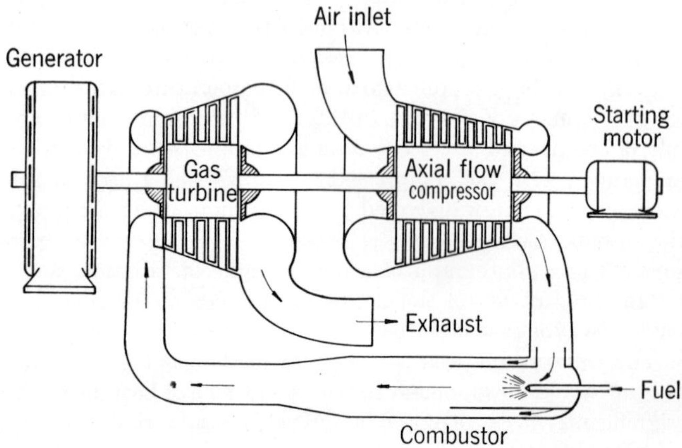

FIG. 1-6. Gas-turbine elements of open cycle.

care must be exercised in regard to setting up rates of diffusion and changing or modifying the direction of flow. Advanced design of air-foil sections for compressor blades in the axial-flow type of compressor and careful adherence to proper rates of diffusion have made it possible, in recent years, to develop compressors which essentially reach the efficiency of the counterpart turbines. In the above discussion the term "diffusion" applies to the process in which the energy of a moving stream of fluid is transformed in such manner that an increase in pressure occurs. The name "diffuser" in common engineering parlance (for subsonic flow) means a passage of increasing cross section in the direction of flow.

Figure 1-6 is a diagrammatic layout of a simple gas-turbine power system of the open type. Air enters the compressor, and its pressure is raised sufficiently so that after heating it can expand in the turbine of the system and produce power. In the combustor, the air is heated by burning fuel in the air stream. The resulting high-temperature air and combustion gases in expanding through the gas turbine drive the compressor and, in addition, produce useful power in a device such as the generator shown. To operate such a system, a starting motor is required to bring the turbine and compressor to one-fourth to one-third of operating speed, following which, under the continued application of fuel, the gas turbine finally is able to bring the system up to full operating speed, drive the compressor, and deliver surplus power as well.

For the gas turbine of the simple system illustrated, and perhaps developing 6000 hp, about 4000 hp would be required by the compressor, leaving 2000 hp available for useful power to the generator. The temperature of the gases supplied to the turbine must be sufficiently high to permit the power developed by the gas turbine to exceed that required by the compressor. A system as pictured here is known as an *open cycle* because the same air which passes through the compressor is used in the combustion process and the air and combustion gases together expand and produce work in the turbine, following which the gases still at high temperature are exhausted to the atmosphere. One advantage of the open system is that the atmosphere acts as the thermodynamic sink of the system and no cooling water is required. The working medium, air, is also available in unlimited quantities.

The temperature of the air and gases leaving the combustor is a critical factor as the efficiency rapidly increases as the temperature of the turbine inlet gas is raised. However, metallurgical considerations limit the maximum temperature that can be used in the rotating turbine parts, and present-day gas-turbine designs call for inlet temperatures ranging from about 1000 F to an upper limit probably not exceeding 1500 F for continuous operation. It will later be shown how the pressure ratio through which the air is compressed also affects the performance of a simple system and must be related to the inlet temperature. The type of compressor used may be axial-flow, centrifugal, or positive-displacement. Compressors will be considered in detail later in the book.

In the case of a jet-operated airplane, a change in the basic cycle is employed. The air inlet is placed in the front of the airplane so that the ramming effect of the air resulting from the forward motion of the airplane helps drive air into the compressor. No generator or other load is attached, the object of the turbine being merely that of supplying sufficient power to drive the compressor. The exhaust from the turbine is led into an expansion cone or outlet duct and leaves the airplane with a high velocity. The useful thrust on the airplane is caused by the change in momentum created because the jet leaves the airplane at a higher velocity than that at which it enters.

1-7. Locomotive Gas-turbine Unit. One of the most promising outlets of the gas turbine is as a possible drive for locomotives, and one arrangement is illustrated in Figs. 1-7 and 1-8, which show a diagrammatic layout and artist's pictorial representation of such a unit. These figures show the air inlet to the first of two centrifugal compressors with the air leaving the first impeller, passing through the diffuser, and entering the second impeller. The compressed air at a pressure of approximately 52 psia then enters the regenerator, where it absorbs some heat from the turbine exhaust gases as it passes through the tubes of the regenerator on its way to the combustor.

In the single combustion chamber, fuel oil is burned in sufficient quantity to raise the temperature of the compressed gases to 1275 F at turbine inlet. The hot gases expand through the four stages of the turbine, dropping in pressure and temperature and delivering work to the turbine-compressor shaft. The exhaust gases leave the turbine at approximately 950 F and, after passing through the regenerator, are exhausted above the locomotive.

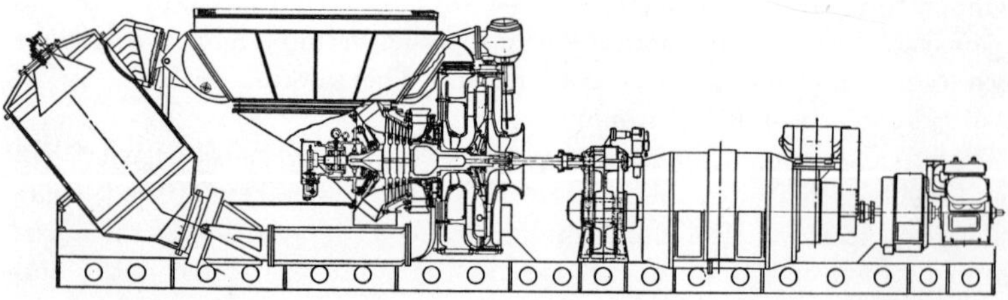

FIG. 1-7. Diagrammatic sectional view of Elliott gas-turbine power plant for locomotive.

This particular unit is designed for an air flow of 67 lb per sec and delivers 3910 hp to the reduction gearing which drives the locomotive generators. The design thermal efficiency of this unit is just under 25 per cent, and the turbine shaft turns at 5910 rpm. The regenerator is purely an economy device and does not increase the load capacity of the plant; in fact, because of pressure drops through it, the power production is decreased. However, because the regenerator can recover up to some 4000 Btu per delivered hp-hr, considerable fuel savings can be expected, and unless fuel is extremely cheap, a regenerator may be a desirable addition. This is particularly true for a unit with a relatively low pressure-compression ratio (4 or under).

It can readily be seen that a unit following the same design could also be utilized for a stationary power plant. As the turbine in either application drives a generator, constant-speed operation would be expected, and under this circumstance control of the unit would be accomplished by variation of the fuel input and slightly varying the air flow in the system. It is conceivable that such a unit could also be operated on powdered coal, provided effective ash-removal facilities were available for eliminating the very fine residual particles resulting from combustion.

Figure 1-9 shows the 4800-hp gas-turbine locomotive unit developed jointly by the American Locomotive Company and the General Electric Company. This is a compact direct-flow unit with the air entering at the left as shown, passing through a 15-stage axial-flow compressor into the six combustors, which are placed circumferentially around the unit. The hot gas leaving the combustors enters the nozzles and is expanded in a

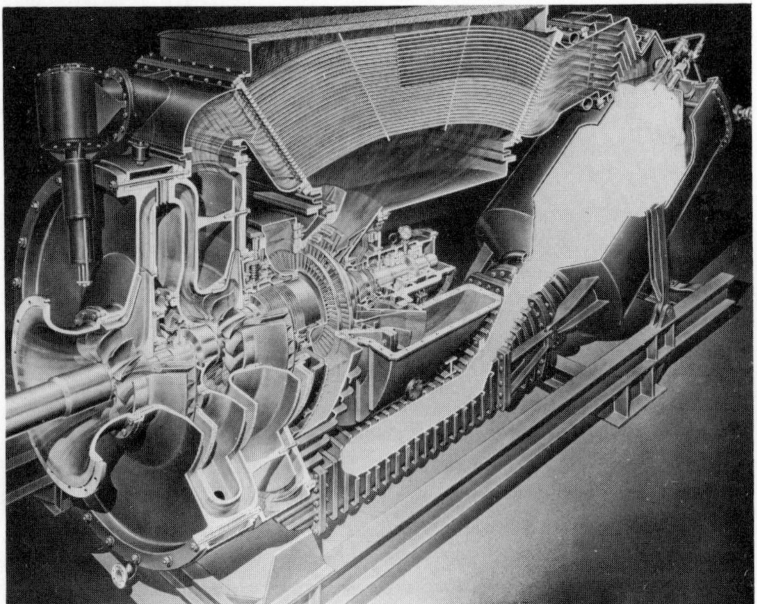

FIG. 1-8. Operational view of Elliott gas-turbine locomotive unit. (*Elliott Co.*)

two-stage turbine unit. The gases then pass into a diffusing section, from which they are directed upward to waste outside of the locomotive. The driven generator connects at the compressor end of the unit.

In over-all length, this gas-turbine unit is 19 ft, with a width of 6.75 ft. Inlet temperature to the turbine is 1400 F, and the pressure leaving the compressor is approximately 75 psig. The turbine turns at 6700 rpm and is designed to develop its rated power at an altitude of 1500 ft. Bunker C fuel is used for running operation, with diesel fuel employed during starting. The air intake rate is 80,000 cfm at 1500 ft altitude, which represents 94 pounds per second. The locomotive is designed to develop a continuous tractive effort of 68,500 lb at 20.5 mph and can develop a maximum speed of 79 mph. It will be noticed that this unit employs one of the simplest possible cycles without using a regenerator or other heat-recovery apparatus. With the moderately high pressure ratio running between 5 and 6, the use of a regenerator is not so necessary. Thermal efficiency is somewhat in excess of 17 per cent, based on lower heating value of the fuel. The compact construction, use of cheap fuel, and light weight of a unit of this type (25,000 lb), all make the unit attractive for locomotive work. This particular unit has been road-tested by several railroads and has given an extremely creditable performance.

It is also possible that such a unit, with little or no change, could be employed in a stationary power plant. However, in a stationary plant it might be desirable to modify the operating cycle to develop greater economy.

FIG. 1-9. 4800-hp gas-turbine locomotive power unit. (*General Electric Co.*)

Much thought is being given to the relative position which the gas turbine might eventually assume in the field of railway transportation. At present, diesel locomotives have established such an outstanding degree of service effectiveness, high economy, and relatively low maintenance cost that most comparisons should be made relative to such units. From a noise viewpoint, the gas turbine presents a problem as under certain conditions the entering air flow and the exhaust from the gas turbine can produce objectionable howling noises. On the other hand, diesel noise has been reduced to an unobjectionable minimum. Although the gas-turbine unit itself is lighter than corresponding diesel equipment, nevertheless the generators, driving motors, and other control equipment of locomotives are little different for either type and the weight factor is not as significant as it first might appear. In addition to this, greater weight on the locomotive drivers helps to provide greater tractive effort.

Both the gas turbine and the diesel engine are clean and do not produce objectionable smoke conditions when operating. The diesel engine has a high lubricating-oil consumption (which is practically eliminated in the case of the gas turbine), and it uses a premium-grade liquid fuel (*i.e.*, highly refined diesel fuel oil). On the other hand, the gas-turbine unit, although frequently started on diesel fuel, can operate on the cheaper residual fuel oils, and at some future time successful operation with pulverized coal is also a possibility.

At this early stage of development, it appears that the diesel locomotive is well entrenched and will continue to hold its position as a

FIG. 1-10. Huey Station 3500-kw gas-turbine power unit. (*General Electric Co.*)

leading motive power of the railroads, with the possibility that the gas-turbine locomotive also will be used in this field and probably in the future will offer significant competition.

1-8. The Gas-turbine Stationary Plant. Perhaps the first specific gas-turbine-unit installation in a central station in the United States was made at the Huey Station of the Oklahoma Gas and Electric Company at Oklahoma City. This gas-turbine unit, which supplements the steam-power units in the station, was put on the line in July, 1949. It was designed by the General Electric Company as a 3500-kw simple open-cycle gas-turbine power plant. Figure 1-10 shows the unit as installed in the station.

This particular unit very closely resembles in its construction the locomotive-gas-turbine unit illustrated in Fig. 1-9. The weight of the complete power plant, including compressor and gas turbine, is 25,000 lb. The complete unit, including the generator and exciter, is less than 50 ft long and 9 ft wide. The gas-turbine exhaust is used in a heat exchanger (recuperator) to supplement heating of the boiler feed water for the main steam plant. This permits additional steam-turbine shaft capacity for the other units in the plant as less bleeding or extraction steam for feed-water heating is required. In fact, some 344,000 lb per hr of exhaust gas from the gas turbine drop in temperature from 780 to 324 F while warming 350,000 lb per hr of feed water from 185 to 295 F.

The gas turbine and its attached alternating current generator turn at 3600 rpm. A 250-hp induction motor is used for starting the gas turbine unit. This can bring the speed up to one-half of normal speed for firing or turn at one-eighth normal speed when purging the unit of gas.

The gas turbine was shipped completely assembled and set up in the plant without requiring field work. An interesting feature of this unit is the fact that it uses natural gas as a fuel. This gas is available in the pipe lines at 150 psig and, consequently, does not require compression. However, in the event that the gas pressure should drop, a rotary gas booster has been made available which can compress 100,000 cfm from 100 to 170 psig while turning at 690 rpm and while powered with a 75-hp motor.

The capacity of a gas turbine decreases as the temperature of the inlet air rises, and in order to maintain unit capacity in hot summer weather, well water can be used for precooling the air entering the compressor of the turbine. This is particularly desirable when the air temperatures approach and exceed 100 F. The well water available at the Huey Station lies in the temperature range of 60 to 65 F. With cool air supply this unit has delivered as much as 5000 kw.

A 15,000-kw gas-turbine generator unit is now under construction by the Westinghouse Electric Corporation to be used in the Bartelsville, Okla., area by the Public Service Company of Oklahoma. This unit, which also will employ natural gas as fuel, will operate on a high-economy complex cycle using high- and low-pressure turbines. These will drive, respectively, high- and low-pressure compressors with the generated power being taken from the high-pressure turbine shaft. Intercoolers will reduce the work of compression by lowering the temperature of the compressed air between compressor units. Gases are expected to enter the turbine at 1350 F, and a regenerator will be used to recover heat from the exhaust gases. The design speed of the turbine is 3600 rpm, and at full load approximately 7½ tons of air per min will flow through the compressors.

1-9. Jet Engines. Because of its historical interest Fig. 1-11 has been included to show a centrifugal-compressor continuous-flow jet-propulsion unit. It employs a single-stage double-entry centrifugal compressor which delivers air under pressure to nine straight-flow combustion chambers placed around the periphery of the unit. Fuel, which is sprayed into the compressed air, burns in the combustion chambers, and the high-temperature products of combustion leave the combustion chambers and enter a nozzle box. In the attached nozzles, as the

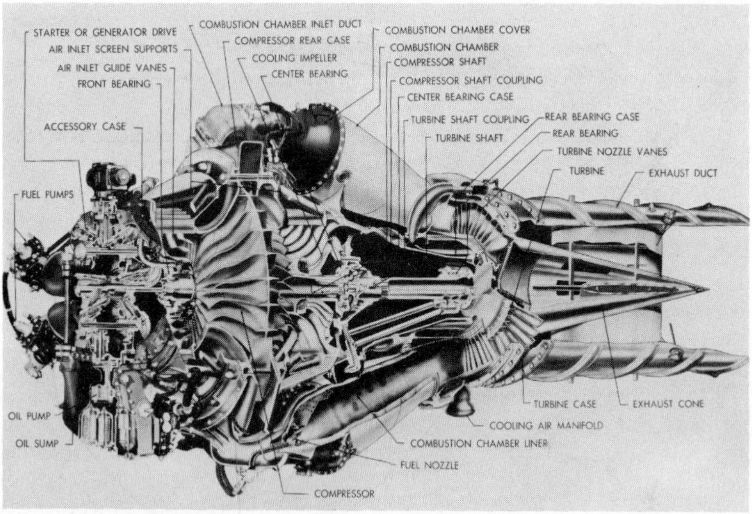

FIG. 1-11. A gas-turbine jet engine employing a double-inlet centrifugal compressor, model JT-6B. (*Pratt and Whitney Division of United Aircraft.*)

gases drop in pressure, thermal energy is changed into kinetic form, which is utilized in the single-stage turbine to produce the power required to drive the compressor. The exhaust gases from the turbine, still at high velocity, flow into the exhaust section and leave the tail pipe as a jet. In a unit of this type, the high-velocity jet is the end desired, as it is the change in momentum of the air between inlet and outlet of the unit which produces the engine thrust.

The turbine itself performs only the function of developing sufficient power to drive the compressor at the desired rate of speed. For the 5000-lb thrust developed under take-off conditions, the unit is extremely light, weighing 1723 lb. The thrust at normal cruising speed amounts to 2700 lb. The fuel consumption in pounds per hour for each pound of thrust is 1.11 lb under cruising conditions. The engine length, without

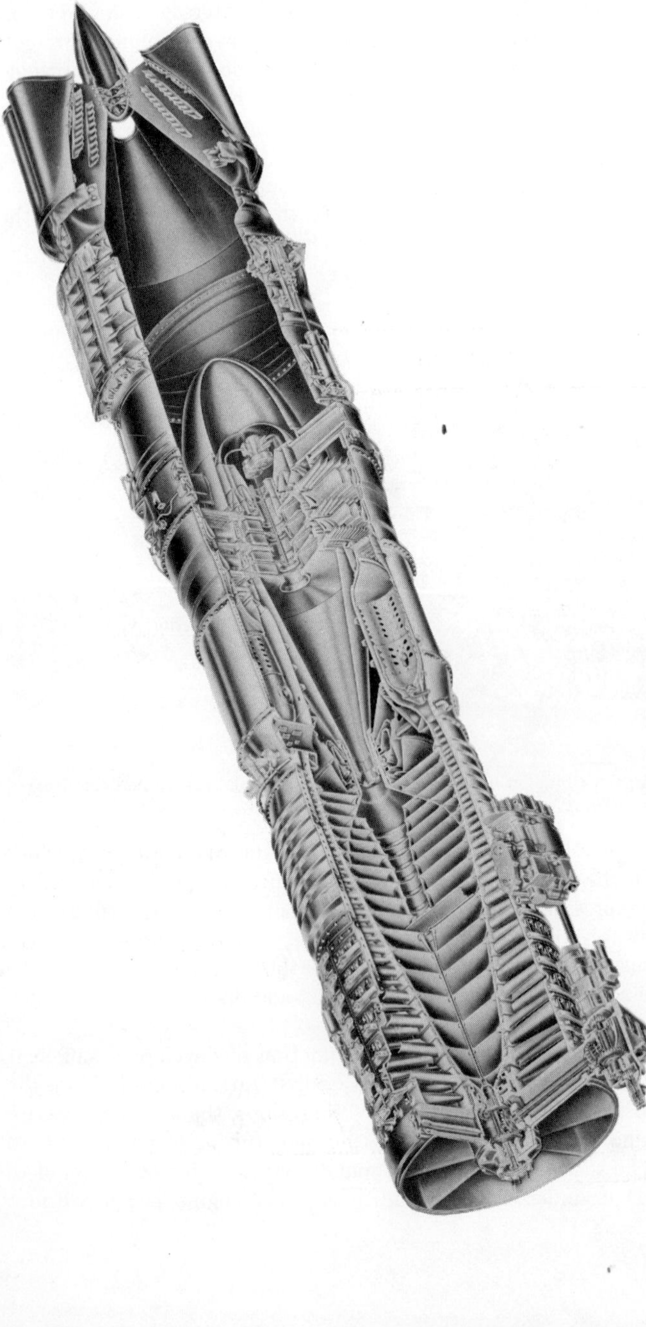

FIG. 1-12. A CJ-805-3 turbojet engine with axial-flow compressor and three-stage turbine. (*General Electric Co.*)

the extension pipe and jet nozzle, is 103 in., and the engine diameter is just under 50 in. The unit turns under normal operating conditions at between 10,000 and 12,300 rpm, with the air flow through the unit ranging from some 60 to 110 lb of air per sec. This engine represents one of the types of jet units which have become extremely important in the field of high-speed flight, where jets have proved themselves superior to propeller-driven airplanes.

It will be noticed that the compressor of this engine is a double-entry centrifugal unit. The standard exhaust-nozzle diameter of this unit is 18¾ in., and it is important that a suitable length-diameter ratio be maintained to give the most effective jet. Either a gasoline- or a kerosene-type fuel can be employed with an engine of this type.

In contrast to the Pratt and Whitney unit of Fig. 1-11 which used a centrifugal compressor, jet engines are now made using axial-flow compressors. Axial-flow compressors do not require as much frontal area as centrifugal compressors, and because they are inherently more efficient have been universally adopted. Fig. 1-12 is an illustration of a General Electric turbojet engine in airline service. It employs a 17-stage axial-flow compressor, which can produce a pressure ratio of 13. The air inlet is shown at the left of the unit. The air from the compressor then passes into ten can-type combustors before entering the first row of nozzles serving the three-stage turbine. Leaving the last turbine-blade row, the still hot gases at about 1100 F further expand in the tail cone and leave at high velocity to produce the required thrust. The conical construction throughout reduces engine weight and the single rotor compressor and turbine require only three main bearings. The length of the unit, including a thrust reverser at the end, is 182 in., and the maximum diameter at the compressor inlet is 31.6 in. The airflow through the compressor-turbine is 171 lb per sec and liquid aviation kerosene is used as fuel.

The unit produces a static thrust on take-off of 11,650 lb and a continuous thrust of 9800 lb can be maintained. At high cruising altitude, 35,000 ft, the thrust reduces to 2190 lb which is sufficient, with the usual complement of four engines, to propel an airplane at 0.8M, 80 per cent of the speed of sound. The unit weighs 2817 lb. Its fuel consumption ranges from 0.81 lb fuel per lb of thrust at take-off to 0.90 lb per lb at 35,000 ft, under the reduced thrust at high altitude. It runs at varying speeds up to about 7500 rpm.

This design of unit has also been modified to provide it with a turbofan. The turbofan or aft-fan arrangement for this unit provides an extra-diameter fan annulus at the rear end of the unit. The blades turning in this annulus accelerate air rearward to supplement the air passing through the basic unit. This aft-fan arrangement is illustrated in Fig. 7-16.

CHAPTER 2

THERMODYNAMICS AND AIR PROPERTIES

2-1. Steady-flow Energy Equation. As the gas turbine and its compressor are continuous-flow machines, for purposes of analysis, the steady-flow equation form of the first law of thermodynamics is extremely useful. In its usual form this equation appears

$$G({}_1q_2) = G\left(h_2 - h_1 + \frac{C_2^2 - C_1^2}{2g_cJ} + \frac{{}_1W_2}{J} + \frac{Z_2 - Z_1}{J}\frac{g_L}{32.17}\right) \quad (2\text{-}1)$$

where G = pound weight (mass)* of fluid under consideration, usually that flowing per unit time, lb per sec

${}_1q_2$ = heat transferred to or from the system, Btu per lb of fluid. When heat is added to the system, the sign of ${}_1q_2$ is +

h_2 = enthalpy of the fluid leaving, Btu per lb

h_1 = enthalpy of the fluid entering, Btu per lb

$C_2^2/2g_cJ$ = kinetic energy of the fluid leaving, Btu per lb

$C_1^2/2g_cJ$ = corresponding kinetic energy of the fluid entering

g_c = 32.17, the dimensionless standard gravitational constant which relates unit mass to unit force.* The value 32.17 is the ratio of

$$\frac{\text{(mass measured in pounds)}}{\text{(mass measured in slugs)}}$$

C_2, C_1 = velocity, fps

J = mechanical equivalent in engineering units, 778.2 ft-lb$_f$ per Btu

${}_1W_2/J$ = work, Btu per lb. When the sign is +, work is delivered by the system

$\left(\frac{Z_2 - Z_1}{J}\right)\left(\frac{g_L}{32.17}\right)$ = change in potential energy, Btu per lb, resulting from a change in elevation of $(Z_2 - Z_1)$ ft. (The term g_L is the local acceleration of gravity, in ft per sec^2.) When the fluid flowing is a gas or vapor the numerical magnitude of this term is so small, relative to the other values in the equation, that it is customary to disregard it

* *Weighing* a quantity of matter by using a balance (not a spring scale) is really a process of comparing the mass of the matter with that of a standard mass. The

Omitting the potential-energy change, Eq. (2-1) becomes

$$G({}_1q_2) = G\left(h_2 - h_1 + \frac{C_2^2 - C_1^2}{2g_cJ} + \frac{{}_1W_2}{J}\right) \tag{2-2}$$

It is sometimes convenient to remember the signs in Eq. (2-2) by the following expression:

Heat added *to* equals *increase in* (enthalpy and kinetic energy) plus work done *by*. By following this system and using a proper interpretation of resulting signs, a quick energy analysis of a *steady-flow* process can be made.

When G is flow in pounds per second, it will be noticed that the equation has the form of a power equation, since pounds per second times Btu per pound becomes Btu per second, which is a *rate* of energy interchange, or power. Recall that 2544 Btu per hr (or 0.7068 Btu per sec) equals 1 hp; and 3413 Btu per hr (or 0.948 Btu per sec) equals 1 kw.

It will be noticed that the heat-flow term, ${}_1q_2$, and the work term, ${}_1W_2$, both represent energy in transition and are not thermodynamic properties as the amount of energy in transition depends, of course, directly on the path and conditions followed during the process. On the other hand, the other energy terms in this equation, such as the enthalpy, are thermodynamic properties or point functions. The magnitude of a property depends solely on the state in which the body exists. It does not depend in any way on the type of path or process used in arriving at that state. It should be recalled that enthalpy h is a composite thermodynamic property defined as

$$h = u + \frac{Pv}{J}$$

number obtained is commonly called the *weight* of the body, although it is actually the *mass* of the body. Both terms—weight and mass—are widely used as synonyms for referring to the mass of a body, and the same practice will be followed in this text. The pound (lb) is the unit of mass in the English system with the slug a unit of mass 32.17 times larger than the pound mass. Whenever abbreviations are needed and a distinction is necessary between a mass pound and a force pound, the abbreviation lb_f will be used for the force pound.

It is necessary to be specific concerning the dimensions of force and mass. Throughout this book, the basic dimensional system employing mass M, length L, and time θ will be used. Force f is defined by Newton's law as $f = MA = ML\theta^{-2}$. Thus the dimensions of force are mass times length divided by time squared. Taking the unit of mass as the slug, the unit of length as the foot, and the unit of time as the second, we arrive at the definition of unit pound force. Thus 1 lb of force is that force magnitude which if acting on 1 slug of mass could produce an acceleration of 1 ft per sec². The *units* of force in pounds are *slug-ft per sec²*.

where u = internal energy, Btu per lb
v = specific volume, cu ft per lb
P = pressure, psfa

It is not important, in most cases, to break enthalpy into its constituents.

2-2. Thermodynamic Relationships for Gases. When the mathematical expressions relating pressure, temperature, volume, and energy values for a medium are complex, it is desirable to use tables of properties. Air, for the working range used in gas turbines, exists entirely in the gaseous phase and approximately follows the simple relationships which hold for the perfect gas and for real gases at very low pressures (practically at less than atmospheric pressure). Because of the simplicity, convenience, and approximate accuracy of these expressions, it seems desirable to review the more significant gas relationships here.

A perfect gas can be defined as one which satisfies the relation

$$PV = wRT \tag{2-3}$$

where P = pressure, psfa (lb per ft^2 abs)
= $144p$, where p is in psia
V = volume, ft^3, occupied by
w = pound weight (mass) of gas
R = a characteristic constant for any gas, ft-lb$_f$ per lb °R. This can be found by dividing the universal gas constant, $R_m = 1545.3$, by the molecular weight m of the particular gas (see Table 2-1)
T = temperature, °F abs, also called degrees Rankine (°R)

$$°\text{R} = 459.7 + t°\text{F}$$

or

$$°\text{R} = 460 + t \qquad \text{approx}$$

Specific heat (per unit weight of a given gas) is defined as the ratio of the heat added (or rejected) to the temperature change

$$c = \frac{dq}{dt} \tag{2-4}$$

The specific heat is not an independent characteristic but depends on the process under which the heat is added. Two common processes are those at constant volume, for which the symbol c_v is used, and at constant pressure, with symbol c_p.

Mathematically

$$c_v = \left(\frac{dq}{dt}\right)_v \tag{2-5}$$

$$c_p = \left(\frac{dq}{dt}\right)_p \tag{2-6}$$

$$c_p = \left(\frac{dh}{dt}\right)_p \tag{2-7}$$

The latter relation states that, with pressure constant, the rate of change of enthalpy with respect to temperature is the specific heat at constant pressure. Each of the preceding specific-heat expressions defines what is more precisely called instantaneous specific heat. As values of specific heats of gases vary significantly with temperature and slightly with pressure, the term mean specific heat, c_m, for a given temperature range has come into use where

$$c_m = \frac{\int_{T_1}^{T_2} dq}{T_2 - T_1} \tag{2-8}$$

Values of c_m become c_{pm} and c_{vm} for constant-pressure and constant-volume processes, respectively. It should be mentioned that Eqs. (2-4) to (2-8) apply to any fluid whether gaseous or not.

Figures 2-1 and 2-2 plot specific heat at constant pressure c_p over a range of temperature, and also plotted is the ratio of specific heats,

$$k = \frac{c_p}{c_v} \tag{2-9}$$

For the perfect gas the relationship holds that

$$c_p - c_v = \frac{R}{J} \tag{2-10}$$

and for a mole of gas

$$m(c_p - c_v) = \frac{mR}{J} = 1.986 \text{ Btu/(lb mole)(°R)} \tag{2-11}$$

where m and R are defined in connection with Eq. (2-3), $J = 778$ ft-lb$_f$ per Btu in the English system of units, with c_p and c_v, Btu/(lb)(°F).

A *pound mole* of gas is a molecular weight of a gas expressed in pounds. Thus a mole of oxygen, molecular weight 32, is 32 lb of oxygen and this weight (mass) occupies 359 ft^3 at 32 F and 14.696 psia. According to Avogadro's law, a mole of any perfect gas would occupy 359 ft^3 at the same pressure and temperature.

Energy relations for the perfect gas show that the change in internal energy u of such a gas, per pound, is given by

$$du = c_v\, dT$$

and, for a gas for which c_v varies with temperature,

$$u_2 - u_1 = \int_{T_1}^{T_2} c_v\, dT = c_{vm}(T_2 - T_1) \tag{2-12}$$

Although changes in internal energy are of major significance, it is convenient, particularly in tabulations of gas properties, to work with values of internal energy referred to an arbitrarily chosen datum T_D, thus

$$u = c_{vm}(T - T_D) \tag{2-13}$$

Similarly, for enthalpy we can write for 1 lb of gas

$$dh = c_p\, dT \tag{2-14}$$

$$h_2 - h_1 = \int_{T_1}^{T_2} c_p\, dT = c_{pm}(T_2 - T_1) \tag{2-15}$$

$$h = c_{pm}(T - T_D) \tag{2-16}$$

where, as in Eq. (2-13), T_D is an arbitrarily chosen datum of temperature. Values which have been used in this connection are 460, 400, and 0 R.

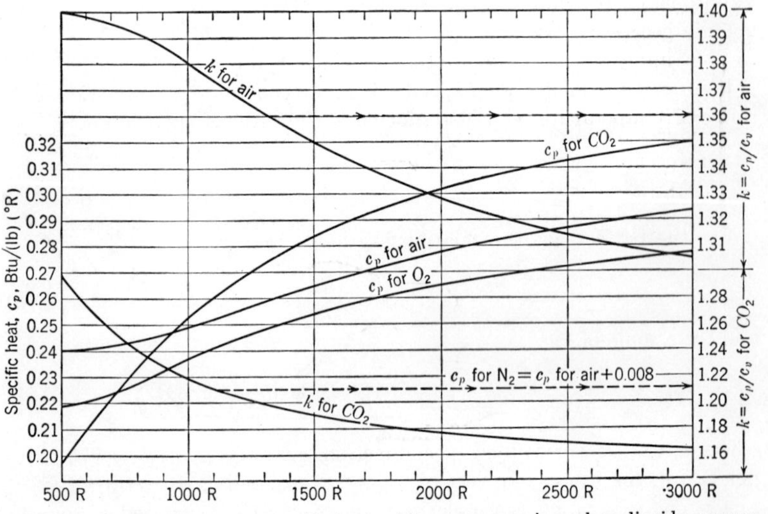

FIG. 2-1. Constant-pressure specific heats of low-pressure air, carbon dioxide, oxygen and k for air and carbon dioxide.

2-3. Adiabatic Process. An *adiabatic* process is defined as one in which no heat is transferred to or from the system during the process. This would mean that in Eq. (2-2), ${}_1q_2$ is zero, and if we were interested in the work term, the equation would appear in Btu per pound as

$$\frac{{}_1W_2}{J} = h_1 - h_2 + \frac{C_1^2 - C_2^2}{2g_cJ} \tag{2-17}$$

When G, the pounds per second flow of working medium, is brought into Eq. (2-17), a power equation results, with P as Btu per second.

$$P = G\left(h_1 - h_2 + \frac{C_1^2 - C_2^2}{2g_cJ}\right) \tag{2-18}$$

$$\text{hp} = \frac{G}{0.707}\left(h_1 - h_2 + \frac{C_1^2 - C_2^2}{2g_cJ}\right) \tag{2-19}$$

where 0.707 (more exactly 0.7068) converts Btu per second to horsepower (hp).

A study of Eq. (2-17) indicates that the work delivered (or absorbed) in an adiabatic-flow process is always measured by the resultant change in enthalpy and in kinetic energy of the fluid. Reasoning developed from the second law of thermodynamics shows that for an expansion the change in enthalpy (which can be converted to work or kinetic energy) has a maximum value when the process followed between a given initial state and final pressure is a reversible one. Similarly in a compression the minimum work is expended to bring the system from state 1 to the pressure at state 2 if the compression is carried out reversibly. In a reversible-adiabatic process the entropy (symbol s or sometimes ϕ) does not change. The criterion of the idealized reversible-adiabatic or constant-entropy process is very important as it gives a basis with which to compare real expansion and compression processes.

During a reversible-adiabatic process for a perfect gas, temperature, pressure, and volume are related by the following equations:

$$\frac{T_2}{T_1} = \left(\frac{P_2}{P_1}\right)^{(k-1)/k} \tag{2-20}$$

$$\frac{T_2}{T_1} = \left(\frac{v_1}{v_2}\right)^{k-1} \tag{2-21}$$

$$P_1v_1^k = P_2v_2^k = Pv^k \tag{2-22}$$

where k = ratio of the specific heats c_p/c_v

The work in reversible-adiabatic expansion or compression of a flowing gas or vapor for negligible changes in kinetic energy appears per pound of gas flowing as

$$\left(\frac{{}_1W_2}{J}\right)_s = (h_1 - h_2)_s \tag{2-23}$$

where the subscript s is used to indicate that the process is considered to take place at constant entropy and is therefore reversibly adiabatic. If the process were limited to a perfect gas, Eq. (2-23) could be transformed to

$$\left(\frac{{}_1W_2}{J}\right)_s = c_{pm}(T_1 - T_2)_s = c_{pm}T_1\left(1 - \left(\frac{P_2}{P_1}\right)^{(k-1)/k}\right)_s \tag{2-24}$$

$$\left(\frac{{}_1W_2}{J}\right)_s = \frac{k}{J(k-1)}(P_1v_1 - P_2v_2)_s \tag{2-25}$$

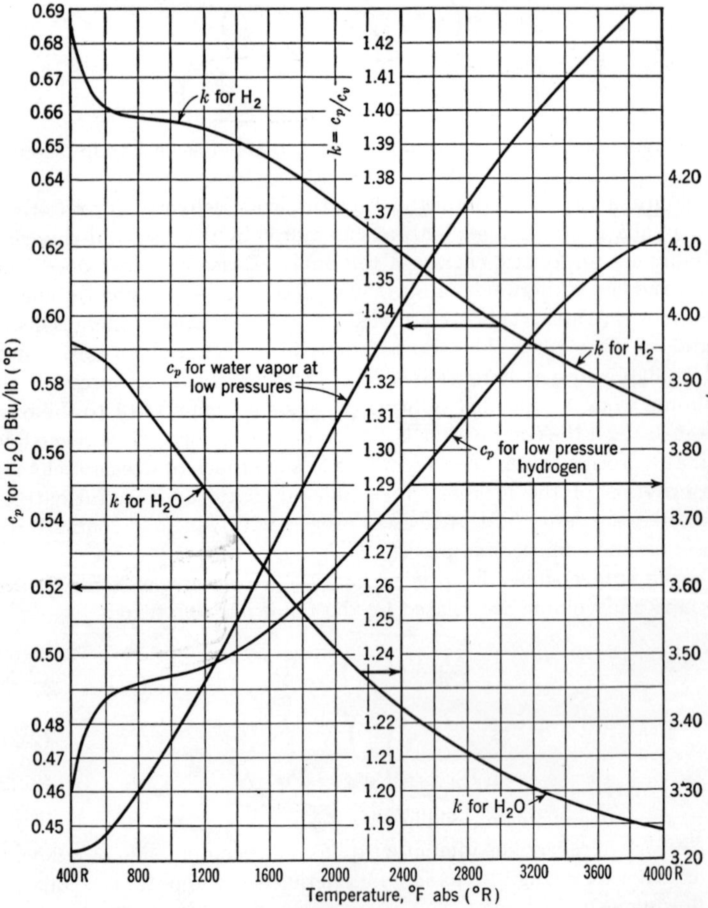

FIG. 2-2. Specific heat and k values for low-pressure water vapor and hydrogen.

A more general form of Eq. (2-23) for any fluid would not disregard kinetic-energy effects and developed from Eq. (2-17) would appear

$$\left(\frac{{}_1W_2'}{J}\right)_s + \frac{C_2^2 - C_1^2}{2g_cJ} = (h_1 - h_2)_s \tag{2-26}$$

Here, even though the process is reversible, $({}_1W_2')_s$ will be less than $({}_1W_2)_s$ in Eq. (2-23) between the same enthalpy limits if additional kinetic energy is stored, as would happen if C_2 is greater than C_1. Also, of course, $({}_1W_2')_s$ may be greater than $({}_1W_2)_s$ if some of the original kinetic energy of the stream is converted into work as would happen if C_1 is greater than C_2.

Consider Eq. (2-17) for the general case of an irreversible adiabatic

$$h_1 - h_2 = \frac{{}_1W_2}{J} + \frac{C_2^2 - C_1^2}{2g_cJ} \tag{2-17}$$

Here h_2 is not the minimum value resulting from isentropic expansion through a given pressure range but can have varying values, even reaching h_1 in the particular case when $C_2 = C_1$ and no work is done. This particular condition goes under the name of *throttling*.

In a moving stream of fluid, kinetic-energy and enthalpy changes are closely related, and it is sometimes convenient to consider the changes in their composite values. Equation (2-17) may be written in the form

$$\frac{{}_1W_2}{J} = \left(h_1 + \frac{C_1^2}{2g_cJ}\right) - \left(h_2 + \frac{C_2^2}{2g_cJ}\right) = H_{o1} - H_{o2} \tag{2-27}$$

It is obvious that the work delivered (or absorbed) by the fluid moving in steady flow depends on the difference in value of the two composite terms in Eq. (2-27). Because of the frequency of appearance of these composite terms, it seems desirable to provide a name for this grouping of kinetic energy and enthalpy. In this text the name *kenthalpy* will be employed.

$$h + \frac{C^2}{2g_cJ} = H_o = \text{kenthalpy in Btu per lb}$$

where C = velocity, fps, of a moving fluid having h for its enthalpy, Btu per lb

g_c = standard gravitational constant = 32.17

J = 778 ft-lb$_f$ per Btu

2-4. Atmospheric Air. It is well known that air is a mechanical mixture of gases of essentially constant composition even at high altitudes. A representative analysis of dry air expressed in per cent by volume shows oxygen (O_2), 20.99; nitrogen (N_2), 78.03; argon (A), 0.94 (including other rare gases, neon, helium, and krypton); carbon dioxide (CO_2), 0.03; and hydrogen (H_2), 0.01.

The water-vapor content, or humidity, in air varies over wide limits from saturation, which implies that the air holds as much water vapor as is possible, to dry air, in which the amount of moisture approaches zero. At 70 F, saturated air (100 per cent relative humidity) at 14.696 atmospheric pressure can have 2.46 per cent water vapor by volume and as temperature increases the percentage rises rapidly, reaching 25.2 per cent at 140 F. In cold weather and at upper altitudes, the amount of water vapor in air is so trivial that it does not affect combustion calculations and can be disregarded. The active constituent of air, as far as combustion is concerned, is oxygen, and all the other gases act as inert materials throughout the combustion process.

The foregoing analysis shows that 20.99 per cent of dry air is oxygen with the remaining 79.01 per cent as inert material. Thus for every cubic foot of oxygen in air 79.01/20.99 = 3.764 ft³ of inert material exist in atmospheric air. Also for every mole of oxygen in air there exist 3.764 moles of inert gases. These inert gases consist largely of nitrogen; and as argon is the second largest constituent, the inert gases in atmospheric air are sometimes called by a synthesized name, *nitargon*. As will be shown, the weight percentages of oxygen and nitargon are, respectively, 23.19 and 76.81, usually rounded to 23.2 and 76.8.

TABLE 2-1. CONSTANTS FOR GASES AND VAPORS

Gas or vapor	Chemical symbol	Mol. wt m	R, ft-lb$_f$/(lb)(°R)	Sp ht at 70°F and low pressure, Btu/(lb)(°R)		$k = \frac{c_p}{c_v}$
				c_p	c_v	
Air	......	28.97	53.34	0.240	0.171	1.40
Argon	A	39.94	38.70	0.124	0.075	1.66
Carbon dioxide	CO_2	44.00	35.12	0.206	0.158	1.30
Carbon monoxide	CO	28.00	55.14	0.243	0.174	1.40
Ethane	C_2H_6	30.05	51.38	0.413	0.339	1.22
Helium	He	4.00	386.00	1.25	0.75	1.66
Hydrogen	H_2	2.016	765.86	3.42	2.44	1.40
Methane	CH_4	16.03	96.31	0.528	0.403	1.31
Nitrogen	N_2	28.02	54.99	0.248	0.176	1.40
Nitargon*	......	28.164	54.82	0.247	0.176	1.40
Octane	C_8H_{18}	114.14	13.55	0.349	0.210	1.66
Oxygen	O_2	32.00	48.25	0.219	0.156	1.40
Propane	C_3H_8	44.06	35.04	0.473	0.411	1.15
Steam	H_2O	18.01	85.6	0.45	0.35	1.28
Sulfur dioxide	SO_2	64.07	24.10	0.154	0.123	1.25

* Nitargon is atmospheric air exclusive of oxygen.

Universal gas constant: 1545.3 ft-lb$_f$/(lb mole)(°R) = 1.986 Btu/(lb mole)(°R) = 1.986 cal/(gram mole)(°C).

In order to demonstrate the method of changing from a volumetric analysis to a weight analysis, a computation will be made to transfer the volumetric analysis of dry air to the weight analysis. A volumetric analysis of gas can also be represented by the moles of each constituent gas in 100 moles of the mixture or the cubic feet of each constituent gas in 100 ft³ of the total mixture (on the assumption that each of these gases might exist under the total pressure of the mixture). A mole of gas is numerically equal in pounds to the molecular weight of that particular gas, and it should be noted that the density of a gas is proportional to its molecular weight.

To carry out the computation, Table 2-2 has been constructed. This lists in the first column the names of the gases, and in the second column the volumetric analysis in per cent or in moles per 100 moles of air. If each of the volumetric percentages is multiplied by its appropriate molecular weight, the resulting products, shown in column (4), represent proportional weights, or the weights of constituents in 100 moles of air. The decimal fraction obtained by dividing each constituent in column (4) by the total of the column gives the resultant weight fraction and weight analysis of the air. The weight fraction, when multiplied by 100, immediately becomes a percentage.

It is of interest to note that the value at the bottom of column (3), 28.967, represents what might be called an equivalent molecular weight of air. This value is usually carried to two decimal places only and

TABLE 2-2. VOLUMETRIC-WEIGHT ANALYSIS OF DRY AIR

Constituents of air (1)	Percentage volumetric analysis, or moles per 100 moles (2)	Mol. wt, or lb per mole (3)	Cols. (2) × (3), weight of constituents in 100 moles air (4)	Weight analysis of dry air, $\frac{\text{col. (4)}}{2896.7}$ (5)
Oxygen	20.99	32.00	671.7	0.2319
Nitrogen	78.03	28.016	2186.1	0.7547
Argon	0.94	39.944	37.6	0.0130
Carbon dioxide	0.03	44.003	1.3	0.0004
Hydrogen	0.01	2.016	0.0	0.0000
Dry air	100.00	(28.967)	2896.7	1.000

appears as 28.97. A figure similarly worked out for the inert materials in air gives the molecular weight of atmospheric nitrogen (nitargon) as 28.161.

When the universal gas constant of 1545.3 is divided by the molecular weight of air, 28.967, the value of R for air results as 53.342, usually considered 53.34.

The United States standard atmosphere is frequently used in aeronautics as a standard of reference. In this standard, 59 F and 29.921 in. Hg represent conditions at sea level, and temperatures are assumed to vary linearly with height up to some 35,000 ft, beyond which the temperature approaches a constant value for a considerable altitude variation before it begins to vary again. Table 2-3 shows the altitude, pressure, and temperature variations for the standard atmosphere.

2-5. Air Tables. The properties of air can be approximated by use of the perfect-gas laws and appropriate constants. However, the deviations of this real gas from perfect-gas relationships may be appreciable, particularly as pressures increase above 200 psi. Air and many similar gases might better be named semiperfect gases, and when perfect-gas relationships are used in connection with them, it should be realized that the relationships are only close approximations. With all gases, the specific heat changes greatly with temperature and consequently, when working with gases, it is extremely convenient to have tabulations of enthalpy and entropy which have taken into account specific-heat variations. In this text air tables have been computed and appear as Table

TABLE 2-3. ALTITUDE, PRESSURE, AND TEMPERATURE FOR THE UNITED STATES STANDARD ATMOSPHERE

Altitude, ft above sea level	Pressure, in. Hg	Temperature °F	Temperature °R
−1,000	31.02	62.6	522.3
−500	30.47	60.8	520.5
0	29.921	59.0	518.7
500	29.38	57.2	516.9
1,000	28.86	55.4	515.1
5,000	24.89	41.2	500.9
10,000	20.58	23.4	483.1
15,000	16.88	5.5	465.2
20,000	13.75	−12.3	447.4
25,000	11.10	−30.1	429.6
30,000	8.88	−47.9	411.8
35,000	7.04	−65.8	393.9
40,000	5.54	−67.0	392.7
45,000	4.36	−67.0	392.7
50,000	3.436	−67.0	392.7

2-4. These tables have been expanded by the authors from tables originally prepared by Prof. Thomas E. Butterfield on 20° intervals and printed in *Steam and Gas Engineering* by Butterfield, Jennings, and Luce, reproduced here by permission of the authors and the publisher, D. Van Nostrand Company. Because of the extensive use which can be made of a table of this type in gas-compressor and turbine work, it seemed necessary to expand it to unit temperature intervals in the commonly used working range.

The enthalpy datum of the tables is taken at 460 R (0.3 F). The datum at which the enthalpy is called zero is an arbitrary decision. The tables use, as their argument, temperature in degrees Rankine and then list enthalpy h in Btu per pound; internal energy u in Btu per pound; entropy s_p based on constant pressure, and relative pressure ratio p_r. Methods of using the tables will be illustrated, but first a brief explanation of how the tables are constructed will be given.

The determination of specific heats of gases by use of spectroscopic readings and summation calculations developed in the field of quantum mechanics has made it possible to give accurate and reliable values for

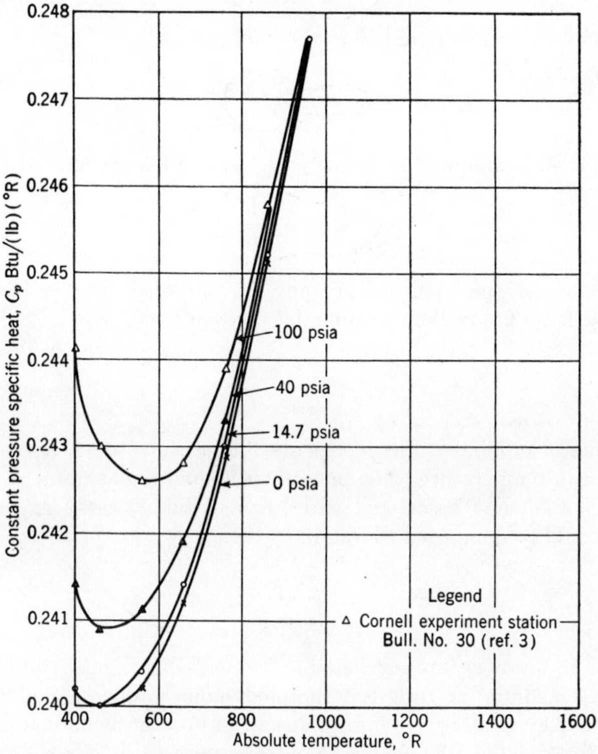

FIG. 2-3. Constant-pressure specific heat of dry air to show effect of pressure.

the specific heats of gases over extensive ranges of temperature. These values are correct only for gases at pressures approaching zero, but it is fortunately true that both at atmospheric pressure and at moderately greater pressures of several atmospheres specific heat is not greatly influenced by pressure. Figure 2-1 is a graph of specific heat of air based on zero pressure, while Fig. 2-3 gives values for specific heats at the higher pressures indicated. It will be noticed that up to 100 psi the variation in specific heat of air with pressure decreases rapidly with temperature and is not of great significance in any event. Consequently, no great inaccuracy develops when tables of gas properties are prepared making use of the zero-pressure specific-heat values. For Table 2-4, the spectroscopic work of Johnson, Davis, and Walker[4,5] was used for the zero-pressure specific-heat values for computing enthalpy following the formulation of Eq. (2-15)

$$h = \int_{460}^{T} c_p \, dT \tag{2-28}$$

The value of u can be found from the enthalpy as

$$u = h - \frac{Pv}{J} \tag{2-29}$$

Further, if air is assumed to follow perfect-gas laws, we can write

$$u = h - \frac{RT}{J} \tag{2-30}$$

In Table 2-4 the value of entropy is computed from a datum of $T_1 = 460$ R by use of the fundamental relation

$$s_p = \int_{460}^{T} \frac{c_p \, dT}{T} \tag{2-31}$$

This value is tabulated and represents the entropy which arises under variations in temperature when pressure is assumed constant. A similar relation for entropy associated with changes taking place at constant volume could be computed as follows:

$$s_v = \int_{460}^{T} \frac{c_v \, dT}{T} \tag{2-32}$$

Values of s_v, however, are not listed in Table 2-4.

For any medium, entropy is a thermodynamic property, and as such its value at a given state is invariant no matter how the medium is brought to that state. Thus, the change in entropy for a gas at state 1 going to state 2 can be computed by any convenient reversible path or any combination of such paths. For example, consider 1 lb of a gas undergoing a constant-pressure change followed by a constant-temperature change from condition 1 to a second condition 2. If for the real gas we can again consider that the perfect-gas relationships hold, this change can be represented by the following equation:

$$s_2 - s_1 = \int_{T_1}^{T_2} c_p \frac{dT}{T} - \frac{R}{J} \log_e \frac{p_2}{p_1} = \Big[s_p \Big]_{T_1}^{T_2} - \Big[s_T \Big]_{p_1}^{p_2} \tag{2-33}$$

Table 2-5 tabulates values for

$$\Big[s_T \Big]_{p_1}^{p_2} = \frac{R}{J} \log_e \frac{p_2}{p_1} \tag{2-34}$$

for various pressure-ratio values of p_2/p_1. Note that using representative values in Eq. (2-34) for air, there results

$$\frac{R}{J} \log_e \frac{p_2}{p_1} = \frac{54.34}{778.2} 2.3026 \log_{10} \frac{p_2}{p_1} = \Big[s_T \Big]_{p_1}^{p_2} = 0.15783 \log_{10} \frac{p_2}{p_1} \tag{2-35}$$

Similarly, for a constant-volume change followed by a constant-temperature change,

$$s_2 - s_1 = \int_{T_1}^{T_2} c_v \frac{dT}{T} + \frac{R}{J} \log_e \frac{v_2}{v_1} = \Big[s_v \Big]_{T_1}^{T_2} + \Big[s_T \Big]_{v_1}^{v_2} \tag{2-36}$$

Let us assume that a reversible-adiabatic (isentropic) process takes place. In such there would be no change in entropy, and

$$\Delta s = s_2 - s_1 = 0$$

Thus in Eq. (2-33)

$$\int_{T_1}^{T_2} c_p \frac{dT}{T} = \frac{R}{J} \log_e \frac{p_2}{p_1}$$

and

$$\frac{R}{J} \log_e \frac{p_2}{p_1} = \int_{T_1}^{T_2} c_p \frac{dT}{T} = \Big[s_p \Big]_{T_1}^{T_2} \tag{2-37}$$

By Eq. (2-34)

$$\Big[s_T \Big]_{p_1}^{p_2} = \Big[s_p \Big]_{T_1}^{T_2} \tag{2-38}$$

$$\Big[s_T \Big]_{p_1}^{p_2} = s_{p2} - s_{p1} \tag{2-39}$$

In the cases where volume ratios of expansion or compression are employed the corresponding form for an isentropic process from Eq. (2-36) is

$$\Big[s_T \Big]_{v_1}^{v_2} = - \Big[s_v \Big]_{T_1}^{T_2} = s_{v1} - s_{v2} \tag{2-40}$$

However, as little use will be made in this text of volume ratios, values of $\Big[s_v \Big]_{T_1}^{T_2}$ are not tabulated. If such a value is required, it could be found from Eq. (2-10) used as follows:

$$\int_{T_1}^{T_2} c_v \frac{dT}{T} = \int_{T_1}^{T_2} c_p \frac{dT}{T} - \frac{R}{J} \log_e \frac{T_2}{T_1}$$

$$\Big[s_v \Big]_{T_1}^{T_2} = \Big[s_p \Big]_{T_1}^{T_2} - \frac{R}{J} \log_e \frac{T_2}{T_1} \tag{2-41}$$

Example 1. By the use of air tables find the final temperature and enthalpy of the air leaving a reversible-adiabatic compressor when air enters at 60.3 F and 14.5 psia at trivial velocity ($C_1 = 0$) and leaves at 34.5 psia with C_2 also essentially zero.

Solution. From Table 2-4 at 60.3 + 459.7 = 520 R, read $h_1 = 14.393$,

$$s_{p1} = 0.02944$$

For a pressure-compression ratio of 34.5/14.5 = 2.38 read in Table 2-5 that

$$s_T = 0.05944$$

By Eq. 2-39,

$$s_{p2} - 0.02944 = 0.05944$$
$$s_{p2} = 0.08888$$

Corresponding to the value of s_{p2} read in Table 2-4 a temperature $T'_2 = 666$ R and an enthalpy = 49.505 Btu per lb.

TABLE 2-4. THERMODYNAMIC PROPERTIES OF AIR AT LOW PRESSURES

°R	°F	h, Btu per lb	u, Btu per lb	s_p	p_r	°R	°F	h, Btu per lb	u, Btu per lb	s_p	p_r
400	−59.7	−14.364	−41.780	−0.03348	0.6140	430	−29.7	−7.182	−36.650	−0.01615	0.7902
401		14.125	41.609	0.03288	0.6193	431		6.943	36.479	0.01559	0.7966
402		13.885	41.438	0.03228	0.6247	432		6.703	36.308	0.01504	0.8031
403		13.646	41.267	0.03169	0.6301	433		6.464	36.137	0.01449	0.8095
404		13.406	41.096	0.03110	0.6356	434		6.224	35.966	0.01394	0.8161
405		−13.167	−40.925	−0.03051	0.6411	435		−5.985	−35.795	−0.01339	0.8227
406		12.928	40.754	0.02992	0.6467	436		5.746	35.624	0.01284	0.8294
407		12.688	40.583	0.02933	0.6522	437		5.506	35.453	0.01229	0.8361
408		12.449	40.412	0.02874	0.6578	438		5.267	35.282	0.01174	0.8428
409		12.209	40.241	0.02815	0.6635	439		5.027	35.111	0.01119	0.8495
410	−49.7	−11.970	−40.070	−0.02756	0.6692	440	−19.7	−4.788	−34.940	−0.01064	0.8563
411		11.731	39.899	0.02699	0.6749	441		4.549	34.769	0.01011	0.8631
412		11.491	39.728	0.02642	0.6806	442		4.309	34.598	0.00958	0.8699
413		11.252	39.557	0.02585	0.6864	443		4.070	34.427	0.00904	0.8768
414		11.012	39.386	0.02527	0.6922	444		3.830	34.256	0.00850	0.8837
415		−10.773	−39.215	−0.02469	0.6981	445		−3.591	−34.085	−0.00796	0.8907
416		10.534	39.044	0.02411	0.7039	446		3.352	33.914	0.00742	0.8977
417		10.294	38.873	0.02353	0.7099	447		3.112	33.743	0.00688	0.9048
418		10.055	38.702	0.02295	0.7158	448		2.873	33.572	0.00634	0.9119
419		9.815	38.531	0.02237	0.7218	449		2.633	33.401	0.00580	0.9190
420	−39.7	− 9.576	−38.360	−0.02179	0.7279	450	− 9.7	−2.394	−33.230	−0.00526	0.9262
421		9.337	38.189	0.02122	0.7340	451		2.155	33.059	0.00473	0.9334
422		9.097	38.018	0.02065	0.7401	452		1.915	32.888	0.00420	0.9406
423		8.858	37.847	0.02008	0.7463	453		1.676	32.717	0.00367	9.9479
424		8.618	37.676	0.01951	0.7525	454		1.436	32.546	0.00314	0.9552
425		− 8.379	−37.505	−0.01895	0.7587	455		−1.197	−32.375	−0.00261	0.9625
426		8.140	37.334	0.01839	0.7649	456		0.958	32.204	0.00208	0.9700
427		7.900	37.163	0.01783	0.7712	457		0.718	32.033	0.00156	0.9775
428		7.661	36.992	0.01727	0.7775	458		0.479	31.862	0.00104	0.9849
429		7.421	36.821	0.01671	0.7838	459		0.240	31.691	0.00052	0.9924
Difference or c_p, c_v		c_p = 0.240	c_v = 0.171	60–56				c_p = 0.240	c_v = 0.171	56–52	

Note that in using this method the s_T value is *added to the s_{p1} value during a compression and subtracted from the s_{p1} value during an expansion.*

By Eq. (2-23), the work is

$$\frac{{}_1W_2}{J} = (h_1 - h_2)_s = 14.393 - 49.505 = -35.112 \text{ Btu per lb}$$

The minus sign merely means that work is done not *by* but *on* the gas in amount +35.112 Btu per lb.

TABLE 2-4. THERMODYNAMIC PROPERTIES OF AIR AT LOW PRESSURES.—(*Continued*)

°R	°F	h, Btu per lb	u, Btu per lb	s_p	p_r	°R	°F	h, Btu per lb	u, Btu per lb	s_p	p_r
460	0.3	0.000	−31.520	0.00000	1.000	495		8.396	−25.534	0.01753	1.292
461		0.240	31.349	0.00050	1.007	496		8.635	25.363	0.01802	1.301
462		0.480	31.178	0.00100	1.015	497		8.875	25.192	0.01852	1.310
463		0.719	31.007	0.00151	1.023	498		9.115	25.020	0.01901	1.319
464		0.959	30.836	0.00201	1.031	499		9.355	24.849	0.01952	1.328
465		1.199	−30.666	0.00252	1.038	500	40.3	9.595	−24.678	0.02003	1.338
466		1.439	30.495	0.00303	1.046	501		9.835	24.507	0.02049	1.347
467		1.679	30.324	0.00353	1.054	502		10.075	24.336	0.02095	1.357
468		1.918	30.153	0.00404	1.062	503		10.315	24.164	0.02142	1.367
469		2.158	29.982	0.00455	1.070	504		10.555	23.993	0.02188	1.376
470	10.3	2.398	−29.811	0.00506	1.078	505		10.795	−23.822	0.02235	1.385
471		2.638	29.640	0.00558	1.086	506		11.034	23.651	0.02281	1.395
472		2.878	29.469	0.00609	1.094	507		11.274	23.480	0.02328	1.404
473		3.118	29.298	0.00660	1.102	508		11.514	23.308	0.02375	1.414
474		3.358	29.127	0.00712	1.110	509		11.754	23.137	0.02422	1.424
475		3.598	−28.957	0.00763	1.118	510	50.3	11.994	−22.966	0.02469	1.434
476		3.837	28.786	0.00815	1.127	511		12.234	22.795	0.02516	1.444
477		4.077	28.615	0.00867	1.135	512		12.474	22.624	0.02563	1.453
478		4.317	28.444	0.00919	1.143	513		12.714	22.452	0.02610	1.463
479		4.557	28.273	0.00970	1.151	514		12.954	22.281	0.02657	1.473
480	20.3	4.797	−28.102	0.01021	1.160	515		13.194	−22.110	0.02705	1.483
481		5.037	27.931	0.01069	1.168	516		13.433	21.939	0.02752	1.494
482		5.277	27.760	0.01117	1.177	517		13.673	21.768	0.02800	1.504
483		5.517	27.588	0.01166	1.185	518		13.913	21.596	0.02847	1.514
484		5.757	27.417	0.01214	1.194	519		14.153	21.425	0.02896	1.524
485		5.997	−27.246	0.01263	1.203	520	60.3	14.393	−21.254	0.02944	1.535
486		6.236	27.075	0.01311	1.212	521		14.633	21.082	0.02989	1.545
487		6.476	26.904	0.01360	1.220	522		14.873	20.911	0.03033	1.555
488		6.716	26.732	0.01409	1.229	523		15.113	20.739	0.03078	1.566
489		6.956	26.561	0.01457	1.238	524		15.353	20.567	0.03123	1.576
490	30.3	7.196	−26.390	0.01506	1.247	525		15.593	−20.396	0.03167	1.587
491		7.436	26.219	0.01555	1.256	526		15.833	20.224	0.03212	1.597
492		7.676	26.048	0.01605	1.265	527		16.075	20.052	0.03257	1.608
493		7.916	25.876	0.01654	1.274	528		16.317	19.880	0.03303	1.619
494		8.156	25.705	0.01703	1.283	529		16.559	19.709	0.03348	1.629
Difference or c_p, c_v		c_p = 0.240	c_v = 0.171	50–49				c_p = 0.240	c_v = 0.171	49–45	

TABLE 2-4. THERMODYNAMIC PROPERTIES OF AIR AT LOW PRESSURES.—(*Continued*)

°R	°F	h, Btu per lb	u, Btu per lb	s_p	p_r	°R	°F	h, Btu per lb	u, Btu per lb	s_p	p_r
530	70.3	16.800	−19.537	0.03393	1.640	570	110.3	26.396	−12.660	0.05141	2.116
531		17.040	19.365	0.03438	1.651	571		26.637	12.483	0.05183	2.129
532		17.280	19.194	0.03484	1.662	572		26.877	12.316	0.05225	2.142
533		17.519	19.022	0.03529	1.673	573		27.117	12.144	0.05267	2.155
534		17.759	18.850	0.03575	1.684	574		27.357	11.972	0.05310	2.169
535		17.999	−18.678	0.03621	1.695	575		27.597	−11.800	0.05352	2.182
536		18.239	18.507	0.03666	1.706	576		27.838	11.628	0.05394	2.195
537		18.479	18.335	0.03712	1.717	577		28.078	11.456	0.05436	2.209
538		18.718	18.163	0.03758	1.728	578		28.318	11.284	0.05479	2.222
539		18.958	17.992	0.03804	1.740	579		28.558	11.112	0.05522	2.236
540	80.3	19.198	−17.820	0.03850	1.751	530	120.3	28.798	−10.940	0.05565	2.249
541		19.438	17.648	0.03892	1.762	581		29.039	10.768	0.05605	2.263
542		19.678	17.476	0.03936	1.774	582		29.279	10.596	0.05645	2.276
543		19.917	17.304	0.03979	1.785	583		29.520	10.424	0.05685	2.290
544		20.157	17.132	0.04022	1.797	584		29.760	10.252	0.05726	2.304
545		20.397	−16.960	0.04065	1.808	585		30.001	−10.080	0.05766	2.318
546		20.637	16.788	0.04109	1.820	586		30.241	9.908	0.05806	2.332
547		20.877	16.616	0.04152	1.832	587		30.482	9.736	0.05847	2.346
548		21.116	16.444	0.04195	1.844	588		30.722	9.564	0.05887	2.360
549		21.356	16.272	0.04239	1.855	589		30.963	9.392	0.05928	2.374
550	90.3	21.596	−16.100	0.04282	1.867	590	130.3	31.203	− 9.220	0.05969	2.388
551		21.836	15.928	0.04326	1.879	591		31.444	9.048	0.06010	2.402
552		22.076	15.756	0.04370	1.891	592		31.684	8.876	0.06051	2.416
553		22.315	15.584	0.04414	1.903	593		31.925	8.704	0.06092	2.431
554		22.555	15.412	0.04457	1.915	594		32.165	8.532	0.06132	2.445
555		22.795	−15.240	0.04501	1.927	595		32.406	− 8.360	0.06173	2.460
556		23.035	15.068	0.04545	1.939	596		32.646	8.188	0.06214	2.474
557		23.275	14.896	0.04589	1.952	597		32.887	8.016	0.06255	2.489
558		23.514	14.724	0.04634	1.964	598		33.127	7.844	0.06296	2.503
559		23.754	14.552	0.04678	1.976	599		33.368	7.672	0.06338	2.518
560	100.3	23.994	−14.380	0.04723	1.989	600	140.3	33.608	− 7.500	0.06379	2.533
561		24.235	14.208	0.04764	2.001	601		33.849	7.328	0.06418	2.547
562		24.475	14.036	0.04806	2.014	602		34.090	7.156	0.06457	2.562
563		24.715	13.864	0.04848	2.026	603		34.330	6.984	0.06496	2.578
564		24.955	13.692	0.04889	2.039	604		34.571	6.812	0.06534	2.593
565		25.195	−13.520	0.04930	2.052	605		34.811	− 6.640	0.06573	2.608
566		25.436	13.348	0.04971	2.064	606		35.052	6.468	0.06612	2.623
567		25.676	13.176	0.05014	2.077	607		35.293	6.296	0.06652	2.638
568		25.916	13.004	0.05056	2.090	608		35.533	6.124	0.06691	2.653
569		26.156	12.832	0.05098	2.103	609		35.774	5.952	0.06730	2.669
Difference or c_p, c_v		c_p = 0.240	c_v = 0.172	45–52				c_p = 0.240	c_v = 0.172	42–39	

TABLE 2-4. THERMODYNAMIC PROPERTIES OF AIR AT LOW PRESSURES.—(*Continued*)

°R	°F	*h*, Btu per lb	*u*, Btu per lb	s_p	p_r	°R	°F	*h*, Btu per lb	*u*, Btu per lb	s_p	p_r
610	150.3	36.014	−5.780	0.06769	2.684	650	190.3	45.646	+1.100	0.08298	3.356
611		36.255	5.608	0.06809	2.699	651		45.887	1.273	0.08335	3.374
612		36.496	5.436	0.06848	2.716	652		46.128	1.446	0.08372	3.393
613		36.736	5.264	0.06887	2.731	653		46.369	1.619	0.08410	3.410
614		36.977	5.092	0.06927	2.748	654		46.610	1.792	0.08447	3.429
615		37.217	−4.920	0.06966	2.763	655		46.851	+1.965	0.08484	3.447
616		37.458	4.748	0.07006	2.778	656		47.092	2.138	0.08522	3.466
617		37.699	4.576	0.07046	2.794	657		47.333	2.311	0.08559	3.485
618		37.939	4.404	0.07086	2.811	658		47.574	2.484	0.08597	3.504
619		38.180	4.232	0.07126	2.826	659		47.815	2.657	0.08634	3.522
620	160.3	38.420	−4.060	0.07166	2.842	660	200.3	48.056	+2.830	0.08672	3.540
621		38.661	3.888	0.07204	2.859	661		48.298	3.003	0.08708	3.559
622		38.902	3.716	0.07241	2.875	662		48.539	3.176	0.08743	3.578
623		39.142	3.544	0.07279	2.890	663		48.781	3.349	0.08779	3.597
624		39.383	3.372	0.07317	2.907	664		49.022	3.522	0.08815	3.616
625		39.623	−3.200	0.07355	2.923	665		49.263	+3.695	0.08851	3.637
626		39.864	3.028	0.07392	2.940	666		49.505	3.868	0.08886	3.655
627		40.105	2.856	0.07430	2.956	667		49.746	4.041	0.08922	3.674
628		40.345	2.684	0.07468	2.974	668		49.988	4.214	0.08958	3.695
629		40.586	2.512	0.07506	2.990	669		50.229	4.387	0.08994	3.714
630	170.3	40.826	−2.340	0.07544	3.007	670	210.3	50.470	+4.560	0.09030	3.734
631		41.067	2.168	0.07583	3.023	671		50.712	4.733	0.09066	3.753
632		41.308	1.996	0.07621	3.039	672		50.953	4.906	0.09102	3.773
633		41.549	1.824	0.07659	3.057	673		51.194	5.079	0.09138	3.792
634		41.790	1.652	0.07697	3.074	674		51.436	+5.252	0.09174	3.812
635		42.031	−1.480	0.07736	3.091	675		51.677	+5.425	0.09210	3.832
636		42.272	1.308	0.07774	3.108	676		51.919	5.598	0.09246	3.853
637		42.513	1.136	0.07812	3.125	677		52.160	5.771	0.09283	3.872
638		42.754	0.964	0.07851	3.143	678		52.402	5.944	0.09319	3.892
639		42.995	0.792	0.07890	3.160	679		52.643	6.117	0.09356	3.912
640	180.3	43.236	−0.620	0.07929	3.177	680	220.3	52.884	+6.290	0.09393	3.932
641		43.477	0.448	0.07966	3.195	681		53.126	6.463	0.09428	3.954
642		43.718	0.276	0.08002	3.213	682		53.368	6.636	0.09462	3.974
643		43.959	0.104	0.08039	3.220	683		53.609	6.809	0.09497	3.994
644		44.200	+0.068	0.08076	3.248	684		53.851	6.982	0.09532	4.014
645		44.441	+0.240	0.08113	3.266	685		54.092	+7.155	0.09566	4.036
646		44.682	0.412	0.08150	3.283	686		54.334	7.328	0.09601	4.056
647		44.923	0.584	0.08186	3.301	687		54.576	7.501	0.09636	4.078
648		45.164	0.756	0.08224	3.319	688		54.817	7.674	0.09671	4.098
649		45.405	0.928	0.08261	3.338	689		55.059	7.847	0.09706	4.119
Difference or c_p, c_v		c_p = 0.241	c_v = 0.172	40–37				c_p = 0.241	c_v = 0.172, 0.173	37–35	

TABLE 2-4. THERMODYNAMIC PROPERTIES OF AIR AT LOW PRESSURES.—(*Continued*)

°R	°F	h, Btu per lb	u, Btu per lb	s_p	p_r	°R	°F	h, Btu per lb	u, Btu per lb	s_p	p_r
690	230.3	55.300	+ 8.020	0.09741	4.142	730	270.3	64.980	+14.960	0.11107	5.057
691		55.542	8.193	0.09776	4.162	731		65.223	15.134	0.11140	5.081
692		55.784	8.366	0.09811	4.184	732		65.466	15.308	0.11173	5.106
693		56.026	8.539	0.09846	4.205	733		65.709	15.482	0.11207	5.130
694		56.268	8.712	0.09881	4.227	734		65.952	15.656	0.11240	5.156
695		56.510	+ 8.885	0.09916	4.248	735		66.195	+15.830	0.11273	5.180
696		56.752	9.058	0.09952	4.270	736		66.438	16.004	0.11306	5.205
697		56.994	9.231	0.09987	4.291	737		66.681	16.178	0.11339	5.230
698		57.236	9.404	0.10022	4.313	738		66.924	16.352	0.11373	5.256
699		57.478	9.577	0.10058	4.336	739		67.167	16.526	0.11406	5.281
700	240.3	57.720	+ 9.750	0.10093	4.357	740	280.3	67.410	+16.700	0.11440	5.306
701		57.962	9.923	0.10127	4.379	741		67.653	16.874	0.11472	5.332
702		58.204	10.096	0.10162	4.401	742		67.895	17.048	0.11505	5.357
703		58.446	10.269	0.10195	4.423	743		68.138	17.222	0.11537	5.382
704		58.688	10.442	0.10229	4.446	744		68.380	17.396	0.11569	5.407
705		58.930	+10.615	0.10264	4.467	745		68.623	+17.570	0.11602	5.434
706		59.172	10.788	0.10298	4.490	746		68.865	17.744	0.11634	5.459
707		59.414	10.961	0.10332	4.513	747		69.108	17.918	0.11666	5.486
708		59.656	11.134	0.10366	4.535	748		69.350	18.092	0.11698	5.511
709		59.898	11.307	0.10400	4.558	749		69.593	18.266	0.11731	5.538
710	250.3	60.140	+11.480	0.10434	4.581	750	290.3	69.835	+18.440	0.11763	5.563
711		60.382	11.654	0.10468	4.603	751		70.078	18.614	0.11795	5.590
712		60.624	11.828	0.10502	4.626	752		70.320	18.788	0.11828	5.616
713		60.866	12.002	0.10536	4.649	753		70.563	18.962	0.11860	5.643
714		61.108	12.176	0.10570	4.672	754		70.805	19.136	0.11892	5.669
715		61.350	+12.350	0.10604	4.696	755		71.048	+19.310	0.11925	5.696
716		61.592	12.524	0.10639	4.719	756		71.290	19.484	0.11957	5.723
717		61.834	12.698	0.10673	4.742	757		71.533	19.658	0.11989	5.749
718		62.076	12.872	0.10707	4.766	758		71.775	19.832	0.12021	5.777
719		62.318	13.046	0.10741	4.790	759		72.018	20.006	0.12054	5.804
720	260.3	62.560	+13.220	0.10775	4.812	760	300.3	72.260	+20.180	0.12087	5.830
721		62.802	13.394	0.10808	4.836	761		72.503	20.354	0.12119	5.858
722		63.044	13.568	0.10841	4.860	762		72.745	20.528	0.12150	5.885
723		63.286	13.742	0.10875	4.883	763		72.988	20.702	0.12192	5.912
724		63.528	13.916	0.10908	4.907	764		73.230	20.876	0.12213	5.940
725		63.770	+14.090	0.10941	4.931	765		73.433	+21.050	0.12245	5.968
726		64.012	14.264	0.10972	4.955	766		73.715	21.224	0.12277	5.995
727		64.254	14.438	0.11007	4.979	767		73.938	21.398	0.12308	6.023
728		64.496	14.612	0.11040	5.003	768		74.200	21.572	0.12340	6.050
729		64.738	14.786	0.11074	5.027	769		74.443	21.746	0.12371	6.078
Difference or c_p, c_v		c_p = 0.242	c_v = 0.173, 0.174	35–34				c_p = 0.243	c_v = 0.174	33–31	

TABLE 2-4. THERMODYNAMIC PROPERTIES OF AIR AT LOW PRESSURES.—(*Continued*)

°R	°F	h, Btu per lb	u, Btu per lb	s_p	p_r	°R	°F	h, Btu per lb	u, Btu per lb	s_p	p_r
770	310.3	74.685	+21.920	0.12403	6.108	810	350.3	84.420	+28.910	0.13637	7.311
771		74.928	22.094	0.12435	6.136	811		84.664	29.085	0.13667	7.343
772		75.170	22.268	0.12466	6.163	812		84.908	29.260	0.13697	7.375
773		75.413	22.442	0.12498	6.193	813		85.152	29.435	0.13727	7.408
774		75.655	22.616	0.12529	6.220	814		85.396	29.610	0.13757	7.440
775		75.898	+22.790	.012561	6.248	815		85.640	+29.785	0.13787	7.473
776		76.140	22.964	0.12593	6.277	816		85.884	29.960	0.13818	7.506
777		76.383	23.138	0.12624	6.306	817		86.128	30.135	0.13848	7.538
778		76.625	23.312	0.12656	6.336	818		86.372	30.310	0.13878	7.571
779		76.868	23.486	0.12687	6.363	819		86.616	30.485	0.13908	7.604
780	320.3	77.110	+23.660	0.12719	6.393	820	360.3	86.860	+30.660	0.13938	7.637
781		77.353	23.835	0.12750	6.422	821		87.104	30.836	0.13967	7.670
782		77.596	24.010	0.12781	6.451	822		87.348	31.012	0.13997	7.703
783		77.839	24.185	0.12811	6.480	823		87.592	31.188	0.14026	7.737
784		78.082	24.360	0.12842	6.510	824		87.836	31.364	0.14056	7.770
785		78.325	+24.535	0.12873	6.539	825		88.080	+31.540	0.14085	7.804
786		78.568	24.710	0.12904	6.568	826		88.324	31.716	0.14114	7.837
787		78.811	24.885	0.12935	6.599	827		88.568	31.892	0.14144	7.871
788		79.054	25.060	0.12965	6.628	828		88.812	32.068	0.14173	7.906
789		79.297	25.235	0.12996	6.658	829		89.056	32.244	0.14203	7.940
790	330.3	79.540	+25.410	0.13027	6.689	830	370.3	89.300	+32.420	0.14232	7.974
791		79.784	25.585	0.13058	6.719	831		89.544	32.596	0.14261	8.008
792		80.028	25.760	0.13089	6.748	832		89.788	32.772	0.14291	8.042
793		80.272	25.935	0.13119	6.779	833		90.032	32.948	0.14320	8.076
794		80.516	26.110	0.13150	6.809	834		90.276	33.124	0.14350	8.112
795		80.760	+26.285	0.13181	6.839	835		90.520	+33.300	0.14379	8.146
796		81.004	26.460	0.13212	6.871	836		90.764	33.476	0.14408	8.180
797		81.248	26.635	0.13243	6.901	837		91.008	33.652	0.14438	8.216
798		81.492	26.810	0.13273	6.932	838		91.252	33.828	0.14467	8.251
799		81.736	26.985	0.13304	6.962	839		91.496	34.004	0.14497	8.285
800	340.3	81.980	+27.160	0.13336	6.994	840	380.3	91.740	+34.180	0.14527	8.321
801		82.224	27.335	0.13366	7.025	841		91.985	34.356	0.14556	8.356
802		82.468	27.510	0.13396	7.056	842		92.230	34.532	0.14585	8.392
803		82.712	27.685	0.13426	7.087	843		92.475	34.708	0.14613	8.427
804		82.956	27.860	0.13456	7.119	844		92.720	34.884	0.14642	8.463
805		83.200	+28.035	0.13487	7.151	845		92.965	+35.060	0.14671	8.499
806		83.444	28.210	0.13517	7.182	846		93.210	35.236	0.14700	8.535
807		83.688	28.385	0.13547	7.214	847		93.455	35.412	0.14729	8.570
808		83.932	28.560	0.13577	7.246	848		93.700	35.588	0.14757	8.607
809		84.176	28.735	0.13607	7.277	849		93.945	35.764	0.14786	8.644
Difference or c_p, c_v		$c_p =$ 0.243, 0.244	$c_v =$ 0.175	32–30				$c_p =$ 0.244, 0.245	$c_v =$ 0.176	30–29	

TABLE 2-4. THERMODYNAMIC PROPERTIES OF AIR AT LOW PRESSURES.—(*Continued*)

°R	°F	*h*, Btu per lb	*u*, Btu per lb	s_p	p_r	°R	°F	*h*, Btu per lb	*u*, Btu per lb	s_p	p_r
850	390.3	94.190	+35.940	0.14815	8.682	885		102.770	+42.125	0.15804	10.032
851		94.435	36.116	0.14844	8.719	886		103.016	42.302	0.15832	10.073
852		94.680	36.292	0.14873	8.756	887		103.262	42.479	0.15859	10.113
853		94.925	36.468	0.14901	8.793	888		103.508	42.656	0.15887	10.154
854		95.170	36.644	0.14930	8.829	889		103.754	42.833	0.15914	10.196
855		95.415	+36.829	0.14959	8.866	890	430.3	104.000	+43.010	0.15942	10.236
856		95.660	36.996	0.14988	8.904	891		104.246	43.187	0.15970	10.278
857		95.905	37.172	0.15017	8.941	892		104.492	43.364	0.15997	10.319
858		96.150	37.348	0.15045	8.979	893		104.738	43.541	0.16025	10.360
859		96.395	37.524	0.15074	9.016	894		104.984	43.718	0.16052	10.402
860	400.3	96.640	+37.700	0.15103	9.053	895		105.230	+43.895	0.16080	10.444
861		96.885	37.877	0.15131	9.090	896		105.478	44.072	0.16108	10.486
862		97.130	38.054	0.15159	9.128	897		105.722	44.249	0.16135	10.528
863		97.375	38.231	0.15187	9.166	898		105.968	44.426	0.16163	10.571
864		97.620	38.408	0.15215	9.204	899		106.214	44.603	0.16190	10.612
865		97.865	+38.585	0.15244	9.242	900	440.3	106.460	+44.780	0.16219	10.654
866		98.110	38.762	0.15272	9.280	901		106.706	44.958	0.16246	10.699
867		98.355	38.939	0.15300	9.319	902		106.952	45.136	0.16273	10.740
868		98.600	39.116	0.15328	9.357	903		107.198	45.314	0.16300	10.783
869		98.845	39.293	0.15356	9.395	904		107.444	45.492	0.16327	10.825
870	410.3	99.090	+39.470	0.15384	9.435	905		107.690	+45.670	0.16354	10.868
871		99.335	39.647	0.15412	9.474	906		107.936	45.848	0.16381	10.911
872		99.580	39.824	0.15440	9.513	907		108.182	46.026	0.16408	10.954
873		99.825	40.001	0.15468	9.551	908		108.428	46.204	0.16435	10.999
874		100.070	40.178	0.15496	9.590	909		108.674	46.382	0.16462	11.042
875		100.315	+40.355	0.15525	9.630	910	450.3	108.920	+46.560	0.16489	11.086
876		100.560	40.532	0.15553	9.669	911		109.166	46.738	0.16516	11.129
877		100.805	40.709	0.15581	9.709	912		109.412	46.916	0.16543	11.174
878		101.050	40.886	0.15609	9.749	913		109.658	47.094	0.16570	11.218
879		101.295	41.063	0.15637	9.788	914		109.904	47.272	0.16597	11.261
880	420.3	101.540	+41.240	0.15666	9.829	915		110.150	+47.450	0.16624	11.305
881		101.786	41.417	0.15694	9.869	916		110.396	47.628	0.16651	11.351
882		102.032	41.594	0.15721	9.908	917		110.642	47.806	0.16678	11.395
883		102.278	41.771	0.15749	9.949	918		110.888	47.984	0.16705	11.440
884		102.524	41.948	0.15776	9.989	919		111.134	48.162	0.16732	11.485
Difference or c_p, c_v		$c_p =$ 0.245, 0.246	$c_v =$ 0.176, 0.177	29–27				$c_p =$ 0.246	$c_v =$ 0.177, 0.178	28–27	

TABLE 2-4. THERMODYNAMIC PROPERTIES OF AIR AT LOW PRESSURES.—(*Continued*)

°R	°F	*h*, Btu per lb	*u*, Btu per lb	s_p	p_r	°R	°F	*h*, Btu per lb	*u*, Btu per lb	s_p	p_r
920	460.3	111.380	+48.340	0.16760	11.530	960	500.3	121.275	+55.480	0.17812	13.442
921		111.627	48.519	0.16787	11.574	961		121.523	55.659	0.17838	13.492
922		111.874	48.698	0.16813	11.619	962		121.770	55.838	0.17863	13.543
923		112.121	48.877	0.16840	11.665	963		122.018	56.017	0.17889	13.594
924		112.368	49.056	0.16866	11.711	964		122.265	56.196	0.17914	13.645
925		112.615	+49.235	0.16893	11.756	965		122.513	+56.375	0.17940	13.696
926		112.862	49.414	0.16919	11.802	966		122.760	56.554	0.17965	13.747
927		113.109	49.593	0.16946	11.847	967		123.008	56.733	0.17991	13.799
928		113.356	49.772	0.16972	11.894	968		123.255	56.912	0.18016	13.850
929		113.603	49.951	0.16999	11.940	969		123.502	57.091	0.18042	13.902
930	470.3	113.850	+50.130	0.17025	11.987	970	510.3	123.750	+57.270	0.18067	13.954
931		114.097	50.309	0.17052	12.032	971		123.998	57.450	0.18093	14.006
932		114.344	50.488	0.17078	12.079	972		124.246	57.630	0.18118	14.059
933		114.591	50.667	0.17105	12.126	973		124.494	57.810	0.18144	14.111
934		114.838	50.846	0.17131	12.173	974		124.742	57.990	0.18169	14.163
935		115.085	+51.025	0.17158	12.220	975		124.990	+58.170	0.18195	14.216
936		115.332	51.204	0.17184	12.267	976		125.238	58.350	0.18220	14.269
937		115.579	51.383	0.17211	12.315	977		125.486	58.530	0.18246	14.315
938		115.826	51.562	0.17237	12.362	978		125.734	58.710	0.18271	14.367
939		116.073	51.741	0.17264	12.409	979		125.982	58.890	0.18297	14.421
940	480.3	116.320	+51.910	0.17291	12.457	980	520.3	126.230	+59.070	0.18323	14.474
941		116.568	52.089	0.17317	12.505	981		126.479	59.250	0.18348	14.528
942		116.816	52.268	0.17343	12.553	982		126.727	59.430	0.18373	14.581
943		117.064	52.447	0.17369	12.601	983		126.976	59.610	0.18398	14.635
944		117.312	52.626	0.17395	12.649	984		127.224	59.790	0.18423	14.689
945		117.560	+52.805	0.17421	12.697	985		127.473	+59.970	0.18449	14.744
946		117.808	52.984	0.17447	12.746	986		127.721	60.150	0.18474	14.798
947		118.056	53.163	0.17473	12.795	987		127.970	60.330	0.18499	14.852
948		118.304	53.342	0.17499	12.843	988		128.218	60.510	0.18524	14.907
949		118.552	53.521	0.17525	12.892	989		128.467	60.690	0.18549	14.962
950	490.3	118.800	+53.700	0.17551	12.941	990	530.3	128.715	+60.870	0.18574	15.016
951		119.048	53.879	0.17577	12.990	991		128.964	61.050	0.18599	15.072
952		119.295	54.058	0.17603	13.039	992		129.212	61.230	0.18624	15.126
953		119.543	54.237	0.17629	13.089	993		129.461	61.410	0.18649	15.182
954		119.790	54.416	0.17655	13.138	994		129.709	61.590	0.18674	15.237
955		120.038	+54.595	0.17681	13.189	995		129.958	+61.770	0.18700	15.293
956		120.285	54.774	0.17707	13.238	996		130.206	61.950	0.18725	15.349
957		120.533	54.953	0.17733	13.289	997		130.455	62.130	0.18750	15.405
958		120.780	55.132	0.17759	13.338	998		130.703	62.310	0.18775	15.460
959		121.028	55.311	0.17785	13.389	999		130.952	62.490	0.18800	15.517
Difference or c_p, c_v		$c_p =$ 0.247, 0.248	$c_v =$ 0.179	27–28				$c_p =$ 0.248, 0.249	$c_v =$ 0.179, 0.180	26–25	

TABLE 2-4. THERMODYNAMIC PROPERTIES OF AIR AT LOW PRESSURES.—(*Continued*)

°R	°F	h, Btu per lb	u, Btu per lb	s_p	p_r	°R	°F	h, Btu per lb	u, Btu per lb	s_p	p_r
1000	540.3	131.200	+62.670	0.18825	15.582	1040	580.3	141.180	+69.910	0.19804	17.972
1001		131.449	62.851	0.18850	15.639	1041		141.430	70.092	0.19828	18.036
1002		131.698	63.032	0.18874	15.696	1042		141.680	70.274	0.19852	18.100
1003		131.947	63.213	0.18899	15.753	1043		141.930	70.456	0.19875	18.163
1004		132.196	63.394	0.18924	15.810	1044		142.180	70.638	0.19899	18.226
1005		132.445	+63.575	0.18949	15.867	1045		142.430	+70.820	0.19923	18.290
1006		132.694	63.756	0.18973	15.924	1046		142.680	71.002	0.19947	18.353
1007		132.943	63.937	0.18998	15.982	1047		142.930	71.184	0.19971	18.418
1008		133.192	64.118	0.19023	16.039	1048		143.180	71.366	0.19994	18.482
1009		133.441	64.299	0.19047	16.097	1049		143.430	71.548	0.20018	18.547
1010	550.3	133.690	+64.480	0.19072	16.156	1050	590.3	143.680	+71.730	0.20042	18.612
1011		133.939	64.661	0.19097	16.213	1051		143.930	71.912	0.20066	18.676
1012		134.189	64.842	0.19121	16.271	1052		144.180	72.094	0.20090	18.741
1013		134.439	65.023	0.19146	16.330	1053		144.430	72.276	0.20113	18.805
1014		134.688	65.204	0.19171	16.389	1054		144.680	72.458	0.20137	18.871
1015		134.938	+65.385	0.19196	16.447	1055		144.930	+72.640	0.20161	18.937
1016		135.187	65.566	0.19220	16.506	1056		145.180	72.822	0.20185	19.003
1017		135.436	65.747	0.19245	16.566	1057		145.430	73.004	0.20209	19.069
1018		135.686	65.928	0.19270	16.624	1058		145.680	73.186	0.20232	19.135
1019		135.935	66.109	0.19294	16.684	1059		145.930	73.368	0.20256	19.201
1020	560.3	136.185	+66.290	0.19319	16.743	1060	600.3	146.180	+73.550	0.20280	19.267
1021		136.434	66.471	0.19343	16.804	1061		146.431	73.732	0.20303	19.334
1022		136.684	66.652	0.19367	16.864	1062		146.682	73.914	0.20327	19.401
1023		136.934	66.833	0.19392	16.923	1063		146.933	74.096	0.20350	19.467
1024		137.183	67.014	0.19416	16.983	1064		147.184	74.278	0.20374	19.534
1025		137.433	+67.195	0.19440	17.044	1065		147.435	+74.460	0.20397	19.601
1026		137.682	67.376	0.19464	17.105	1066		147.686	74.642	0.20420	19.670
1027		137.931	67.577	0.19488	17.165	1067		147.937	74.824	0.20444	19.737
1028		138.181	67.738	0.19513	17.226	1068		148.188	75.006	0.20467	19.804
1029		138.430	67.910	0.19537	17.287	1069		148.439	75.188	0.20491	19.873
1030	570.3	138.680	+68.100	0.19561	17.348	1070	610.3	148.690	+75.370	0.20514	19.940
1031		138.930	68.281	0.19582	17.410	1071		148.941	75.552	0.20537	20.008
1032		139.179	68.462	0.19609	17.472	1072		149.192	75.734	0.20561	20.077
1033		139.429	68.643	0.19634	17.533	1073		149.443	75.916	0.20584	20.145
1034		139.678	68.824	0.19658	17.595	1074		149.694	76.098	0.20608	20.213
1035		139.928	+69.005	0.19682	17.656	1075		149.945	+76.280	0.20631	20.283
1036		140.177	69.186	0.19706	17.718	1076		150.196	76.462	0.20654	20.353
1037		140.427	69.367	0.19730	17.781	1077		150.447	76.644	0.20678	20.421
1038		140.676	69.548	0.19755	17.843	1078		150.698	76.826	0.20701	20.491
1039		140.926	69.729	0.19779	17.905	1079		150.949	77.008	0.20725	20.561
Difference or c_p, c_v		c_p = 0.249, 0.250	c_v = 0.181	25–24				c_p = 0.250, 0.251	c_v = 0.182	24	

TABLE 2-4. THERMODYNAMIC PROPERTIES OF AIR AT LOW PRESSURES.—(*Continued*)

°R	°F	h, Btu per lb	u, Btu per lb	s_p	p_r	°R	°F	h, Btu per lb	u, Btu per lb	s_p	p_r
1080	620.3	151.200	+77.190	0.20749	20.630	1115		160.010	+83.610	0.21552	23.193
1081		151.451	77.373	0.20772	20.699	1116		160.262	83.794	0.21574	23.269
1082		151.702	77.556	0.20795	20.770	1117		160.514	83.978	0.21597	23.346
1083		151.953	77.739	0.20818	20.841	1118		160.766	84.162	0.21620	23.423
1084		152.204	77.922	0.20841	20.912	1119		161.018	84.346	0.21642	23.501
1085		152.455	+78.105	0.20865	20.983	1120	660.3	161.270	+84.530	0.21665	23.578
1086		152.706	78.288	0.20888	21.054	1121		161.523	84.714	0.21687	23.655
1087		152.957	78.471	0.20911	21.125	1122		161.776	84.898	0.21710	23.734
1088		153.208	78.654	0.20934	21.196	1123		162.029	85.082	0.21732	23.811
1089		153.459	78.837	0.20957	21.267	1124		162.282	85.266	0.21754	23.890
1090	630.3	153.710	+79.020	0.20980	21.339	1125		162.535	+85.450	0.21777	23.968
1091		153.962	79.203	0.21003	21.412	1126		162.788	85.634	0.21799	24.047
1092		154.214	79.386	0.21026	21.484	1127		163.041	85.818	0.21821	24.125
1093		154.466	79.569	0.21049	21.556	1128		163.294	86.002	0.21843	24.204
1094		154.718	79.752	0.21072	21.628	1129		163.547	86.186	0.21866	24.283
1095		154.970	+79.935	0.21096	21.702	1130	670.3	163.800	+86.370	0.21888	24.362
1096		155.222	80.118	0.21119	21.774	1131		164.053	86.554	0.21910	24.442
1097		155.474	80.301	0.21142	21.848	1132		164.306	86.738	0.21933	24.522
1098		155.726	80.484	0.21165	21.921	1133		164.559	86.922	0.21955	24.602
1099		155.978	80.667	0.21188	21.995	1134		164.812	87.106	0.21977	24.682
1100	640.3	156.230	+80.850	0.21211	22.068	1135		165.065	+87.290	0.22000	24.762
1101		156.482	81.034	0.21234	22.142	1136		165.318	87.474	0.22022	24.843
1102		156.734	81.218	0.21256	22.215	1137		165.571	87.658	0.22044	24.924
1103		156.986	81.402	0.21279	22.289	1138		165.824	87.842	0.22066	25.005
1104		157.238	81.586	0.21302	22.362	1139		166.077	88.026	0.22089	25.086
1105		157.490	+81.770	0.21325	22.437	1140	680.3	166.330	+88.210	0.22112	25.167
1106		157.742	81.954	0.21347	22.512	1141		166.583	88.395	0.22133	25.248
1107		157.994	82.138	0.21370	22.586	1142		166.836	88.580	0.22156	25.331
1108		158.246	82.322	0.21393	22.661	1143		167.089	88.765	0.22178	25.412
1109		158.498	82.506	0.21415	22.736	1144		167.342	88.950	0.22200	25.494
1110	650.3	158.750	+82.690	0.21438	22.812	1145		167.595	+89.135	0.22222	25.577
1111		159.002	82.874	0.21461	22.888	1146		167.848	89.320	0.22244	25.659
1112		159.254	83.058	0.21483	22.964	1147		168.101	89.505	0.22266	25.741
1113		159.506	83.242	0.21506	23.040	1148		168.354	89.690	0.22288	25.825
1114		159.758	83.426	0.21529	23.116	1149		168.607	89.875	0.22310	25.909
Difference or c_p, c_v		$c_p =$ 0.251, 0.252	$c_v =$ 0.183, 0.184	23				$c_p =$ 0.252, 0.253	$c_v =$ 0.184, 0.185	22	

TABLE 2-4. THERMODYNAMIC PROPERTIES OF AIR AT LOW PRESSURES.—(*Continued*)

°R	°F	*h*, Btu per lb	*u*, Btu per lb	s_p	p_r	°R	°F	*h*, Btu per lb	*u*, Btu per lb	s_p	p_r
1150	690.3	168.860	+90.060	0.22332	25.994	1190	730.3	179.020	+97.470	0.23201	29.507
1151		169.114	90.245	0.22354	26.070	1191		179.275	97.656	0.23222	29.596
1152		169.367	90.430	0.22376	26.159	1192		179.530	97.842	0.23244	29.685
1153		169.621	90.615	0.22398	26.248	1193		179.785	98.028	0.23265	29.774
1154		169.874	90.800	0.22420	26.324	1194		180.040	98.214	0.23287	29.875
1155		170.128	+90.985	0.22442	26.413	1195		180.295	+98.400	0.23308	29.964
1156		170.381	91.170	0.22464	26.501	1196		180.550	98.586	0.23329	30.052
1157		170.634	91.355	0.22486	26.577	1197		180.805	98.772	0.23351	30.154
1158		170.888	91.540	0.22508	26.666	1198		181.060	98.958	0.23372	30.243
1159		171.142	91.725	0.22530	26.755	1199		181.315	99.144	0.23394	30.344
1160	700.3	171.395	+91.910	0.22553	26.844	1200	740.3	181.570	+99.330	0.23415	30.433
1161		171.649	92.095	0.22575	26.920	1201		181.825	99.517	0.23436	30.521
1162		171.902	92.280	0.22596	27.008	1202		182.080	99.704	0.23457	30.623
1163		172.156	92.465	0.22618	27.097	1203		182.335	99.891	0.23478	30.712
1164		172.409	92.650	0.22640	27.186	1204		182.590	100.078	0.23499	30.813
1165		172.663	+92.835	0.22662	27.262	1205		182.845	+100.265	0.23521	30.902
1166		172.916	93.020	0.22683	27.350	1206		183.100	100.452	0.23542	31.003
1167		173.169	93.205	0.22705	27.439	1207		183.355	100.639	0.23563	31.092
1168		173.423	93.390	0.22727	27.528	1208		183.610	100.826	0.23584	31.193
1169		173.676	93.575	0.22748	27.617	1209		183.865	101.013	0.23605	31.295
1170	710.3	173.930	+93.760	0.22770	27.705	1210	750.3	184.120	+101.200	0.23626	31.383
1171		174.184	93.945	0.22792	27.794	1211		184.375	101.387	0.23647	31.485
1172		174.438	94.130	0.22813	27.883	1212		184.630	101.574	0.23668	31.573
1173		174.692	94.315	0.22835	27.972	1213		184.885	101.761	0.23689	31.675
1174		174.946	94.500	0.22857	28.060	1214		185.140	101.948	0.23710	31.776
1175		175.200	+94.685	0.22879	28.149	1215		185.395	+102.135	0.23732	31.865
1176		175.454	94.870	0.22900	28.238	1216		185.650	102.322	0.23753	31.966
1177		175.708	95.055	0.22922	28.326	1217		185.905	102.509	0.23774	32.068
1178		175.962	95.240	0.22944	28.415	1218		186.160	102.696	0.23795	32.169
1179		176.216	95.425	0.22965	28.504	1219		186.415	102.883	0.23816	32.271
1180	720.3	176.470	+95.610	0.22987	28.593	1220	760.3	186.670	+103.070	0.23837	32.359
1181		176.725	95.796	0.23008	28.681	1221		186.926	103.257	0.23858	32.461
1182		176.980	95.982	0.23030	28.770	1222		187.182	103.444	0.23879	32.562
1183		177.235	96.168	0.23051	28.859	1223		187.438	103.631	0.23899	32.663
1184		177.490	96.354	0.23073	28.947	1224		187.694	103.818	0.23920	32.765
1185		177.745	+96.540	0.23094	29.036	1225		187.950	+104.005	0.23941	32.854
1186		178.000	96.726	0.23115	29.138	1226		188.206	104.192	0.23962	32.955
1187		178.255	96.912	0.23137	29.226	1227		188.462	104.379	0.23982	33.056
1188		178.510	97.098	0.23158	29.315	1228		188.718	104.566	0.24003	33.158
1189		178.765	97.284	0.23180	29.404	1229		188.974	104.753	0.24024	33.259
Difference or c_p, c_v		c_p = 0.254, 0.255	c_v = 0.185, 0.186	22				c_p = 0.255, 0.256	c_v = 0.186, 0.187	21	

TABLE 2-4. THERMODYNAMIC PROPERTIES OF AIR AT LOW PRESSURES.—(*Continued*)

°R	°F	h, Btu per lb	u, Btu per lb	s_p	p_r	°R	°F	h, Btu per lb	u, Btu per lb	s_p	p_r
1230	770.3	189.230	+104.940	0.24045	33.363	1270	810.3	199.500	+112.470	0.24866	37.613
1231		189.486	105.127	0.24066	33.465	1271		199.757	112.659	0.24886	37.727
1232		189.742	105.314	0.24087	33.566	1272		200.014	112.848	0.24907	37.828
1233		189.998	105.501	0.24107	33.667	1273		200.271	113.037	0.24927	37.942
1234		190.254	105.688	0.24128	33.769	1274		200.528	113.226	0.24947	38.056
1235		190.510	+105.875	0.24149	33.870	1275		200.785	+113.415	0.24968	38.170
1236		190.766	106.062	0.24170	33.972	1276		201.042	113.604	0.24988	38.285
1237		191.022	106.249	0.24191	34.086	1277		201.299	113.793	0.25008	38.399
1238		191.278	106.436	0.24211	34.187	1278		201.556	113.982	0.25028	38.513
1239		191.534	106.623	0.24232	34.289	1279		201.813	114.171	0.25049	38.627
1240	780.3	191.790	+106.810	0.24253	34.390	1280	820.3	202.070	+114.360	0.25069	38.728
1241		192.047	106.998	0.24274	34.491	1281		202.328	114.549	0.25089	38.842
1242		192.304	107.186	0.24294	34.593	1282		202.586	114.738	0.25109	38.956
1243		192.561	107.374	0.24315	34.707	1283		202.844	114.927	0.25129	39.070
1244		192.818	107.562	0.24335	34.808	1284		203.102	115.116	0.25149	39.185
1245		193.075	+107.750	0.24356	34.910	1285		203.360	+115.305	0.25169	39.311
1246		193.332	107.938	0.24376	35.011	1286		203.618	115.494	0.25188	39.425
1247		193.589	108.126	0.24397	35.125	1287		203.876	115.683	0.25208	39.540
1248		193.846	108.314	0.24417	35.227	1288		204.134	115.872	0.25228	39.654
1249		194.103	108.502	0.24438	35.328	1289		204.392	116.061	0.25248	39.768
1250	790.3	194.360	+108.690	0.24458	35.442	1290	830.3	204.650	+116.250	0.25268	39.882
1251		194.617	108.879	0.24479	35.544	1291		204.908	116.440	0.25288	39.996
1252		194.874	109.068	0.24499	35.645	1292		205.166	116.630	0.25308	40.123
1253		195.131	109.257	0.24520	35.759	1293		205.424	116.820	0.25328	40.237
1254		195.388	109.446	0.24540	35.860	1294		205.682	117.010	0.25348	40.351
1255		195.645	+109.635	0.24561	35.974	1295		205.940	+117.200	0.25368	40.465
1256		195.902	109.824	0.24581	36.076	1296		206.198	117.390	0.25387	40.592
1257		196.159	110.013	0.24602	36.190	1297		206.456	117.580	0.25407	40.706
1258		196.416	110.202	0.24622	36.291	1298		206.714	117.770	0.25427	40.820
1259		196.673	110.391	0.24643	36.405	1299		206.972	117.960	0.25447	40.947
1260	800.3	196.930	+110.580	0.24663	36.507	1300	840.3	207.230	+118.150	0.25468	41.061
1261		197.187	110.769	0.24683	36.621	1301		207.488	118.340	0.25488	41.175
1262		197.444	110.958	0.24704	36.722	1302		207.746	118.530	0.25507	41.302
1263		197.701	111.147	0.24724	36.836	1303		208.004	118.720	0.25527	41.416
1264		197.958	111.336	0.24744	36.951	1304		208.262	118.910	0.25547	41.530
1265		198.215	+111.525	0.24765	37.052	1305		208.520	+119.100	0.25567	41.657
1266		198.472	111.714	0.24785	37.166	1306		208.778	119.290	0.25586	41.783
1267		198.729	111.903	0.24805	37.280	1307		209.036	119.480	0.25606	41.897
1268		198.986	112.092	0.24825	37.382	1308		209.294	119.670	0.25626	42.024
1269		199.243	112.281	0.24846	37.496	1309		209.552	119.860	0.25645	42.138
Difference or c_p, c_v		c_p = 0.256, 0.257	c_v = 0.187, 0.189	21				c_p = 0.257, 0.258	c_v = 0.189, 0.190	20	

TABLE 2-4. THERMODYNAMIC PROPERTIES OF AIR AT LOW PRESSURES.—(*Continued*)

°R	°F	*h*, Btu per lb	*u*, Btu per lb	s_p	p_r	°R	°F	*h*, Btu per lb	*u*, Btu per lb	s_p	p_r
1310	850.3	209.810	+120.050	0.25665	42.268	1345		218.890	+126.715	0.26347	46.694
1311		210.068	120.240	0.25685	42.383	1346		219.150	126.906	0.26366	46.821
1312		210.326	120.430	0.25704	42.509	1347		219.410	127.097	0.26385	46.948
1313		210.584	120.620	0.25724	42.623	1348		219.670	127.288	0.26405	47.087
1314		210.842	120.810	0.25744	42.750	1349		219.930	127.479	0.26424	47.214
1315		211.101	+121.000	0.25764	42.877	1350	890.3	220.190	+127.670	0.26443	47.353
1316		211.360	121.190	0.25783	43.000	1351		220.450	127.861	0.26462	47.480
1317		211.619	121.380	0.25803	43.121	1352		220.710	128.052	0.26481	47.620
1318		211.878	121.570	0.25823	43.245	1353		220.970	128.243	0.26501	47.746
1319		212.137	121.760	0.25842	43.369	1354		221.230	128.434	0.26520	47.886
1320	860.3	212.396	+121.950	0.25862	43.490	1355		221.490	+128.625	0.26539	48.013
1321		212.656	122.140	0.25881	43.612	1356		221.750	128.816	0.26558	48.152
1322		212.916	122.331	0.25901	43.739	1357		222.010	129.007	0.26577	48.292
1323		213.176	122.522	0.25920	43.866	1358		222.270	129.198	0.26597	48.418
1324		213.436	122.712	0.25940	43.993	1359		222.530	129.389	0.26616	48.558
1325		213.695	+122.903	0.25959	44.119	1360	900.3	222.790	+129.590	0.26636	48.697
1326		213.954	123.093	0.25978	44.246	1361		223.050	129.782	0.26655	48.824
1327		214.213	123.284	0.25998	44.373	1362		223.310	129.974	0.26674	48.964
1328		214.472	123.474	0.26017	44.500	1363		223.570	130.166	0.26693	49.103
1329		214.731	123.665	0.26037	44.627	1364		223.830	130.358	0.26712	49.243
1330	870.3	214.990	+123.855	0.26056	44.753	1365		224.090	+130.550	0.26731	49.369
1331		215.250	124.046	0.26075	44.880	1366		224.350	130.742	0.26750	49.509
1332		215.510	124.236	0.26095	45.007	1367		224.610	130.934	0.26769	49.648
1333		215.770	124.427	0.26114	45.134	1368		224.870	131.126	0.26788	49.788
1334		216.030	124.617	0.26134	45.260	1369		225.130	131.318	0.26807	49.927
1335		216.290	+124.807	0.26153	45.387	1370	910.3	225.390	+131.510	0.26826	50.067
1336		216.550	124.998	0.26172	45.514	1371		225.650	131.702	0.26845	50.206
1337		216.810	125.189	0.26192	45.641	1372		225.910	131.894	0.26864	50.346
1338		217.070	125.379	0.26211	45.780	1373		226.170	132.086	0.26883	50.485
1330		217.330	125.569	0.26231	45.907	1374		226.430	132.278	0.26902	50.624
1340	880.3	217.590	+125.760	0.26251	46.034	1375		226.690	+132.470	0.26921	50.764
1341		217.850	125.951	0.26270	46.161	1376		226.950	132.662	0.26940	50.903
1342		218.110	126.142	0.26289	46.300	1377		227.210	132.854	0.26959	51.043
1343		218.370	126.333	0.26309	46.427	1378		227.470	133.046	0.26978	51.182
1344		218.630	126.524	0.26328	46.554	1379		227.730	133.238	0.26997	51.322
Difference or c_p, c_v		c_p = 0.258, 0.260	c_v = 0.190, 0.191	19				c_p = 0.260	c_v = 0.191, 0.192	19	

TABLE 2-4. THERMODYNAMIC PROPERTIES OF AIR AT LOW PRESSURES.—(*Continued*)

°R	°F	h, Btu per lb	u, Btu per lb	s_p	p_r	°R	°F	h, Btu per lb	u, Btu per lb	s_p	p_r
1380	920.3	227.990	+133.430	0.27016	51.464	1420	960.3	238.450	+141.140	0.27763	57.394
1381		228.251	133.622	0.27035	51.616	1421		238.712	141.334	0.27781	57.546
1382		228.512	133.814	0.27054	51.756	1422		238.974	141.528	0.27800	57.711
1383		228.773	134.006	0.27072	51.895	1423		239.236	141.722	0.27818	57.863
1384		229.034	134.198	0.27091	52.035	1424		239.498	141.916	0.27836	58.016
1385		229.295	+134.390	0.27110	52.187	1425		239.760	+142.110	0.27855	58.168
1386		229.556	134.582	0.27129	52.326	1426		240.022	142.304	0.27873	58.333
1387		229.817	134.774	0.27148	52.466	1427		240.284	142.498	0.27891	58.485
1388		230.078	134.966	0.27166	52.618	1428		240.546	142.692	0.27909	58.637
1389		230.339	135.158	0.27185	52.757	1429		240.808	142.886	0.27928	58.802
1390	930.3	230.600	+135.350	0.27204	52.909	1430	970.3	241.070	+143.080	0.27946	58.954
1391		230.861	135.542	0.27223	53.049	1431		241.332	143.274	0.27964	59.119
1392		231.122	135.734	0.27242	53.188	1432		241.594	143.468	0.27983	59.271
1393		231.383	135.926	0.27260	53.341	1433		241.856	143.662	0.28001	59.436
1394		231.644	136.118	0.27279	53.480	1434		242.118	143.856	0.28019	59.588
1395		231.905	+136.310	0.27298	53.632	1435		242.380	+144.050	0.28038	59.753
1396		232.166	136.502	0.27317	53.772	1436		242.642	144.244	0.28056	59.905
1397		232.427	136.694	0.27336	53.924	1437		242.904	144.438	0.28074	60.070
1398		232.688	136.886	0.27354	54.076	1438		243.166	144.632	0.28092	60.222
1399		232.949	137.078	0.27373	54.215	1439		243.428	144.826	0.28111	60.387
1400	940.3	233.210	+137.280	0.27392	54.368	1440	980.3	243.690	+145.020	0.28130	60.552
1401		233.472	137.473	0.27411	54.520	1441		243.953	145.214	0.28148	60.704
1402		233.734	137.666	0.27429	54.659	1442		244.216	145.408	0.28168	60.869
1403		233.996	137.859	0.27448	54.811	1443		244.479	145.602	0.28184	61.034
1404		234.258	138.052	0.27466	54.963	1444		244.742	145.796	0.28202	61.199
1405		234.520	+138.245	0.27485	55.116	1445		245.005	+145.990	0.28221	61.363
1406		234.782	138.438	0.27503	55.255	1446		245.268	146.184	0.28239	61.516
1407		235.044	138.631	0.27522	55.407	1447		245.531	146.378	0.28257	61.680
1408		235.306	138.824	0.27540	55.559	1448		245.794	146.572	0.28275	61.845
1409		235.568	139.017	0.27559	55.712	1449		246.057	146.766	0.28293	62.010
1410	950.3	235.830	+139.210	0.27577	55.864	1450	990.3	246.320	+146.960	0.28311	62.175
1411		236.092	139.403	0.27596	56.016	1451		246.583	147.154	0.28329	62.340
1412		236.354	139.596	0.27614	56.168	1452		246.846	147.348	0.28347	62.505
1413		236.616	139.789	0.27633	56.320	1453		247.109	147.542	0.28365	62.670
1414		236.878	139.982	0.27651	56.472	1454		247.372	147.736	0.28383	62.834
1415		237.140	+140.175	0.27670	56.624	1455		247.635	+147.930	0.28402	62.999
1416		237.402	140.368	0.27688	56.777	1456		247.898	148.124	0.28420	63.177
1417		237.664	140.561	0.27707	56.929	1457		248.161	148.318	0.28438	63.342
1418		237.926	140.754	0.27725	57.081	1458		248.424	148.512	0.28456	63.506
1419		238.188	140.947	0.27744	57.233	1459		248.687	148.706	0.28474	63.671
Difference or c_p, c_v		c_p = 0.261, 0.262	c_v = 0.192, 0.193	19				c_p = 0.262, 0.263	c_v = 0.194	18	

TABLE 2-4. THERMODYNAMIC PROPERTIES OF AIR AT LOW PRESSURES.—(*Continued*)

°R	°F	*h*, Btu per lb	*u*, Btu per lb	s_p	p_r	°R	°F	*h*, Btu per lb	*u*, Btu per lb	s_p	p_r
1460	1000.3	248.950	+148.900	0.28493	63.844	1495		258.180	+155.740	0.29117	69.943
1461		249.213	149.095	0.28511	64.009	1496		258.444	155.936	0.29134	70.121
1462		249.476	149.290	0.28529	64.186	1497		258.708	156.132	0.29152	70.311
1463		249.739	149.485	0.28547	64.351	1498		258.972	156.328	0.29170	70.489
1464		250.002	149.680	0.28565	64.516	1499		259.236	156.524	0.29187	70.666
1465		250.265	+149.875	0.28583	64.693	1500	1040.3	259.500	+156.720	0.29206	70.857
1466		250.528	150.070	0.28600	64.858	1501		259.765	156.916	0.29224	71.034
1467		250.791	150.265	0.28618	65.023	1502		260.030	157.112	0.29241	71.212
1468		251.054	150.460	0.28636	65.201	1503		260.295	157.308	0.29259	71.402
1469		251.317	150.655	0.28654	65.366	1504		260.560	157.504	0.29276	71.580
1470	1010.3	251.580	+150.850	0.28672	65.543	1505		260.825	+157.700	0.29294	71.770
1471		251.844	151.045	0.28690	65.721	1506		261.090	157.896	0.29311	71.948
1472		252.108	151.240	0.28708	65.886	1507		261.355	158.092	0.29329	72.138
1473		252.372	151.435	0.28726	66.050	1508		261.620	158.288	0.29346	72.315
1474		252.636	151.630	0.28744	66.228	1509		261.885	158.484	0.29364	72.506
1475		252.900	+151.825	0.28762	66.406	1510	1050.3	262.150	+158.680	0.29381	72.683
1476		253.164	152.020	0.28779	66.570	1511		262.415	158.876	0.29399	72.874
1477		253.428	152.215	0.28797	66.748	1512		262.680	159.072	0.29416	73.064
1478		253.692	152.410	0.28815	66.926	1513		262.945	159.268	0.29434	73.241
1479		253.956	152.605	0.28833	67.103	1514		263.210	159.464	0.29451	73.432
1480	1020.3	254.220	+152.800	0.28851	67.270	1515		263.475	+159.660	0.29469	73.622
1481		254.484	152.996	0.28869	67.446	1516		263.740	159.856	0.29486	73.812
1482		255.748	153.192	0.28886	67.623	1517		264.005	160.052	0.29504	74.003
1483		255.012	153.388	0.28904	67.801	1518		264.270	160.248	0.29521	74.193
1484		255.276	153.584	0.28922	67.978	1519		264.535	160.444	0.29539	74.370
1485		255.540	+153.780	0.28940	68.143	1520	1060.3	264.800	+160.640	0.29556	74.561
1486		255.804	153.976	0.28957	68.321	1521		265.065	170.837	0.29573	74.751
1487		256.068	154.172	0.28975	68.498	1522		265.330	161.034	0.29591	74.941
1488		256.332	154.368	0.28993	68.676	1523		265.595	161.231	0.29608	75.131
1489		256.596	154.564	0.29010	68.853	1524		265.860	161.428	0.29625	75.322
1490	1030.3	256.860	+154.760	0.29028	69.031	1525		266.125	+161.625	0.29643	75.512
1491		257.124	154.956	0.29046	69.221	1526		266.390	161.822	0.29660	75.715
1492		257.388	155.125	0.29063	69.399	1527		266.655	162.019	0.29677	75.905
1493		257.652	155.348	0.29081	69.576	1528		266.920	162.216	0.29694	76.096
1494		257.916	155.544	0.29099	69.754	1529		267.185	162.413	0.29712	76.286
Difference or c_p, c_v		c_p = 0.263, 0.264	c_v = 0.195, 0.196	18				c_p = 0.264, 0.265	c_v = 0.196, 0.197	17	

TABLE 2-4. THERMODYNAMIC PROPERTIES OF AIR AT LOW PRESSURES.—(*Continued*)

°R	°F	h, Btu per lb	u, Btu per lb	s_p	p_r	°R	°F	h, Btu per lb	u, Btu per lb	s_p	p_r
1530	1070.3	267.450	+162.610	0.29729	76.478	1570	1110.3	278.090	+170.500	0.30416	84.525
1531		267.716	162.807	0.29746	76.681	1571		278.367	170.699	0.30433	84.741
1532		267.982	163.004	0.29764	76.871	1572		278.634	170.897	0.30450	84.957
1533		268.248	163.201	0.29781	77.061	1573		278.901	171.096	0.30467	85.160
1534		268.514	163.398	0.29798	77.264	1574		279.168	171.294	0.30484	85.375
1535		268.780	+163.595	0.29816	77.455	1575		279.435	+171.493	0.30501	85.578
1536		269.046	163.792	0.29833	77.645	1576		279.702	171.691	0.30517	85.794
1537		269.312	163.989	0.29850	77.848	1577		279.969	171.890	0.30534	86.010
1538		269.578	164.186	0.29867	78.038	1578		280.236	172.088	0.30551	86.225
1539		269.844	164.383	0.29885	78.241	1579		280.503	172.287	0.30568	86.428
1540	1080.3	270.110	+164.580	0.29903	78.431	1580	1120.3	280.760	+172.485	0.30586	86.644
1541		270.376	164.777	0.29920	78.634	1581		281.027	172.684	0.30603	86.860
1542		270.642	164.974	0.29937	78.825	1582		281.294	172.882	0.30620	87.075
1543		270.908	165.171	0.29955	79.028	1583		281.561	173.081	0.30636	87.291
1544		271.174	165.368	0.29972	79.231	1584		281.828	173.279	0.30653	87.507
1545		271.440	+165.565	0.29989	79.421	1585		282.095	+173.478	0.30670	87.722
1546		271.706	165.762	0.30006	79.624	1586		282.362	173.676	0.30687	87.938
1547		271.972	165.959	0.30023	79.827	1587		282.629	173.875	0.30704	88.154
1548		272.238	166.156	0.30041	80.030	1588		282.896	174.073	0.30720	88.369
1549		272.504	166.353	0.30058	80.220	1589		283.163	174.272	0.30737	88.585
1550	1090.3	272.770	+166.550	0.30075	80.423	1590	1130.3	283.430	+174.470	0.30754	88.801
1551		273.036	166.747	0.30092	80.626	1591		283.697	174.669	0.30771	89.016
1552		273.302	166.944	0.30109	80.829	1592		283.964	174.868	0.30788	89.232
1553		273.568	167.141	0.30127	81.032	1593		284.231	175.067	0.30804	89.448
1554		273.834	167.338	0.30144	81.235	1594		284.498	175.266	0.30821	89.670
1555		274.100	+167.535	0.30161	81.438	1595		284.765	+175.465	0.30838	89.886
1556		274.366	167.732	0.30178	81.641	1596		285.032	175.664	0.30855	90.106
1557		274.632	167.929	0.30195	81.844	1597		285.299	175.863	0.30872	90.330
1558		274.898	168.126	0.30213	82.047	1598		285.566	176.062	0.30888	90.546
1559		275.164	168.323	0.30230	82.250	1599		285.833	176.261	0.30905	90.766
1560	1100.3	275.430	+168.520	0.30247	82.453	1600	1140.3	286.100	+176.460	0.30922	90.990
1561		275.697	168.718	0.30264	82.655	1601		286.368	176.659	0.30939	91.209
1562		275.964	168.916	0.30281	82.871	1602		286.636	176.858	0.30955	91.433
1563		276.231	169.114	0.30298	83.074	1603		286.904	177.057	0.30972	91.652
1564		276.498	169.312	0.30315	83.277	1604		287.172	177.256	0.30988	91.877
1565		276.765	+169.510	0.30332	83.480	1605		287.440	+177.455	0.31005	92.106
1566		277.032	169.708	0.30348	83.696	1606		287.708	177.654	0.30122	92.321
1567		277.299	169.906	0.30365	83.899	1607		287.976	177.853	0.31038	92.550
1568		277.566	170.104	0.30382	84.114	1607		288.244	178.052	0.31055	92.778
1569		277.833	170.302	0.30399	84.317	1609		288.512	178.251	0.31071	92.994
Difference or c_p, c_v		$c_p =$ 0.266	$c_v =$ 0.197, 0.198	17				$c_p =$ 0.267, 0.268	$c_v =$ 0.198, 0.199	16	

TABLE 2-4. THERMODYNAMIC PROPERTIES OF AIR AT LOW PRESSURES.—(*Continued*)

°R	°F	h, Btu per lb	u, Btu per lb	s_p	p_r
1610	1150.3	288.780	+178.450	0.31088	93.244
1611		289.048	178.649	0.31105	93.472
1612		289.316	178.848	0.31121	93.701
1613		289.584	179.047	0.31138	93.929
1614		289.852	179.246	0.31154	94.158
1615		290.120	+179.445	0.31171	94.386
1616		290.388	179.644	0.31188	94.614
1617		290.656	179.843	0.31204	94.843
1618		290.924	180.042	0.31221	95.071
1619		291.192	180.241	0.31237	95.300
1620	1160.3	291.460	+180.440	0.31255	95.528
1621		291.728	180.640	0.31271	95.756
1622		291.996	180.840	0.31288	95.985
1623		292.264	181.040	0.31304	96.226
1624		292.532	181.240	0.31321	96.454
1625		292.800	+181.440	0.31337	96.683
1626		293.068	181.640	0.31353	96.911
1627		293.336	181.840	0.31370	97.152
1628		293.604	182.040	0.31386	97.380
1629		293.872	182.240	0.31403	97.621
1630	1170.3	294.140	+182.440	0.31419	97.850
1631		294.408	182.640	0.31435	98.091
1632		294.676	182.840	0.31452	98.319
1633		294.944	183.040	0.31468	98.560
1634		295.212	183.240	0.31485	98.801
1635		295.480	+183.440	0.31501	99.030
1636		295.748	183.640	0.31517	99.271
1637		296.016	183.840	0.31534	99.639
1638		296.284	184.040	0.31550	99.740
1639		296.552	184.240	0.31567	99.981
1640	1180.3	296.830	+184.440	0.31584	100.223
1641		297.099	184.640	0.31600	100.464
1642		297.367	184.840	0.31617	100.705
1643		297.636	185.040	0.31633	100.946
1644		297.904	185.240	0.31649	101.187
1645		298.173	+185.440	0.31666	101.428
1646		298.441	185.640	0.31682	101.669
1647		298.710	185.840	0.31698	101.910
1648		298.978	186.040	0.31714	102.151
1649		299.247	186.240	0.31731	102.392
Difference or c_p, c_v		c_p = 0.268, 0.269	c_v = 0.199, 0.200	17	

°R	°F	h, Btu per lb	u, Btu per lb	s_p	p_r
1650	1190.3	299.515	+186.440	0.31747	102.647
1651		299.784	186.640	0.31763	102.888
1652		300.052	186.840	0.31780	103.129
1653		300.321	187.040	0.31796	103.383
1654		300.589	187.240	0.31812	103.624
1655		300.858	+187.441	0.31829	103.865
1656		301.126	187.643	0.31845	104.119
1657		301.395	187.845	0.31861	104.360
1658		301.663	188.047	0.31877	104.614
1659		301.932	188.249	0.31894	104.855
1660	1200.3	302.200	+188.450	0.31910	105.109
1661		302.469	188.651	0.31926	105.350
1662		302.738	188.852	0.31942	105.604
1663		303.007	189.053	0.31958	105.857
1664		303.276	189.254	0.31974	106.099
1665		303.545	+189.455	0.31991	106.352
1666		303.814	189.656	0.32007	106.606
1667		304.083	189.857	0.32023	106.860
1668		304.352	190.058	0.32039	107.114
1669		304.621	190.259	0.32055	107.355
1670	1210.3	304.890	+190.460	0.32071	107.609
1671		305.160	190.661	0.32087	107.862
1672		305.430	190.862	0.32103	108.116
1673		305.700	191.063	0.32119	108.370
1674		305.970	191.264	0.32135	108.624
1675		306.240	+191.465	0.32152	108.878
1676		306.510	191.666	0.32168	109.144
1677		306.780	191.867	0.32184	109.398
1678		307.050	192.068	0.32200	109.652
1679		307.320	192.269	0.32216	109.905
1680	1220.3	307.590	+192.470	0.32233	110.172
1681		307.860	192.671	0.32249	110.426
1682		308.130	192.872	0.32265	110.680
1683		308.400	193.073	0.32281	110.946
1684		308.670	193.274	0.32297	111.200
1685		308.940	+193.475	0.32313	111.466
1686		309.210	193.676	0.32328	111.720
1687		309.480	193.877	0.32344	111.987
1688		309.750	194.078	0.32360	112.240
1689		310.020	194.279	0.32376	112.507
Difference or c_p, c_v		c_p = 0.269, 0.270	c_v = 0.200, 0.201	16	

TABLE 2-4. THERMODYNAMIC PROPERTIES OF AIR AT LOW PRESSURES.—(*Continued*)

°R	°F	h, Btu per lb	u, Btu per lb	s_p	p_r	°R	°F	h, Btu per lb	u, Btu per lb	s_p	p_r
1690	1230.3	310.290	+194.480	0.32392	112.794	1730	1270.3	321.110	+202.560	0.33025	123.703
1691		310.560	194.682	0.32408	113.048	1731		321.381	202.763	0.33041	123.982
1692		310.830	194.884	0.32424	113.312	1732		321.652	202.966	0.33056	124.262
1693		311.100	195.086	0.32440	113.581	1733		321.923	203.169	0.33072	124.554
1694		311.370	195.288	0.32456	113.847	1734		322.194	203.372	0.33087	124.833
1695		311.640	+195.490	0.32472	114.114	1735		322.465	+203.575	0.33103	125.125
1696		311.910	195.692	0.32487	114.368	1736		322.736	203.778	0.31119	125.404
1697		312.180	195.894	0.32503	114.634	1737		323.007	203.981	0.33134	125.683
1698		312.450	196.096	0.32519	114.901	1738		323.278	204.184	0.33150	125.975
1699		312.720	196.298	0.32535	115.167	1739		323.549	204.387	0.33165	126.267
1700	1240.3	312.990	+196.500	0.32552	115.434	1740	1280.3	323.820	+204.590	0.33182	126.546
1701		313.261	196.702	0.32568	115.700	1741		324.092	204.793	0.33198	126.838
1702		313.532	196.904	0.32584	115.967	1742		324.364	204.996	0.33213	127.118
1703		313.803	197.106	0.32599	116.246	1743		324.636	205.199	0.33229	127.410
1704		314.074	197.308	0.32615	116.513	1744		324.908	205.402	0.32244	127.702
1705		314.345	+197.510	0.32631	116.780	1745		325.180	+205.605	0.33260	127.994
1706		314.616	197.712	0.32647	117.046	1746		325.452	205.808	0.33275	128.286
1707		314.887	197.914	0.32663	117.325	1747		325.724	206.011	0.33291	128.578
1708		315.158	198.116	0.32678	117.591	1748		325.996	206.214	0.33306	128.869
1709		315.429	198.318	0.32696	117.858	1749		326.268	206.417	0.33322	129.161
1710	1250.3	315.700	+198.520	0.32710	118.137	1750	1290.3	326.540	+206.620	0.33337	129.453
1711		315.971	198.722	0.32726	118.404	1751		326.812	206.823	0.33353	129.745
1712		316.241	198.924	0.32742	118.683	1752		327.084	207.026	0.33368	130.037
1713		316.512	199.126	0.32757	118.949	1753		327.356	207.229	0.33384	130.329
1714		316.782	199.328	0.32773	119.229	1754		327.628	207.432	0.33399	130.621
1715		317.053	+199.530	0.32789	119.508	1755		327.900	+207.635	0.33415	130.913
1716		317.323	199.732	0.32805	119.774	1756		328.172	207.838	0.33430	131.218
1717		317.594	199.934	0.32821	120.054	1757		328.444	208.041	0.33446	131.510
1718		317.864	200.136	0.32836	120.333	1758		328.716	208.244	0.33461	131.802
1719		318.135	200.338	0.32852	120.599	1759		328.988	208.447	0.33477	132.106
1720	1260.3	318.405	+200.540	0.32869	120.879	1760	1300.3	329.260	+208.650	0.33493	132.398
1721		318.676	200.742	0.32885	121.158	1761		329.532	208.853	0.33508	132.703
1722		318.946	200.944	0.32900	121.437	1762		329.804	209.056	0.33524	132.995
1723		319.217	201.146	0.32916	121.716	1763		330.076	209.259	0.33539	133.300
1724		319.487	201.348	0.32931	121.996	1764		330.348	209.462	0.33554	133.592
1725		319.758	+201.550	0.32947	122.274	1765		330.620	+209.665	0.33570	133.897
1726		320.028	201.752	0.32963	122.554	1766		330.892	209.868	0.33585	134.201
1727		320.299	201.954	0.32978	122.833	1767		331.164	210.071	0.33600	134.493
1728		320.569	202.156	0.32994	123.112	1768		331.436	210.274	0.33615	134.798
1729		320.840	202.358	0.33009	123.404	1769		331.708	210.477	0.33631	135.102
Difference or c_p, c_v		c_p = 0.270, 0.271	c_v = 0.201, 0.202	16–15				c_p = 0.271, 0.272	c_v = 0.203	16	

TABLE 2-4. THERMODYNAMIC PROPERTIES OF AIR AT LOW PRESSURES.—(*Continued*)

°R	°F	h, Btu per lb	u, Btu per lb	s_p	p_r	°R	°F	h, Btu per lb	u, Btu per lb	s_p	p_r
1770	1310.3	331.980	+210.680	0.33646	135.417	1810	1350.3	342.880	+218.850	0.34256	147.986
1771		332.252	210.884	0.33661	135.722	1811		343.154	219.055	0.34271	148.316
1772		332.524	211.088	0.33677	136.027	1812		343.428	219.260	0.34286	148.633
1773		332.796	211.292	0.33692	136.332	1813		343.702	219.465	0.34301	148.963
1774		333.068	211.496	0.33707	136.636	1814		343.976	219.670	0.34316	149.293
1775		333.340	+211.700	0.33723	136.941	1815		344.250	+219.875	0.34332	149.623
1776		333.612	211.904	0.33738	137.246	1816		344.524	220.080	0.34347	149.953
1777		333.884	212.108	0.33753	137.550	1817		344.798	220.285	0.34362	150.238
1778		334.156	212.312	0.33768	137.868	1818		345.072	220.490	0.34377	150.613
1779		334.428	212.516	0.33784	138.172	1819		345.346	220.695	0.34392	150.944
1780	1320.3	334.700	+212.720	0.33800	138.477	1820	1360.3	345.620	+220.900	0.34407	151.274
1781		334.973	212.924	0.33815	138.794	1821		345.894	221.105	0.34422	151.604
1782		335.246	213.128	0.33830	139.099	1822		346.168	221.310	0.34437	151.934
1783		335.519	213.332	0.33846	139.416	1823		346.442	221.515	0.34452	152.264
1784		335.792	213.536	0.33861	139.721	1824		346.716	221.720	0.34467	152.594
1785		336.065	+213.740	0.33876	140.026	1825		346.990	+221.925	0.34482	152.937
1786		336.338	213.944	0.33891	140.343	1826		347.264	222.130	0.34496	153.267
1787		336.611	214.148	0.33906	140.661	1827		347.538	222.335	0.34511	153.597
1788		336.884	214.352	0.33922	140.965	1828		347.812	222.540	0.34526	153.940
1789		337.157	214.556	0.33937	141.283	1829		348.086	222.745	0.34541	154.270
1790	1330.3	337.430	+214.760	0.33952	141.600	1830	1370.3	348.360	+222.950	0.34556	154.612
1791		337.703	214.964	0.33967	141.905	1831		348.634	223.155	0.34571	154.942
1792		337.975	215.168	0.33982	142.222	1832		348.908	223.360	0.34586	155.285
1793		338.248	215.372	0.33998	142.539	1833		349.182	223.565	0.34601	155.628
1794		338.520	215.576	0.34013	142.857	1834		349.456	223.770	0.34616	155.958
1795		338.793	+215.780	0.34028	143.174	1835		349.730	+223.975	0.34631	156.301
1796		339.065	215.984	0.34043	143.492	1836		350.004	224.180	0.34645	156.644
1797		339.338	216.188	0.34058	143.809	1837		350.278	224.385	0.34660	156.986
1798		339.610	216.392	0.34074	144.126	1838		350.552	224.590	0.34675	157.329
1799		339.883	216.596	0.34089	144.444	1839		350.826	224.795	0.34690	157.672
1800	1340.3	340.155	+216.800	0.34105	144.761	1840	1380.3	351.100	+225.000	0.34706	158.015
1801		340.428	217.005	0.34120	145.078	1841		351.374	225.205	0.34721	158.357
1802		340.700	217.210	0.34135	145.409	1842		351.648	225.410	0.34736	158.700
1803		340.973	217.415	0.34150	145.726	1843		351.922	225.615	0.34750	159.043
1804		341.245	217.620	0 34165	146.043	1844		352.196	225.820	0.34765	159.386
1805		341.518	+217.825	0.34181	146.373	1845		352.470	+226.025	0.34780	159.728
1806		341.790	218.030	0.34196	146.691	1846		352.744	226.230	0.34795	160.071
1807		342.063	218.235	0.34211	147.008	1847		353.018	226.435	0.34810	160.414
1808		342.335	218.440	0.34226	147.338	1848		353.292	226.640	0.34824	160.769
1809		342.608	218.645	0.34241	147.668	1849		353.566	226.845	0.34839	161.112
Difference or c_p, c_v		c_p = 0.272, 0.273	c_v = 0.204, 0.205	15				c_p = 0.274	c_v = 0.204, 0.205	15	

TABLE 2-4. THERMODYNAMIC PROPERTIES OF AIR AT LOW PRESSURES.—(*Continued*)

°R	°F	h, Btu per lb	u, Btu per lb	s_p	p_r	°R	°F	h, Btu per lb	u, Btu per lb	s_p	p_r
1850	1390.3	353.840	+227.050	0.34854	161.468	1890	1430.3	364.810	+235.300	0.35441	175.917
1851		354.114	227.256	0.34869	161.823	1891		365.085	235.507	0.35456	176.285
1852		354.388	227.462	0.34884	162.166	1892		365.360	235.714	0.35470	176.666
1853		354.662	227.668	0.34898	162.521	1893		365.635	235.921	0.35485	177.034
1854		354.936	227.874	0.34913	162.877	1894		365.910	236.128	0.35499	177.415
1855		355.210	+228.080	0.34928	163.220	1895		366.185	+236.335	0.35514	177.796
1856		355.484	228.286	0.34943	163.575	1896		366.460	236.542	0.35529	178.164
1857		355.758	228.492	0.34958	163.931	1897		366.735	236.749	0.35543	178.545
1858		356.032	228.698	0.34972	164.286	1898		367.010	236.956	0.35558	178.926
1859		356.306	228.904	0.34987	164.629	1899		367.285	237.163	0.35572	179.307
1860	1400.3	356.580	+229.110	0.35002	164.985	1900	1440.3	367.560	+237.370	0.35587	179.675
1861		356.854	229.316	0.35017	165.340	1901		367.836	237.577	0.35601	180.056
1862		357.128	229.522	0.35031	165.695	1902		368.112	237.784	0.35616	180.437
1863		357.402	229.728	0.35046	166.051	1903		368.388	237.991	0.35630	180.818
1864		357.676	229.934	0.35060	166.406	1904		368.664	238.198	0.35645	181.199
1865		357.950	+230.140	0.35075	166.762	1905		368.940	+238.405	0.35659	181.580
1866		358.224	230.346	0.35090	167.130	1906		369.216	238.612	0.35673	181.961
1867		358.498	230.552	0.35104	167.486	1907		369.492	238.819	0.35688	182.342
1868		358.772	230.758	0.35119	167.841	1908		369.768	239.026	0.35702	182.723
1869		359.046	230.964	0.35133	168.209	1909		370.044	239.233	0.35717	183.116
1870	1410.3	359.320	+231.170	0.35148	168.565	1910	1450.3	370.320	+239.440	0.35731	183.510
1871		359.594	231.376	0.35163	168.920	1911		370.596	239.647	0.35745	183.891
1872		359.868	231.582	0.35177	169.288	1912		370.872	239.854	0.35760	184.284
1873		360.142	231.788	0.35192	169.657	1913		371.148	240.061	0.35774	184.678
1874		360.416	231.994	0.35206	170.012	1914		371.424	240.268	0.35789	185.059
1875		360.690	+232.200	0.35221	170.380	1915		371.700	+240.475	0.35803	185.452
1876		360.964	232.406	0.35236	170.749	1916		371.976	240.682	0.35817	185.846
1877		361.238	232.612	0.35250	171.117	1917		372.252	240.889	0.35832	186.227
1878		361.512	232.818	0.35265	171.485	1918		372.528	241.096	0.35846	186.621
1879		361.786	233.024	0.35279	171.840	1919		372.804	241.303	0.35861	187.014
1880	1420.3	362.060	+233.230	0.35295	172.209	1920	1460.3	373.080	+241.510	0.35876	187.395
1881		362.335	233.437	0.35310	172.577	1921		373.356	241.717	0.35890	187.789
1882		362.610	233.644	0.35324	172.945	1922		373.632	241.924	0.35905	188.182
1883		362.885	233.851	0.35339	173.313	1923		373.908	242.131	0.35919	188.576
1884		363.160	234.058	0.35353	173.681	1924		374.184	242.338	0.35933	188.969
1885		363.435	+234.265	0.35368	174.062	1925		374.460	+242.545	0.35948	189.363
1886		363.710	234.472	0.35383	174.430	1926		374.736	242.752	0.35962	189.757
1887		363.985	234.679	0.35397	174.786	1927		375.012	242.959	0.35976	190.150
1888		364.260	234.886	0.35412	175.154	1928		375.288	243.166	0.35990	190.544
1889		364.535	235.093	0.35426	175.538	1929		375.564	243.373	0.36005	190.950
Difference or c_p, c_v		$c_p =$ 0.274, 0.275	$c_v =$ 0.206, 0.207	15				$c_p =$ 0.275, 0.276	$c_v =$ 0.207	15	

TABLE 2-4. THERMODYNAMIC PROPERTIES OF AIR AT LOW PRESSURES.—(*Continued*)

°R	°F	h, Btu per lb	u, Btu per lb	s_p	p_r	°R	°F	h, Btu per lb	u, Btu per lb	s_p	p_r
1930	1470.3	375.840	243.580	0.36019	191.344	1970	1510.3	386.900	251.900	0.36585	207.903
1931		376.116	243.788	0.36033	191.750	1971		387.177	252.109	0.36599	208.322
1932		376.392	243.996	0.36048	192.144	1972		387.454	252.318	0.36613	208.754
1933		376.668	244.204	0.36062	192.550	1973		387.731	252.527	0.36627	209.185
1934		376.944	244.412	0.36076	192.944	1974		388.008	252.736	0.36641	209.617
1935		377.220	244.620	0.36091	193.337	1975		388.285	252.945	0.36655	210.036
1936		377.496	244.828	0.36105	193.743	1976		388.562	253.154	0.36669	210.468
1937		377.772	245.036	0.36119	194.150	1977		388.839	253.363	0.36683	210.900
1938		378.048	245.244	0.36133	194.556	1978		389.116	253.572	0.36697	211.332
1939		378.324	245.452	0.36148	194.962	1979		389.393	253.781	0.36711	211.764
1940	1480.3	378.600	245.660	0.36162	195.369	1980	1520.3	389.670	253.990	0.36726	212.196
1941		378.876	245.868	0.36176	195.775	1981		389.947	254.199	0.36740	212.627
1942		379.152	246.076	0.36190	196.181	1982		390.224	254.408	0.36754	213.059
1943		379.428	246.284	0.36204	196.588	1983		390.501	254.617	0.36768	213.491
1944		379.704	246.492	0.36218	196.994	1984		390.778	254.826	0.36782	213.923
1945		379.980	246.700	0.36233	197.413	1985		391.055	255.035	0.36796	214.355
1946		380.256	246.908	0.36247	197.832	1986		391.332	255.244	0.36809	214.799
1947		380.532	247.116	0.36261	198.238	1987		391.609	255.453	0.36823	215.231
1948		380.808	247.324	0.36275	198.645	1988		391.886	255.662	0.36837	215.676
1949		381.084	247.532	0.36289	199.064	1989		392.163	255.871	0.36851	216.120
1950	1490.3	381.360	247.740	0.36303	199.470	1990	1530.3	392.440	256.080	0.36865	216.552
1951		381.637	247.948	0.36317	199.889	1991		392.718	256.289	0.36879	216.997
1952		381.914	248.156	0.36331	200.308	1992		392.996	256.498	0.36893	217.441
1953		382.191	248.364	0.36345	200.714	1993		393.274	256.707	0.36907	217.873
1954		382.468	248.572	0.36359	201.133	1994		393.552	256.916	0.36921	218.317
1955		382.745	248.780	0.36374	201.539	1995		393.830	257.125	0.36935	218.775
1956		383.022	248.988	0.36388	201.958	1996		394.108	257.334	0.36948	219.219
1957		383.299	249.196	0.36402	202.377	1997		394.386	257.543	0.36962	219.664
1958		383.576	249.404	0.36416	202.796	1998		394.664	257.752	0.36976	220.108
1959		383.853	249.612	0.36430	203.203	1999		394.942	257.961	0.36990	220.553
1960	1500.3	384.130	249.820	0.36445	203.622	2000	1540.3	395.220	258.170	0.37005	220.997
1961		384.407	250.028	0.36459	204.041	2001		395.498	258.379	0.37019	221.442
1962		384.684	250.236	0.36473	204.460	2002		395.776	258.588	0.37033	221.886
1963		384.961	250.444	0.36487	204.879	2003		396.054	258.797	0.37046	222.331
1964		385.238	250.652	0.36501	205.298	2004		396.332	259.006	0.37060	222.788
1965		385.515	250.860	0.36515	205.717	2005		396.610	259.215	0.37074	223.245
1966		385.792	251.068	0.36529	206.148	2006		396.888	259.424	0.37088	223.690
1967		386.069	251.276	0.36543	206.567	2007		397.166	259.633	0.37102	224.147
1968		386.346	251.484	0.36557	206.986	2008		397.444	259.842	0.37115	224.592
1969		386.623	251.692	0.36571	207.418	2009		397.722	260.051	0.37129	225.049
Difference or c_p, c_v		$c_p =$ 0.276, 0.277	$c_v =$ 0.208	14				$c_p =$ 0.277, 0.278	$c_v =$ 0.209	14	

TABLE 2-4. THERMODYNAMIC PROPERTIES OF AIR AT LOW PRESSURES.—(*Continued*)

°R	°F	h, Btu per lb	u, Btu per lb	s_p	p_r	°R	°F	h, Btu per lb	u, Btu per lb	s_p	p_r
2010	1550.3	398.000	260.260	0.37143	225.524	2050	1590.3	409.130	268.650	0.37692	244.311
2011		398.278	260.469	0.37157	225.981	2051		409.409	268.860	0.37706	244.793
2012		398.556	260.678	0.37171	226.439	2052		409.688	269.070	0.37719	245.276
2013		398.834	260.887	0.37184	226.896	2053		409.967	269.280	0.37733	245.772
2014		399.112	261.096	0.37198	227.353	2054		410.246	269.490	0.37746	246.254
2015		399.390	261.305	0.37212	227.810	2055		410.525	269.700	0.37760	246.750
2016		399.668	261.514	0.37226	228.268	2056		410.804	269.910	0.37774	247.245
2017		399.946	261.723	0.37240	228.725	2057		411.083	270.120	0.37787	247.728
2018		400.224	261.932	0.37253	229.182	2058		411.362	270.330	0.37801	248.223
2019		400.502	262.141	0.37267	229.652	2059		411.641	270.540	0.37814	248.719
2020	1560.3	400.780	262.350	0.37282	230.109	2060	1600.3	411.920	270.750	0.37828	249.202
2021		401.058	262.560	0.37296	230.567	2061		412.199	270.961	0.37842	249.697
2022		401.336	262.770	0.37309	231.037	2062		412.478	271.172	0.37855	250.193
2023		401.614	262.980	0.37323	231.494	2063		412.757	271.383	0.37869	250.675
2024		401.892	263.190	0.37337	231.964	2064		413.036	271.594	0.37882	251.171
2025		402.170	263.400	0.37351	232.434	2065		413.315	271.805	0.37896	251.679
2026		402.448	263.610	0.37364	232.891	2066		413.594	272.016	0.37909	252.174
2027		402.726	263.820	0.37378	233.348	2067		413.873	272.227	0.37923	252.657
2028		403.004	264.030	0.37392	233.818	2068		414.152	272.438	0.37936	253.153
2029		403.282	264.240	0.37405	234.276	2069		414.431	272.649	0.37950	253.661
2030	1570.3	403.560	264.450	0.37419	234.746	2070	1610.3	414.710	272.860	0.37963	254.156
2031		403.838	264.660	0.37433	235.216	2071		414.989	273.071	0.37977	254.652
2032		404.116	264.870	0.37446	235.686	2072		415.268	273.282	0.37990	255.147
2033		404.394	265.080	0.37460	236.156	2073		415.547	273.493	0.38004	255.655
2034		404.672	265.290	0.37474	236.626	2074		415.826	273.704	0.38017	256.151
2035		404.950	265.500	0.37488	237.108	2075		416.105	273.915	0.38031	256.659
2036		405.228	265.710	0.37501	237.578	2076		416.384	274.126	0.38044	257.167
2037		405.506	265.920	0.37515	238.061	2077		416.663	274.337	0.38058	257.663
2038		405.784	266.130	0.37529	238.531	2078		416.942	274.548	0.38071	258.171
2039		406.062	266.340	0.37542	239.001	2079		417.221	274.759	0.38085	258.679
2040	1580.3	406.340	266.550	0.37556	239.484	2080	1620.3	417.500	274.970	0.38098	259.187
2041		406.619	266.760	0.37570	239.953	2081		417.780	275.181	0.38111	259.695
2042		406.898	266.970	0.37583	240.423	2082		418.060	275.392	0.38125	260.203
2043		407.177	267.180	0.37597	240.906	2083		418.340	275.603	0.38138	260.711
2044		407.456	267.390	0.37610	241.389	2084		418.620	275.814	0.38151	261.220
2045		407.735	267.600	0.37624	241.871	2085		418.900	276.025	0.38165	261.741
2046		408.014	267.810	0.37638	242.354	2086		419.180	276.236	0.38178	262.249
2047		408.293	268.020	0.37651	242.837	2087		419.460	276.447	0.38191	262.757
2048		408.572	268.230	0.37665	243.320	2088		419.740	276.658	0.38204	263.278
2049		408.851	268.440	0.37678	243.802	2089		420.020	276.869	0.38218	263.799
Difference or c_p, c_v		c_p = 0.278, 0.279	c_v = 0.209, 0.210	14				c_p = 0.279, 0.280	c_v = 0.210, 0.211	14	

TABLE 2-4. THERMODYNAMIC PROPERTIES OF AIR AT LOW PRESSURES.—(*Continued*)

°R	°F	h, Btu per lb	u, Btu per lb	s_p	p_r	°R	°F	h, Btu per lb	u, Btu per lb	s_p	p_r
2090	1630.3	420.300	+277.080	0.38231	264.319	2130	1670.3	431.490	+285.540	0.38762	285.631
2091		420.580	277.291	0.38244	264.828	2131		431.771	285.752	0.38775	286.139
2092		420.859	277.502	0.38258	265.348	2132		432.052	285.964	0.38788	286.774
2093		421.139	277.713	0.38271	265.869	2133		432.333	286.176	0.38801	287.283
2094		421.418	277.924	0.38284	266.390	2134		432.614	286.388	0.38814	287.791
2095		421.698	+278.135	0.38298	266.911	2135		432.895	+286.600	0.38828	288.299
2096		421.977	278.346	0.38311	267.419	2136		433.176	286.812	0.38841	288.934
2097		422.257	278.557	0.38324	267.940	2137		433.457	287.024	0.38854	289.443
2098		422.536	278.768	0.38337	268.461	2138		433.738	287.236	0.38867	290.078
2099		422.816	278.979	0.38351	268.982	2139		434.019	287.448	0.38880	290.586
2100	1640.3	423.095	+279.190	0.38365	269.552	2140	1680.3	434.300	+287.660	0.38894	291.194
2101		423.375	279.401	0.38378	270.087	2141		434.581	287.872	0.38907	291.730
2102		423.654	279.612	0.38392	270.595	2142		434.862	288.084	0.38920	292.288
2103		423.934	279.823	0.38405	271.103	2143		435.143	288.296	0.38933	292.873
2104		424.213	280.034	0.38418	271.612	2144		435.424	288.508	0.38946	293.382
2105		424.493	+280.245	0.38432	272.120	2145		435.705	+288.720	0.38959	293.890
2106		424.772	280.456	0.38445	272.628	2146		435.986	288.932	0.38972	294.505
2107		425.052	280.667	0.38458	273.136	2147		436.267	289.144	0.38985	295.083
2108		425.331	280.878	0.38471	273.644	2148		436.548	289.356	0.38998	295.669
2109		425.611	281.089	0.38485	274.279	2149		436.829	289.568	0.39011	296.177
2110	1650.3	425.890	+281.300	0.38498	274.788	2150	1690.3	437.110	+289.780	0.39024	296.685
2111		426.170	281.512	0.38511	275.296	2151		437.391	289.992	0.39037	297.320
2112		426.450	281.724	0.38525	275.804	2152		437.672	290.204	0.39050	297.829
2113		426.730	281.936	0.38538	276.312	2153		437.953	290.416	0.39063	298.464
2114		427.010	282.148	0.38551	276.947	2154		438.234	290.628	0.39076	298.972
2115		427.290	+282.360	0.38565	277.455	2155		438.515	+290.840	0.39089	299.607
2116		427.570	282.572	0.38578	277.964	2156		438.796	291.052	0.39102	300.116
2117		427.850	282.784	0.38591	278.472	2157		439.077	291.264	0.39115	300.751
2118		428.130	282.996	0.38604	279.040	2158		439.358	291.476	0.39128	301.259
2119		428.410	283.208	0.38618	279.615	2159		439.639	291.688	0.39141	301.768
2120	1660.3	428.690	+283.420	0.38631	280.123	2160	1700.3	439.920	+291.900	0.39155	302.403
2121		428.970	283.632	0.38644	280.691	2161		440.201	292.112	0.39168	303.038
2122		429.250	283.844	0.38657	281.267	2162		440.482	292.324	0.39181	303.546
2123		429.530	284.056	0.38670	281.775	2163		440.763	292.536	0.39194	304.182
2124		429.810	284.268	0.38683	282.283	2164		441.044	292.748	0.39207	304.690
2125		430.090	+284.480	0.38697	282.791	2165		441.325	+292.960	0.39220	305.325
2126		430.370	284.692	0.38710	283.356	2166		441.606	293.172	0.39232	305.833
2127		430.650	284.904	0.38723	283.934	2167		441.887	293.384	0.39245	306.469
2128		430.930	285.116	0.38736	284.493	2168		442.168	293.596	0.39258	306.977
2129		431.210	285.328	0.38749	285.078	2169		442.449	293.808	0.39271	307.612
Difference or c_p, c_v		c_p = 0.280	c_v = 0.211, 0.212	13				c_p = 0.280, 0.281	c_v = 0.212	13	

Table 2-4. Thermodynamic Properties of Air at Low Pressures.—(*Continued*)

°R	°F	h, Btu per lb	u, Btu per lb	s_p	p_r	°R	°F	h, Btu per lb	u, Btu per lb	s_p	p_r
2170	1710.3	442.730	+294.020	0.39284	308.296	2210	1750.3	453.990	+302.540	0.39798	332.266
2171		443.011	294.233	0.39297	308.804	2211		454.272	302.754	0.39811	332.901
2172		443.292	294.446	0.39310	309.440	2212		454.554	302.968	0.39823	333.537
2173		443.573	294.659	0.39323	309.948	2213		454.836	303.182	0.39836	334.172
2174		443.854	294.872	0.39336	310.584	2214		455.118	303.396	0.39849	334.808
2175		444.135	+295.085	0.39349	311.219	2215		455.400	+303.610	0.39862	335.443
2176		444.416	295.298	0.39361	311.727	2216		455.682	303.824	0.39874	336.079
2177		444.697	295.511	0.39374	312.363	2217		455.964	304.038	0.39887	336.587
2178		444.978	295.724	0.39387	312.871	2218		456.246	304.252	0.39900	337.223
2179		445.259	295.937	0.39400	313.506	2219		456.528	304.466	0.39912	337.858
2180	1720.3	445.540	+296.150	0.39414	314.142	2220	1760.3	456.810	+304.680	0.39926	338.494
2181		445.821	296.363	0.39427	314.650	2221		457.092	304.894	0.39939	339.129
2182		446.102	296.576	0.39440	315.285	2222		457.374	305.108	0.39951	339.765
2183		446.383	296.789	0.39452	315.921	2223		457.656	305.322	0.39964	340.401
2184		446.664	297.002	0.39465	316.429	2224		457.938	305.536	0.39977	340.909
2185		446.945	+297.215	0.39478	317.065	2225		458.220	+305.750	0.39990	341.672
2186		447.226	297.428	0.39491	317.700	2226		458.502	305.964	0.40002	342.307
2187		447.507	297.641	0.39504	318.208	2227		458.784	306.178	0.40015	342.943
2188		447.788	297.854	0.39516	318.844	2228		459.066	306.392	0.40028	343.578
2189		448.069	298.067	0.39529	319.479	2229		459.348	306.606	0.40040	344.214
2190	1730.3	448.350	+298.280	0.39542	320.115	2230	1770.3	459.630	+306.820	0.40053	344.849
2191		448.632	298.493	0.39555	320.623	2231		459.912	307.034	0.40066	345.485
2192		448.914	298.706	0.39568	321.258	2232		460.194	307.248	0.40078	346.121
2193		449.196	298.919	0.39580	321.894	2233		460.476	307.462	0.40091	346.756
2194		449.478	299.132	0.39593	322.402	2234		460.758	307.676	0.40104	347.392
2195		449.760	+299.345	0.39606	323.037	2235		461.040	+307.890	0.40117	348.027
2196		450.042	299.558	0.39619	323.673	2236		461.322	308.104	0.40129	348.663
2197		450.324	299.771	0.39632	324.308	2237		461.604	308.318	0.40142	349.298
2198		450.606	299.984	0.39644	324.816	2238		461.886	308.532	0.40155	349.934
2199		450.888	300.197	0.39657	325.452	2239		462.168	308.746	0.40167	350.569
2200	1740.3	451.170	+300.410	0.39671	326.087	2240	1780.3	462.450	+308.960	0.40180	351.205
2201		451.452	300.623	0.39684	326.723	2241		462.733	309.174	0.40193	351.840
2202		451.734	300.836	0.39696	327.358	2242		463.016	309.388	0.40205	352.475
2203		452.016	301.049	0.39709	327.866	2243		463.299	309.602	0.40218	353.169
2204		452.298	301.262	0.39722	328.502	2244		463.582	309.816	0.40230	353.864
2205		452.580	+301.475	0.39735	329.137	2245		463.865	+310.030	0.40243	354.510
2206		452.862	301.688	0.39747	329.773	2246		464.148	310.244	0.40255	355.145
2207		453.144	301.901	0.39760	330.408	2247		464.431	310.458	0.40268	355.781
2208		453.426	302.114	0.39773	331.043	2248		464.714	310.672	0.40280	356.416
2209		453.708	302.327	0.39785	331.679	2249		464.997	310.886	0.40293	357.052
Difference or c_p, c_v		$c_p =$ 0.281, 0.282	$c_v =$ 0.213	12				$c_p =$ 0.282, 0.283	$c_v =$ 0.214	13	

TABLE 2-4. THERMODYNAMIC PROPERTIES OF AIR AT LOW PRESSURES.—(*Continued*)

°R	°F	h, Btu per lb	u, Btu per lb	s_p	p_r	°R	°F	h, Btu per lb	u, Btu per lb	s_p	p_r
2250	1790.3	465.280	+311.100	0.40305	357.716	2290	1830.3	476.600	+319.670	0.40804	384.823
2251		465.563	311.314	0.40318	358.418	2291		476.883	319.885	0.40816	385.458
2252		465.846	311.528	0.40330	359.114	2292		477.166	320.100	0.40829	386.154
2253		466.129	311.742	0.40343	359.750	2293		477.449	320.315	0.40841	386.857
2254		466.412	311.956	0.40355	360.385	2294		477.732	320.530	0.40854	387.492
2255		466.695	+312.170	0.40368	361.021	2295		478.015	+320.745	0.40866	388.255
2256		466.978	312.384	0.40380	361.696	2296		478.298	320.960	0.40878	388.891
2257		467.261	312.598	0.40393	362.419	2297		478.581	321.175	0.40891	389.653
2258		467.544	312.812	0.40405	363.055	2298		478.864	321.390	0.40903	390.289
2259		467.827	313.026	0.40418	363.690	2299		479.147	321.605	0.40916	391.052
2260	1800.3	468.110	+313.240	0.40431	364.326	2300	1840.3	479.430	+321.820	0.40928	391.688
2261		468.393	313.454	0.40443	364.962	2301		479.714	322.035	0.40940	392.450
2262		468.676	313.668	0.40456	365.724	2302		479.998	322.250	0.40952	393.213
2263		468.959	313.882	0.40468	366.360	2303		480.282	322.465	0.40965	393.849
2264		469.242	314.096	0.40481	366.995	2304		480.566	322.680	0.40977	394.612
2265		469.525	+314.310	0.40493	367.758	2305		480.850	+322.895	0.40989	395.247
2266		469.808	314.524	0.40505	368.394	2306		481.134	323.110	0.41001	396.010
2267		470.091	314.738	0.40518	369.029	2307		481.418	323.325	0.41013	396.646
2268		470.374	314.952	0.40530	369.665	2308		481.702	323.540	0.41026	397.408
2269		470.657	315.166	0.40543	370.428	2309		481.986	323.755	0.41038	398.171
2270	1810.3	470.940	+315.380	0.40555	371.063	2310	1850.3	482.270	+323.970	0.41050	398.807
2271		471.223	315.594	0.40567	371.699	2311		482.554	324.185	0.41062	399.570
2272		471.506	315.808	0.40580	372.462	2312		482.838	324.400	0.41074	400.275
2273		471.789	316.022	0.40592	373.097	2313		483.122	324.615	0.41087	400.968
2274		472.072	316.236	0.40605	373.733	2314		483.406	324.830	0.41099	401.731
2275		472.355	+316.450	0.40617	374.496	2315		483.690	+325.045	0.41111	402.366
2276		472.638	316.664	0.40629	375.131	2316		483.974	325.260	0.41123	403.129
2277		472.921	316.878	0.40642	375.767	2317		484.258	325.475	0.41135	403.892
2278		473.204	317.092	0.40654	376.529	2318		484.542	325.690	0.41148	404.528
2279		473.487	317.306	0.40667	377.165	2319		484.826	325.905	0.41160	405.290
2280	1820.3	473.770	+317.521	0.40680	377.801	2320	1860.3	485.110	+326.120	0.41173	406.053
2281		474.053	317.736	0.40692	378.563	2321		485.394	326.336	0.41185	406.816
2282		474.336	317.951	0.40705	379.299	2322		485.678	326.552	0.41197	407.452
2283		474.619	318.166	0.40717	379.962	2323		485.962	326.768	0.41210	408.214
2284		474.902	318.381	0.40730	380.597	2324		486.246	326.984	0.41222	408.977
2285		475.185	+318.596	0.40742	381.233	2325		486.530	+327.200	0.41234	409.673
2286		475.468	318.811	0.40754	381.996	2326		486.814	327.416	0.41246	410.376
2287		475.751	319.026	0.40767	382.631	2327		487.098	327.632	0.41258	411.138
2288		476.034	319.241	0.40779	383.394	2328		487.382	327.848	0.41271	411.801
2289		476.317	319.456	0.40792	384.130	2329		487.666	328.064	0.41283	412.537
Difference or c_p, c_v		$c_p =$ 0.283	$c_v =$ 0.214, 0.215	13				$c_p =$ 0.283, 0.284	$c_v =$ 0.215, 0.216	12	

TABLE 2-4. THERMODYNAMIC PROPERTIES OF AIR AT LOW PRESSURES.—(*Continued*)

°R	°F	*h*, Btu per lb	*u*, Btu per lb	s_p	p_r	°R	°F	*h*, Btu per lb	*u*, Btu per lb	s_p	p_r
2330	1870.3	487.950	+328.280	0.41295	413.332	2370	1910.3	499.320	+336.920	0.41779	443.591
2331		488.234	328.496	0.41307	414.095	2371		499.605	337.136	0.41791	444.354
2332		488.518	328.712	0.41319	414.808	2372		499.890	337.352	0.41803	445.117
2333		488.802	328.928	0.41332	415.494	2373		500.175	337.568	0.41815	445.880
2334		489.086	329.144	0.41344	416.256	2374		500.460	337.784	0.41827	446.643
2335		489.370	+329.360	0.41356	417.019	2375		500.745	+338.000	0.41839	447.406
2336		489.654	329.576	0.41368	417.782	2376		501.030	338.216	0.41851	448.199
2337		489.938	329.792	0.41380	418.545	2377		501.315	338.432	0.41863	449.058
2338		490.222	330.008	0.41393	419.308	2378		501.600	338.648	0.41875	449.821
2339		490.506	330.224	0.41405	419.943	2379		501.885	338.864	0.41887	450.584
2340	1880.3	490.790	+330.440	0.41417	420.706	2380	1920.3	502.170	+339.080	0.41899	451.347
2341		491.074	330.656	0.41429	421.469	2381		502.455	339.297	0.41911	452.110
2342		491.358	330.872	0.41441	422.232	2382		502.740	339.513	0.41923	452.950
2343		491.642	331.088	0.41453	422.995	2383		503.025	339.730	0.41935	453.763
2344		491.926	331.304	0.41465	423.708	2384		503.310	339.946	0.41947	454.526
2345		492.210	+331.520	0.41478	424.420	2385		503.595	+340.163	0.41959	455.288
2346		492.494	331.736	0.41490	425.156	2386		503.880	340.379	0.41970	456.091
2347		492.778	331.952	0.41502	425.919	2387		504.165	340.596	0.41982	456.901
2348		493.062	332.168	0.41514	426.682	2388		504.450	340.812	0.41994	457.704
2349		493.346	332.384	0.41526	427.445	2389		504.735	341.029	0.42006	458.467
2350	1890.3	493.630	+332.600	0.41538	428.208	2390	1930.3	505.020	+341.245	0.42018	459.357
2351		493.914	332.816	0.41550	428.970	2391		505.305	341.462	0.42030	460.120
2352		494.198	333.032	0.41562	429.733	2392		505.590	341.678	0.42042	460.883
2353		494.482	333.248	0.41574	430.496	2393		505.875	341.895	0.42054	461.645
2354		494.766	333.464	0.41586	431.259	2394		506.160	342.111	0.42066	462.505
2355		495.050	+333.680	0.41599	432.022	2395		506.445	+342.328	0.42078	463.298
2356		495.334	333.896	0.41611	432.785	2396		506.730	342.544	0.42089	464.061
2357		495.618	334.112	0.41623	433.547	2397		507 015	342.761	0.42101	464.951
2358		495.902	334.328	0.41635	434.310	2398		507.300	342.977	0.42113	465.741
2359		496.186	334.544	0.41647	435.073	2399		507.585	343.194	0.42125	466.477
2360	1900.3	496.470	+334.760	0.41659	435.836	2400	1940.3	507.870	+343.410	0.42138	467.367
2361		496.755	334.976	0.41671	436.599	2401		508.155	343.626	0.42150	468.129
2362		497.040	335.192	0.41683	437.362	2402		508.440	343.842	0.42162	468.892
2363		497.325	335.408	0.41695	438.124	2403		508.725	344.058	0.42173	469.782
2364		497.610	335.624	0.41707	438.887	2404		509.010	344.274	0.42185	470.545
2365		497.895	+335.840	0.41719	439.650	2405		509.295	+344.490	0.42197	471.308
2366		498.180	336.056	0.41731	440.413	2406		509.580	344.706	0.42209	472.198
2367		498.465	336.272	0.41743	441.176	2407		509.865	344.922	0.42221	472.961
2368		498.750	336.488	0.41755	441.989	2408		510.150	345.138	0.42232	473.851
2369		499.035	336.704	0.41767	442.829	2409		510.435	345.354	0.42244	474.714
Difference or c_p, c_v		$c_p =$ 0.284, 0.285	$c_v =$ 0.215, 0.216	12				$c_p =$ 0.285	$c_v =$ 0.216	12	

TABLE 2-4. THERMODYNAMIC PROPERTIES OF AIR AT LOW PRESSURES.—(*Continued*)

°R	°F	h, Btu per lb	u, Btu per lb	s_p	p_r	°R	°F	h, Btu per lb	u, Btu per lb	s_p	p_r
2410	1950.3	510.720	345.570	0.42256	475.541	2445		520.720	353.172	0.42668	504.91
2411		511.005	345.787	0.42268	476.304	2446		521.006	353.390	0.42680	505.74
2412		511.292	346.004	0.42280	477.194	2447		521.292	353.608	0.42691	506.57
2413		511.578	346.221	0.42291	477.957	2448		521.578	353.825	0.42703	507.46
2414		511.864	346.438	0.42303	478.847	2449		521.864	354.043	0.42714	508.35
2415		512.150	346.655	0.42315	479.610	2450	1990.3	522.150	354.260	0.42726	509.24
2416		512.436	346.872	0.42327	480.500	2451		522.436	354.477	0.42738	510.13
2417		512.722	347.089	0.42339	481.263	2452		522.722	354.694	0.42749	510.89
2418		513.008	347.306	0.42350	482.153	2453		523.008	354.911	0.42761	511.78
2419		513.294	347.523	0.42362	482.966	2454		523.294	355.128	0.42772	512.67
2420	1960.3	513.580	347.740	0.42375	483.806	2455		523.580	355.345	0.42784	513.56
2421		513.866	347.957	0.42387	484.569	2456		523.866	355.562	0.42796	514.45
2422		514.151	348.174	0.42398	485.459	2457		524.152	355.779	0.42807	515.34
2423		514.437	348.391	0.42410	486.222	2458		524.438	355.996	0.42819	516.23
2424		514.722	348.608	0.42422	487.112	2459		524.724	356.213	0.42830	517.12
2425		515.008	348.825	0.42434	488.002	2460	2000.3	525.010	356.430	0.42843	517.95
2426		515.293	349.042	0.42445	488.825						
2427		515.579	349.259	0.42457	489.655						
2428		515.864	349.476	0.42469	490.418	2500	2040.3	536.45	365.08	0.43305	553.99
2429		516.150	349.693	0.42480	491.308	2550	2090.3	550.76	375.96	0.43872	601.80
								$c_p =$	$c_v =$		
2430	1970.3	516.435	349.910	0.42492	492.071			0.286	0.217		
2431		516.721	350.127	0.42504	492.961	2600	2140.3	565.08	386.84	0.44431	652.92
2432		517.006	350.345	0.42515	493.851	2650	2190.3	579.45	397.86	0.44980	707.33
2433		517.292	350.563	0.42527	494.674			$c_p =$	$c_v =$		
2434		517.577	350.780	0.42539	495.500			0.287	0.220		
						2700	2240.3	593.85	408.94	0.45520	765.32
2435		517.863	350.998	0.42551	496.39	2750	2290.3	608.28	419.99	0.46052	826.98
2436		518.148	351.215	0.42562	497.20			$c_p =$	$c_v =$		
2437		518.434	351.433	0.42574	498.05			0.288	0.221		
2438		518.719	351.650	0.42586	498.94	2800	2340.3	622.74	431.08	0.46575	892.60
2439		519.005	351.868	0.42597	499.83	2850	2390.3	637.28	442.20	0.47090	962.15
								$c_p =$	$c_v =$		
2440	1980.3	519.290	352.085	0.42610	500.59			0.290	0.222		
2441		519.576	352.303	0.42622	501.48	2900	2440.3	651.85	453.34	0.47596	1036.03
2442		519.862	352.520	0.42633	502.37	2950	2490.3	666.45	464.52	0.48096	1114.35
2443		520.148	352.738	0.42645	503.13	3000	2540.3	681.08	475.72	0.48587	1196.99
2444		520.434	352.956	0.42656	504.02			$c_p =$	$c_v =$		
								0.292	0.224		
Difference or c_p, c_v		$c_p =$ 0.285, 0.286	$c_v =$ 0.216, 0.217	12							

TABLE 2-5. PRESSURE-RATIO ENTROPY CHANGE FOR AIR, s_T

$$\Delta s = s_T = \frac{R}{J} \log_e \frac{p_2}{p_1} = 0.15783 \log_{10} \frac{p_2}{p_1}$$

$\frac{p_2}{p_1}$	s_T	Diff.	$\frac{p_2}{p_1}$	s_T	Diff.	$\frac{p_2}{p_1}$	s_T	Diff.	$\frac{p_2}{p_1}$	s_T	Diff.
1.00	0.00000	68									
1.01	0.00068	68	1.36	0.02108	50	1.71	0.03677	40	2.06	0.04954	33
1.02	0.00136	67	1.37	0.02158	50	1.72	0.03717	40	2.07	0.04987	33
1.03	0.00203	66	1.38	0.02208	49	1.73	0.03757	40	2.08	0.05020	33
1.04	0.00269	65	1.39	0.02257	49	1.74	0.03797	39	2.09	0.05053	33
1.05	0.00334	65	1.40	0.02306	49	1.75	0.03836	39	2.10	0.05086	33
1.06	0.00399	65	1.41	0.02355	49	1.76	0.03875	39	2.11	0.05119	32
1.07	0.00464	64	1.42	0.02404	48	1.77	0.03914	39	2.12	0.05151	32
1.08	0.00528	63	1.43	0.02452	48	1.78	0.03953	38	2.13	0.05183	32
1.09	0.00591	62	1.44	0.02500	47	1.79	0.03991	38	2.14	0.05215	32
1.10	0.00653	62	1.45	0.02547	47	1.80	0.04029	38	2.15	0.05247	32
1.11	0.00715	62	1.46	0.02594	47	1.81	0.04067	38	2.16	0.05279	32
1.12	0.00777	61	1.47	0.02641	46	1.82	0.04105	38	2.17	0.05311	31
1.13	0.00838	60	1.48	0.02687	46	1.83	0.04143	37	2.18	0.05342	31
1.14	0.00898	60	1.49	0.02733	46	1.84	0.04180	37	2.19	0.05373	31
1.15	0.00958	59	1.50	0.02779	46	1.85	0.04217	37	2.20	0.05404	62
1.16	0.01017	59	1.51	0.02825	45	1.86	0.04254	37	2.22	0.05466	62
1.17	0.01076	58	1.52	0.02870	45	1.87	0.04291	36	2.24	0.05528	61
1.18	0.01134	58	1.53	0.02915	45	1.88	0.04327	36	2.26	0.05589	60
1.19	0.01192	58	1.54	0.02960	44	1.89	0.04363	36	2.28	0.05649	60
1.20	0.01250	57	1.55	0.03004	44	1.90	0.04399	36	2.30	0.05709	60
1.21	0.01307	56	1.56	0.03048	44	1.91	0.04435	36	2.32	0.05769	59
1.22	0.01363	56	1.57	0.03092	44	1.92	0.04471	36	2.34	0.05828	58
1.23	0.01419	55	1.58	0.03136	43	1.93	0.04507	35	2.36	0.05886	58
1.24	0.01474	55	1.59	0.03179	43	1.94	0.04542	35	2.38	0.05944	57
1.25	0.01529	55	1.60	0.03222	42	1.95	0.04578	35	2.40	0.06001	57
1.26	0.01584	54	1.61	0.03265	42	1.96	0.04613	35	2.42	0.06058	56
1.27	0.01638	54	1.62	0.03307	42	1.97	0.04648	35	2.44	0.06114	56
1.28	0.01692	53	1.63	0.03349	42	1.98	0.04683	34	2.46	0.06170	56
1.29	0.01745	53	1.64	0.03391	42	1.99	0.04717	34	2.48	0.06226	55
1.30	0.01798	53	1.65	0.03433	41	2.00	0.04751	34	2.50	0.06281	54
1.31	0.01851	52	1.66	0.03474	41	2.01	0.04785	34	2.52	0.06335	54
1.32	0.01903	52	1.67	0.03515	41	2.02	0.04819	34	2.54	0.06389	54
1.33	0.01955	51	1.68	0.03556	41	2.03	0.04853	34	2.56	0.06443	54
1.34	0.02006	51	1.69	0.03597	40	2.04	0.04887	33	2.58	0.06497	52
1.35	0.02057		1.70	0.03637		2.05	0.04920		2.60	0.06549	

TABLE 2-5. PRESSURE-RATIO ENTROPY CHANGE FOR AIR, s_T.—(*Continued*)

$\frac{p_2}{p_1}$	s_T	Diff.	$\frac{p_2}{p_1}$	s_T	Diff.	$\frac{p_2}{p_1}$	s_T	Diff.	$\frac{p_2}{p_1}$	s_T	Diff.
2.62	0.06602	52	3.32	0.08225	41	4.05	0.09587	84	5.80	0.12049	59
2.64	0.06654	52	3.34	0.08266	41	4.10	0.09671	84	5.85	0.12108	58
2.66	0.06706	51	3.36	0.08307	41	4.15	0.09755	82	5.90	0.12166	58
2.68	0.06757	51	3.38	0.08348	40	4.20	0.09837	81	5.95	0.12224	57
2.70	0.06808		3.40	0.08388		4.25	0.09918		6.00	0.12281	
		51			40			80			57
2.72	0.06859	50	3.42	0.08428	40	4.30	0.09998	79	6.05	0.12338	57
2.74	0.06909	50	3.44	0.08468	40	4.35	0.10077	78	6.10	0.12395	56
2.76	0.06959	49	3.46	0.08508	40	4.40	0.10155	78	6.15	0.12451	55
2.78	0.07008	49	3.48	0.08548	39	4.45	0.10233	76	6.20	0.12506	55
2.80	0.07057		3.50	0.08587		4.50	0.10309		6.25	0.12561	
		49			39			76			55
2.82	0.07106	49	3.52	0.08626	39	4.55	0.10385	75	6.30	0.12616	54
2.84	0.07155	48	3.54	0.08665	39	4.60	0.10460	74	6.35	0.12670	54
2.86	0.07203	48	3.56	0.08704	38	4.65	0.10534	74	6.40	0.12724	53
2.88	0.07251	47	3.58	0.08742	38	4.70	0.10608	72	6.45	0.12777	53
2.90	0.07298		3.60	0.08780		4.75	0.10680		6.50	0.12830	
		47			38			72			53
2.92	0.07345	47	3.62	0.08818	38	4.80	0.10752	71	6.55	0.12883	52
2.94	0.07392	46	3.64	0.08856	38	4.85	0.10823	70	6.60	0.12935	52
2.96	0.07438	46	3.66	0.08894	37	4.90	0.10893	70	6.65	0.12987	51
2.98	0.07484	46	3.68	0.08931	37	4.95	0.10963	69	6.70	0.13038	51
3.00	0.07530		3.70	0.08968		5.00	0.11032		6.75	0.13089	
		46			37			68			51
3.02	0.07576	45	3.72	0.09005	37	5.05	0.11100	67	6.80	0.13140	50
3.04	0.07621	45	3.74	0.09042	36	5.10	0.11167	67	6.85	0.13190	50
3.06	0.07666	45	3.76	0.09078	36	5.15	0.11234	67	6.90	0.13240	49
3.08	0.07711	44	3.78	0.09114	36	5.20	0.11301	65	6.95	0.13289	49
3.10	0.07755		3.80	0.09150		5.25	0.11366		7.00	0.13338	
		44			36			65			97
3.12	0.07799	44	3.82	0.09186	36	5.30	0.11431	64	7.10	0.13435	96
3.14	0.07843	44	3.84	0.09222	36	5.35	0.11495	64	7.20	0.13531	95
3.16	0.07887	43	3.86	0.09258	36	5.40	0.11559	63	7.30	0.13626	93
3.18	0.07930	43	3.88	0.09294	35	5.45	0.11622	63	7.40	0.13719	92
3.20	0.07973		3.90	0.09329		5.50	0.11685		7.50	0.13811	
		43			35			62			91
3.22	0.08016	42	3.92	0.09364	35	5.55	0.11747	62	7.60	0.13902	89
3.24	0.08058	42	3.94	0.09399	35	5.60	0.11809	61	7.70	0.13991	89
3.26	0.08100	42	3.96	0.09434	34	5.65	0.11870	60	7.80	0.14080	87
3.28	0.08142	42	3.98	0.09468	33	5.70	0.11930	60	7.90	0.14167	86
3.30	0.08184		4.00	0.09501		5.75	0.11990		8.00	0.14253	

TABLE 2-5. PRESSURE-RATIO ENTROPY CHANGE FOR AIR, s_T.—(*Continued*)

$\frac{p_2}{p_1}$	s_T	Diff.	$\frac{p_2}{p_1}$	s_T	Diff.	$\frac{p_2}{p_1}$	s_T	Diff.
8.10	0.14339		11.60	0.16800		15.10	0.18608	
		84			59			45
8.20	0.14423		11.70	0.16859		15.20	0.18653	
		83			58			45
8.30	0.14506		11.80	0.16917		15.30	0.18698	
		82			58			45
8.40	0.14588		11.90	0.16975		15.40	0.18743	
		81			58			44
8.50	0.14669		12.00	0.17033		15.50	0.18787	
		80			57			44
8.60	0.14749		12.10	0.17090		15.60	0.18831	
		79			56			44
8.70	0.14828		12.20	0.17146		15.70	0.18875	
		79			56			43
8.80	0.14907		12.30	0.17202		15.80	0.18918	
		77			55			43
8.90	0.14984		12.40	0.17257		15.90	0.18961	
		77			55			43
9.00	0.15061		12.50	0.17312		16.00	0.19004	
		75			55			43
9.10	0.15136		12.60	0.17367		16.10	0.19047	
		75			54			43
9.20	0.15211		12.70	0.17421		16.20	0.19090	
		74			54			42
9.30	0.15285		12.80	0.17475		16.30	0.19132	
		74			53			42
9.40	0.15359		12.90	0.17528		16.40	0.19174	
		72			53			42
9.50	0.15431		13.00	0.17581		16.50	0.19216	
		72			53			41
9.60	0.15503		13.10	0.17634		16.60	0.19257	
		71			52			41
9.70	0.15574		13.20	0.17686		16.70	0.19298	
		71			52			41
9.80	0.15645		13.30	0.17738		16.80	0.19339	
		69			51			41
9.90	0.15714		13.40	0.17789		16.90	0.19380	
		69			51			40
10.00	0.15783		13.50	0.17840		17.00	0.19420	
		68			51			40
10.10	0.15851		13.60	0.17891		17.10	0.19460	
		68			50			40
10.20	0.15919		13.70	0.17941		17.20	0.19500	
		67			50			40
10.30	0.15986		13.80	0.17991		17.30	0.19540	
		66			49			40
10.40	0.16052		13.90	0.18040		17.40	0.19580	
		65			49			39
10.50	0.16117		14.00	0.18089		17.50	0.19619	
		65			49			39
10.60	0.16182		14.10	0.18138		17.60	0.19658	
		65			49			39
10.70	0.16247		14.20	0.18187		17.70	0.19697	
		64			48			39
10.80	0.16311		14.30	0.18235		17.80	0.19736	
		63			47			38
10.90	0.16374		14.40	0.18282		17.90	0.19774	
		62			47			38
11.00	0.16436		14.50	0.18329		18.00	0.19812	
		62			47			188
11.10	0.16498		14.60	0.18376		18.50	0.20000	
		62			47			92
11.20	0.16560		14.70	0.18423		18.75	0.20092	
		61			47			90
11.30	0.16621		14.80	0.18470		19.00	0.20182	
		60			46			178
11.40	0.16681		14.90	0.18516		19.50	0.20360	
		60			46			174
11.50	0.16741		15.00	0.18562		20.00	0.20534	

Example 2. In a real compressor air enters at 60.3 F and 14.5 psia with a velocity of 25 fps and leaves at 34.5 psia, $t = 242.3$ F and at 125 fps velocity. If the compressor can be considered adiabatic, find the work required for the compression per pound of air. Compare the result with Example 1.

Solution. By Table 2-4 read $h_1 = 14.393$, $h_2 = 58.204$. The kenthalpy values, entry and exit, are, respectively,

$$H_{o1} = h_1 + \frac{C_1^2}{2g_cJ} = 14.393 + \frac{25^2}{(64.34)(778)} = 14.406 \text{ Btu per lb}$$

and

$$H_{o2} = 58.204 + \frac{125^2}{(64.34)(778)} = 58.516 \text{ Btu per lb}$$

By Eq. (2-27), work $= 14.406 - 58.516 = -44.110$, or the work input is $+44.110$ Btu per lb.

Compressor efficiency can be defined as

$$\eta_c = \frac{\text{ideal isentropic work required}}{\text{actual work required}}$$

For Examples 2 and 1

$$\eta_c = \frac{35.112}{43.998} = 0.798 = 79.8\%$$

Pressure Ratios and Relative Pressure. Equation (2-37), which applies to isentropic processes, can be expanded into the form

$$\frac{R}{J} \log_e \frac{p_2}{p_1} = \Big[s_p \Big]_{T_1}^{T_2} = \Big[s_p \Big]_{460}^{T_2} - \Big[s_p \Big]_{460}^{T_1}$$

to apply to the air tables of this text, for which the method of obtaining s_p has already been explained. For any temperature T referred to the datum 460 R let us take

$$\frac{R}{J} \log_e \frac{p_2}{p_1} = \Big[s_p \Big]_{460}^{T} = s_p$$

where s_p is the value listed. Corresponding to each temperature in the table, it is possible to substitute proper values for air and solve for p_2/p_1 as follows:

$$\log_{10} \frac{p_2}{p_1} = \frac{J}{(R)(2.3026)} s_p = \frac{778.2}{(53.34)(2.3026)} s_p$$

$$\log_{10} \frac{p_2}{p_1} = 6.3359 s_p$$

$$\frac{p_2}{p_1} = \text{antilog } 6.3359 s_p$$

For example, at 480 R

$$\frac{p_2}{p_1} = \text{antilog } (6.3359 \times 0.01021) = 1.160$$

and this value is seen tabulated in the column marked p_r. The use of the symbol p_r was employed by Keenan and Kaye,[1,6] while Π was used by the NACA.[8]

It will be noted from the derivation that the values found as above are associated with pressure ratios that exist in isentropic compressions and expansions. These p_r values are commonly called *relative pressures.* Note that p_r is no longer in logarithmic form, which simplifies arithmetic operations.

$$\frac{p_{\text{final}}}{p_{\text{initial}}} = r = \text{pressure ratio}$$

Also,

$$\frac{p_{\text{final}}/p_1}{p_{\text{initial}}/p_1} = r$$

where p_1 is any arbitrary reference pressure. This can be written as

$$\frac{p_{r2}}{p_{r1}} = r$$

or

$$p_{r2} = rp_{r1} \tag{2-42}$$

As an example, consider an isentropic compression from 480 R through a pressure ratio of $r = 6$. We would find, after using the p_r value previously found, that

$$p_{r2} = 6p_{r1}$$
$$p_{r2} = (6)(1.16) = 6.96$$

Opposite 6.962 in Table 2-4 read a final temperature after isentropic compression of 799 R. Note that in a compression with the final pressure p_2 and the original pressure p_1

$$r = \frac{p_2}{p_1} = \frac{p_{r2}}{p_{r1}} \qquad p_2 = \frac{p_{r2}}{p_{r1}} p_1$$

Similarly for an expansion from p_1 to a final value p_2 with $r_e = p_1/p_2$ it follows that

$$p_{r2} = \frac{p_{r1}}{r_e} = \frac{p_2}{p_1} p_{r1} \tag{2-43}$$

Example 3. Solve Example 1 by use of relative-pressures tabulations for the case of reversible-adiabatic compression of air from 60.3 F and 14.5 psia ($C_1 \equiv 0$) to a final pressure of 34.5 psia ($C_2 \equiv 0$).

Solution. From Table 2-4 at 60.3 F (520 R) read $h_1 = 14.393$ Btu per lb and $p_{r1} = 1.535$. The pressure-compression ratio $r = 34.5/14.5 = 2.38$. By Eq. (2-42)

$$p_{r2} = 2.38p_{r1} = (2.38)(1.535) = 3.654$$

Corresponding to this value of p_{r2} read in Table 2-4 that $T_2 = 666$ R, $h_2 = 49.505$. By Eq. (2-23) the work is

$$\frac{{}_1W_2}{J} = (h_1 - h_2)_s = 14.393 - 49.505 = -35.112 \text{ Btu per lb}$$

with the minus sign indicating that work is done *on* and not *by* the air in the amount 35.112 Btu per lb.

2-6. Keenan-Kaye Gas Tables. These are complete tables covering air, air and combustion products, and compressible-flow functions.[6] They are carefully prepared, very complete, and convenient to use. Extracts from two representative tables of this book are shown as Table 2-6 of this chapter. These tables use a datum temperature of 0 R in

TABLE 2-6. EXTRACT FROM KEENAN-KAYE GAS TABLES*

T	t	h	p_r	u	v_r	ϕ
From Table 1. Air at Low Pressures (for 1 Lb)						
530	70.3	126.66	1.2983	90.34	151.22	0.59630
531		126.90	1.3068	90.51	150.52	0.59675
532		127.14	1.3155	90.68	149.81	0.59720
533		127.38	1.3242	90.85	149.10	0.59765
534		127.62	1.3329	91.02	148.41	0.59810
535	75.3	127.86	1.3416	91.19	147.72	0.59855
1000	540.3	240.98	12.298	172.43	30.12	0.75042
1001		241.23	12.343	172.61	30.04	0.75067
1002		241.48	12.388	172.79	29.96	0.75092
1003		241.73	12.433	172.97	29.88	0.75117
1004		241.98	12.478	173.15	29.81	0.75142
1005	545.3	242.23	12.523	173.33	29.73	0.75166
From Table 4. Products, 400% Theoretical Air (for 1 Lb Mole)						
530	70.3	3697.6	1.2856	2645.1	4424	46.225
531		3704.6	1.2942	2650.1	4403	46.239
532		3711.7	1.3029	2655.2	4382	46.252
533		3718.7	1.3116	2660.2	4361	46.265
534		3725.8	1.3204	2665.3	4340	46.278
535	75.3	3732.8	1.3291	2670.4	4320	46.291
1000	540.3	7072.1	12.694	5086.2	845.3	50.773
1001		7079.3	12.742	5091.4	843.0	50.780
1002		7086.7	12.789	5096.9	840.7	50.787
1003		7094.1	12.837	5102.3	838.5	50.795
1004		7101.6	12.884	5107.8	836.4	50.802
1005	545.3	7109.0	12.932	5113.2	834.1	50.809

* By permission John Wiley & Sons, Inc., New York, 1948.

contrast to the 460 R used for Table 2-4 of this text. It will also be noted that the symbol ϕ is used for entropy. Relative isentropic volume ratios v_r are also tabulated. These v_r values can be used for isentropic compression or expansion ratios expressed in terms of volumes in the same way that the p_r values are used for pressure ratios. Prior to the publication of these tables, the same authors compiled a volume entitled

"Thermodynamic Properties of Air."[7] The previous volume did not contain the many additional functions of the more recent book.

Example 4. A mixture which can be considered air is compressed reversibly and adiabatically from 14.7 psia and 73.3 F through a compression ratio (by volume) of 5. Find the final conditions and the work of compression per pound of mixture (*a*) if a continuous flow of mixture moving at low velocity is compressed (steady flow), (*b*) if a simple charge (nonflow) is compressed.

Solution. By use of v_r and the Keenan-Kaye tables. Refer to the abbreviated values in Table 2-6 and read, for 533 R (73.3 F), $h = 127.38$ and $v_r = 149.10$.

$$v_{r2} = \frac{v_{r1}}{5} = \frac{149.10}{5} = 29.82$$

Corresponding to this value of v_r, the temperature and enthalpy after compression are 1004 R and 241.98 Btu per lb.

(*a*) In steady-flow adiabatic compression the work input is

$${}_1W_2 = h_2 - h_1 = 241.98 - 127.38 = 114.6 \text{ Btu per lb}$$

An alternate solution can be had by use of Eq. (2-21).

$$\frac{T_2}{T_1} = \left(\frac{v_1}{v_2}\right)^{k-1} \qquad \frac{T_2}{533} = (5)^{0.394} = 1.885$$

where the value of k is read as 1.394 from Fig. 2-1 in the range between, say, 530 to 1000. Thus $T_2 = (1.885)(533) = 1005$ R. This is in reasonable agreement with the other solution, but the result depends largely on properly selecting the value of k, and great inaccuracy can result if the compression- or expansion-ratio range is extreme.

$${}_1W_2 = h_2 - h_1 = c_{pm}(T_2 - T_1)$$

but as enthalpy values are preferable, reference to Table 2-4 shows

$$W = 242.23 - 127.38 = 114.85 \text{ Btu per lb}$$

This differs from the previous answer because of the slightly different final temperatures indicated by the two methods.

(*b*) In a nonflow adiabatic compression the work appears as increased internal energy of the medium and using tabular values

$$\begin{aligned} {}_1W_2 &= u_2 - u_1 = 173.15 - 90.85 \\ &= 82.30 \text{ Btu per lb} \end{aligned}$$

For a gaseous medium, Eq. (2-30) could also be used if desired

$${}_1W_2 = u_2 - u_1 = \left(h_2 - \frac{RT_2}{J}\right) - \left(h_1 - \frac{RT_1}{J}\right)$$

Moreover for a gas it follows that

$$u_2 - u_1 = c_{vm}(T_2 - T_1)$$

and this is equal to

$$u_2 - u_1 = \left(c_{pm} - \frac{53.34}{778}\right)(T_2 - T_1)$$

This gives an alternate method of solution for part (*b*).

2-7. Cycles. For the continuous production of power, the thermodynamic working medium employed goes through a series of events in an essentially cyclical manner. In a power cycle, the working medium would go through at least these minimum processes, *viz.*, *heat is added*, *power* or *work is developed*, and *heat is rejected*, with the medium finally arriving at the same condition it had at the beginning of the series of processes. In the steam power plant, which is an external-combustion engine, the working medium starts in the liquid phase and is vaporized in a boiler by the addition of heat; the steam expands in a turbine developing useful work or power, then at lower pressure it is condensed as heat is rejected to the cooling medium, and finally the condensate has its pressure increased by a feed pump to exactly the same state the water had at the beginning of the cycle. A similar cyclical arrangement exists for the closed gas turbine. The air or working gas is heated, power is produced as expansion occurs, heat is rejected, and the medium is compressed to its original temperature and pressure. The air or other gas constituting the medium carries out a true cycle of operations, and this is also an external-combustion cycle.

However, in the true internal-combustion engine or internal-combustion gas turbine, the working medium is a mixture of air and fuel which, during combustion, releases energy from the chemical form and transfers this energy to internal energy of the products of combustion and uncombined air. Thus, even though the products are brought back to the same pressure and temperature as existed originally, they are now a different species of matter and consequently a true cycle does not exist. In spite of this, it is sometimes convenient to disregard the change in the nature of the working medium and to analyze the process as though the same medium were carried through the events of a cycle and brought back exactly to the original condition. For example, if an air-fuel ratio is very high before combustion, it may be that the medium after combustion is still primarily air.

From an over-all viewpoint, the performance of the gas turbine or internal-combustion engine is directly associated with a flow of air and fuel to the engine and a flow of exhaust products from the engine. This resembles what thermodynamically could be called a steady-flow process. Such a condition is fully true in the case of a gas turbine and is approximated in the case of the reciprocating engine. However, in the reciprocating engine, many of the individual processes are not steady flow. For example, compression, which takes place after the charge has been drawn into the cylinder, is obviously not of itself steady flow. Thus many types of analyses concerned with events in the whole cycle or process have to be analyzed specifically as nonflow or steady-flow processes as the case may be.

Otto Cycle. This cycle represents a prototype cycle which is approximated in spark-ignition reciprocating engines. In the idealized spark-ignition engine, a charge of air and fuel is brought into the engine cylinder. It is then compressed reversibly and adiabatically. Ignition is caused by a suitable spark, and burning takes place in the cylinder of the engine essentially at constant volume, with combustion ideally being carried to completion. Expansion of the products of combustion then takes place, with work being developed, and finally the products are released to atmospheric pressure and exhausted from the cylinder. External to the engine in the atmosphere, the hot products cool down and arrive finally at the initial temperature under which the cycle started. The theoretical Otto cycle which resembles this process is assumed to consist of (1) isentropic compression, (2) heat addition at constant volume, (3) isentropic expansion, and (4) rejection of heat at constant volume. The medium (air) does not change composition and remains in the cylinder at all times.

Diesel Cycle. In the idealized Diesel cycle the events consist of (1) isentropic compression of the working medium, air; (2) heat addition at constant pressure; (3) isentropic expansion following heat addition; and finally (4) heat rejection at constant volume. In real diesel-engine operation, instead of heat being added from an external source, it is furnished by a spray of fuel oil which ignites when it is injected into the air which has been compressed. The air as a result of compression is at a sufficiently high temperature so that ignition of the fuel spray entering it takes place automatically. The hot gases then expand and finally, with release occurring as the piston reaches the end of its stroke, the gases exhaust into the atmosphere.

Brayton Cycle. For the gas turbine, the Brayton cycle (Fig. 3-1) most closely represents a prototype cycle. In the ideal Brayton cycle the working medium, air or gas, (1) is compressed isentropically and (2) is heated at constant pressure. Following this, (3) the medium expands, dropping in pressure and delivering work until the low pressure of the cycle is reached, following which (4) heat is rejected at constant pressure to bring the medium back to its original temperature and state. This cycle could be followed in a closed-type gas-turbine system. However in the open gas-turbine system, compression takes place in an axial or centrifugal-type compressor, and following this the fuel is injected into the air stream. The fuel in burning at constant pressure raises the products of combustion and excess air to the desired temperature, following which the hot gases expand through the power turbine, finally being exhausted to the atmosphere after they have reached the low pressure of the system. The final cooling of these gaseous products to the original temperature actually takes place external to the machine.

Cycles will be discussed in greater detail at other places in this text.

At this point it should merely be emphasized that the cyclical process is only approximated in internal-combustion engines. However, the resemblance of the real process to a prototype cycle makes it convenient to analyze the prototype cycle and to use this information for predicting what results can be expected in the real operation under actual performance conditions. The term cycle will be used in describing these real sequences of processes although it may be true that a cycle does not in fact exist.

Example 5. *Ideal closed cycle.* In a closed cycle, air at 70.3 F at 100 psia is isentropically compressed to 400 psia, then heated in an externally fired heater to 1140.3 F, after which it expands isentropically in a power turbine to 100 psia. The hot air leaving the turbine then enters a water-cooled heat exchanger, from which it leaves at a temperature of 70.3 F. For this ideal cycle, which represents the events of a simple closed gas-turbine system with the working medium continuously under pressure, compute: (*a*) the compressor work, (*b*) heat added, (*c*) turbine work, (*d*) heat rejected, (*e*) efficiency of the internal cycle (air alone exclusive of fuel and combustion effectiveness). Consider the system adiabatic in regard to heat losses to surroundings.

Solution. At 70.3 F (530 R) from Table 2-4 the enthalpy of the air is 16.800 Btu per lb, and $p_r = 1.640$. These values apply with trivial error whether the air is at 100 psia or at a lower or higher pressure. As the air is compressed through a pressure ratio of 4, the value of p_{r2} after compression is found.

$$p_{r2} = rp_{r1} = (4)(1.640) = 6.560$$

For this value of p_{r2} from Table 2-4 read $h_2 = 78.501$ Btu per lb, $T_2 = 785.72$ R.

(*a*) $$\text{Compressor work} = (1)(78.501 - 16.800) = 61.701 \text{ Btu per lb}$$

(*b*) The enthalpy at 1140.3 F (1600 R) of the air at 400 psia leaving the heater by Table 2-4 is 286.100.

$$Q_A = (1)(286.100 - 78.501) = 207.599 \text{ Btu per lb, heat added}$$

(*c*) At 1600 R, which we can call point 3 of the cycle, the value of $p_{r3} = 90.990$. Use the pressure-expansion ratio $^{400}/_{100} = 4$ to find the p_{r4} value after isentropic expansion, *viz.*,

$$p_{r4} = \frac{p_{r3}}{r_e} = \frac{90.990}{4} = 22.748$$

Corresponding to this value read from Table 2-4 for conditions leaving the turbine

$$h_4 = 158.538 \text{ Btu per lb}, \quad T_4 = 1109.16 \text{ R.}$$

$$\text{Turbine work} = (1)(286.100 - 158.538) = 127.562 \text{ Btu per lb}$$

(*d*) Heat is rejected in cooling the gases leaving the turbine back to 70.3 F.

$$Q_R = (1)(h_4 - h_1) = (1)(158.538 - 16.800) = 141.738 \text{ Btu per lb}$$

(*e*)
$$\begin{aligned}\text{Eff} &= \frac{\text{net work}}{\text{heat added}} = \frac{\text{heat added} - \text{heat rejected}}{\text{heat added}} \\ &= \frac{127.562 - 61.701}{207.599} = \frac{207.599 - 141.738}{207.599} = \frac{65.861}{207.599} \\ &= 0.317 = 31.7\%\end{aligned}$$

The efficiency of the actual cycle is less than the above because of losses in the compressor and turbine, pressure losses in the system, heat losses from equipment, and a combustion-stack loss in such a closed-cycle unit.

Power expressions can be developed for parts (*a*) and (*c*) when a design flow rate is chosen. For example, if the flow rate is 1 lb of air per sec,

$$\text{Compressor hp} = \frac{(1)(61.701)}{550/778} = 87.3 \text{ hp}$$

$$\text{Turbine hp} = \frac{(1)(127.562)}{0.707} = 180.4 \text{ hp}$$

$$\text{Net hp} = 180.4 - 87.3 = 93.1 \text{ hp}$$

2-8. Nozzle Relationships. A nozzle is a flow passage especially shaped to produce kinetic energy at the expense of other forms of energy. Nozzles are used extensively in gas turbines and also in reciprocating combustion engines in such places as the spray nozzle of the diesel fuel jet and at the carburetor throat, where gasoline is mixed with air passing to the engine. In general the fluid flows through a nozzle so rapidly that the heat loss from the fluid to the nozzle walls is trivial and the nozzle process can be considered as adiabatic. Applying the steady-flow equation (2-2) to a nozzle carrying gas or vapor, and with ${}_1q_2 = 0$ and the ${}_1W_2$ term irrelevant, there results

$$\frac{C_2^2 - C_1^2}{2g_cJ} = h_1 - h_2 \tag{2-44}$$

$$C_2^2 = 2g_cJ(h_1 - h_2) + C_1^2 \tag{2-45}$$

$$C_2 = \sqrt{(64.34)(778.16)(h_1 - h_2) + C_1^2}$$

$$= \sqrt{(50{,}070)(h_1 - h_2) + C_1^2} \tag{2-46}$$

$$C_2 = \sqrt{50{,}070 c_{pm}(T_1 - T_2) + C_1^2} \tag{2-46a}$$

where C_2 = velocity, at any point in nozzle, fps
C_1 = velocity at entry to nozzle, fps
h_1 = enthalpy at entry to nozzle, Btu per lb
h_2 = enthalpy at same location in nozzle where C_2 exists, Btu per lb
c_{pm} = mean specific heat at constant pressure, Btu/(lb)(°R), over the temperature range $T_1 - T_2$

If the velocity, C_1, of the medium entering the nozzle is very small by comparison with leaving velocity, C_1 can be disregarded and Eq. (2-46) becomes

$$C_2 = 223.8\sqrt{h_1 - h_2} \tag{2-47}$$

Equations (2-46) and (2-47) apply to a nozzle whether the flow is frictionless (reversible) or not. If the flow is reversible, the value of $h_1 - h_2$ in a nozzle will be the maximum for a given pressure drop from a given initial state to a given final pressure and this corresponds to an isentropic change between these limits, namely, $(h_1 - h_2')_s$. Carefully designed nozzles are extremely efficient in developing kinetic energy, and efficiencies in the range of 95 to 98 per cent can be realized. However, for nozzles with rough walls and for those with wide angles of divergence, particularly on the downstream side of supersonic nozzles, the efficiencies may drop below 90 per cent. In a nozzle the fluid flows in the direction of the pressure gradient (falling pressure), and the boundary layer, while moving more slowly than the main stream and thus retarding the flow, nevertheless flows in the stream direction. In a diffuser, with the flow in the direction of increasing pressure, lower efficiencies can be expected than for nozzles in conserving available mechanical energy.

Nozzle efficiency compares the actual kinetic energy produced on discharge (or between any two points in a nozzle) to that obtainable by assuming an isentropic expansion in the nozzle.

$$\eta_n = \frac{C_2^2/2g_cJ}{(C_1^2/2g_cJ) + (h_1 - h_2')_s} \tag{2-48}$$

$$C_2 = \sqrt{\eta_n}\sqrt{C_1^2 + 2g_cJ(h_1 - h_2')_s}$$

$$= \sqrt{\eta_n}\sqrt{50{,}070(h_1 - h_2')_s + C_1^2} \tag{2-49}$$

For a nozzle efficiency of unity the complete isentropic velocity C_{s2} would be realized and Eq. (2-49) would show

$$C_{s2} = \sqrt{C_1^2 + 2g_cJ(h_1 - h_2')_s} \tag{2-49a}$$

If the actual discharge velocity C_2 is compared with the isentropic discharge velocity C_{s2}, we obtain the nozzle velocity coefficient k_{nv}, where

$$k_{nv} = \frac{C_2}{C_{s2}} \tag{2-50}$$

Substituting Eqs. (2-49) and (2-49*a*) in (2-50), it can be seen that

$$k_{nv} = \sqrt{\eta_n} \tag{2-51}$$

and

$$C_2 = \sqrt{\eta_n}\, C_{s2} \tag{2-52}$$

It should not be thought that Eqs. (2-46) and (2-49) are incompatible. In any adiabatic process with no external work, enthalpy and kinetic energy are always interchangeable and the kenthalpy is constant. The import of the two equations can be brought out by comparing them for the case where the approach velocity is trivial ($C_1 \to 0$). By Eqs. (2-46) and (2-49)

$$C_2 = \sqrt{2gJ(h_1 - h_2)}$$

$$C_2 = \sqrt{\eta_n 2g_cJ(h_1 - h_2')_s}$$

$$\eta_n = \frac{h_1 - h_2}{(h_1 - h_2')_s}$$

Here it is obvious that, with zero approach velocity, η_n merely measures the fraction of the isentropic enthalpy drop which is actually converted into kinetic form.

Continuity of Flow. In a nozzle or duct as soon as a steady flow rate is established the mass rate of flow (pounds per second) remains constant at any location in the nozzle. Expressed as an equation for various locations in a nozzle, there results

$$G = \frac{A_1C_1}{v_1} = \frac{A_2C_2}{v_2} = \frac{AC}{v} \tag{2-53}$$

where G = mass flow rate, lb per sec
A = area, ft²
C = velocity, fps
v = sp vol, ft³ per lb

As G for a given flow rate is a constant, Eq. (2-53) can be written in differential form as

$$d\left(\frac{AC}{v}\right) = 0 \tag{2-54}$$

and differentiating

$$\frac{dA}{A} + \frac{dC}{C} - \frac{dv}{v} = 0 \tag{2-55}$$

For a reversible (isentropic) case Eq. (2-44) can be written*

$$\frac{C_2^2 - C_1^2}{2g_cJ} = (h_1 - h_2')_s = -\frac{1}{J}\int v\,dp$$

$$d\left(\frac{C^2}{2g_c}\right) = -v\,dp \tag{2-56}$$

and

$$C\,dC = -g_cv\,dp \tag{2-57}$$

Substitute (2-57) in (2-55), and rearrange.

$$\frac{dA}{A} = -\frac{dC}{C} + \frac{dv}{v} = -\frac{dC}{C}\left(1 - \frac{dv\,C^2}{v\,C\,dC}\right)$$

$$\frac{dA}{A} \quad -\frac{dC}{C}\left(1 + \frac{C^2\,dv}{g_cv^2\,dp}\right) \tag{2-58}$$

$$g_c\rho = \frac{1}{v} \qquad \text{where } \rho = \text{density in slugs per ft}^3$$

$$g_c\,d\rho = -\frac{dv}{v^2} \tag{2-59}$$

Substitute in Eq. (2-58).

$$\frac{dA}{A} = -\frac{dC}{C}\left(1 - \frac{C^2}{dp/d\rho}\right) \tag{2-60}$$

* Keenan, "Thermodynamics," Chap. XVIII.

Examination of this equation reveals that $\sqrt{dp/d\rho}$ is a significant parameter. When C is less than $\sqrt{dp/d\rho}$, the fluid speeds up in a convergent passage (dA is minus). Further, the maximum value of velocity that can be reached in the converging passage occurs when $C = \sqrt{dp/d\rho}$, that is, at the throat. In a converging passage it is of course not necessarily true that C reaches this value. Past the throat in the diverging section of a nozzle two conditions are possible for full-flow conditions: (1) the value of $\sqrt{dp/d\rho}$ may be such that C decreases in the diverging section which corresponds to subsonic flow conditions with that part of the nozzle acting as a diffuser or (2) $\sqrt{dp/d\rho}$ is such that C increases in the divergent section and supersonic velocity exists. When this latter condition holds, sonic velocity must have been previously reached at the throat.

Formal proof will not be supplied here (see Refs. 9 and 10) that

$$a = \sqrt{\frac{dp}{d\rho}} \tag{2-61}$$

where a is the velocity of a sound wave. For a perfect gas undergoing an isentropic process Eq. (2-22) applies, pv^k = constant.

After differentiation and use of Eq. (2-59) in Eq. (2-61), it appears that

$$a = \sqrt{kg_cPv} = \sqrt{\frac{kP}{\rho}} \tag{2-62}$$

Apply $Pv = RT$, the characteristic gas equation with v the specific volume, and there results

$$a = \sqrt{kg_cRT} \tag{2-63}$$

$$a = \sqrt{kg_c \frac{1545}{m} T} = 223 \sqrt{\frac{k}{m} T} \tag{2-64}$$

For air in the range where k is closely 1.4, Eq. (2-64) simplifies to

$$a = 49\sqrt{T} \tag{2-64a}$$

where a = sonic velocity, fps
$k = c_p/c_v$ (see Figs. 2-1 and 2-2)
g_c = 32.17 as before
P = pressure, psfa
v = sp vol, ft³ per lb
ρ = density, slugs per ft³
R = gas constant, $1545/m$
m = molecular wt of a gas
T = temperature, °R

Standard texts on thermodynamics develop a perfect-gas relation for the pressure ratio with acoustic flow at the throat of a frictionless nozzle, *viz.*,

$$\frac{P_t}{P_1} = \left(\frac{2}{k+1}\right)^{k/(k-1)} \tag{2-65}$$

It is also possible for a given pressure ratio to show that the velocity at any point in a frictionless nozzle with small velocity of approach is

$$C_{s2} = \sqrt{2g_c P_1 v_1 \frac{k}{k-1}\left[1 - \left(\frac{P_2}{P_1}\right)^{(k-1)/k}\right]} \tag{2-66}$$

With air tables available, however, it is probably easier to use Eq. (2-46) or 2-47.

Example 6. A gas turbine is supplied with gas, which can be considered as air, at 1340.3 F and at 45 psia. The gas enters the turbine nozzles with negligible initial velocity where it expands to 15 psia. If the nozzle efficiency is 94 per cent, find the leaving velocity and the total mouth and throat area of the nozzle bank to pass 10 lb per sec.

Solution. From Table 2-4 read $h_1 = 340.15$ at 1800 R and $p_{r1} = 144.761$. For $r = 45/15 = 3$, $p_{r2} = 144.761/3 = 48.254$, $h_{2s} = 221.94$.
By Eqs. (2-49) and (2-46)

$$h_1 - h_2 = (0.94)(340.15 - 221.94) = 111.1$$
$$C_m = \sqrt{(64.3)(778)(111.1) + 0} = 2360 \text{ fps}$$

The temperature at the mouth of the nozzle can be found by making use of the fact that the kenthalpy is constant in a nozzle.

$$h_1 + \frac{C_1^2}{2g_cJ} = h_m + \frac{C_m^2}{2g_cJ} = H_o$$
$$340.15 + 0 = h_m + \frac{(2360)^2}{2g_cJ}$$
$$h_m = 229.05 \text{ Btu per lb at mouth}$$
$$T_m = 1384.1 \text{ R temperature at mouth (Table 2-4)}$$

By Eq. (2-3) under mouth conditions $p_m = 15$ psia and 1384.1 R.

$$(144)(15)(v_m) = (1)(53.34)(1384.1)$$
$$v_m = 34.2 \text{ ft}^3 \text{ per lb}$$

By Eq. (2-53) for $G = 10$ lb per sec

$$G = \frac{AC}{v} \qquad 10 = \frac{(A)(2360)}{34.2}$$
$$A_m = 0.1448 \text{ ft}^2 = 20.9 \text{ in.}^2 \text{ mouth area}$$

Ideal throat pressure by Eq. (2-65) and k value from Fig. 2-1:

$$p_t = (45)\left(\frac{2}{1.342+1}\right)^{1.342/0.342} = (45)(0.538) = 24.2 \text{ psia}$$
$$r = \frac{45}{24.2} = 1.86 \text{ pressure expansion ratio}$$

Using the same inlet conditions $h_1 = 340.15$ and $p_{r1} = 144.761$,

$$p_{rt} = \frac{144.761}{1.86} = 77.80$$
$$h_{ts} = 269.25$$
$$h_1 - h_t = (0.94)(340.15 - 269.25) = 66.65$$
$$C_t = \sqrt{(50{,}070)(66.65)} = 1826 \text{ fps}$$
$$h_t = 340.15 - 66.65 = 273.50 \text{ Btu per lb}$$

for which $T_t = 1552.7$ R.

$$(144)(24.2)(v_t) = (1)(53.34)(1552.7)$$
$$v_t = 23.8 \text{ ft}^3 \text{ per lb}$$
$$G = \frac{AC}{v} \qquad 10 = \frac{(A_t)(1826)}{23.8}$$
$$A_t = 0.130 \text{ ft}^2 \doteq 18.7 \text{ in.}^2 \text{ throat area}$$

Although idealized conditions with no friction were assumed in determining the throat pressure, sufficiently close design results for friction can usually be obtained by this method. However, if more precision is required, another check point at a slightly lower pressure should be computed.

2-9. Static and Total Pressure. Consider a stream of fluid moving under adiabatic conditions with no work interchange to or from the stream. If in addition potential-energy changes are negligible, Eq. (2-2) simplifies to

$$h_1 + \frac{C_1^2}{2g_cJ} = h_2 + \frac{C_2^2}{2g_cJ} = H_o \tag{2-67}$$

Break enthalpy into its constituents (Art. 2-1) in the preceding equation.

$$\left(\frac{P_1v_1}{J} + u_1\right) + \frac{C_1^2}{2g_cJ} = \left(\frac{P_2v_2}{J} + u_2\right) + \frac{C_2^2}{2g_cJ}$$
$$\frac{P_2v_2 - P_1v_1}{J} = (u_1 - u_2) + \frac{C_1^2 - C_2^2}{2g_cJ} \tag{2-68}$$

For a perfect gas, Eq. (2-67) can be modified by the following substitution:

$$u_1 - u_2 = (1)(c_v)(T_1 - T_2) = \frac{R}{(k-1)(J)}(T_1 - T_2)$$
$$= \frac{R}{(k-1)(J)}\left(\frac{P_1v_1}{R} - \frac{P_2v_2}{R}\right) = \frac{P_1v_1 - P_2v_2}{(k-1)(J)}$$

Employing this in Eq. (2-68),

$$\frac{P_2v_2 - P_1v_1}{J} = \frac{P_1v_1 - P_2v_2}{(k-1)(J)} + \frac{C_1^2 - C_2^2}{2g_cJ}$$
$$\frac{k}{k-1}(P_2v_2 - P_1v_1) = \frac{C_1^2 - C_2^2}{2g_c} \tag{2-69}$$

Equations (2-68) and (2-69) are useful in that pressure is introduced as a variable; however, pressure change cannot be predicted in most cases as the relationships between the other variables are not uniquely known.

In the particular case of reversible-adiabatic compression (expansion) with a semiperfect gas, substitution of Eq. (2-22), $P_1v_1^k = P_2v_2^k$, makes it possible to determine the final pressure for any final velocity.

$$\frac{k}{k-1} P_1v_1 \left[\left(\frac{P_2}{P_1} \right)^{(k-1)/k} - 1 \right] = \frac{C_1^2 - C_2^2}{2g_c} \tag{2-70}$$

Introduce by means of Eq. (2-62) the acoustic velocity at initial conditions in the fluid.

$$a_1 = \sqrt{kg_cP_1v_1} \qquad \text{and} \qquad P_1v_1 = \frac{a_1^2}{kg_c}$$

$$\frac{a_1^2}{k-1} \left[\left(\frac{P_2}{P_1} \right)^{(k-1)/k} - 1 \right] = \frac{C_1^2 - C_2^2}{2} \tag{2-71}$$

$$\frac{P_2}{P_1} = \left[1 + \frac{(k-1)(C_1^2 - C_2^2)}{2a_1^2} \right]^{k/(k-1)} \tag{2-72}$$

This expression represents for semiperfect gases the pressure change, in a ratio form, when the velocity changes from C_1 to C_2 under isentropic conditions.

Consider the case of the velocity C_2 approaching zero, then P_1 represents the static pressure of the stream at velocity C_1 before it is stopped and P_2 becomes the total, or stagnation, pressure P_T. For this condition ($C_2 = 0$) Eq. (2-72) becomes

$$\frac{P_T}{P_1} = \left[1 + \frac{(k-1)C_1^2}{2a_1^2} \right]^{k/(k-1)} \tag{2-73}$$

$$\frac{P_T}{P_1} = \left[1 + \frac{k-1}{2} M_1^2 \right]^{k/(k-1)} \tag{2-74}$$

where M = Mach number, or ratio of the stream velocity to the local acoustic velocity

$M_1 = C_1/a_1$

P_T = total, or stagnation, pressure for isentropic realization of pressure build-up, psia

P_1 = static pressure of the moving stream, psia

k = ratio of specific heats, c_p/c_v

2-10. Diffusers. How effectively enthalpy can be converted into kinetic energy for a given pressure decrease, or the reverse problem of how effectively a decrease in kinetic energy to enthalpy results in pressure rise, is related to the irreversibilities which exist in the system. Consider the case of a moving body, such as an airplane, passing through a gaseous medium at velocity C_1. In essence, this problem is exactly the same as though the airplane were still and the air approached it with the same relative velocity. Under these conditions if entry flow reduces to trivially low velocity, the air entering the nose of the airplane could be considered as stopped and stagnation conditions would result. Under isentropic stagnation the resulting pressure rise would be from point 1 to point T in Fig. 2-4, and the corresponding pressure at T would be called the theoretical stagnation pressure P_T. However, in general, reversible conditions do not hold, and a pressure lower than P_T is reached, namely, P_B. The enthalpy increase in either case, however, must be the same because of the constancy of the kenthalpy H_o in Eq. (2-67).

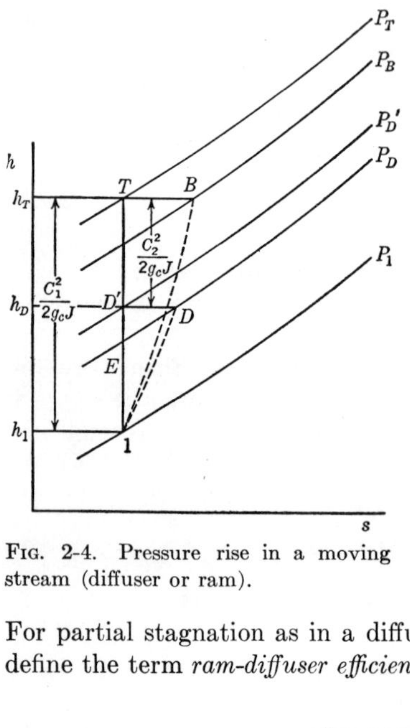

FIG. 2-4. Pressure rise in a moving stream (diffuser or ram).

In the case of a diffusing section in which the velocity C_1 is reduced to a value C_2 the maximum pressure increase would be from P_1 to $P_{D'}$ in Fig. 2-4. However, the probable actual pressure rise is to some lower value P_D because of irreversibilities (entropy increase), while in either case the kinetic energy remaining is $C_2^2/2g_cJ$.

Ram efficiency η_R is defined as the actual pressure rise compared with the pressure rise possible under reversible conditions. Referring to the stagnation condition in Fig. 2-4,

$$\eta_R = \frac{P_B - P_1}{(P_T - P_1)_s} \tag{2-75}$$

For partial stagnation as in a diffuser with the velocity $C_2 > 0$ we can define the term *ram-diffuser efficiency* η_{RD} in similar manner as

$$\eta_{RD} = \frac{P_D - P_1}{(P_{D'} - P_1)_s} \tag{2-76}$$

In Eq. (2-76) η_{RD} as defined is also known as *diffusion effectiveness*. For subsonic velocities, ram efficiencies under stagnation conditions can reach 90 per cent or more although lower values can often be expected. Diffuser efficiencies rarely exceed 85 per cent and are frequently in the range of 60 to 80 per cent.

It is frequently convenient and desirable to analyze diffuser performance on an energy basis and only indirectly in terms of pressures. Referring to Fig. 2-4, consider a stream of fluid decreasing in velocity from

C_1 to a final velocity C_2. The final fluid condition is represented as having an enthalpy value h_D and a temperature of T_D. T_D and $T_{D'}$ are the same, whether the diffusion takes place reversibly or not. Write an energy balance for this adiabatic-diffusion process, and by Eq. (2-67) there results

$$\frac{C_1^2 - C_2^2}{2g_cJ} = h_D - h_1 = (h_{D'} - h_1)_s \tag{2-77}$$

For the enthalpy increase $(h_{D'} - h_1 = h_D - h_1)$ the maximum pressure rise to $p_{D'}$ takes place only if the diffusion process occurs under reversible conditions. The actual pressure reached in the diffuser for the final-state point D is the lower pressure p_D (Fig. 2-4). Under reversible diffusion this actual static pressure increase could be attained if the enthalpy increased only in amount $(h_E - h_1)_s$. For the actual pressure increase of $p_D - p_1$ let us compare the minimum (isentropic) enthalpy increase with the kinetic energy converted to enthalpy in the diffuser. This gives an energy ratio on which to base diffuser efficiency, *viz.*,

$$\eta_D = \frac{(h_E - h_1)_s}{(h_{D'} - h_1)_s} = \frac{(h_E - h_1)_s}{(C_1^2 - C_2^2)/2g_cJ} \tag{2-78}$$

Where gas-property tables are available such as those for air (Table 2-4), the pressure increase can readily be determined by use of Eq. (2-78). It is also possible to start with Eq. (2-78) and develop explicit expressions involving pressure by the use of the gas relationships.

$$\eta_D = \frac{(h_E - h_1)_s}{(h_{D'} - h_1)_s} = \frac{c_{pm}(T_E - T_1)_s}{c_{pm}(T_{D'} - T_1)_s} = \frac{(T_E/T_1)_s - 1}{(T_{D'}/T_1)_s - 1} \tag{2-79}$$

$$\eta_D = \frac{(p_D/p_1)^{(k-1)/k} - 1}{(p_{D'}/p_1)^{(k-1)/k} - 1} = \frac{1 - (p_D/p_1)^{(k-1)/k}}{1 - (p_{D'}/p_1)^{(k-1)/k}} \tag{2-80}$$

$$\eta_D = \frac{1 - (p_D/p_1)^{(k-1)/k}}{1 - (T_{D'}/T_1)} = \frac{1 - (p_D/p_1)^{(k-1)/k}}{1 - (T_D/T_1)} \tag{2-81}$$

In Eqs. (2-77) to (2-81)

$p_D = p_E$ = realized pressure, psia, in a diffuser, for a final velocity of C_2 fps

$p_{D'}$ = pressure, psia, which could be realized in an isentropic diffuser for the same velocity change to C_2 fps

p_1 = pressure at inlet to diffuser, psia

$h_D = h_{D'}$; h_E, h_1 enthalpy values at various conditions for the diffuser

C_1, C_2 = velocities at inlet and at any other stated point in a diffuser, fps

k = ratio of specific heats

Example 7. The diffusing section following the last stage of a gas turbine receives gases, considered as air, at 600 fps at 1060 R and at 14.5 psia. The gases are reduced in velocity to 150 fps on exit from the section. Compute the pressure increase and the diffuser area ratio, (*a*) if the diffuser efficiency η_D is 100 per cent, (*b*) if the diffuser efficiency is 80 per cent.

Solution. (*a*) Using the air table 2-4 read at 1060 R that $h_1 = 146.180$ Btu per lb, $p_{r1} = 19.267$. By Eq. (2-78)

$$1.00 = \frac{h_E - 146.180}{\dfrac{(600)^2 - (150)^2}{50{,}070}} \qquad h_E = 152.918 \text{ Btu per lb}$$

For this value of enthalpy $p_{rE} = 21.114$ and $T_E = 1086.8 \text{ R} = T_{D'} = T_D$ because $\eta_D = 100$ per cent.

$$\frac{p_E}{p_1} = \frac{p_{rE}}{p_{r1}} \qquad \frac{p_E}{14.5} = \frac{21.114}{19.267} \qquad p_E = 15.9 \text{ psia}$$

$$\Delta p = 15.9 - 14.5 = 1.4 \text{ psi}$$

$$\text{lb per sec flow} = \frac{A_2C_2}{v_2} = \frac{A_1C_1}{v_1}$$

$$\frac{A_1}{A_2} = \frac{C_2v_1}{C_1v_2}$$

By Eq. (2-3)

$$(144)(14.5)(v_1) = (1)(53.34)(1060)$$

$$v_1 = 27.08 \text{ ft}^3 \text{ per lb}$$

$$(144)(15.9)(v_2) = (1)(53.34)(1086.8)$$

$$v_2 = 25.31 \text{ ft}^3 \text{ per lb}$$

$$\frac{A_1}{A_2} = \frac{(150)(27.08)}{(600)(25.31)} = 0.267 \qquad \frac{A_2}{A_1} = 3.74$$

(*b*) If $\eta_D = 0.80$, by Eq. (2-78),

$$0.8 = \frac{h_E - 146.180}{\dfrac{(600)^2 - (150)^2}{50{,}070}} \qquad h_E = 151.570$$

For this value of h_E read $p_{rE} = 20.732$.

$$\frac{p_E}{14.5} = \frac{20.732}{19.267} \qquad p_E = 15.6 \text{ psia}$$

$$\Delta p = 15.6 - 14.5 = 1.1 \text{ psi}$$

To find v at exit, it should be realized that the temperature is still $1086.8 \text{ R} = T_{D'}$ irrespective of the pressure rise.

$$(144)(15.6)(v_2) = (1)(53.34)(1086.8)$$

$$v_2 = 25.80 \text{ ft}^3 \text{ per lb}$$

$$\frac{A_1}{A_2} = \frac{(150)(27.08)}{(600)(25.80)} = 0.262 \qquad \frac{A_2}{A_1} = 3.81$$

Equations (2-79) and (2-80) could also be employed as an alternate solution if desired.

It has previously been mentioned that the efficiency of a diffuser is less than the efficiency of a nozzle. In the diffuser, flow is in the direction of increasing pressure, which is an unnatural condition for flow. At the diffuser walls the boundary layer of gas moves at much slower velocity than the main stream of the diffuser. Because of this, in extreme conditions it is even possible for the main stream to tear away from the boundary layer and move through the diffuser almost as a jet. This undesirable condition is illustrated in Fig. 2-5, where a central core is shown passing through at only slightly reduced velocity with a region of high turbulence developed in the surrounding area of the downstream part of the diffuser. If design pressure rise is too rapid in a diffuser, it is possible

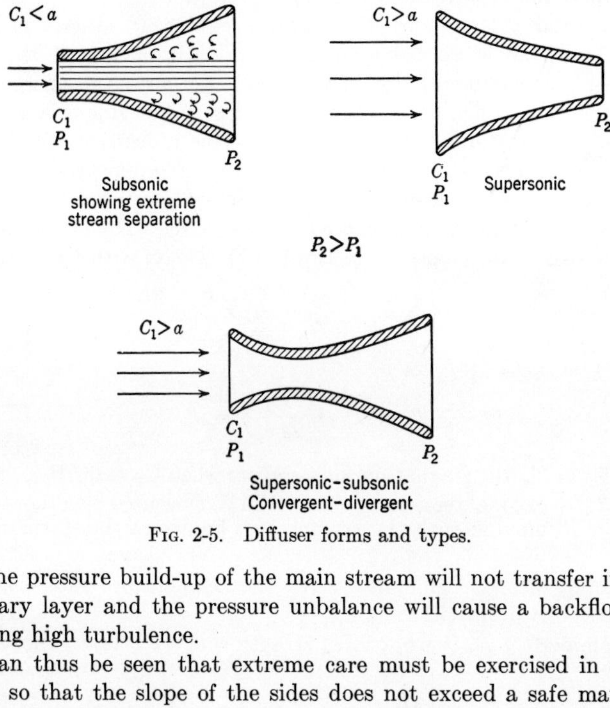

FIG. 2-5. Diffuser forms and types.

that the pressure build-up of the main stream will not transfer into the boundary layer and the pressure unbalance will cause a backflow with resulting high turbulence.

It can thus be seen that extreme care must be exercised in diffuser design so that the slope of the sides does not exceed a safe maximum. On the other hand, if the slope is made too shallow, a long diffusing section results, which can result in high wall friction and inefficiency from this cause. It is customary to design subsonic diffusers with the included angle ranging from 4 to 6° along the length of the diffuser. The subsonic diffuser is of diverging cross section in the direction of flow. On the other hand, the supersonic diffuser must converge in the direction of flow. These types are illustrated in Fig. 2-5 along with a convergent-divergent type which can build up pressure when velocities are reduced from supersonic to subsonic values. Diffuser efficiencies, except with very carefully designed units, seldom exceed 85 per cent.

For subsonic diffusion with either compressible or incompressible fluids, the shape of the section is always divergent. With supersonic velocities of the fluid undergoing diffusion, the cross section converges from large to smaller area in the direction of flow; except that, in a pipe of constant cross-sectional area carrying a stream moving at supersonic velocity, the pressure builds up in the direction of flow, provided friction is present, and such a pipe acts as a diffuser. It is, of course, realized that with a subsonic flow in a pipe of constant cross-sectional area the pressure drops in the direction of flow when friction is present.

2-11. Total (Stagnation) Temperature. When a stream of fluid is stopped without work being done, the kinetic energy of the stream appears as increased enthalpy of the fluid. For a gas this energy change is evidenced by a significant temperature rise. For adiabatic conditions the final temperature is the same whether the process of kinetic-energy transformation takes place reversibly (with maximum pressure increase being realized) or irreversibly. The final temperature reached by the gas is known as the *total*, or *stagnation, temperature.* The stagnation temperature can be developed from Eq. (2-67), where, with $C_2 = 0$, we can write

$$h_2 = h_1 + \frac{C_1^2}{2g_cJ} \tag{2-82}$$

and as $h_2 - h_1 = c_p(T_2 - T_1)$

$$T_2 = T_1 + \frac{C_1^2}{2g_cc_pJ} = T_o \tag{2-83}$$

where $T_2 = T_o$, the stagnation temperature when $C_2 = 0$, °R

T_1 = original temperature as would be measured by a thermometer moving with the stream, also known as the static temperature, °R

c_p = mean specific heat over the temperature range $T_0 - T_1$, Btu/(lb)(°R)

For a semiperfect gas it is possible to transform Eq. (2-83) by the substitution of

$$c_p = \frac{kR}{(k-1)J} \qquad \text{and} \qquad a = \sqrt{g_ckRT}$$

into

$$T_o = T_1 + \frac{C_1^2(k-1)T_1J}{2g_ckRJT_1} = T_1 + \frac{T_1(k-1)C_1^2}{2a_1^2}$$

$$T_o = T_1\left(1 + \frac{k-1}{2}\frac{C_1^2}{a_1^2}\right) = T_1\left(1 + \frac{k-1}{2}M_1^2\right) \tag{2-84}$$

A thermometer placed in a moving stream indicates a temperature greater than the static temperature because the moving stream in slowing down against the bulb produces a temperature rise. Although ideally a thermometer could indicate the full stagnation temperature, this is seldom reached because only a fraction of the kinetic energy is dissipated at the thermometer bulb. The shape and location of the bulb are important factors in determining how close the thermometer reading approaches the stagnation temperature. From 40 to 85 per cent of the kinetic energy of the moving stream can be expected to appear as measurable temperature rise above the static temperature. Assuming that a portion n of the initial kinetic energy is realized as a temperature increase, Eq. (2-83) could be written to indicate the temperature reading T_t as follows:

$$T_t = T_1 + n \frac{C_1^2}{2c_p g_c J} \tag{2-85}$$

The real static stream temperature is very often a desired value, and this can be found if definite knowledge of n values for the particular thermometric element is had. To transform Eq. (2-85) into a form which is sometimes more convenient, solve it for C_1^2, and equate to the C_1^2 value found in Eq. (2-83).

$$C_1^2 = \frac{(T_t - T_1)2g_c c_p J}{n} \tag{2-86}$$

$$C_1^2 = T_1\left(\frac{T_0}{T_1} - 1\right) 2g_c c_p J \tag{2-87}$$

Equate the right-hand parts of the above two equations and solve for T_1.

$$T_1 = \frac{T_t}{n\left(\dfrac{T_0}{T_1} - 1\right) + 1} \tag{2-88}$$

Substitute T_0/T_1 from Eq. (2-84), giving

$$T_1 = \frac{T_t}{n\left(1 + \dfrac{k-1}{2} M_1^2 - 1\right) + 1} = \frac{T_t}{1 + nM_1^2 \dfrac{k-1}{2}} \tag{2-89}$$

where T_t = temperature of a gas stream as indicated by a thermometric element, °R

T_1 = static temperature of gas stream, °R

n = temperature recovery factor of the thermometric element

Example 8. A thermometric element in a wind tunnel has a recovery factor of 0.74. Air at 20 psia moving at 400 mph shows an indicated temperature of 97.3 F. What is the probable static temperature in the tunnel?

Solution. Either Eq. (2-89) or Eq. (2-85) can be used. Here (2-85) appears more convenient as the local acoustic velocity is not known with certainty because the static temperature is not known.

$$T_1 = T_t - n\frac{C_1^2}{2c_p g_c J} = (97.3 + 459.7) - 0.74\left\{\frac{[(400)(1.466)]^2}{(2)(0.240)(32.2)(778)}\right\}$$

In the above, mph $\times$ 1.466 = fps, 0.240 is c_p from Table 2-4.

$$T_1 = 557.0 - 21.1 = 535.9 \text{ R} = 76.2 \text{ F}$$

2-12. Momentum Changes in a Flowing Fluid. Newton's second law of motion applied to a mass particle or system of such particles states: *The resultant external force acting on a system of particles is equal to the rate of change of momentum of the particles.* Expressed mathematically,

$$F = \frac{d\mathbf{M}}{dt} \tag{2-90}$$

$$F_x = \frac{d\mathbf{M}_x}{dt} \qquad F_y = \frac{d\mathbf{M}_y}{dt} \qquad F_z = \frac{d\mathbf{M}_z}{dt} \tag{2-91}$$

where F_x, F_y, and F_z are the components of external force along any coordinate axes. The analysis made here will be limited to the xy plane, but the conclusions apply equally rigorously to three dimensions.

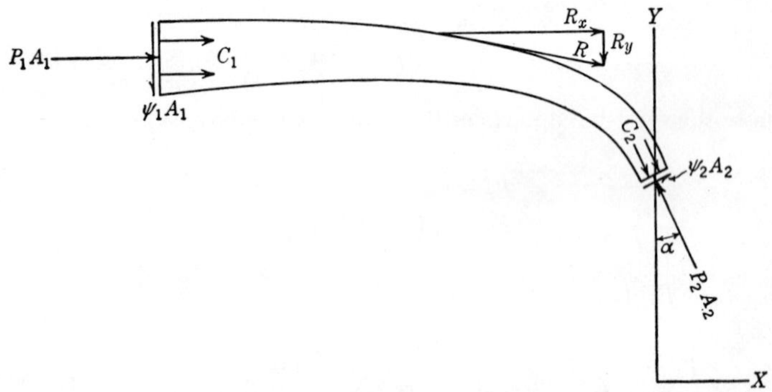

FIG. 2-6. Flow stream of a fluid, indicating pressure and momentum changes.

Consider the stream tube of Fig. 2-6, stationary in space, the side boundaries of which are formed either by a solid wall or by adjacent fluid flowing in the direction of the side boundaries shown. Within this boundary there is a steady flow of a system of mass particles (the fluid) which is acted upon by forces external to the system. The resultant of these external forces causes a momentum change of the mass particles, according to Newton's laws.

$$\sum \begin{matrix}\text{external forces}\\ \text{on the system}\end{matrix} = \sum \begin{matrix}\text{time rate of momentum change}\\ \text{for the particles of the system}\end{matrix} \tag{2-92}$$

Forces and momentum are vector quantities, and so *vectorial* summations must be used.

$$\sum \begin{matrix}\text{external forces}\\ \text{on the system}\end{matrix} = \mathbf{P}_1\mathbf{A}_1 +\!\!\!\!\to \mathbf{P}_2\mathbf{A}_2 +\!\!\!\!\to \mathbf{R} +\!\!\!\!\to \boldsymbol{\psi}_1 A_1 +\!\!\!\!\to \boldsymbol{\psi}_2\mathbf{A}_2 \qquad (2\text{-}93)$$

In Eq. (2-93) and throughout this section, standard vector notation has been employed. With this notation, vectorial addition is indicated by the symbol $+\!\!\!\!\to$ and vectorial subtraction by $\rightarrow$.

In Eq. (2-93), **PA** and $\boldsymbol{\psi}$**A** are the normal and shear forces on the fluid, exerted on the ends of the stream tube. The term **R** represents the net force exerted by the side boundaries on the enclosed fluid. This may be a solid wall reaction or a pressure and shear-force reaction, depending upon the nature of the boundary.

Fluid entering the stream tube at section 1 and leaving at section 2 is acted upon by the foregoing set of forces during its passage through the tube. During that time, the momentum of the fluid changes from that at 1 to that at 2, or

$$\sum \begin{matrix}\text{time rate of momentum change for}\\ \text{the particles of the system}\end{matrix} = \sum \frac{(m_2\mathbf{C}_2/g_c) \rightarrow (m_1\mathbf{C}_1/g_c)}{\Delta t} \qquad (2\text{-}94)$$

where m_2, m_1 = lb flowing across the respective boundaries of the system in time Δt sec

$\mathbf{C}_1$, $\mathbf{C}_2$ = initial and final velocities, respectively

This summation can also be written

$$\begin{aligned}\sum &= \sum \frac{m_2}{g_c\,\Delta t}\mathbf{C}_2 \rightarrow \sum \frac{m_1}{g_c\,\Delta t}\mathbf{C}_1\\ &= \sum \frac{G_2}{g_c}\mathbf{C}_2 \rightarrow \sum \frac{G_1}{g_c}\mathbf{C}_1 \qquad (2\text{-}95)\end{aligned}$$

where G is the mass flow rate, in pounds per second. If the velocity is not constant across section 2, then the leaving momentum per second, which is $\Sigma\,(G_2/g_c)C_2$, is obtained by integration or summation of the momentum leaving across the elemental areas of the section. Expressed mathematically,

$$\sum \frac{G_2}{g_c}\mathbf{C}_2 = \int_0^{A_2} \frac{G_{A2}}{g_c}\,dA\;\mathbf{C}_{A2} \qquad (2\text{-}96)$$

where G_{A2} = local mass flow rate, lb/(sec)(ft²), at any point in the cross section for which the fluid velocity is $\mathbf{C}_{A2}$

dA = small elemental area, ft², through which the local mass flow G_{A2} takes place

A similar expression can be written for the fluid entering at section 1.

The total rate of change of momentum becomes

$$\int_0^{A_2} \frac{G_{A2}}{g_c}\, dA_2\, \mathbf{C}_{A2} \rightarrow \int_0^{A_1} \frac{G_{A1}}{g_c}\, dA_1\, \mathbf{C}_{A1} \tag{2-97}$$

The basic equation (2-92) can now be written as follows:

$$\mathbf{P}_1\mathbf{A}_1 \nrightarrow \mathbf{P}_2\mathbf{A}_2 \nrightarrow \mathbf{R} \nrightarrow \boldsymbol{\psi}_1\mathbf{A}_1 \nrightarrow \boldsymbol{\psi}_2\mathbf{A}_2 = \int_0^{A_2} \frac{G_{A2}\, dA_2\, \mathbf{C}_{A2}}{g_c} \rightarrow \int_0^{A_1} \frac{G_{A1}\, dA_1\, \mathbf{C}_{A1}}{g_c} \tag{2-98}$$

If the velocities are uniform across sections 1 and 2, having values $\mathbf{C}_1$ and $\mathbf{C}_2$, the equation reduces to

$$\mathbf{P}_1\mathbf{A}_1 \nrightarrow \mathbf{P}_2\mathbf{A}_2 \nrightarrow \mathbf{R} \nrightarrow \boldsymbol{\psi}_1\mathbf{A}_1 \nrightarrow \boldsymbol{\psi}_2\mathbf{A}_2 = \frac{G}{g_c}\,\mathbf{C}_2 \rightarrow \frac{G}{g_c}\,\mathbf{C}_1 \tag{2-99}$$

Usually the magnitude of the shearing forces on the ends is trivially small so that the $\boldsymbol{\psi}\mathbf{A}$ terms can be eliminated. This leaves

$$\mathbf{P}_1\mathbf{A}_1 \nrightarrow \mathbf{P}_2\mathbf{A}_2 \nrightarrow \mathbf{R} = \frac{G}{g_c}\,\mathbf{C}_2 \rightarrow \frac{G}{g_c}\,\mathbf{C}_1 \tag{2-100}$$

This vector equation can be replaced by algebraic equations by taking components of forces and momentum in the x and y directions (for two-dimensional flow). In Fig. 2-6, let the positive directions be chosen to the right (for x components) and downward (for y components). These positive directions correspond to the directions of flow.

For the x direction,

$$(P_1A_1)_x - (P_2A_2)_x + R_x = \frac{G}{g_c}(C_{2x} - C_{1x}) \tag{2-101}$$

For the y direction,

$$(P_1A_1)_y - (P_2A_2)_y + R_y = \frac{G}{g_c}(C_{2y} - C_{1y}) \tag{2-102}$$

At any section, the mass rate of flow, G/g_c slugs per sec, can be expressed as

$$\frac{G}{g_c} = \frac{A_1C_1}{g_c v_1} = \frac{\rho_1 A_1 C_1}{g_c} = \frac{\rho_2 A_2 C_2}{g_c} \tag{2-103}$$

where ρ = fluid density in pounds per cubic foot. Substituting in Eqs. (2-101) and (2-102),

$$(P_1A_1)_x - (P_2A_2)_x + R_x = \frac{1}{g_c}(\rho_2 A_2 C_2 C_{2x} - \rho_1 A_1 C_1 C_{1x}) \tag{2-104}$$

$$(P_1A_1)_y - (P_2A_2)_y + R_y = \frac{1}{g_c}(\rho_2 A_2 C_2 C_{2y} - \rho_1 A_1 C_1 C_{1y}) \tag{2-105}$$

One practical problem of importance is the flow of fluid through a straight symmetrical duct, such as a pipe or a jet engine. If the duct center line (or axis of symmetry) is chosen as the x axis, Eq. (2-101) is all that is needed completely to specify the dynamic equilibrium of the fluid within the duct. (The assumption of uniform velocity over inlet and outlet sections is necessary for this equation, of course.)

PROBLEMS

2-1. The combustion products from a gas turbine after analysis on a volume basis were found to have a composition of $2.7CO_2$, $16.6O_2$, $77.9N_2$, $2.8H_2O$. For these products at high temperature and before condensation of any water vapor has taken place determine (*a*) the analysis in percentage by weight, (*b*) the composite (equivalent) molecular weight, (*c*) the value of R, (*d*) the density of this gas in pounds per cubic foot at 440.3 F and at 14.7 psia. *Ans.* (*a*) 4.10, 18.44, 75.71, 1.75; (*b*) 28.82; (*c*) 53.6; (*d*) 0.0439.

2-2. A fuel gas consists of 91% CH_4, 3.5% CO_2, 5.5% N_2 by volume. Find (*a*) the analysis by weight, (*b*) composite molecular weight, (*c*) the value of R, (*d*) the density of this gas in pounds per cubic foot at 14.7 and 60.3 *R*. *Ans.* (*a*) 82.54, 8.73, 8.73 per cent; (*b*) 17.64; (*c*) 87.5, (*d*) 0.0464.

2-3. Use the air tables to compute the work required to compress air reversibly in an adiabatic compressor from 15 psia at 0.3 F (460 R) to a final leaving pressure of 90 psia. Make use of the entropy function, and disregard kinetic energy changes. *Ans.* 73.75 Btu per lb.

2-4. Air expands in a gas turbine from 1600 R and 75 psia to 15 psia. If the expansion is reversible and adiabatic, compute the work delivered per pound of air by use of the air tables and the entropy function. *Ans.* 144.014 Btu per lb.

2-5. For temperatures of 500 and 1000 R check the values of relative pressure p_r found in Table 2-4.

2-6. Work Prob. 2-3 by making use of the p_r column in the air table 2-4. *Ans.* 73.758 Btu per lb.

2-7. Work Prob. 2-4 by making use of the p_r column in the air table 2-4. *Ans.* 144.031 Btu per lb.

2-8. A gas which can be considered to have the properties of air enters a cooler at 300 fps at a temperature of 800 R and a pressure of 400 psia. It is cooled to 600 R while transferring heat to water-cooled tubes and leaves with a velocity of 100 fps at slightly less than 400 psia pressure. (*a*) Compute the kenthalpy of the entering and leaving air. (*b*) Find the heat absorbed by the water per pound of air passing through the cooler. (The pressure of the air does not enter in the solution.) *Ans.* (*a*) 83.778 Btu per lb; 33.808. (*b*) 49.97.

2-9. Air is heated in a cross-tube gas-fired unit from 90.3 to 490.3 F by combustion gases. The entering air has a velocity of 150 fps and the leaving air a velocity of 50 fps. Find (*a*) the kenthalpy of the entering and leaving air and (*b*) the heat absorbed per pound of air. (*c*) By what per cent would the answer to part (*b*) be in error if kinetic-energy effects were disregarded? *Ans.* (*a*) 22.046, 118.850; (*b*) 96.804; (*c*) 0.41 per cent.

2-10. Find the velocity at exit from a nozzle supplied with air at 800 R, 30 psia at 100 fps and expanding to 20 psia if the nozzle efficiency is considered unity. Find the inlet and exit areas required to pass 20 lb per sec. *Ans.* 1029 fps, 1.97 ft^2, 0.257 ft^2 at exit.

2-11. Solve Prob. 2-10 with no other change in data except to allow for a nozzle efficiency of 94 per cent. *Ans.* 998 fps, 1.97 ft^2, 0.267 ft^2 at exit.

2-12. A nozzle is to be designed to pass 20 lb per sec of helium supplied at 100 fps at 800 R, 30 psia and discharging at 20 psia. For a frictionless nozzle find the exit velocity and the required inlet and exit areas. [Suggestion: Use Table 2-1 and Eqs. (2-20), and (2-3).] *Ans.* 2729 fps, 14.3 ft^2, 0.67 ft^2 at exit.

2-13. Find the acoustic velocity of dry air at 30 psia at temperatures of 0.3, 60.3, 120.3 F. Express in feet per second and in miles per hour. (Note that the acoustic velocity of a perfect gas is independent of pressure.) *Ans.* At 60.3 F, $a = 1117$ fps and 752 mph.

2-14. Find the acoustic velocity of helium at 250 psia and at 60.3 F. (See Table 2-1.) *Ans.* 3273 fps.

2-15. Air is flowing in a duct at 59.3 F (519 R) at a velocity of 600 fps and under a static pressure of 30 psia. Use the methods of Art. 2-9 to compute the total (stagnation) pressure. *Ans.* 36.5 psia.

2-16. Air is flowing in a duct at 59.3 F (519 R) at a velocity of 600 fps and under a static pressure of 30 psia. The air is slowed down under reversible-adiabatic conditions to 100 fps. Use the methods of Art. 2-9 to compute the resultant static pressure. *Ans.* 36.3 psia.

2-17. Solve Prob. 2-15 by use of the air tables. *Ans.* $p_T = 36.5$ psia.

2-18. Solve Prob. 2-16 by use of the air tables.

2-19. Compute the kenthalpy and find the total (stagnation) temperature of the air flowing in the duct of Prob. 2-15. *Ans.* $H_o = 21.346$, $T_o = 548.96$.

2-20. *Diffuser.* An airplane flying 400 mph at 15,000 ft (16.88 in. Hg absolute pressure) inducts air at 5.5 F (465.2 R). The induction system has a diffuser efficiency of 0.78, and the air velocity is reduced to 200 fps on leaving the diffuser and at entry to the compressor. (*a*) Compute the air temperature at inlet to the compressor. (*b*) Compute the actual pressure leaving the diffuser. (*c*) Find the area ratio required. (*d*) What is the ram diffuser efficiency? *Ans.* (*a*) 490.5 R, (*b*) 19.52 in. Hg. (*c*) 2.67, (*d*) 0.77. Work by use of the air tables.

2-21. *Diffuser.* Work Prob. 2-20 by use of conventional gas relationships.

2-22. *Diffuser.* An eight-stage gas turbine receives air at 1350.3 F, 75 psia, and develops 6000 shaft hp when turning at 9200 rpm. The stages of the turbine are designed for equal Δh values over the stationary and rotating blades at the mean diameter, and the blades are tapered and twisted. The rotor of 19-9 stainless steel is 14½ in. in diameter and 24 in. long. The blades themselves taper from 1.98 in. height of the first fixed blade to 4.72 in. mean height of the last moving blade. The gases leave the last row of blades with a maximum velocity of 500 fps and enter the diffuser section at 470 fps and 850.3 F, where the velocity decreases as the passage changes from an annular section 5 in. deep over the 14½-in. rotor core to 13.7 in. deep over a 1.85-in. fixed core. Velocity at exit from the diffuser is 210 fps.

a. Verify that the diffuser area ratio is 2.13 and $A_1 = 2.12$ ft,^{2} $A_2 = 4.52$ ft.^{2}

b. Compute the ideal exit pressure from the diffuser if reversible adiabatic flow is considered and at inlet to the diffuser the pressure is 14.32 psia. *Ans.* 14.97 psia.

c. Work part (*b*) if the diffuser efficiency is 0.8. *Ans.* 14.80 psia.

d. Find the pounds per second flow based on inlet conditions to the diffuser. *Ans.* 29.6.

2-23. *Diffuser.* A supersonic diffuser receives an air supply at 2360 fps at 1384.3 R and 15 psia. Assume diffuser efficiency is 85 per cent. (*a*) For two velocity values in the diffuser, namely, 1888 fps and 1400 fps, find both theoretical and realized pressure values. (*b*) Find for 1 lb per sec flow the diffuser areas at inlet, at 1888 fps, and at 1400 fps. (*c*) Find the ram diffuser efficiency at the given velocities. *Ans.* (*a*) 22.32 and 21.0 psia, 29.89 and 27.2 psia; (*b*) 0.01448, 0.01432, and 0.0161 ft^2; (*c*) 82% and 81.8%.

2-24. An impact thermometric element in the duct of Prob. 2-15 indicates a temperature of 540.0 R. What is the recovery factor of this particular element? *Ans.* 0.7.

REFERENCES

1. Keenan and Kaye: A Table of Thermodynamic Properties of Air, *Trans.* ASME, vol. 10, pp. A123–A130, September, 1943.
2. Gerhart, Brunner, Mickley, Sage, and Lacey: Thermodynamic Properties of Air, *Mech. Eng.*, vol. 64, pp. 270–272, 1942.
3. Ellenwood, Kulik, and Gay: Specific Heats at High Pressures, *Cornell Univ. Expt. Sta. Bull.* 30, 1942.
4. Johnson and Davis: Heat Capacity Curves of the Simpler Gases, *J. Am. Chem. Soc.*, vol. 56, pp. 271–276, 1934.
5. Johnson and Walker: Heat Capacity Curves of the Simpler Gases, *J. Am. Chem. Soc.*, vol. 55, pp. 172–187, 1933, vol. 57, pp. 682–684, 1935.
6. Keenan and Kaye: "Gas Tables," John Wiley & Sons, Inc., New York, Inc., 1948.
7. Keenan and Kaye: "Thermodynamic Properties of Air," John Wiley & Sons, Inc., New York, 1945.
8. *NACA Tech. Note* 1138, Standard Procedures for Rating and Testing Multistage Axial Flow Compressors, September, 1946.
9. Hunsacker and Rightmire: "Engineering Applications of Fluid Mechanics," McGraw-Hill Book Company, Inc., New York, 1947.
10. Sauer, Robert: "Introduction to Theoretical Gas Dynamics," translated by Hill and Alpher, Edwards Bros, Inc., Ann Arbor, Mich., 1947.
11. Shapiro and Hawthorne: Mechanics and Thermodynamics of Steady One-dimensional Gas Flow, *ASME J. Applied Mechanics*, vol. 14, no. 4, pp. A317–A336, December, 1947.
12. Shapiro: Nozzles for Supersonic Flow without Shock Fronts, *ASME J. Applied Mechanics*, vol. 11, no. 2, June, 1944.
13. von Mises, R.: On the Thickness of a Steady Shock Wave, *J. Aeronaut. Sci.*, vol. 17, no. 9, pp. 551–555, September, 1950.
14. Courant and Friedrichs: "Supersonic Flow and Shock Waves," Interscience Publishers, Inc., New York, 1948.
15. Dahl and Fiock: Shielded Thermocouples for Gas Turbines, *Trans. ASME*, vol. 71, pp. 153–161, 1949.

CHAPTER 3

GAS-TURBINE CYCLES AND THERMODYNAMIC ANALYSES

3-1. Brayton Cycle. This prototype cycle for the gas turbine, which also goes under the name of the Joule cycle, is shown in Fig. 3-1 on the *pv* and *Ts* planes. It will be noted that the cycle consists of two isentropics, 1 to 2 and 3 to 4, and two constant-pressure lines, 2 to 3 and 4 to 1. For this cycle operating with a perfect gas it is possible to derive an expression for the efficiency in simple form.

$$\eta_t = \frac{\text{work}}{\text{heat added}} = \frac{\text{heat added} - \text{heat rejected}}{\text{heat added}} \tag{3-1}$$

$$\eta_{tB} = \frac{c_{pm}(T_3 - T_2) - c_{pm}(T_4 - T_1)}{c_{pm}(T_3 - T_2)} = 1 - \frac{T_4 - T_1}{T_3 - T_2} \tag{3-2}$$

It is assumed above that the mean specific heats are not appreciably different over the two temperature ranges.

Introduce the pressure ratio, $r = p_2/p_1 = p_3/p_4$, and it follows by Eq. (2-20) that

$$T_2 = T_1 r^{(k-1)/k} \qquad T_3 = T_4 r^{(k-1/k)}$$

Thus

$$\frac{T_2}{T_1} = \frac{T_3}{T_4} \quad \text{and} \quad \frac{T_2}{T_3} = \frac{T_1}{T_4} \tag{3-3}$$

$$\eta_{tB} = 1 - \frac{T_4 - T_1}{T_3 - T_2} = 1 - \frac{T_1 \dfrac{T_3}{T_2} - T_1}{T_2 \dfrac{T_3}{T_2} - T_2} = 1 - \frac{T_1\left(\dfrac{T_3}{T_2} - 1\right)}{T_2\left(\dfrac{T_3}{T_2} - 1\right)}$$

$$\eta_{tB} = 1 - \frac{T_1}{T_2} \tag{3-4}$$

$$\eta_{tB} = 1 - \frac{1}{r^{(k-1)/k}} \tag{3-5}$$

On the *Ts* diagram for this reversible cycle, the area 2-3-5-8 represents the heat added, and the area 1-4-5-8 represents the heat rejected, with the work area 2-3-4-1 appearing as the difference. On the *pv* diagram the net work area is correspondingly 2-3-4-1. With the initial temperature of the cycle fixed at T_1 it is obvious that more work is realized from the heat supplied as the spread between T_2 and T_1 is increased, *i.e.*, as the pressure ratio is increased. Whereas in the theoretical cycle it is true that performance is independent of the top temperature T_3, it should be emphasized that this is not the case in the real gas-turbine power cycle, as performance is related closely not only to pressure ratio but also to the top or turbine inlet temperature T_3, and for every pressure ratio there exists an optimum T_3.

This is true because the working medium, although largely air, is of variable specific heat and the fuel added from 2 to 3 supplies products of combustion to the expanding gases. In addition the irreversibilities and losses of the real cycle even more greatly reduce the actual thermal

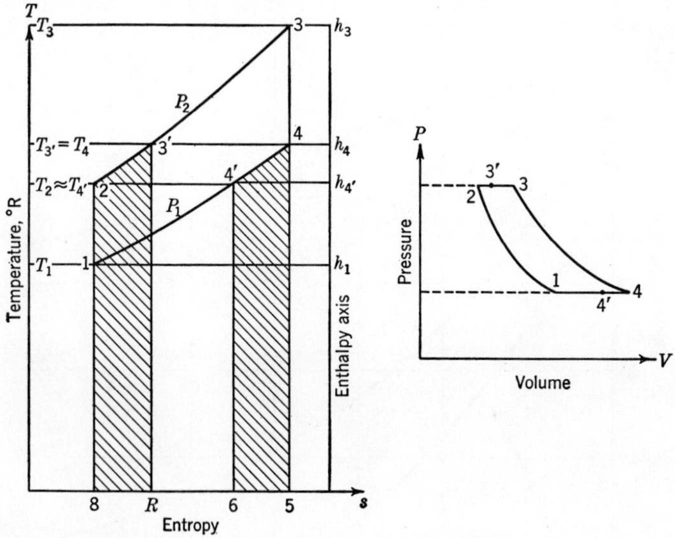

FIG. 3-1. Temperature-entropy (and Pv) diagrams of Brayton cycle with and without regeneration.

efficiency. The expression of Eq. (3-5) is of interest, however, in indicating that pressure ratio is a significant and limiting factor in efficiency. The efficiency of the basic Brayton for air appears in Fig. 3-2 as the increasing graph from left to right.

It will later be shown how actual thermal efficiency of the cycle is greatly affected by the efficiencies of its two major components, the compressor and the turbine. Efficiencies of these components range between 75 and 90 per cent, with the turbine running usually 1 to 3 percentage points higher in efficiency than the compressor. Typical comparative values for an axial-flow compressor and a multistage turbine might run 84 and 86 per cent, respectively. In addition there are pressure losses, radiation-conduction losses, and imperfect combustion, all tending to reduce over-all efficiency.

3-2. Regenerative Brayton Cycle. In the cycle illustrated in Fig. 3-1 gases leave the power unit at temperature T_4 and the compressor at temperature T_2. By proper use of regenerative heat-exchanger surface the compressed gas can be warmed to a higher temperature $T_{3'}$ by cooling down the hot gases to $T_{4'}$. By using counterflow heat exchangers the hot gases could be cooled in the limit until $T_{4'}$ reached the value T_2 while

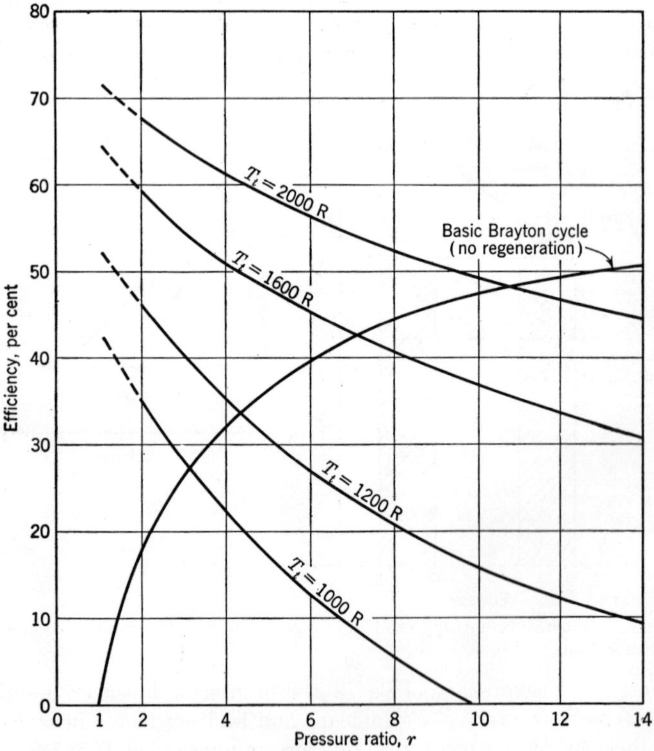

FIG. 3-2. Brayton efficiencies for basic cycle and for regenerative cycle. Computed for air with k at 1.39, T_1 = 530 R; top temperatures are indicated for the regenerative cycle.

the compressed gases warmed to T_4. This conclusion must assume equal values of mean specific heat for the hot and cold gases and the use of infinite heat-transfer surface. With these premises, it furnishes a basis for analysis of an ideal regenerative cycle.

On the Ts diagram (Fig. 3-1) the area representing heat addition from the fuel is 3′-3-5-R, the heat rejected to waste is the area 1-4′-6-8, and the work area is unchanged and is 2-3-4-1.

With the regenerative Brayton cycle $T_{3'}$ essentially reaches T_4, and $T_{4'}$ approaches T_2. The efficiency for the cycle appears

$$\eta_{tBR} = \frac{\text{work}}{\text{heat added}} = \frac{c_{pm}(T_3 - T_2) - c_{pm}(T_4 - T_1)}{c_{pm}(T_3 - T_4)}$$

$$= \frac{T_3 - T_2 - T_4 + T_1}{T_3 - T_4} = 1 - \frac{T_2 - T_1}{T_3 - T_4} \tag{3-6}$$

Refer to Eq. (2-20), and set $T_2/T_1 = r^{(k-1)/k} = \delta$ and from Eq. (3-3), $T_2/T_1 = T_3/T_4 = \delta$. Thus Eq. (3-6) becomes

$$\eta_{tBR} = 1 - \frac{(\delta T_1 - T_1)}{T_3 - (T_3/\delta)} = 1 - \frac{T_1}{T_3}\,\delta = 1 - \frac{T_1}{T_4} \tag{3-7}$$

$$\eta_{tBR} = 1 - \frac{T_1}{T_3}\,r^{(k-1)/k} \tag{3-8}$$

Note that the efficiency of the regenerative cycle depends not only on the pressure ratio but also on both the inlet and top temperatures of the theoretical cycle. This is true in real cycles as well.

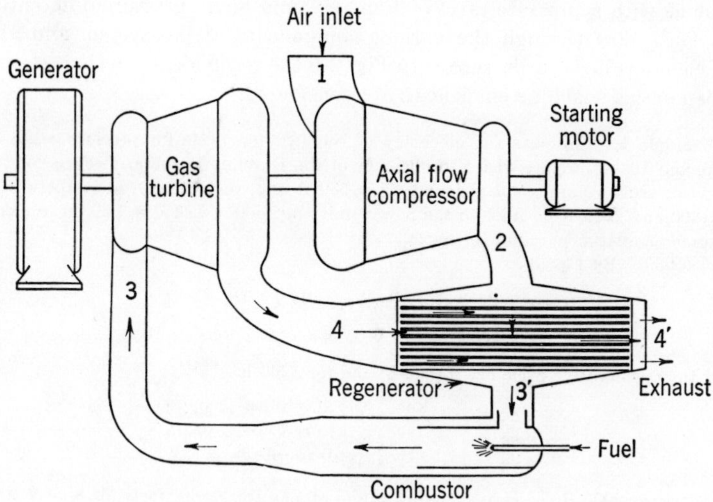

FIG. 3-3. Open-cycle system arranged for regenerative heating.

Figure 3-3 shows diagrammatically how a real gas-turbine power cycle could be arranged for regeneratively heating the gases leaving the compressor utilizing the hot exhaust gases from the turbine.

It is of interest to study the conditions under which regenerative heating is desirable. To visualize trends, the graphs of Fig. 3-2 have been computed by use of Eq. (3-8). Although these are drawn for an idealized air cycle (with $k = 1.39$), they clearly indicate trends for the real cycle. Constant turbine inlet-temperature lines are plotted in terms of pressure ratio and thermal efficiency. It will be noted that the efficiency of the regenerative cycle falls with increased pressure ratio and increases with temperature at inlet to the turbine. These curves are reversed in trend to that representing the efficiency of the nonregenerative cycle. For a constant turbine inlet temperature the reason for the decrease in efficiency with increasing pressure ratio is that the turbine exhaust-gas temperature becomes increasingly lower as the pressure ratio increases, finally reaching a temperature lower than the temperature at which air leaves the compressor. Heat would then actually be absorbed by the exhaust gases from cooling the compressed air before it entered the combustor. Following this reasoning, it is obvious that the best performance with regeneration occurs with low compression ratios and high combustion temperatures. In all real gas-turbine cycles, whether regeneration is used or not, efficiency increases rapidly with maximum temperature of the cycle. In Eq. (3-8), with a compression ratio of 1, a positive finite value of efficiency is indicated. This is only theoretically true as with a pressure ratio of 1 there would be no pressure differential to cause flow through the various components of the system and the efficiency is necessarily zero. In Fig. 3-9 the regenerative cycle is shown when actual machine efficiencies are considered.

Example 1. Compute the efficiency of the Brayton cycle for pressure ratios of 2, 6, and 10. Compare with the efficiency of the Brayton regenerative cycle for the same pressure compression ratios at $T_1 = 530$ R and, respectively, at temperatures of 1200 and 2000 R on inlet to the power unit (turbine). Use $k = 1.39$ for air, and assume it invariant around the cycle.

Solution. By Eq. (3-5)

$$\eta_{tB} = 1 - \frac{1}{(2)^{(1.39-1)/1.39}} = 17.7\% \text{ for } r = 2$$

$$\eta_{tB} = 39.6 \text{ for } r = 6 \qquad \eta_{tB} = 47.6 \text{ for } r = 10$$

For the regenerative cycle use Eq. (3-8) and for 1200 R

$$\eta_{tBR} = 1 - \tfrac{530}{1200}(2)^{(1.39-1)/1.39} = 46.3$$
$$\eta_{tBR} = 1 - \tfrac{530}{1200}(6)^{(1.39-1)/1.39} = 26.9$$
$$\eta_{tBR} = 1 - \tfrac{530}{1200}(10)^{(1.39-1)/1.39} = 15.6$$

Similarly at 2000 R the corresponding values of η_{tBR} are, respectively, 67.8, 56.2, and 49.4 per cent.

These and additional data are plotted in Fig. 3-2.

3-3. Open Gas-turbine Cycle. The real gas turbine operating through the same pressure ratio as the Brayton cycle shows much poorer performance. This is true largely because real turbines and compressors can only approach theoretically perfect performance. For example, compressor efficiencies until recent years did not exceed 80 per cent, and turbine efficiencies were but little better. In fact in small sizes (below

100 hp) it is still extremely difficult to build compressors which reach 75 per cent efficiency. This means that the power input required for the compressor is 1/0.75, or 1.33 times as great as that needed for a theoretical compressor operating through the same pressure ratio. Similarly a 75 per cent efficient turbine develops only 75 per cent as much power as an ideal turbine. Also, in the real system, undesired pressure drops occur as the gas flows from the compressor through the regenerator and combustor to the turbine, and on exit from the turbine additional pressure is required to force the waste gases through the regenerator and exhaust passages to discharge. Thus the pressure expansion ratio is less than the pressure compression ratio. Combustion is also not perfect although with liquid and gas fuels values of 95 to 99 per cent utilization are possible. Good insulation reduces heat losses, but no system is actually adiabatic.

To give good performance with the limitations mentioned, it has been necessary to improve mechanical performance and in addition to work continuously toward satisfactory operation at higher and higher inlet temperatures. The latter procedure is of course limited almost entirely by metallurgical considerations.

In Fig. 3-4 a real gas-turbine cycle is illustrated on an *hs* diagram. The compressor operating through a pressure ratio $r = \frac{p_2}{p_1}$ delivers air having enthalpy h_2, a higher value than for point 2*s* because of compressor irreversibilities. Heating in the regenerator may raise the temperature and enthalpy to point 3′ with the combustor bringing the temperature finally to the value at 3. Pressure losses from 2 to 3 reduce the pressure available at the turbine inlet to the lower value (p_t). Expansion now proceeds in an irreversible turbine to the turbine exhaust pressure p_e and a final state, point 4. The exhaust gases from the turbine pass through the regenerator, cooling to 4′, or they can be discharged direct to atmosphere. The turbine exhaust pressure is necessarily higher than free discharge pressure (that is, $p_e > p_1$).

The compressed gas (air) from the compressor experiences a pressure loss in the combustor ranging from 0.5 to some 5 psi. If a regenerator is used, an additional pressure loss of some 0.5 to 3 psi occurs. Pressure losses also exist in connecting piping. Care in design can reduce pressure losses somewhat, but a range of loss values varying from 1 to 7 psi has to be considered in any carefully made real-cycle analysis. Often the effect of pressure loss is considered by expressing the pressure loss as a fraction of the pressure at a given inlet point in the system.

$$\lambda = \frac{\Delta p_l}{p_{\text{inlet}}} \tag{3-9}$$

$$\Delta p_l = \lambda p_{\text{inlet}}$$

$$p_{\text{outlet}} = (1 - \lambda)p_{\text{inlet}} \tag{3-10}$$

where Δp_l = psi pressure loss in a component or group of gas-turbine components

λ = ratio of pressure loss to the pressure at inlet to a gas-turbine component or components

p_{inlet} = psi pressure at inlet to any component such as the regenerator-combustor parts of system (*i.e.*, leaving compressor)

The value of $1 - \lambda$ ranges from 0.90 to 0.98. For a straight-flow burner a very low loss exists, with $1 - \lambda$ ranging from 0.97 to 0.995; for the full-circumference annular-type burner, $1 - \lambda$ ranges from some 0.93 to 0.99; for reverse-flow-type combustors, $1 - \lambda$ ranges from 0.90 to 0.99. Pressure loss results from the misdirection of momentum and from fluid friction. The regenerator when used offers additional pressure loss with λ amounting to 0.003 to 0.04.

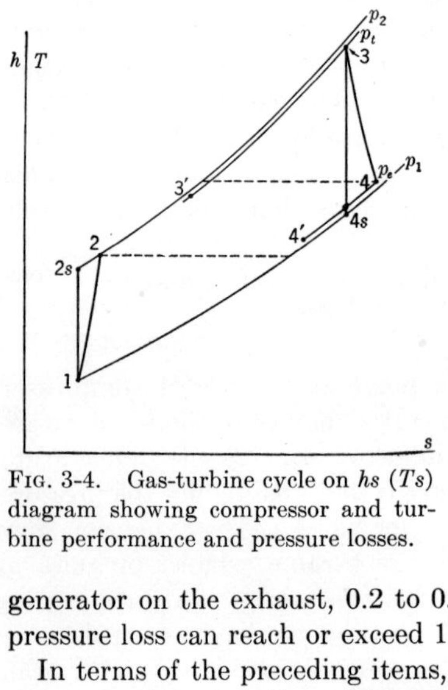

FIG. 3-4. Gas-turbine cycle on *hs* (*Ts*) diagram showing compressor and turbine performance and pressure losses.

When the compressor discharge pressures are low, more care must be exercised in design to minimize pressure losses. The above ranges are representative though actual pressure losses are variable in terms of the absolute pressure of the flowing medium. On the turbine exhaust side, pressure losses are much more critical in reducing capacity and performance and must be kept to an absolute minimum. Without a regenerator on the exhaust, 0.2 to 0.5 psi pressure drop may occur. The pressure loss can reach or exceed 1.0 psi with a restricted regenerator.

In terms of the preceding items, the pressure expansion ratio r_e might appear in the symbolism of Fig. 3-4 as

$$r_e = \frac{(1 - \lambda)p_2}{p_e} = \frac{(1 - \lambda)p_2}{p_1 + \Delta p_{\text{exhaust}}} \tag{3-11}$$

This is in contrast to the pressure compression ratio, which in terms of Fig. 3-4 is

$$r = \frac{p_2}{p_1}$$

With the help of the pressure ratios r and r_e it is possible to find the appropriate enthalpy values and then determine

$$\text{Power}_{\text{compressor}} = G_c(h_2 - h_1) \tag{3-12}$$

$$\text{Power}_{\text{turbine}} = G_t(h_3 - h_4) \tag{3-13}$$

where G_c = weight flow through compressor, lb per sec

G_t = lb per sec weight flow through turbine equal to $G_c(1 + f)$, where f is the ratio of fuel to air expressed as a decimal

h_2, h_1, h_3, h_4 = respective enthalpy values, Btu per lb. Power is expressed in Btu per second for these equations, which assume adiabatic conditions and a trivial difference in the kinetic energy of the stream at inlet to and exit from each machine

The fuel ratio f, which is the weight of fuel used to heat unit weight of compressed air to turbine inlet temperature, varies directly with the temperature increase required. However, it is also dependent upon the temperature range through which the heating takes place, as the specific heats of most gases increase with temperature. The lower heating value of the fuel is the important consideration, as this energy alone contributes to the temperature rise under conditions of gas-turbine operation.

Combustion calculations are considered in detail in a later chapter, but for the present, in problem solution, use can be made of Figs. 9-19 to 9-21, which have been prepared for various types of hydrocarbon fuels. Values from the appropriate figure can be used as required. Suppose, for example, compressed air at 900 R (440 F) is to be heated to 1900 R, or through a temperature rise of 1000°, using a light fuel oil. For this case, Fig. 9-20 shows a ratio of 0.0160 lb of fuel per lb of air heated. Thus 0.0160 lb of fuel is now associated with each original pound of air to constitute 1.0160 lb of products. The fuel, of course, has been transformed to constitute part of the products of combustion and appears in the CO_2 and H_2O along with unburned hydrocarbon traces and rarely with CO. Unless the temperature rise is excessive, f seldom amounts to more than 0.04 so that the difference between G_t and G_c is of the order of 1 to 4 per cent. Also, as the combustion products expanding through the turbine are not exclusively air, their thermal characteristics are not accurately represented by the air tables. For most calculations, however, it is sufficiently accurate to consider the expanding products to be air, after modification to account for the proper value of G_t in terms of G_c.

3-4. Regenerator Effectiveness. The effect of regenerative warming of the compressed gas can be illustrated either on the *hs* or on the *Ts* diagram. Using Fig. 3-4, compression is shown from 1 to 2 with expansion from 3 to 4 in a real cycle. The gases at 4 are sufficiently hot to warm the compressed gas at 2 through an appreciable temperature rise to point 3′, with the hot gases at the same time cooling in counterflow to 4′. In the limit, the hot gases from 4 could be cooled until they essentially reach the temperature at 2. This would, of course, require an excessive amount of surface and also assumes that the weights per second of the turbine gases and of compressed gases are the same, which is not true in the open cycle as the products of combustion appear in the turbine gases. However, because cooling from 4 to 2 represents a possible limiting condition, it is frequently used as a datum by which to compare regenerator effectiveness.

A heat balance of a perfectly insulated regenerator using the symbols of Fig. 3-4 and with the same terminology used in Eqs. (3-12) and (3-13) appears as

$$G_c(h_{3'} - h_2) = G_t(h_4 - h_{4'}) = G_c(1 + f)(h_4 - h_{4'})$$

$$h_{3'} - h_2 = (1 + f)(h_4 - h_{4'}) \tag{3-14}$$

and

$$c_{pm}(T_{3'} - T_2) = c_{pm}(1 + f)(T_4 - T_{4'}) \tag{3-15}$$

which becomes

$$T_{3'} - T_2 = (1 + f)(T_4 - T_{4'}) \tag{3-16}$$

provided the mean specific heats of the hot and cold gases are not appreciably different.

If, for a basis of comparison, it is assumed $T_{4'}$ in the limit could essentially reach T_2, an expression for regenerator effectiveness ϵ would appear if written on a heat-balance basis as

$$\epsilon = \frac{G_c(h_{3'} - h_2)}{G_t(h_4 - h_2)} = \frac{h_{3'} - h_2}{(1 + f)(h_4 - h_2)} \tag{3-17}$$

This is closely approximated by

$$\epsilon = \frac{T_{3'} - T_2}{(1 + f)(T_4 - T_2)} \tag{3-18}$$

and if the difference in weight rate of flow of G_c and G_t is disregarded,

$$\epsilon = \frac{T_{3'} - T_2}{T_4 - T_2} \tag{3-19}$$

Form (3-19) is probably most frequently used in stating regenerator-effectiveness values.

In actual heat-transfer equipment, regenerator effectiveness for commercial design ranges from 0.50 to 0.75. However, it is possible to design for higher values of effectiveness approaching 0.85 to 0.90. In general, counterflow design is preferable, and terminal temperature differences of not less than 50° are employed. Where the regenerator is not adiabatic, a portion of the thermal energy of the warm gases is dissipated to the surroundings. Energy lost in this way produces essentially the equivalent of a unit operating with a lower effectiveness ratio.

3-5. Stage Efficiency. Before going into the details of turbine operation, it is desirable to develop an expression for the efficiency of a turbine or compressor stage. A stage of a turbine consists of the fixed nozzle row (or fixed blade row) and the moving row of blades which receives the gases. In the moving row a portion of the energy from the gases is absorbed and flows as power to the turbine shaft. In the axial-flow compressor a similar working pair, consisting of a moving row and a fixed row of blades, constitutes a stage. Here power flows from the shaft to the moving blade row, where the blades do work accelerating the moving gases being passed through.

Figure 3-5 is an enthalpy-entropy diagram of a turbine stage. Hot gases at temperature T_0 and pressure p_0 enter the fixed blades or nozzles with an initial velocity C_0. In the fixed blades pressure decreases to p_1 and velocity of the stream increases. The enthalpy decreases to h_{1i}, if the process is reversible (isentropic), but in the actual case arrives at

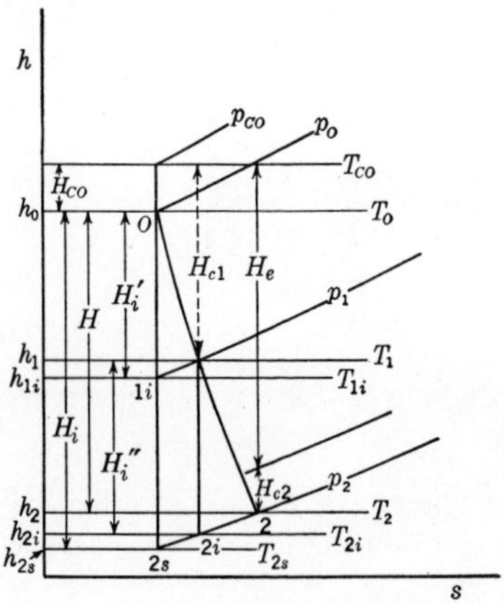

$\mathbf{H}_i$ = reversible-adiabatic enthalpy change in a stage (isentropic)
$\mathbf{H}_i'$ = reversible-adiabatic enthalpy change in a fixed row (isentropic)
$\mathbf{H}_i''$ = reversible-adiabatic enthalpy change in a moving row (isentropic)
$\mathbf{H}$ = real enthalpy change in a stage
$\mathbf{H}_e$ = effective delivered work in a stage
$\mathbf{H}_{C0}$ = enthalpy equivalent to entering absolute velocity = $C_0^2/2g_cJ$
$\mathbf{H}_{C1}$ = enthalpy equivalent to absolute gas velocity entering moving row, $C_1^2/2g_cJ$
$\mathbf{H}_{C2}$ = enthalpy equivalent to leaving absolute velocity = $C_2^2/2g_cJ$
C_0 = absolute velocity, fps, entering inlet guides or nozzles of turbine
C_1 = absolute inlet velocity, fps, into a moving row
C_2 = absolute outlet velocity, fps, from a moving row

FIG. 3-5. *hs* diagram of a reaction turbine stage.

the leaving value $h_1 > h_{1i}$. The energy ideally available for work in the fixed row is represented by the length $\mathbf{H}_i'$. (In the fixed row, the decrease in enthalpy can exist only as increased velocity of the stream.) In the moving row the gases which enter with enthalpy h_1 (and the velocity C_1) expand to pressure p_2, under reversible expansion arriving at h_{2i}, but under actual expansion reaching $h_2 > h_{2i}$. The ideally available work from expansion in the moving row is $\mathbf{H}_i''$. The real enthalpy change for the whole stage is represented by the length $\mathbf{H}$, and $\mathbf{H}$ also represents the work per pound of gas delivered from the stage if C_0 entering and C_2 leaving the stage are essentially equal. When this is not the case, the work delivered is accurately represented by the length $\mathbf{H}_e$. The temperature and pressure are sufficient to locate the enthalpy of a gas on the *hs* plane. To represent the fact that kinetic energy is also associated with a gas, it is possible to add the lengths $\mathbf{H}_{Co} = C_0^2/2g_cJ$ and

$\mathbf{H}_{C2} = C_2^2/2g_cJ$ onto the proper h values on the chart. Using the kenthalpy points thus found, $\mathbf{H}_e$, the effective delivered work of the stage, can be shown.

Turbine stage efficiency η_s can now be indicated as the ratio of work delivered to the thermally available energy of the whole stage. If it is assumed that the kinetic energy entering is closely equivalent to that leaving the stage, the efficiency takes the form

$$\eta_s = \frac{H}{H_i} = \frac{h_0 - h_2}{h_0 - h_{2s}} = \frac{1 - (h_2/h_0)}{1 - (h_{2s}/h_0)} \tag{3-20}$$

This form of the equation can be used directly in the solution of problems if gas or air tables are available. However, it may be desirable to obtain a solution where, over a small range, deviations from the gas relationships are small. To do this, substitute in Eq. (3-20), using Eqs. (2-16) and (2-20).

$$\eta_s = \frac{h_0 - h_2}{h_0 - h_{2s}} = \frac{c_{pm}(T_0 - T_2)}{c_{pm}(T_0 - T_{2s})} = \frac{T_0\left(1 - \dfrac{T_2}{T_0}\right)}{T_0\left(1 - \dfrac{T_{2s}}{T_0}\right)} \tag{3-21}$$

$$\eta_s = \frac{1 - (T_2/T_0)}{1 - (p_2/p_0)^{(k-1)/k}} \tag{3-22}$$

This form, Eq. (3-22), is very commonly employed for stage efficiency as it brings pressure into the equation.

Example 2. Gases enter a turbine at 1240.3 F at 75 psia and expand to 62 psia in the first reaction-type stage, which consists of a fixed and moving row. Assume that the KE (kinetic energy) entering and leaving the stage are essentially the same. Find the leaving temperature if the stage efficiency is 85 per cent.

Solution. $T_0 = 1240.3 + 459.7 = 1700$ R, $p_1 = 75$ psia, $p_2 = 62$ psia. Select the value of k from Fig. 2-1 for an estimated mean temperature somewhat below 1700 R, giving $k = 1.342$. By Eq. (3-22)

$$\eta_s = 0.85 = \frac{1 - (T_2/1700)}{1 - (62/75)^{(1.342-1)/1.342}} = \frac{1 - (T_2/1700)}{1 - 0.9528}$$

from which

$$T_2 = 1631.7 \text{ R}$$

An alternate but preferred solution using Table 2-4 for air properties requires the use of Eq. (3-20). Read from the table at 1700 R, $h_0 = 312.99$ Btu per lb and $p_{r0} = 115.434$. The pressure expansion ratio is

$$r_e = 75/62 = 1.21$$

As this is an expansion, divide the initial p_r value by the pressure ratio to find the p_{r2} value which can indicate the enthalpy resulting after isentropic expansion. Thus

$$p_{r2} = \frac{p_{r0}}{r_e} = \frac{115.434}{1.21} = 95.40$$

This value corresponds to a temperature of 1619.4 R and an enthalpy of 291.30. Substituting in Eq. (3-20)

$$\eta_s = 0.85 = \frac{1 - (h_2/312.99)}{1 - (291.30/312.99)}$$

$$h_2 = 294.63 \text{ Btu per lb}$$

for which the corresponding temperature by Table 2.4 is 1631.8 R.

For this small temperature range a close agreement between the gas-equation solution and the tabular solution can be observed. The tabular solution is preferred wherever it can be employed.

Internal efficiency of a turbine is the ratio of the work (power) delivered to the rotor of the turbine by the gaseous medium compared with the ideal energy available for work (power) from the medium in expanding through the same pressure range and ratio. A portion of the work (power) delivered to the rotor is absorbed in the glands and bearings so that the useful shaft work (power) is less than the internal (rotor) work (power). Over-all turbine efficiency η_{ot} compares the delivered shaft work (power) with the ideal energy available from the medium.

3-6. Reheat Factor. The fact that internal turbine efficiency is higher than average stage efficiency is a characteristic of a multistage turbine. The waste energy not converted into work in an early stage is added to the supply of energy in a following stage and makes more work available in that stage. Referring to Fig. 3-5, if a slight change in interpretation is considered which shows p_0 to p_1 as being one complete stage and p_1 to p_2 a second complete stage, then the available energy in these two stages consists of $\mathbf{H}_i' + \mathbf{H}_i''$. It should be recalled that the pressure lines on the *hs* chart diverge with increasing entropy; consequently $h_1 - h_{2i} > h_{1i} - h_{2s}$. Because of this when the terms $\mathbf{H}_i' + \mathbf{H}_i''$ are compared with H_i, the result is greater than unity.

$$\frac{\mathbf{H}_i' + \mathbf{H}_i''}{\mathbf{H}_i} > 1$$

A term called *reheat factor*, R_f, is defined in this connection as

$$R_f = \frac{\mathbf{H}_i' + \mathbf{H}_i''}{\mathbf{H}_i} \tag{3-23}$$

or, in general,

$$R_f = \frac{\Sigma \mathbf{H}_i'}{\mathbf{H}_i} \tag{3-24}$$

The reheat factor of a multistage gas or steam turbine is greater than 1, reaching a maximum value in the neighborhood of 1.05. The reheat factor is related to stage efficiency in the following way:

$$\eta_{it} = R_f \eta_s \tag{3-25}$$

where η_s is an average stage efficiency in contrast to η_{it}, the internal efficiency of the turbine. In the case of a compressor, this effect is backward from a use viewpoint, as progressively with increasing entropy the effect of reheat in an earlier stage of a compressor adds to the amount of work which must be done in the following stages of the compressor. In the case of a turbine and a compressor, each of the same design type and each carefully constructed, the turbine will have the higher internal efficiency. One of the factors contributing to this lower efficiency is the previously mentioned deleterious effect of reheat in the compressor. Equation (3-25) is not applicable to a compressor.

Example 3. The (internal efficiency) of the turbine of Example 2 is 87.2 per cent. When the expansion is carried to 15 psia from the 75-psia and 1700 R inlet conditions, what is the final leaving temperature, assuming entering and leaving velocities essentially equivalent? What work is delivered per pound of air flowing?

Solution. $r_e = 75/15 = 5$. Select a value of k equal to 1.355 from Fig. 2-1 at about 1400 R. By use of Table 2-4

$$p_{r2} = \frac{p_{r0}}{r_e} = \frac{115.434}{5} = 23.087$$

For this value of p_{r2}, $h_{2s} = 159.66$. Using Eq. (3-20),

$$\eta_{it} = 0.872 = \frac{1 - (h_2/312.99)}{1 - (159.66/312.99)}$$

$$h_2 = 179.24 \text{ Btu per lb}$$

for which

Corresponding $T_2 = 1190.9$ R

$$\text{Work} = h_0 - h_2 = 312.99 - 179.24 = 133.75 \text{ Btu per lb}$$

An alternate solution would be to use Eq. (3-22) to find T_2.

$$\eta_{it} = 0.872 = \frac{1 - (T_2/1700)}{1 - (15/75)^{(1.355-1)/1.355}} = \frac{1 - (T_2/1700)}{1 - (1/1.524)}$$

$$T_2 = 1190 \text{ R}$$

To find the work by this method, use

$$W = c_{pm}(T_o - T_2) \qquad \text{Btu per lb}$$

or read the proper enthalpy values from a table of air properties.

3-7. Compressor Performance. Figure 3-6 is an hs diagram drawn for two compressor stages. Here gas is shown entering the compressor with enthalpy h_A, pressure p_A, temperature T_A, and velocity C_A. In the first stage the pressure of the gas increases to p_1 with the work input supplied by the compressor shaft amounting to $\mathbf{H}_1$. If compression had occurred under reversible-adiabatic (isentropic) conditions, the work input would have been the lesser value of $\mathbf{H}_i'$. In the second stage, gas at temperature T_1 and pressure p_1 has $\mathbf{H}_2$ units of work added to it in raising the pressure to p_2. If the second-stage compression could have been carried through reversibly, only $\mathbf{H}_i''$ units of work would have been absorbed from the shaft. The result of compression inefficiency is progressively to move the state point of the compressed gas in the direction of increasing entropy. As the pressure lines diverge with increasing entropy, more shaft work for a given pressure lift occurs as a result of inefficiencies in previous stages.

If a term similar to the reheat factor used with the turbine is employed in compressor analysis, it will show that the internal efficiency, η_{ic} of a

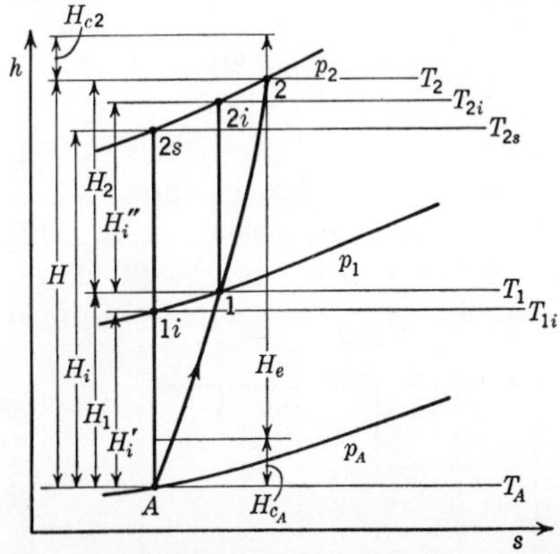

$\mathbf{H}_i'$ = isentropic enthalpy increase in first stage
$\mathbf{H}_1$ = enthalpy increase in first stage
$\mathbf{H}_i''$ = isentropic enthalpy increase in second stage
$\mathbf{H}_2$ = enthalpy increase in second stage
$\mathbf{H}_i$ = isentropic enthalpy increase for compressor
$\mathbf{H}_{CA}$ = enthalpy equivalent of gas velocity entering compressor $C_A^2/2g_cJ$
$\mathbf{H}_{C2}$ = enthalpy equivalent of gas velocity leaving compressor $C_2^2/2g_cJ$
$\mathbf{H}$ = enthalpy increase in whole compressor
$\mathbf{H}_e$ = kenthalpy increase in compressor or work input

FIG. 3-6. *hs* diagram of a compressor having two stages.

multistage compressor is less than average individual stage efficiency η_s; thus

$$\eta_{ic} = \frac{\eta_s}{R_c} \tag{3-26}$$

where R_c for a multistage compressor has a value greater than 1 and usually less than 1.05. Also,

$$R_c = \frac{\mathbf{H}_i' + \mathbf{H}_i''}{\mathbf{H}_i} \tag{3-27}$$

Over-all efficiency η_c of the compressor is less than internal efficiency by virtue of the power absorbed in bearings, glands, etc. If a term mechanical efficiency η_m is introduced, then

$$\eta_c = \eta_m \eta_{ic} \tag{3-28}$$

Using the notation of Fig. 3-6, an expression for compressor internal efficiency can be developed for compression between p_A and p_2:

$$\text{Compressor eff} = \frac{\text{isentropic work}}{\text{actual work on gas}} = \eta_{ic} \tag{3-29}$$

$$\eta_{ic} = \frac{(h_{2s} - h_A)_s}{(h_2 - h_A)} = \frac{c_{pm}(T_{2s} - T_A)}{c_{pm}(T_2 - T_A)} = \frac{T_A\left(\dfrac{T_{2s}}{T_A} - 1\right)}{T_A\left(\dfrac{T_2}{T_A} - 1\right)} \tag{3-30}$$

$$\eta_{ic} = \frac{(p_2/p_A)^{(k-1)/k} - 1}{(T_2/T_A) - 1} \tag{3-31}$$

where η_{ic} = compressor efficiency internal
p_2 = discharge pressure, psia
p_A = inlet pressure, psia
T_2 = temperature of compressed gas on leaving, °R
T_A = temperature of inlet gas, °R

In Eq. (3-31), η_s may be substituted for η_{ic} if the pressures and temperatures are referred to a stage instead of to the whole compressor.

Mechanical efficiencies η_m of turbines and compressors are comparable and range in the neighborhood of 97 to 99.5 per cent when based on full- (or rated-) load capacity. Equation (3-28) for the compressor would

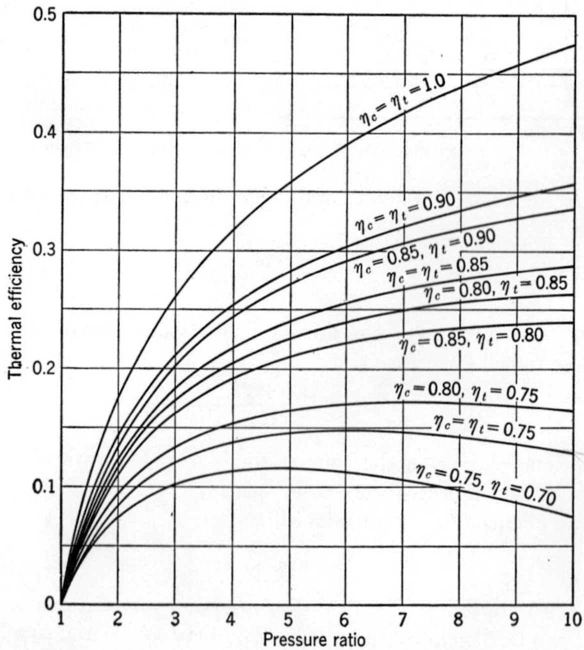

FIG. 3-7. Effect of compressor and turbine efficiencies on cycle performance. Compressor inlet at 14.7 psia, 530 R; inlet turbine temperature 2000 R, no pressure losses, and no regeneration.

take the following form when applied to the turbine:

$$\eta_{ot} = \eta_m \eta_{it} \tag{3-32}$$

Note that the losses in bearings, glands, etc., for turbines and compressors are of relatively fixed magnitude essentially independent of load for a constant-speed machine. For example, a turbine of 5000-hp rated output may have bearing and gland losses amounting to 30 hp. When delivering 2500 hp, the losses are still essentially 30 hp. Consequently a rated-load mechanical efficiency of 5000/5030 = 99.4 per cent is only 2500/2530 = 98.8 per cent when producing half load. It should be observed that for a turbine mechanical efficiency is the ratio of shaft horsepower to rotor (internal) horsepower, while for a compressor mechanical efficiency is the ratio of rotor horsepower to shaft horsepower.

3-8. Open-cycle Performance. Mention has been made of other variables, besides pressure ratio, which affect the performance of the open cycle. In particular, the inlet temperature to the turbine, individual

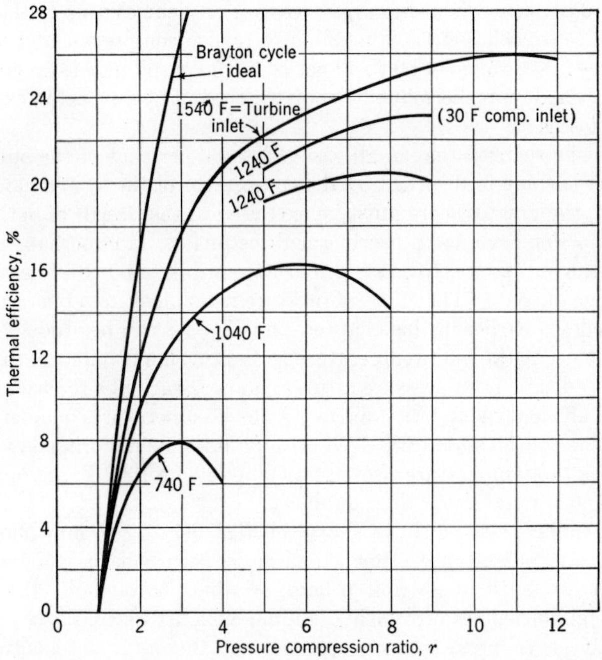

FIG. 3-8. Efficiency curves for simple open cycle at various inlet turbine temperatures. p_1 = 14.7 psia, T_1 = 530 R, η_t = 0.86, η_c = 0.84, 3 per cent pressure loss combustor to turbine.

efficiencies of the turbine and compressor, and ambient-supply-air temperature are significant. Also, pressure losses in the cycle make the expansion (pressure) ratio of the turbine appreciably less than the compression (pressure) ratio of the compressor and thereby reduce both efficiency and available power. Combustion effectiveness likewise must be maintained at high levels with radiation-heat losses controlled within desired limits.

Various heat-exchange devices, such as the regenerator which utilizes some of the energy in the turbine exhaust or intercoolers between groups of compressor stages, may lead to improvement in over-all performance. The effect of different variables on performance can be seen if a few of them are changed while the remainder are kept constant. Figure 3-7 shows specifically the effect of varying compressor and turbine efficiencies with both inlet temperatures to the compressor and to the turbine fixed at stated values. Study of the curves will show how at a particular pressure ratio, for example 6, the thermal efficiency of nearly 40 per cent with ideal compressors and turbines decreases to a value in the neighborhood of 10 per cent when the respective unit efficiencies drop to the range of 70 per cent each. As it is possible to realize compressor and turbine efficiencies in the range of 80 to 90 per cent, the graph in this range might represent maximum thermal efficiency which could be expected exclusive of other losses.

It must be realized that in all cases 60 to 80 per cent of the output of the power turbine is absorbed by the compressor of the system, and consequently the greatest care must be exercised in the design of both these elements. For a value to use in rough estimates, it is perhaps safe to think of the net power of the system as amounting to 30 to 40 per cent of the turbine output. The effect of pressure ratio on system operation has been discussed earlier in this chapter, and Fig. 3-8 can be studied in this connection. As the inlet temperature to the turbine is reduced, the compression and expansion pressure ratios must necessarily be reduced as well if highest efficiency is to be attained. At low temperatures on inlet to the turbine, the compression ratio need be only in the neighborhood of 2 or 3, whereas as the temperature rises appreciably above 1000 F, the optimum pressure-ratio range more than doubles.

Performance plots, such as shown in Figs. 3-7 to 3-9, may show certain amounts of variance. Such deviations occur because of inclusion or not of pressure losses and other variations throughout the cycle. Mechanical efficiencies of rotating equipment must be considered in any final analysis. Several terms concerned with the output of gas turbines have come into use and will here be defined. *Power ratio* (work ratio) is the ratio of useful or net horsepower of the cycle compared with the power developed by the turbine of the system. *Specific power* is the net horsepower developed per 1 lb per sec of air flow. *Air rate* is the pounds per second of air flow required per net horsepower developed. This can also be called flow rate.

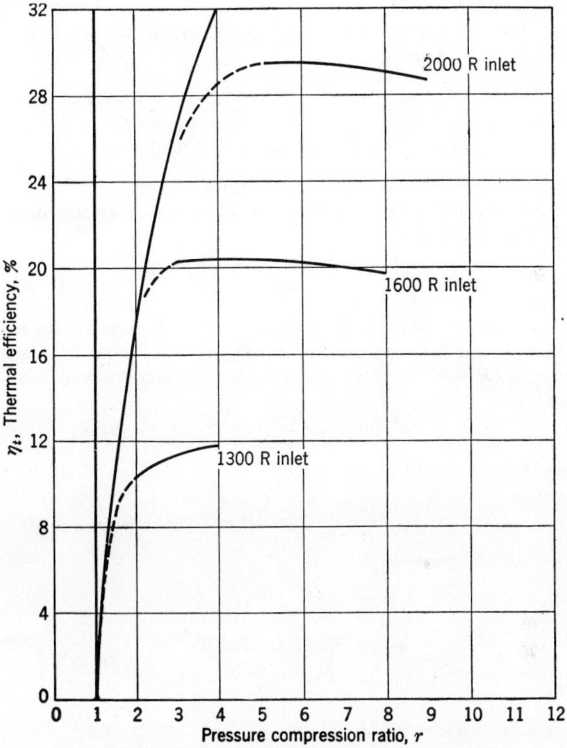

FIG. 3-9. Efficiency curves for regenerative open cycle at various turbine inlet temperatures.

Example 4. An open gas-turbine system follows a cycle similar to that shown in Fig. 1-6 and is illustrated on the *hs* plane of Fig. 3-4. The axial-flow compressor has an internal efficiency of 0.84 and the turbine an internal efficiency of 0.86. Atmospheric air is at 70.3 F (530 R), 14.7 psia, and the combustors deliver air to the turbine at 1040.3 F (1500 R). The pressure of the air leaving the compressor suffers a 3 per cent decrease by the time it reaches the turbine. The gases leave the turbine at 15 psia. Consider the system components adiabatic, and take the mechanical efficiencies of turbine and compressor as unity in an over-all analysis of the system for a compressor pressure ratio of 8 and an air flow of 1 lb per sec. The fuel used is a kerosene type with a hydrogen-carbon ratio of 0.15 (see Fig. 9-20). The subscript notation of Fig. 3-4 will be used.

Solution. Compressor: $r = 8$, $T_1 = 530$ R, $p_1 = 14.7$, $p_2 = (14.7)(8) = 117.6$ psia. By Table 2-4,

$$h_1 = 16.800 \qquad p_{r1} = 1.640$$
$$p_{r2} = (1.640)(8) = 13.12 \qquad \text{and} \qquad h_{2s} = 119.699$$

corresponding to this value of p_{r2}.

Using Eq. (3-30) with the nomenclature referred to Fig. 3-4 and with a compressor efficiency of 0.84, there results

$$\eta_{ic} = \frac{h_{2s} - h_1}{h_2 - h_1} \qquad 0.84 = \frac{119.699 - 16.800}{h_2 - 16.800}$$
$$h_2 = 139.30 \text{ Btu per lb leaving compressor}$$

The corresponding leaving temperature is 1032.5 R.

Compressor horsepower for gas compression, with kinetic-energy changes considered negligible, by Eq. (2-19) is

$$hp_c = (1)\,\frac{139.30 - 16.80}{0.707} = 173.3$$

Combustor takes air at 1032.5 R, $h_2 = 139.30$, and delivers products, largely air, at 1500 R with $h_3 = 259.500$ Btu per lb. By Fig. 9-20 for a 467.5° temperature rise the fuel-air ratio f is 0.0067. The approximate heat added per second in the combustor is thus

$${}_2Q_3 = (1.0067)(259.50) - (1.0)(139.30) = 121.93$$

The pressure at turbine inlet is

$$p_t = (117.6)(1 - 0.03) = 114.1 \text{ psia}$$

The pressure ratio of expansion is

$$\frac{p_t}{p_e} = \frac{114.1}{15} = 7.61$$
$$p_{r3} = 70.857 \text{ at } 1500 \text{ R}$$
$$p_{re} = \frac{p_{r3}}{r} = \frac{70.857}{7.61} = 9.311$$

The enthalpy at 4*s* is $h_{4s} = 98.305$.

$$h_4 = 259.50 - 0.86(259.50 - 98.305) = 120.88$$
$$T_4 = 958.5 \text{ R}$$

The turbine output is thus

$$hp_t = \frac{(1.0067)(259.50 - 120.88)}{0.707} = 197.4$$
$$\text{Net hp} = hp_t - hp_c = 197.4 - 173.3 = 24.1$$
$$\text{Cycle thermal efficiency} = \frac{(2545)(24.1)}{(3600)(121.93)} = 0.14 = 14.0\%$$
$$\text{Power ratio} = \frac{\text{net hp}}{\text{turbine hp}} = \frac{24.1}{197.4} = 0.12$$
$$\text{Sp power} = \frac{24.1}{1} = 24.1 \text{ hp per 1 lb per sec flow}$$
$$\text{Air rate} = \frac{1}{24.1} = 0.041 \text{ lb per sec per net hp}$$

3-9. Open-cycle Generalizations. Mention has already been made of the effectiveness of regenerative heating in conserving fuel energy input to the cycle. The benefits of this are much greater at lower pressure ratios than at high because the difference in temperature between the gases leaving the compressor and the gases leaving the turbine is greatest at low pressure ratios. Comparison of the efficiency found in Example 4 with the proper graph of Fig. 3-7 will show that the value found is lower than the peak for that temperature. This can be directly attributed to the high pressure ratio of 8. Note that the power ratio is also low for the same reason.

When it is desirable to work with high pressure ratios, it becomes necessary to reduce the compressor horsepower in every way possible. This, in general, will call for intercooling between stages of the compressor. For example, Fig. 3-10 shows a complete cycle in which intercooling between groups of stages of the compressor is considered. There is necessarily some pressure loss for the air passing through the intercoolers; nevertheless the reduction in temperature of the air with consequent reduction in compressor work more than offsets the effects of the pressure loss, and intercoolers are desirable when high economy is required. A study of Example 4 would show that the gas temperature leaving the turbine is lower than the air temperature leaving the compressor so that a regenerator for the conditions of this problem could serve no useful purpose. On the other hand, with intercooling, the compressed-air temperature would be so reduced that the benefits of regenerative heating could be realized. If operating pressure ratios are kept low, intercooling is not particularly important and much of the energy in the exhaust can be saved through use of a regenerator. Cycles leading to extremely high efficiencies will utilize high turbine temperatures, intercooling between stage groups in the compressor, regenerative heating, and possibly reheating between multiple turbine units or groups of turbine stages.

A comparison might be made between the gas-turbine and the steam-turbine power plant. Particularly with the open-system cycle operating at low pressures, the specific power based on unit air flow is small. Also because of the low pressures, specific power on a volume-flow basis is even more unsatisfactory. In contrast, steam turbines operate with steam of low specific volume and liquid water throughout the greater portion of the cycle. Steam-plant sizes have not yet reached a top limit and already exceed 100,000 kw in single units. On the other hand, there appears to be a definite limit to the practical size for which open-cycle gas-turbine units can be built. Present-day designs apparently put this limit at about 25,000 kw. The big advantage of the open-cycle gas-turbine system is that heat-transfer surfaces are not required (except in intercoolers and regenerators) as the combustion energy is added directly to the working

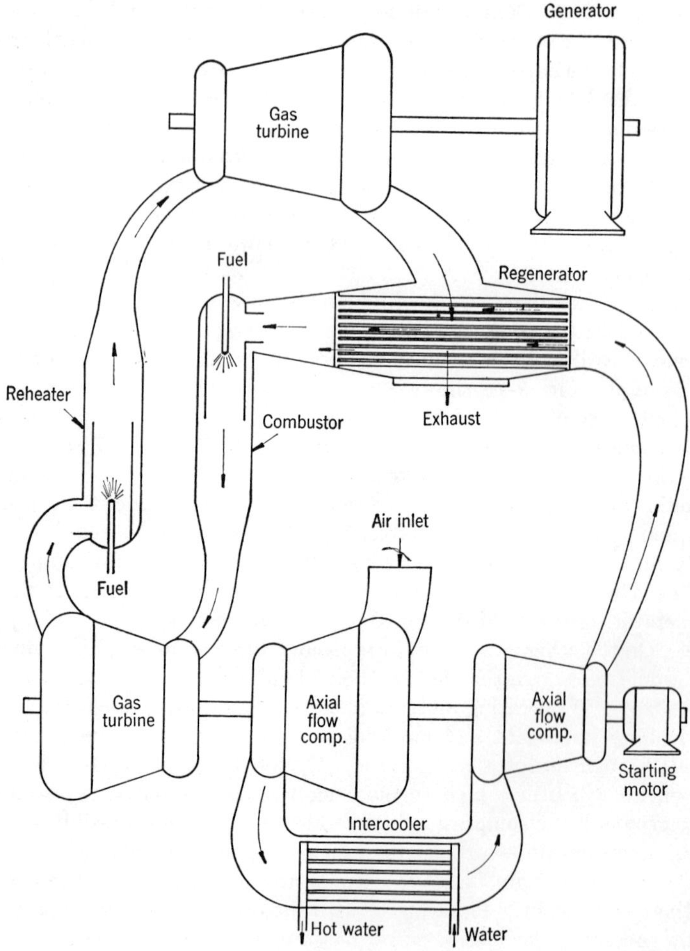

FIG. 3-10. Diagrammatic layout of open-cycle system with regeneration, intercooling, and reheating.

medium. Where a clean fuel such as a petroleum product is available, this is particularly advantageous.

In terms of economy, metallurgical advances have now reached a point where the gas turbine can be considered very competitive to small- and medium-size steam turbines. This is particularly true of the efficiency of gas turbines at full load. However, at partial loads the efficiency of the open-cycle gas-turbine system may fall off rapidly unless specific compensation features are added to the cycle. The useful service life of a gas-turbine system can be of long duration except where extremely high temperatures are used for some of the operating elements. The metals must be carefully selected for the combustors and the blading in early turbine stages. The fact that the open-system gas turbine does not require a source of cooling water greatly broadens its field of usefulness for such things as outlying or stand-by power plants and for railway transportation units. This is in addition to the present field of greatest utility of the gas turbine, *viz.*, aircraft propulsion.

Example 5. An open gas-turbine system operates under the conditions of Example 4, but in addition an intercooler and regenerator are added. The intercooler has a 3 per cent pressure drop based on the pressure of the air on entry to it and it is designed to receive its air at $\sqrt{8}$ times the compressor inlet pressure. The air is intercooled to 530 R. The regenerator effectiveness is 0.6 based on temperatures. The air pressure drop through the regenerator is 1.7 psi and on the gas side 0.7 psi pressure drop occurs to the atmosphere. Pressure loss in the combustors is 3.3 psi. Carry out an analysis of the system.

Solution. $r = 8$, $T_1 = 530$ R, $p_1 = 14.7$, $h_1 = 16.80$, $p_{r1} = 1.640$,

$$p_2 = (14.7)(8) = 117.6 \text{ psia.}$$

$p_{\text{int}} = (14.7)(\sqrt{8}) = 41.6$ psia, pressure leaving first group of compressor stages to enter intercooler

$p_{r\,\text{int}} = (1.64)(\sqrt{8}) = 4.174$

$h_{\text{int}\,s} = 55.674$ Btu per lb

$$0.84 = \frac{55.674 - 16.80}{h_{\text{int}} - 16.80}$$

$h_{\text{int}} = 63.05$ Btu per lb for air on leaving first group of compressor stages at 41.6 psia

In the intercooler the pressure drops to

$41.6(1.00 - 0.03) = 40.35$ psia, and the air is cooled to 530 R at 16.80 Btu per lb

$$r = \frac{117.6}{40.35} = 2.914 \text{ pressure ratio for second group of compressor stages}$$

For 530 R and 40.35 psia, $p_r = 1.64$.

$p_{r2} = (1.64)(2.914) = 4.78$, for this value

$h_{2s} = 62.217$ Btu per lb for air

$$0.84 = \frac{62.217 - 16.80}{h_2 - 16.80}$$

$h_2 = 70.87$ Btu per lb leaving second group of compressor stages at 754.3 R

$$hp_c = (1)\left(\frac{1}{0.707}\right)[(70.87 - 16.80) + (63.05 - 16.80)] = 141.8$$

Air leaves the compressor at 117.6 psia, and pressure losses of 1.7 psi and 3.3 psi occur before entry to the turbine with $p_t = 112.6$ psia.

The pressure leaving the turbine is $14.7 + 0.7 = 15.4$ psia $= p_e$. The ratio of expansion is thus

$$r_e = \frac{112.6}{15.4} = 7.31$$

$$p_{r3} = 70.857 \text{ at } 1500 \text{ R}$$

$$p_{re} = \frac{70.857}{7.31} = 9.69 \qquad h_{4s} = 100.69$$

Inlet conditions to the turbine as before are at 1500 R and 112.6 psi, and $h_3 = 259.50$.

$$h_4 = h_e = 259.50 - 0.86(259.50 - 100.69) = 122.90 \text{ Btu per lb}$$

Gases thus leave the turbine at 122.90 Btu per lb at a corresponding temperature of 966.6 R.

The regenerator is supplied with hot gases at 966.6 R flowing counter- or cross flow to the air entering at 754.3 R. Use Eq. (3-19), and for an effectiveness of 0.6 there results

$$0.6 = \frac{T_{3'} - T_2}{T_4 - T_2} = \frac{T_{3'} - 754.3}{966.6 - 754.3}$$
$$T_{3'} = 881.7 \text{ R}$$

Corresponding $h_3' = 101.96$

From Fig. 9-20 for a temperature rise of 1500 − 881.7, or 618.3° the value of f is 0.0094.

The approximate heat added in the combustor is

$${}_{3'}Q_3 = (1 + f)h_3 - (1)h_{3'} = (1.0094)(259.50) - (1)(101.96) = 159.98 \text{ Btu per sec}$$

The turbine output per 1 lb per sec of air compressed is

$$hp_t = \frac{(1.0094)(259.50 - 122.90)}{0.707} = 194.9$$

$$\text{Net hp} = \text{hp}_t - \text{hp}_c = 194.9 - 141.8 = 53.1$$

$$\text{Cycle thermal efficiency} = \frac{(2545)(53.1)}{(3600)(159.98)} = 0.235 = 23.5\%$$

$$\text{Power ratio} = \frac{53.1}{194.9} = 0.272$$

$$\text{Sp power} = \frac{53.1}{1.0} = 53.1 \text{ hp per 1 lb per sec of compressed air flow}$$

$$\text{Air flow} = \frac{1}{53.1} = 0.0188 \text{ lb compressed air per sec per net hp}$$

The annulus area of the last turbine stage must pass the gas flow at 966.6 R at 15.4 psia. This gas has a specific volume by Eq. (2-3) of

$$v = \frac{RT}{144p} = \frac{(53.34)(966.6)}{(144)(15.4)} = 23.2 \text{ ft}^3 \text{ per lb}$$

If the axial velocity is arbitrarily taken as 500 fps, the required annulus area A can be found, thus:

$$(\text{Area})(\text{velocity}) = \text{cfs} = (G_t)(v)$$
$$(A)(500) = (1.0094)(23.2)$$
$$A = 0.0468 \text{ ft}^2$$

This area is for 1 lb per sec of compressed air flow and for 53.1 net hp. For each 1000 net hp the net annulus area required is

$$A_{1000} = 0.0468 \times \frac{1000}{53.1} = 0.881 \text{ ft}^2$$

This area was selected for computation as it shows one of the critical flow areas which can limit gas-turbine capacity. This area must be available through the final row of blades of the turbine. These blades, which move in the gas-flow field, are subject to centrifugal stresses which limit their length and the permissible rotational speed.

Another critical spot from a size-limitation viewpoint is the entry stage of the compressor. However, here the temperature is low so that high-temperature alloys are not required for blading material. At the compressor inlet row, with $p = 14.7$ psia and $T = 530$ R, by Eq. (2-3), $v = 13.4$ ft^3 per lb. With the axial-air-flow velocity assumed at 500 fps the annulus area required is 0.0268 ft^2 for 1 lb per sec air flow. For each 1000 net hp developed, the net annulus area required is

$$A_{1000} = 0.0268 \times \frac{1000}{53.1} = 0.504 \text{ ft}^2$$

3-10. Closed-cycle System. The closed-cycle system is an external-combustion engine and as such uses an external supply of heat and external heat rejection. In some ways it resembles a steam power plant, the fundamental difference being that the gaseous working medium does not change its phase, *i.e.*, liquefy, at any part of the cycle.

Figure 3-11 is a diagrammatic arrangement of a simple closed cycle. The compressed gases, heated by an external source, flow to the turbines where they expand and produce power for driving the compressors and also for delivering useful power. The gases, at a reduced but still high pressure, leave the turbine to pass through a regenerative-type heat exchanger, following which they are precooled to a low temperature for entry to the compressor. In the compressor the pressure of the gases is raised sufficiently so that the cycle events can again be carried out.

It should be observed that heat rejection from the cycle in the precooler requires utilization of an external cooling medium such as circulating water. As in the open cycle, pressure ratios of expansion and compression are directly related to the temperature range available in the cycle. Pressure ratios of 3 to 4 or even higher have been used.

One of the advantages of the closed cycle is the fact that it can be operated at higher pressures than are feasible for the open cycle. This means that the flow passages throughout the turbine and compressor for a given weight rate of flow can be made smaller for the same velocities. Pressures as high as 900 psi have been considered, and commercial designs in the neighborhood of 400 psi are in actual operation. The efficiency is, of course, independent of the actual pressures in the system, although it is necessarily related to the pressure ratio. The actual pressure used is one of design convenience in terms of the over-all limitations of physical size and in contrast to the heavier walls required as the pressure increases. If, for example, a low pressure of 100 psia is set for the system with a compression ratio of 4, the high pressure of the system would necessarily be 400 psi. If this appeared too high, a base pressure of, let us say, 50 psi might be used, giving a top pressure of 200 psi. This latter arrangement would mean that in the low-pressure parts of the system the specific volume of the gases would be double that required for the pressure of 100 psi if the temperatures were the same in both cases.

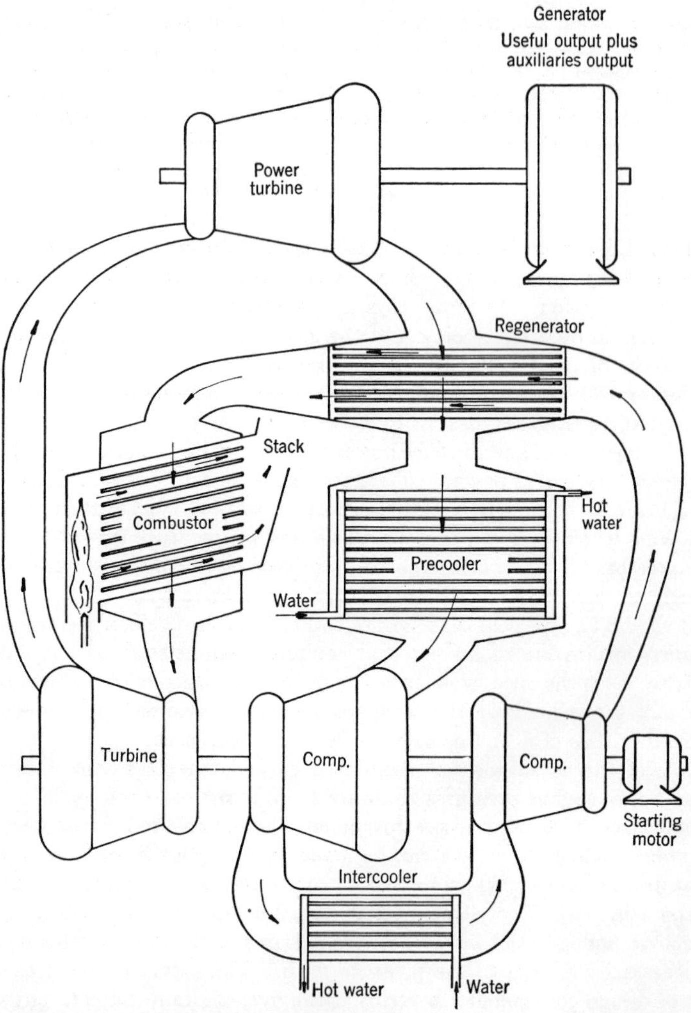

Fig. 3-11. Diagrammatic layout of closed-cycle system for a gas turbine.

In regard to maximum temperatures, the same limits apply to closed-cycle units as apply to open-cycle units, with commercial installations probably not exceeding 1300 F for continuous operation and discharge temperatures ranging from 950 down to 750 F. Because of the relatively low temperatures on compressor outlet and in the low-pressure parts of the system, heat exchangers need not necessarily be made of expensive alloys.

The basic development work on the closed-cycle gas turbine was largely done in Switzerland. In this connection the name of the Escher-Wyss Engineering Works is closely associated with early and continuing developments. In 1940 one of its test units produced 1000 kw in a 7-hr run. Since that time several plants have been built, with a 2000-kw plant having furnished more than 12 years of operating experience. Design and construction have been completed on two 12,500-kw plants.

3-11. Advantages and Disadvantages of the Closed Cycle. Because the pressure in the system is not related to atmospheric pressure, it can be arbitrarily fixed. This results in a particular advantage in that through regulation of the pressure in the cycle it is possible to vary the work produced in terms of the gas density within the system. Expressed

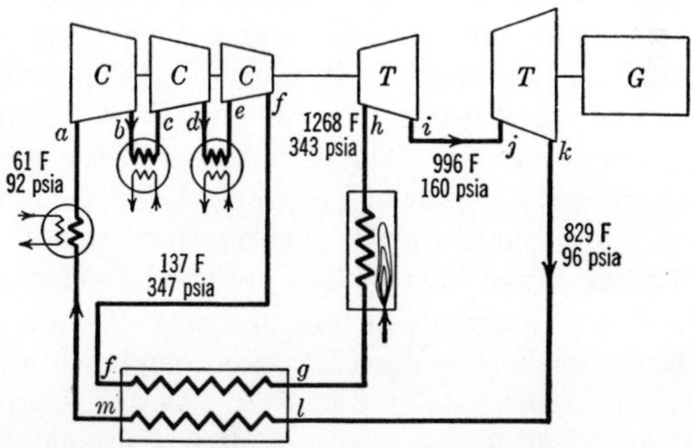

FIG. 3-12. Diagram of one-stage-heated closed gas turbine with data for full-load operation. (*Escher-Wyss, Ref.* 6.)

in another way, governing can be partly accomplished by variation of the pressure levels of the gas in the system.

The thermal efficiency varies but slightly under changes in system density. The power developed is proportional to the weight of gas flowing and to gas density in so far as the flow rate changes with variations in density. In order to increase power, by system-pressure regulation the gas density is increased by raising the operating pressure of the system. To accomplish this end, gas can be supplied to the system at an elevated pressure from an auxiliary reservoir or from a compressor operating at higher pressure. In order to reduce the pressure, it is possible to let a portion of the gas escape either directly to air or into a reservoir held at lower pressure. A particular advantage of this type of control lies in the fact that operating temperatures in the system can remain unchanged independent of the load. Moreover no regulating valves are required in the main gas-flow circuits of the machine.

The decrease in size of the machine because of higher operating pressures has already been noted, and the closed system holds promise for plants in the range of 15,000 to 25,000 kw. High-pressure systems also operate at higher Reynolds numbers, which in some cases results in lower pressure losses. Higher values of heat-transfer coefficients can also be expected as the range of Reynolds-number values increases.

The closed cycle can operate with gases other than air as the working medium. Helium and carbon dioxide have been considered, and even mixtures of the two. Hydrogen has also been considered as another possibility, but its low density partially offsets its advantages of high specific heat and low viscosity. Because the fuel does not enter the flow stream of the working medium, cheaper fuels and ash-carrying fuels can be used in this system.

On the other hand, definite disadvantages exist in the closed cycle, the main one of which is the difficulty of heating and cooling gases through metallic surfaces. For effective heat transfer elaborate heat-transfer units are required. In addition to this, because of the pressures involved, the regenerator and other heat exchangers must be sufficiently strong to resist the pressures developed. The performance of the cycle is directly related to the temperature of air or gas supply to the compressor so that for best efficiency low-temperature circulating water should be available.

3-12. The Escher-Wyss Plants. The previously mentioned 2000 kw-output plant, shown diagrammatically in Fig. 3-12, was designed by Ackeret and Keller for Escher-Wyss. It was completed in 1939 and has operated since that time. As it was set up as an experimental unit, various minor modifications have been made in its design from time to time to investigate the desirability of indicated changes. The compressor, which is powered by the compressor turbine, is built in three sections so that the benefits of intercooling can be realized. The air leaving the compressor passes through a regenerative heat exchanger and thence into the combustor and to the compressor turbine. The temperature leaving the compressor turbine is sufficiently high to serve the power turbine. From this, the gases leave to pass through the heat exchanger and precooler before reentering the compressor. In the figure, representative operating data are shown which for this unit have given efficiencies based on lower heating value of the fuel in the neighborhood of 30 per cent.

Various possibilities exist as to how the efficiency and capacity of a closed system can be increased, and such a modification is indicated in Fig. 3-13 with possible operating pressures and temperatures indicated. It will here be noted that two stages of heating are provided, and to utilize these effectively the pressure ratio has been increased to over 10, in contrast to the ratio of less than 4 used in the cycle of Fig. 3-12. Computation will show that for the conditions of Fig. 3-12 about 20 lb per sec of working air must flow in the circuit for each 1000 kw developed. In contrast to this, Fig. 3-13 would show that for each 1000 kw developed only about half as big a flow rate is required. For proper relative intercooling in the equipment of each design and for the same temperature difference in their respective regenerators, the two-stage-heated unit will show higher efficiency. To illustrate this the theoretical curves from Ref. 6 have been drawn in Fig. 3-14. These are arbitrarily drawn for t_1 to the turbine from a heater as 1200 F, t_2 after the intercoolers or precoolers as 80 F, with the pressure loss $\lambda = \Delta P/P_{in}$, as 10 per cent for each exchanger and the regenerator terminal Δt, 36°. The turbine efficiency of

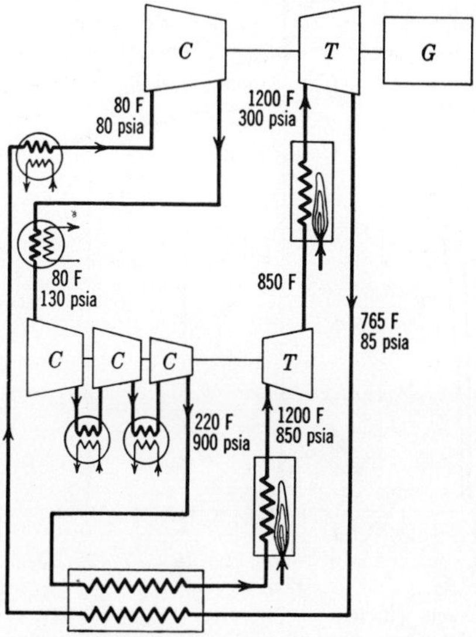

FIG. 3-13. Diagram of two-stage-heated closed gas-turbine system. (*Escher-Wyss, Ref.* 6.)

91 per cent used by the author (Ref. 6) appears higher than would probably be realized in operation.

A 12,500-net-kw closed-cycle plant of Escher-Wyss design went into service in 1950 in the St. Denis station at Paris, France. This plant is oil-fired and is purported to have a heat rate of 10,000 Btu per kwhr and a thermal efficiency of 34.1 per cent based on lower heating value of the fuel. The higher pressure of the system is 750 psia, and the high-pressure turbine employs a temperature of 1230 F. The reheat turbine receives air at 1250 F at a pressure of 255 psia and exhausts at 65 psia. Because of the high pressures, sizes of the power elements are small; one of the turbines developing 18,500 kw total power and running at 3000 rpm has a tip diameter of only 42 in. maximum.

The ultimate position of large-size closed-cycle plants in the United States is one of pure speculation. Units ranging in size from 10,000 to 20,000 kw are competitive with steam plants on a first-cost basis and have about the same efficiency under present-day allowable operating temperatures. Larger steam plants are appreciably lower in first cost per kilowatt of installed capacity. Heat transfer from gas to gas even under

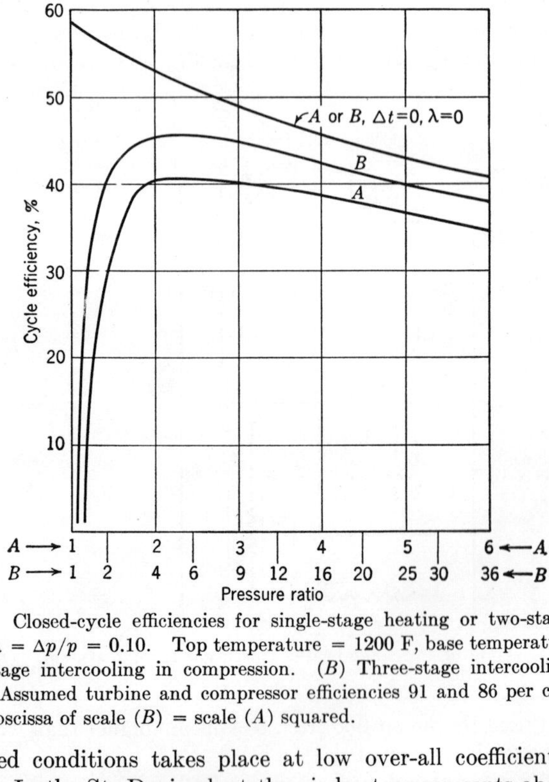

FIG. 3-14. Closed-cycle efficiencies for single-stage heating or two-stage heating. $\Delta t = 36°$, $\lambda = \Delta p/p = 0.10$. Top temperature = 1200 F, base temperature = 80 F. (A) One-stage intercooling in compression. (B) Three-stage intercooling in compression. Assumed turbine and compressor efficiencies 91 and 86 per cent, respectively. Abscissa of scale (B) = scale (A) squared.

compressed conditions takes place at low over-all coefficients of heat transfer. In the St. Denis plant the air heater represents about 30 per cent of the total weight and cost.

Example 6. A closed-cycle gas turbine employing the arrangement shown in Fig. 3-12 on a particular test indicated the following operating data, which differ in some respects from the values shown in the figure.

$p_a = 92$ psia, $T_a = 521$ R
$p_f = 347$, $T_a = T_c = T_e$
Internal compressor eff = 0.85; disregard pressure losses in intercoolers, and divide work equally.
$p_h = 343$ psia, $T_h = 1728$ R

Mechanical efficiency of compressor and turbine is 0.984 for each; internal turbine efficiency is 0.87.

Compressor turbine must supply sufficient power to drive compressor.

$p_j = 179$ psia, $p_k = 96$ psia, $T_m = 760$ R

Disregard radiation losses.

Fuel: 18,450 Btu per lb LHV, 19,650 Btu per lb HHV

Ratio of heat absorption from combustion based on LHV is 0.905.

Compute missing cycle temperatures, heat transferred, work quantities, and appropriate efficiencies. For an electrical output of 2050 kw, compute the air flow and the required fuel per hour (generator efficiency is 0.967).

Solution. The compressor increases the pressure from 92 to 347 psia, a pressure ratio of $^{347}/_{92} = 3.77$. To produce equal work in the three compressors, the ratio per unit will be $r = \sqrt[3]{3.77} = 1.556$. This will be used, and the 1 psi or so pressure loss in the intercoolers will be disregarded in this calculation to reduce the illustrative arithmetic.

$p_a = 92$ psia, $p_b = (92)(1.556) = 143.2$ psia

$T_a = 521$ R, and, by Table 2-4, $h_a = 14.633$ Btu per lb and $p_{ra} = 1.545$,

$(p_{rb})_s = p_{ra}r = (1.545)(1.556) = 2.404$

Corresponding $(h_b)_s = 31.478$

$$\text{Work} = \frac{31.478 - 14.633}{\eta_c} = \frac{16.845}{0.85} = 19.82 \text{ Btu per lb added in each of the three compressors}$$

$h_b = h_a + 19.82 = 34.45$ Btu per lb, and 603.5 R is the temperature T_b.

In the intercooler from b to c, the heat taken out by the water is

${}_bQ_c = 34.45 - 14.63 = 19.82$ Btu per lb as the temperature is brought from 603.5 R back to 521 R

Because the pressure ratio and the temperature ranges are the same, the work in both the intermediate and final compressors is

$${}_aW_b = {}_cW_d = {}_eW_f = 19.82 \text{ Btu per lb}$$

Also

$${}_bQ_c = {}_dQ_e = 19.82 \text{ Btu per lb}$$

Also

$T_f = T_b = 603.5$ R and $h_f = h_b = 34.45$ Btu per lb

Total compression work per lb of air flowing $= 19.82 + 19.82 + 19.82 = 59.46$ Btu

For a mechanical efficiency of the compressor of $\eta_m = 0.984$ based on rated load conditions

$$\eta_m = \frac{\text{compression work}}{\text{shaft supply work}} = \frac{59.46}{\text{shaft work}} = 0.984$$

Shaft work $= 60.43$ Btu per lb of air compressed

Thus work dissipated in external compressor bearings (*i.e.*, oil coolers, etc.) $= 60.43 - 59.46 = 0.97$ Btu on basis of each lb of air flowing through compressors.

Consider the mechanical efficiency of the turbine as 0.984 based on rated load conditions.

$$\eta_m = \frac{\text{shaft work output}}{\text{internal work from rotor}} = \frac{60.43}{W_{ti}} = 0.984$$

$W_{ti} = 61.41$ Btu work required from each lb of air flowing through the turbine which drives the compressor

Air enters the turbine for the compressor drive at 343 psia and at $T_h = 1728$ R, for which

$h_h = 320.569$ Btu per lb, and $p_{rh} = 123.112$

$h_i = h_h - 61.41 = 320.569 - 61.41 = 259.159$ and corresponding $T_i = 1498.7$ R

Isentropic work in turbine $= \dfrac{61.41}{0.87} = 70.58$ Btu per lb for expansion through actual pressure ratio

$(h_i)_s = 320.569 - 70.58 = 249.989$ value of enthalpy if expansion in turbine was isentropic, for which

Corresponding $p_r = 64.508$, and pressure ratio for turbine is

$$r = \frac{p_{rh}}{p_r} = \frac{123.112}{64.508} = 1.908$$

Thus

$p_i = \dfrac{343}{1.908} = 179.7$ psia; this indicates that a pressure loss of 0.7 psi occurs before reaching the power turbine as $p_j = 179$ psia.

$r_{jk} = 179/96 = 1.865$

$p_{rj} = 70.633$ at 1498.7 R

$p_{rk} = \dfrac{70.633}{1.865} = 37.87$

Corresponding value $(h_k)_s = 200.109$ Btu per lb

Work in power turbine $= 0.87(259.159 - 200.109) = 51.373$ Btu per lb of air

Net shaft work $= (51.373)(0.984) = 50.551$ Btu per lb

$h_k = 259.159 - 51.373 = 207.786$ Btu per lb, for which corresponding temperature is 1302.2 R.

Regenerator:

$T_l = 1302.2$ R, $T_f = 603.5$ R, $T_m = 760$ R

If radiation losses are disregarded and specific heats are equal,

$c_{pm}(T_l - T_m) = c_{pm}(T_g - T_f)$

$1302.2 - 760 = T_g - 603.5, \qquad T_g = 1145.7$ R

$h_g = 167.772$ Btu per lb

${}_gQ_h = h_h - h_g = 320.569 - 167.772 = 152.797$ Btu per lb of air flowing

${}_aQ_m = h_a - h_m = 14.66 - 72.260 = -57.627$ Btu per lb of air flowing

Efficiency of the system based on shaft power, assuming perfect combustion and utilization of all energy in fuel, is

$$\frac{{}_jW_k}{{}_gQ_h} = \frac{50.55}{152.797} = 0.331$$

If 90.5 per cent of the energy in the lower heating value of the fuel is absorbed by the air in the furnace, the thermal efficiency based on lower heating value (LHV) is

$(0.331)(0.905) = 0.299$, or 29.9 per cent on net shaft power

and

$(29.9)(0.967) = 28.9$ per cent based on electrical power

Based on higher heating value (HHV), the efficiency is

$$0.299 \frac{\text{LHV}}{\text{HHV}} = (0.299)\left(\frac{18{,}450}{19{,}650}\right) = 0.281 \text{ based on net shaft power}$$

$$\text{Air flow} = \frac{2545}{(60)(60)(50.551)} = 0.01398 \text{ lb per sec per net shaft hp}$$

$$\text{Air flow} = \left(\frac{0.01398}{0.746}\right)\left(\frac{1}{0.967}\right) = 0.01938 \text{ lb/(sec)(kw) delivered}$$

Air flow = (0.01938)(2050) = 39.7 lb per sec for 2050 kw delivered

(Air flow per sec per kw)($_gQ_h$)(3600) = (0.01938)(152.797)(3600) = 10,660 Btu per kwhr delivered

$$\text{Fuel used} = \frac{(10{,}660)(2050)}{(18{,}450)(0.905)} = 1308 \text{ lb per hr for 2050 kw delivered}$$

Specific volume of air at inlet to compressor at 92 psia at 521 R by Eq. (2-3) is

$$pv = (1)RT$$
$$(144)(92)(v) = (1)(53.34)(521)$$
$$v = 2.10 \text{ ft}^3 \text{ per lb}$$

Annular area required at inlet to compressor, if velocity is 500 fps, can be found:

ft^3 per sec = (lb per sec)(sp vol) = (area)(velocity) = (39.7)(2.10) = (area)(500)

Annular area = 0.1666 ft^2 = 24.0 in.^{2}

3-13. Gas-turbine Control. Mention has previously been made of the relative ease by which closed-cycle systems can be governed by varying the pressure in the system. Figure 3-15 indicates diagrammatically a method of accomplishing this result. As power load decreases with consequent rise in speed, the centrifugal governing device causes the discharge valve to release air into the low-pressure accumulator. When an increased power load falls on the system, causing a decrease in speed, the governor responds so as to admit air to the system from the high-pressure accumulator. For small reductions in load, it is possible to control the output without making use of the accumulators and changing the air pressure in the system. This is accomplished by operating the air by-pass from the high-pressure side of the circuit to the low-pressure side of the circuit so that useful output of the plant is reduced when a slight variation in load occurs or for small swings in capacity. Although this is wasteful of the compressed-air supply, it is less expensive than utilizing (and wasting) large quantities of high-pressure air from the accumulator, as this air also requires compression work. Acting on this basic principle, many variations of the governing arrangement can be developed for more complex arrangements such as occur when two turbines are used, one of which drives the compressor and the other the generator. Of course, it is also necessary to decrease the thermal (fuel) input when the load reduces. This can be done, however, without to any appreciable extent changing the inlet temperature to the turbine.

Control of system capacity essentially is always related to a change in fuel input. Means of operating the equipment so as to permit governing and control can be varied in many cases. In the open-cycle system we can think of (1) constant-temperature–varying-speed operations; (2) constant-speed–varying-temperature operation, or (3) varying-temperature–varying-speed operation together. For each installation, one of these possibilities or a combination of them must be employed. With constant-temperature–varying-speed operation, a high thermal efficiency can result for loads running from one-quarter to full load. However, such an arrangement would not be possible with a single turbine driving both a compressor and a generator if the latter were of the synchronous

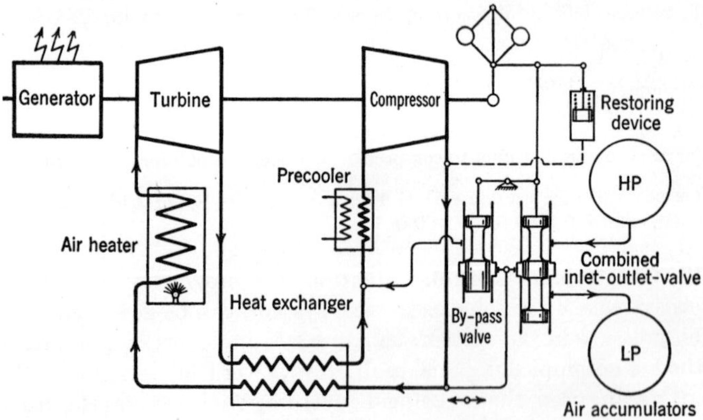

FIG. 3-15. Schematic diagram of centrifugally controlled regulating device for single-shaft Escher-Wyss closed-cycle unit. (*C. Keller, Ref.* 6.)

type. For this case, it would be necessary to operate at constant speed and variable temperature. This, unfortunately, does not give the best performance as the efficiency rapidly drops and at partial loads the low turbine inlet temperature is not conducive to economical use of fuel. This brings out the desirability of having separate turbines for driving the compressor and the generator as under this arrangement it is possible to operate the compressor at variable speed while the turbogenerator always turns at synchronous speed. Where the synchronous-speed aspect is not inherent in the design, modification 3 above offers certain advantages and has been used in one gas-turbine locomotive design as illustrated in Fig. 8-10. Reference to this figure will show that in the high-horsepower range of the turbine the temperature is kept constant and the speed is reduced, with the efficiency holding at a good value over the range of operation. At extremely light loads, down to idling, a point is reached at which the speed cannot be reduced further, and in this range the temperature is reduced. A study of the figure will show the very abrupt decrease in thermal efficiency, but inasmuch as operation in this range normally would not be for long periods, the arrangement is satisfactory. Under any of these variations, as output-power requirements decrease, the fuel input must also be decreased. Throughout the text, other examples of control arrangements will be indicated.

PROBLEMS

3-1. A Brayton cycle operates with air through a temperature range of 70.3 to 1240.3 F. (*a*) Compute thermal efficiencies of the cycle for pressure ratios of 4, 8, and 12, respectively. (*b*) Compute thermal efficiencies for a regenerative Brayton cycle operating under the same conditions. Use gas relations with a suggested value of $k = 1.39$ for air.

3-2. For a Brayton cycle under a pressure-compression ratio of $r = 8$, draw a Ts diagram to scale for limits of $T_1 = 530$ R and $T_3 = 2000$ R. Indicate the areas representing heat added, heat rejected, and work done. Superpose on this diagram a regenerative Brayton cycle, and indicate the areas representing heat added and rejected and work done. Use air tables for finding values. Note that any arbitrary datum can be used for entropy on the diagram.

3-3. A gas turbine operating on a simple cycle compresses air supplied at 13.5 psia and 70.3 F through a pressure compression ratio of 4. Find for a compressor efficiency of 100 per cent (*a*) the exit pressure and temperature from the compressor, (*b*) the work of compression per pound of air, and (*c*) the power absorbed for each pound per second of air compressed. Disregard changes in kinetic energy on inlet and exit from compressor.

3-4. Work Prob. 3-3 for the data given except that the compressor efficiency is 84 per cent.

3-5. A gas turbine, which operates on a simple power cycle, is supplied with air at 30.3 F at 14.7 psia. The air is compressed in an axial-flow compressor having an internal efficiency of 87 per cent to a pressure of 117.6 psia. (*a*) Find the isentropic work of compression and the actual work of compression per pound of air. (*b*) State the air temperature leaving the compressor (disregard kinetic contribution). (*c*) For an air flow of 25 lb per sec, find the internal compressor horsepower. (*d*) When the compressor bearing and gland losses indicate a mechanical efficiency of 97.5 per cent, find the shaft horsepower input required from the driving turbine. *Ans.* (*a*) 95.247, 109.48; (*b*) 941.44 R; (*c*) 3871 hp; (*d*) 3970 hp.

3-6. The compressor of Prob. 3-5 supplies air to combustors in which the temperature is raised to 1360.3 F (1820 R) while the pressure drops 3.2 psi. The turbine exhausts into a regenerator at a pressure of 15.5 psia. The internal efficiency of the turbine is 85 per cent, and its mechanical efficiency 97.5 per cent. (*a*) Find the pressure expansion ratio of the turbine and the isentropic work per pound of medium. (*b*) Find the actual internal work of the turbine and the leaving temperature. (*c*) For an air flow of 25 lb per sec, compute the internal turbine horsepower and the delivered shaft horsepower. (*d*) If it is known that the fuel-air ratio is 0.014 with the regenerator not in use, recompute the turbine horsepower values for the resulting medium flow of (25) (1.014) lb per sec. *Ans.* (*a*) 7.38, 194.897; (*b*) 165.662 Btu per lb, 1193.68 R; (*c*) 5858 hp, 5712 shaft hp; (*d*) 5937 hp, 5790 shaft hp.

3-7. For Probs. 3-6 and 3-5, find the heat added per second and the internal cycle efficiency provided the regenerator is not being used (*i.e.*, by-passed), (*a*) disregarding mechanical efficiencies; (*b*) considering mechanical efficiencies of the units. *Ans.* 5844.6 Btu per sec; (*a*) 25.02; (*b*) 22.05.

3-8. A gas turbine operates on a simple open cycle with gases entering the turbine at 1140.3 F (1600 R). Consider the turbine efficiency as 0.86, the compressor efficiency 0.84, and disregard mechanical-efficiency correction for each unit. The air from the compressor passing through the combustor undergoes a 3 per cent pressure drop based on compressor pressure. Barometer is 14.7 psia, and gases leave the turbine at 15 psia. Consider the system units each to be adiabatic. Bunker C fuel oil with an HHV of 18,540 Btu per lb and LHV of 17,540 Btu per lb is used (Fig. 9-21). Correct for the additional weight of gas passing through the turbine in contrast to the air compressed, but consider the gases to have the properties of air.

The ambient-air temperature is 70.3 F (530 R). Work for a compression pressure ratio of 3 and for a flow of 10 lb per sec of air. Compute or confirm the items called for in the tabulation for the appropriate pressure ratio.

SUMMARY OF RESULTS FOR PROBLEMS 3-8 TO 3-11

	3-8	3-9	3-10	3-11	Problem number
	3	4	6	8	Pressure ratio
(*a*)	761.3				Compressor leaving temp, °R
(*b*)					Compressor hp
(*c*)					Temp rise in combustor
(*d*)	.0132				Fuel-air ratio *f*
(*e*)					Energy added in combustor, Btu per sec
(*f*)					Pressure entering turbine, psia
(*g*)					Expansion pressure ratio
(*h*)					Turbine leaving temp, °R
(*i*)					Turbine hp
(*j*)					Net hp
(*k*)					Cycle thermal eff
(*l*)					Air rate, lb per net hp-sec
(*m*)					Power ratio, net hp/turbine hp
(*n*)					Sp power, net hp/(lb)/(sec) flow

3-9. Use the data given in Prob. 3-8, and for a pressure compression ratio of 4 compute or confirm the items called for in the summary following Prob. 3-8.

3-10. Use the data given in Prob. 3-8, and for a pressure compression ratio of 6 compute or confirm the items called for in the summary following Prob. 3-8.

3-11. Use the data given in Prob. 3-8, and for a pressure compression ratio of 8 compute or confirm the items called for in the summary following Prob. 3-8.

3-12. A gas turbine operates on a simple open cycle with gases entering the turbine at 1440.3 F (1900 R). Consider the turbine efficiency as 0.86, the compressor efficiency 0.84, and disregard mechanical-efficiency correction for each unit. The air from the compressor passing through the combustor undergoes a 3 per cent pressure drop based on compressor pressure. Barometer is 14.7 psia, and gases leave the turbine at 15 psia. Consider the system units each to be adiabatic. Bunker C fuel oil with an HHV of 18,540 Btu per lb and LHV of 17,540 Btu per lb is used (Fig. 9-21). Correct for the additional weight of gas passing through the turbine in contrast to the air compressed, but consider the gases to have the properties of air.

The ambient-air temperature is 70.3 F (530 R). Work for a compression pressure ratio of 4, and for a flow of 10 lb per sec of air. Compute or confirm the items called for in the tabulation for the appropriate pressure ratio.

Summary of Results for Problems 3-12 to 3-15

	3-12	3-13	3-14	3-15	Problem number
	4	6	8	11	Pressure ratio
(*a*)					Compressor leaving temp, °R
(*b*)					Compressor hp
(*c*)					Temp rise in combustor
(*d*)					Fuel-air ratio f
(*e*)					Energy added in combustor, Btu per sec
(*f*)					Pressure entering turbine, psia
(*g*)					Expansion pressure ratio
(*h*)					Turbine leaving temp, °R
(*i*)					Turbine hp
(*j*)					Net hp
(*k*)					Cycle thermal eff
(*l*)					Air rate, lb per net hp-sec
(*m*)					Power ratio, net hp/turbine hp
(*n*)					Sp power, net hp/(lb)/(sec) flow

3-13. Use the data given in Prob. 3-12, and for a pressure compression ratio of 6 compute or confirm the items called for in the summary which follows Prob. 3-12.

3-14. Use the data given in Prob. 3-12, and for a pressure compression ratio of 8.0 compute or confirm the items called for in the summary which follows Prob. 3-12.

3-15. Use the data given in Prob. 3-12, and for a pressure compression ratio of 11compute or confirm the items called for in the summary which follows Prob. 3-12.

3-16. A gas turbine operates on a simple power cycle except that a regenerator is used. Consider a pressure loss of 3.2 psi from the compressor through the regenerator

SUMMARY OF RESULTS FOR PROBLEMS 3-16 TO 3-18

	3-16	3-17	3-18	Problem number
	1800	1800	1800	Inlet turbine temp, °R
	4	6	8	Compression pressure ratio
(*a*)				Compressor leaving temp, °R
(*b*)			1733	Compressor hp
(*c*)				Pressure entering turbine, psia
(*d*)			7.29	Expansion pressure ratio
(*e*)				Turbine exhaust temp, °R
(*f*)				Air temp entering combustor, °R
(*g*)				Fuel-air ratio *f*
(*h*)				Turbine hp
(*i*)				Net hp
(*j*)				Energy added in combustor, Btu per sec
(*k*)				Cycle thermal eff, %
(*l*)				Thermal eff based on LHV at 95% combustion eff
(*m*)				Corresponding thermal eff based on HHV
(*n*)				Air rate, lb per net hp-sec
(*o*)				Power ratio, net hp/turbine hp

and combustor to the turbine inlet. On exhaust, the pressure leaving the turbine is 1 psi greater than atmospheric at 15.7 psia. Consider the compressor and turbine as adiabatic devices, although there is necessarily some heat loss from these units. Consider the working medium as 10 lb of air per sec compressed. Correct for the additional gas weight resulting from combustion and passing through the combustor and turbine, by use of Fig. 9-21. Use a turbine efficiency of 86 per cent and compressor efficiency of 84 per cent. Air supply is at 70.3 F and 14.7 psia. The fuel is Bunker C oil with a HHV of 18,540 Btu per lb and a LHV of 17,540. Assume that the air leaving the regenerator is warmed to within 60° of the gas temperature leaving the turbine.

For a pressure compression ratio of 4, compute or confirm the items called for in the summary on page 128.

3-17. Use the data given in Prob. 3-16, and for a pressure compression ratio of 6 compute or confirm the items called for in the summary which follows Prob. 3-16.

3-18. Use the data given in Prob. 3-16, and for a pressure compression ratio of 8 compute or confirm the items called for in the summary which follows Prob. 3-16.

3-19. An intercooler is used on the compressor of an open power cycle when the pressure compression ratio is 8 and the inlet temperature to the turbine is 1240.3 F. Consider that the intercooler pressure is equal to the square root of the product of inlet and outlet pressure and that air leaves the intercooler at 80.3 F. Disregard pressure loss in the intercooler, and consider each compressor as adiabatic and having an efficiency of 84 per cent. Pressure loss from the compressor through the regenerator and combustor is 3.2 psi to turbine inlet. The turbine can be considered adiabatic and 86 per cent efficient based on delivered power. The exhaust gases enter the regenerator at 15.7 psi. Atmospheric air is at 70.3 F at 14.7 psi. Flow through the compressor is 10 lb per sec. Modify the flow through the turbine for combustion products from light fuel oil (see Fig. 9-20). Find (*a*) temperatures leaving low-pressure and high-pressure compressors, horsepower absorbed by each, and heat rejected in intercooler; (*b*) pressure entering turbine, expansion pressure ratio, gas temperature at exit from turbine. (*c*) Considering the regenerator to be 75 per cent effective, based on temperatures, find the leaving air and gas temperatures and the energy per second required in the combustor to raise the total products to turbine inlet temperature, considering these to have the properties of air. (*d*) Find the turbine gas-flow rate in pounds per second, delivered horsepower, net horsepower, power ratio, and air rate in pounds per second per net horsepower. (*e*) Compute the annular area into the turbine if a gas velocity of 400 fps is employed.

3-20. Find the results asked for in Prob. 3-19 if the pressure compression ratio is 6 instead of 8, with no other changes in the problem.

3-21. For the gas-turbine unit discussed in Probs. 3-11 and 3-8 with a turbine efficiency of 0.86 and a compressor efficiency of 0.84, it is planned to add an intercooler between two units of the compressor and interstage reheating between two sections of the turbine. The pressure compression ratio is 8, and the inlet temperature to each turbine is set at 1600 R. Air enters the low-pressure compressor at 70.3 F and is intercooled to the same temperature for entry to the high-pressure compressor.

Disregard pressure loss in the intercooler, and arrange the intercooler pressure to be equal to $p_1 \sqrt{r}$, where p_1 is 14.7 psia, the inlet pressure of the system. Under this condition, equal work is done on the air in the low-pressure and high-pressure compressors.

Compute the intermediate pressure and the discharge pressure from the high-pressure unit, the discharge temperature and enthalpy from each compressor, and the work required in each compressor; also, the heat rejected in the intercooler. Assume that a 3 per cent pressure loss occurs in the burner piping system for the air entering the high-pressure turbine. The intermediate pressure at inlet to the low-pressure reheating burner should be computed as equal to $\sqrt{p_{1t} \times p_e}$, where p_{1t} is the pressure at entry to the high-pressure turbine and p_e is the pressure (15 psia) at exhaust from the low-pressure turbine. A 3 per cent pressure loss is considered to take place in the reheating burner. *Ans.* 41.6 psia, 117.6 psia; 747.1 R, $h = 69.135$,

$$W_1 = W_2 = 52.335 \text{ Btu per lb} = Q_{ic};$$

$p_{1t} = 114.07$ psia, $p_{2t} = (0.97)(41.36)$.

Compute the temperatures and enthalpies at exhaust from each turbine. These will be slightly different for the two units because of pressure losses. *Ans.* $h = 201.989$ and 204.196 Btu per lb.

Find the fuel-air ratio for the high-pressure turbine, and then find the fuel-air ratio for the low-pressure turbine. It should be realized that the exhaust products from the first turbine also pass through the second turbine. On the basis of 10 lb per sec flow through the compressor, find the power produced by each of the turbines. Bunker C fuel oil with a LHV of 17,540 Btu per lb is used with 95 per cent combustion efficiency. (Refer to Fig. 9-21 for fuel-air ratios.) *Ans.* 0.0134, 0.0051; 852.5 and 834.2 Btu per sec.

Summarize the problem, indicating total compressor horsepower, total turbine horsepower, net horsepower, total heat added by the fuel, cycle thermal efficiency based on heat added, air rate, power ratio, and specific power. *Ans.* 1480.5 compressor hp, 2385.7 turbine hp, 3075.2 Btu per sec heat added.

3-22. A closed-cycle gas-turbine layout similar to that of the Escher Wyss AK Plant has the following operating data:

Compressor, eff of each part 85 per cent

1. Entry: 89 psia, 70.3 F
   Leaving: 141 psia
2. Entry: 140 psia, 70.3 F
   Leaving: 220 psia
3. Entry: 219 psia, 70.3 F
   Leaving: 346 psia

Temperature of air leaving preheater will be found by heat-balance computation

Air enters turbine at 340 psia, 1270.3 F

Turbine eff is 87 per cent. Use with compressor requirement to find condition leaving this turbine and entering the power turbine

Pressure leaving power turbine is 95 psia

a. Find the work supply per pound of air to each part of the compressor, and compute the final leaving temperature and enthalpy. *Ans.* Work = 21.00, 20.59, 20.87 Btu per lb; 616.87 R, 37.668.

b. Use this information to compute the leaving temperature and pressure from the compressor turbine. *Ans.* 1496.79 R, 176.8 psia.

c. Find the work developed per pound of air in the power turbine and leaving conditions when the discharge pressure is 95 psia. *Ans.* 51.2 Btu per lb, 1300.8 R.

d. In the regenerator the turbine air is cooled to 210.3 F giving up 94 per cent of its energy to warming of the compressed air and 6 per cent to radiation; find the leaving temperature and enthalpy of the compressed air. *Ans.* 1214.3 R, 185.2 Btu per lb.

e. Find the energy per pound air to cooling water in the precooler to bring the 210.3 F air to 70.3 F. Find also the energy per pound air to cooling water in the compressor intercoolers. *Ans.* 33.67 Btu per lb; 41.59 Btu per lb air.

f. When the power turbine delivers 2110 kw to the generator, compute the air flow required in pounds per second to create this power. *Ans.* 39.1 lb per sec.

g. The generator efficiency is 96.7 per cent, and of the power generated 3.8 per cent is used for auxiliaries (water pump, combustion-air fan, flue-gas fan, circulating-gas fan, fuel-injection pumps). Find the net kilowatts generated. *Ans.* 1963 kw.

h. Find the kilowatts required to drive the compressor. *Ans.* 2576 kw.

i. Find the heat required for the combustor in Btu per hour and find the pound per hour of fuel oil at 18,540 Btu per lb HHV (17,540 LHV) if 85 per cent of the LHV is transferred to useful heating. *Ans.* 19,210,000 Btu per hr; 1288 lb fuel per hr.

j. What is the thermal eff. based on net kw output and higher heating value of the fuel? *Ans.* 28.05 per cent

REFERENCES

1. Fischer and Meyer: The Combustion Gas Turbine Cycle, *Westinghouse Eng.*, vol. 4, pp. 78–84, May, 1944.
2. Streid: Gas Turbine Fundamentals, *Mech. Eng.*, vol. 68, pp. 127–133, February, 1946.
3. Meyer, A.: Recent Developments in Gas Turbines, *Mech. Eng.*, vol. 69, pp. 273–277, April, 1947.
4. Puffer and Alford: The Gas Turbine in Aviation, *Mech. Eng.*, vol. 67, pp. 803–812, December, 1945.
5. Keller, C.: Closed Cycle Gas Turbine, *ASME Trans.*, 1950, vol. 72, pp. 835–850.
6. Keller, C.: The Escher Wyss-AK Closed-Cycle Turbine, Its Actual Development and Future Prospects, *ASME Trans.*, vol. 68, pp. 791–822, November, 1946.
7. L'Application de la turbine á gaz á la production d'énergie électrique, extract from *Mém. soc. roy. belge ing. ind.*, series B, no. 5, 1949.
8. Salisbury, J. K.: The Basic Gas Turbine Plant and Some of Its Variants, *Mech. Eng.*, vol. 66, pp. 373–383, June, 1944.
9. 1000-KW Gas-turbine Plant Features New Centrifugal Compressor Design, *Power*, vol. 91, pp. 851–853, December, 1947.

CHAPTER 4

GAS TURBINE TYPES AND DESIGN FEATURES

4-1. Impulse-turbine Types. The impulse turbine in its simplest form consists of a nozzle or group of nozzles with a following row of blades or buckets (Fig. 4-1). In the nozzle the gaseous working medium expands and, during the process, converts available thermal energy into kinetic form. The jet of gas at high velocity then impinges on blades (buckets) of suitable form. These buckets are capable of converting a large portion of the kinetic energy of the moving jet into work or power at the turbine shaft. The impulse turbine is characterized by the fact that theoretically no pressure drop or expansion as such occurs in the blades themselves.

Velocity-compounded Impulse Turbine. In order to absorb a large fraction of the kinetic energy from the moving jet in the blades, it is necessary that the blades move at about half the velocity of the gas jet. Inasmuch as blade velocities are limited by centrifugal stresses, it may happen that for maximum utilization of kinetic energy the optimum blade speed cannot be utilized. Velocity compounding permits the use of a low blade speed by absorbing the energy in two or more rows of moving blades. Figure 4-2 is a diagrammatic section through such a turbine. In the first row, the jet from the nozzle impinges on the blades as in an ordinary impulse turbine. However, the gas leaving this row enters a row of blades fixed to the casing of the turbine, which merely redirects the gas onto a second moving row, where an additional amount of kinetic energy is absorbed. In round figures, a blade speed of approximately 25 per cent of the jet velocity represents an optimum speed for absorbing the kinetic energy of the jet. This is in contrast to a simple impulse turbine, where the optimum blade speed is approximately 49 per cent. Velocity-compounded turbine units are often designated by the name of Curtis turbines.

Pressure Compounding. In pressure compounding the pressure range available for the gas expansion is broken into a series of steps or stages. Each stage consists of a nozzle or bank of nozzles followed by a row of turbine blades or buckets. In each nozzle-stage row, the kinetic energy of the jet is increased, and in each following row of moving blades the greater portion of the kinetic energy of the jet is absorbed as useful work on the turbine rotor. The gas leaving a moving row of blades is then led into a following row of nozzles, where the pressure drops a further amount, the velocity increases, and, in its associated row of moving blades, the greater portion of this kinetic energy is absorbed in useful work. Figure 4-3 shows pressure-compounded stages.

Turbines of this type are impulse turbines in that theoretically there is no pressure drop within the moving blades, the pressure drop all taking place in the nozzles. Turbines using pressure-compounded stages are sometimes referred to as Rateau turbines.

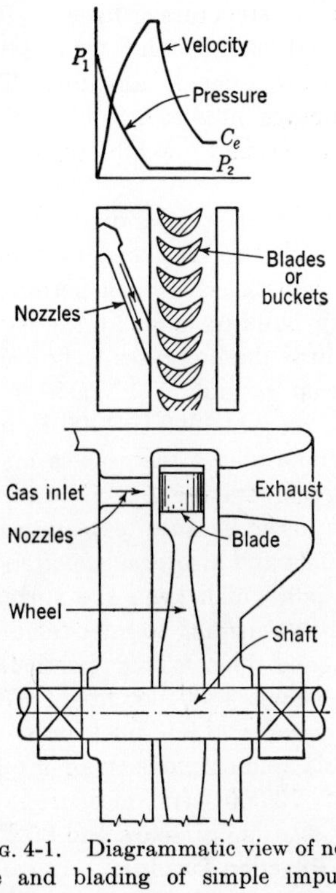

FIG. 4-1. Diagrammatic view of nozzle and blading of simple impulse turbine.

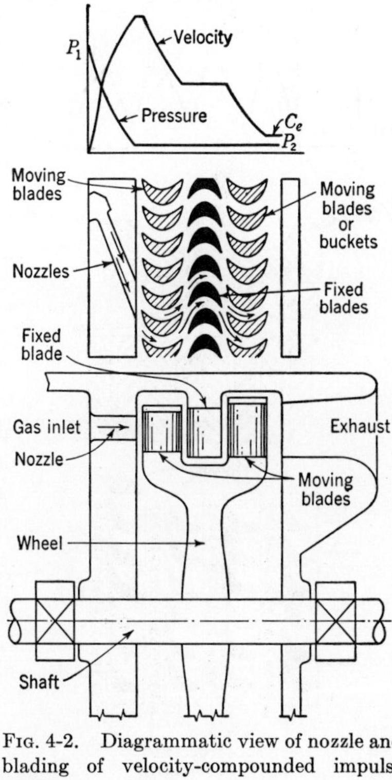

FIG. 4-2. Diagrammatic view of nozzle and blading of velocity-compounded impulse turbine.

Figures 4-4 and 4-5 give details of the turbine for the General Electric 4800-hp railway locomotive which is illustrated as a complete power unit in Fig. 1-9. It can be noticed that the turbine is a two-stage Rateau-type impulse unit. The blades (buckets) are fastened to the wheels by means of the fir-tree root form (Fig. 10-13). The rim of the wheel which holds the blades is of stainless material suitable for high-temperature operation, while the main portion of the wheel is of low-alloy high-strength steel suitable for operation below 700 F. Figure 4-5 is a photograph of the turbine rotor, the first-stage buckets being the short, stubby blading at the left of the photograph. The shape of the blading, in particular the twist of the second-stage blades from hub to tip, should be observed.

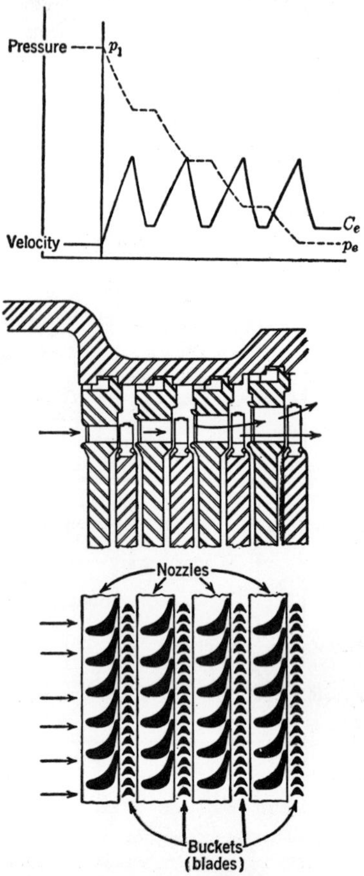

FIG. 4-3. Diagrammatic view of nozzles and blading of a four-stage pressure-compounded (Rateau) impulse turbine.

The first-stage nozzles consist of partitions welded into an outer ring and projecting through closely fitting holes into an inner ring. This permits a structure which is free to expand radially and reduces the stresses to a small amount. The second-stage nozzles are of similar construction and are split to permit assembly.

The gases, after expanding through the second-stage buckets, pass into a diffusing passage which permits a pressure build-up and, at the same time, turns the hot gases to flow upward from the unit. The gases enter the first stage at 1300 to 1400 F, but the turbine blades themselves operate at a temperature several hundred degrees lower because of boundary-layer effect and also because of cooling-air jets which keep the turbine rotor from warming to inlet-temperature conditions. This particular unit develops 4800 net hp at 6700 rpm, operates at an inlet pressure of 75 psia, and employs an air-intake rate of 70,000 cfm measured at 1500 ft altitude pressure and 80 F.

4-2. Reaction Turbines. In reaction turbines the nozzles and moving blades are each made in the same general form with the cross section in the direction of gas flow reduced so that both the fixed and moving blades act as expanding nozzles.

Figure 4-6 is a diagrammatic section through a reaction turbine. The pressure decreases in both the fixed and moving blades, and gas velocity is increased relative to the blade passageway in both types of blading. In passing through the stationary blades, the gas is directed into the moving blades, usually at a velocity higher than that at which the blades are

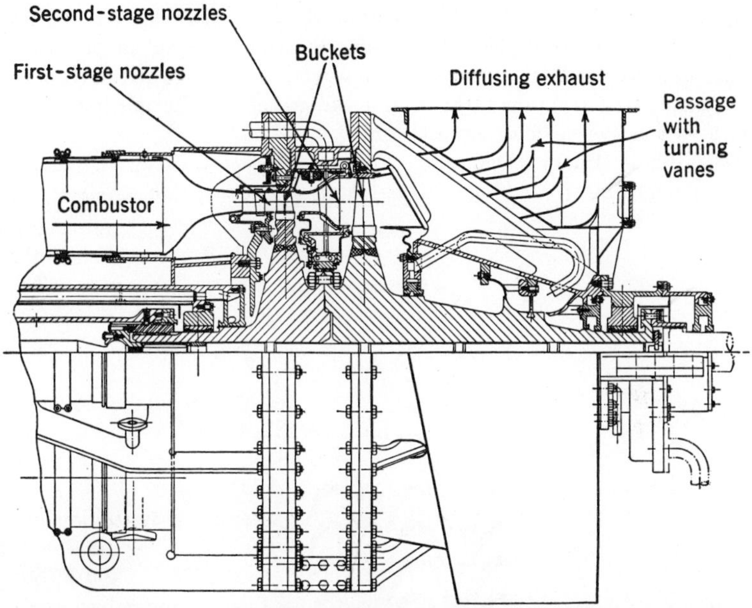

FIG. 4-4. Detail of two-stage General Electric gas turbine for the 4800-hp locomotive of Fig. 1-9.

FIG. 4-5. Turbine rotor for General Electric gas-turbine power plant for 4800-hp locomotive drive; inlet row at left.

moving. In this respect the reaction turbine resembles an impulse turbine, with the fixed blades performing the same function as the nozzles of an impulse turbine. It happens, however, that the velocities used are lower with the reaction turbine, and the gas enters the moving blades with its relative velocity almost axial.

4-3. Blade (Bucket) Diagram. Figure 4-7 shows the velocity triangles for an impulse-turbine stage. Gas leaves the nozzle with a velocity C_1

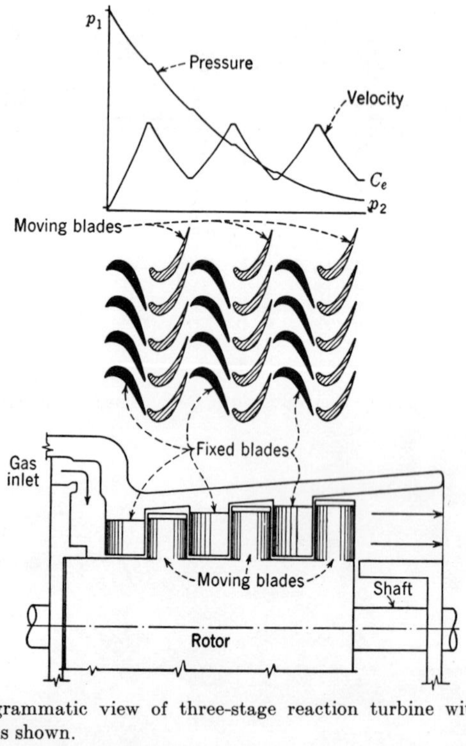

FIG. 4-6. Diagrammatic view of three-stage reaction turbine with pressure and velocity patterns shown.

and is directed into the blade passageway. The blade itself is shown having a tangential speed of u_1 as it turns around the axis of the turbine. The relative velocity w_1 with which the gas enters the blade passage is the vectorial difference between these two velocities, and its direction lies at an angle β_1 measured from the direction of blade motion. In an impulse turbine, if there is no pressure change in the blade passageway (*i.e.*, no reaction), the relative gas velocity w_1 will remain unchanged except for some diminution because of friction and the gas will leave with relative velocity w_2 in a direction β_2 which is set by the outlet blade angle β_2', that is, $180 - \beta_2$. After laying off the blade speed u_2, it is then possible to complete the outlet-velocity triangle which shows the absolute leaving velocity to have a magnitude of C_2 and direction of θ_2. In passing through the blade passageway the absolute velocity has been reduced in magnitude from C_1 to C_2 and also changed in direction.

The tangential change of velocity in the direction of blade motion, which goes under the name *velocity of whirl*, creates the driving force acting on the blades. This velocity change is obviously

$$\Delta V = C_1 \cos \theta_1 - C_2 \cos \theta_2 \tag{4-1}$$
$$\Delta V = C_{1u} - C_{2u} \tag{4-2}$$
$$\Delta V = w_1 \cos \beta_1 - w_2 \cos \beta_2 \tag{4-3}$$

When the angles are measured consistently, *i.e.*, both clockwise from the direction of u as shown in Fig. 4-7, the terms C_{1u} and C_{2u} have the proper trigonometric sign and will automatically indicate the proper value of ΔV. Notice that in this case the total velocity change is C_{1u} plus C_{2u}. This can be shown in Eq. (4-1) for Fig. 4-7, as here θ_2 lies between 90 and 180°. Its cosine is minus so that $-C_2 \cos \theta_2$ becomes

$$-C_2[-\cos(180 - \theta_2)] = +C_2 \cos \theta_2'$$

If the angles θ_2' and $180 - \theta_2$ are used in drawing diagrams, care must always be observed to interpret the proper sign for C_{2u}.

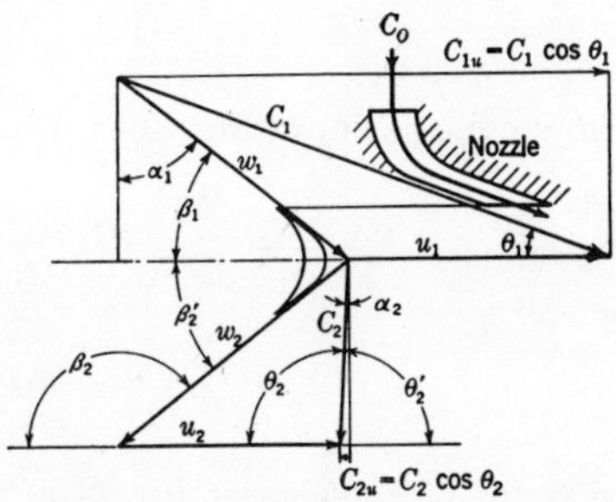

FIG. 4-7. Velocity triangles for impulse-turbine stage.

From Newton's second law, by proper substitution

$$F = \frac{W}{g_c} a = \frac{W}{g_c} \frac{\Delta V}{\Delta t} = \frac{W}{g_c\, \Delta t} \Delta V \tag{4-4}^*$$

Let the term $W/\Delta t$, the pounds mass (weight) flow per second, be represented by a single term, G lb per sec. Then

$$F = \frac{G}{g_c} \Delta V \tag{4-5}$$

and the force on the blade element measured in pounds force (lb_f) is

$$F = \frac{G}{g_c} (C_{1u} - C_{2u}) \tag{4-6}$$

* In this expression if W is measured in *pounds mass (weight)* units, then g_c is a numerical dimensionless ratio 32.17 which changes W/g_c to a mass (weight) unit measured in slugs.

Power developed can be expressed as

$$P = Fu \tag{4-7}$$

where P = power, ft-lb$_f$ per sec
F = pounds force (lb$_f$)
u = velocity in the line of action of the force, fps

For a turbine blade element through which a gas flow of G lb per sec is passing with the blade velocity at the leading edge u_1 fps and with u_2 fps at the trailing edge,

$$\begin{aligned} P &= Fu \\ &= \frac{G}{g_c}(C_{1u}u_1 - C_{2u}u_2) \end{aligned} \tag{4-8}$$

In most turbines for a gas-stream filament passing through the passageway, u_1 and u_2 are little different as can be seen from Fig. 4-8. If such is the case, Eq. (4-8) becomes

$$P = \frac{G}{g_c}(C_{1u} - C_{2u})u \tag{4-9}$$

Employing gas-flow directions, Eq. (4-9) appears from Fig. 4-7 or 4-9 as

$$P = \frac{G}{g_c}(C_1 \cos \theta_1 + C_2 \cos \theta_2')u \tag{4-10}$$

or

$$P = \frac{G}{g_c}(w_1 \cos \beta_1 + w_2 \cos \beta_2')u \tag{4-11}$$

In terms of horsepower Eq. (4-9) appears

$$\text{hp} = \frac{(G/g_c)(C_{1u} - C_{2u})u}{550} \tag{4-12}$$

In blade analyses for many applications the blade-velocity diagram at the mean radius and the mean blade speed in connection with the total flow through the blade are used to find diagram horsepower. However, for more precise analyses, diagrams are made for the tip, mean radius, and hub (base) of the blade.

From Fig. 4-8, the mean blade speed for a turbine turning at N rpm, with a blade height (length) of h in., can be developed as

$$u_m = \frac{2\pi r_m}{12}\frac{N}{60} = \frac{2\pi\left(r_h + \dfrac{h}{2}\right)N}{12 \times 60}$$

$$u_m = \frac{\pi(2r_h + h)N}{720} \quad \text{fps} \tag{4-13}$$

The root (hub) speed is

$$u_h = \frac{2\pi r_h N}{12 \times 60} = \frac{\pi r_h N}{360} \quad \text{fps} \tag{4-14}$$

The tip speed is

$$u_t = \frac{2\pi(r_h + h)N}{12 \times 60} = \frac{\pi(r_h + h)N}{360} \quad \text{fps} \tag{4-15}$$

$$u_t = \frac{\pi r_t N}{360}$$

The passages through turbine blading are guiding channels for conducting the gas in such manner as to meet the objectives of the turbine designer and to produce the power required. It is obvious that the flow area per inch of blade height (length) increases as the distance from the center of rotation increases. Variations in flow patterns along the blade occur not only because of area change but also because the blade velocity varies along the blade length. To meet design objectives, it has been found desirable not only to twist the blades but to taper them as well.

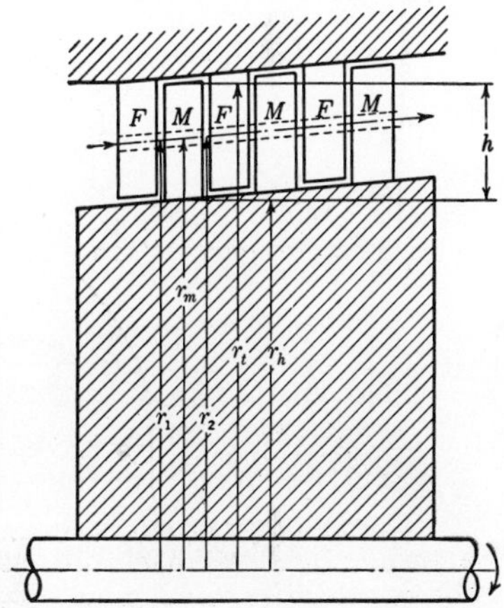

FIG. 4-8. Flow pattern through turbine blading.

Figure 4-10 shows moving-blade profiles employed by one manufacturer. In this figure it will be noticed that the inlet and outlet angles of the blade vary from base (hub) to tip. Figures 4-17, 4-18, and 4-25 are photographs of turbine blades in each of which the twist of the blades can also be observed. The blades taper (decrease in depth) from base to tip. Decreased blade volume outward from the center diminishes the centrifugal stress which would otherwise have to be carried at the base or hub sections of the blade.

Figure 4-24 showing the blades mounted in the rotor illustrates the use of the fir-tree type of attachment. With this the blades are slipped into position sideways and held in place by a supporting side member on the rotor. It has been found desirable to use a relatively loose fit at the base so that vibrational energy can be damped out with the friction resulting from relative motion between the blade and its mount. This is not possible with a rigid mounting.

The use of round-nose blades at the inlet edge allows for variations in the direction at which the fluid enters without causing undue shock when the angle is appreciably different from the design angle. It should be realized that turbine blading designed for one load on the turbine will necessarily pass fluids which enter and turn through different angles both at partial load and at overload conditions. The thickness of the trailing edge of the blade is normally set by manufacturing convenience, with one manufacturer using a minimum thickness of approximately 1 per cent of the blade chord. The blading for impulse turbines is normally thicker at the center-chord position, and such blades are relatively thin at inlet and leaving edges.

It should be observed that in drawing the blade-velocity diagram (Fig. 4-7) the absolute-velocity vectors C and u were drawn with their direction arrows meeting at a common point and the relative velocity w then closed the triangle. This system is convenient and will be followed throughout

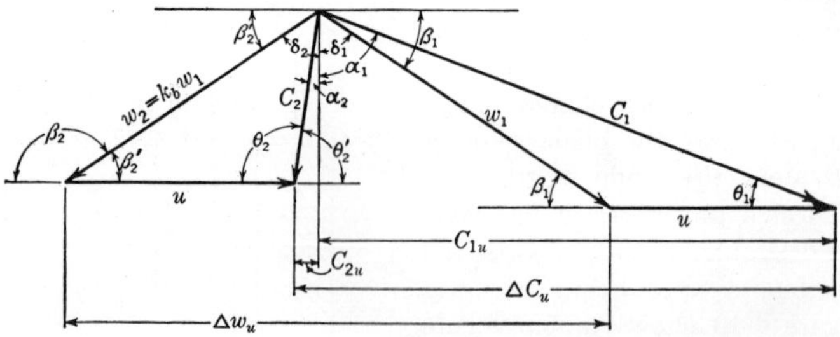

$$\Delta V = V_m = \Delta C_u = \Delta w_u$$

FIG. 4-9. Axial (pole) representation for blade velocity triangles.

the text. Observation of this simple rule of meeting arrowheads for absolute velocities can eliminate doubt in constructing velocity triangles. A system in which the tails of the absolute-velocity vectors are caused to meet could also be employed, but the two systems should not be confused.

In the diagram analysis it was tacitly assumed that the gas flow followed the passage direction indicated by the blade angles. This is reasonably true for a machine working under design conditions but is of course not true over the whole range of operating conditions. Actually under the force and energy analyses just given the angles used must refer to the direction in which the gas streams are moving and not necessarily to the configuration and twist of the passage.

The notation employed in Fig. 4-7 has been used extensively in steam-turbine usage, whereas with gas turbines and axial-flow compressors it is more usual to indicate nozzle and blading angles at a pole point in a plane normal to that passing through the axis of rotation. In Fig. 4-9 the diagrams of Fig. 4-7 have been relocated and marked to indicate this practice. In the latter system the nozzle angle θ_1 is discarded in favor of α_1, while the relative velocity flow angles become δ_1 and δ_2.

The axial velocity of the gas entering the moving blade is

$$C_{a1} = C_1 \cos \alpha_1 = C_1 \sin \theta_1$$
$$C_{a1} = w_1 \cos \delta_1 = w_1 \sin \beta_1 \quad (4\text{-}16)$$

The axial velocity leaving the moving blade row is

$$C_{a2} = C_2 \cos \alpha_2 = C_2 \sin \theta_2'$$
$$C_{a2} = w_2 \cos \delta_2 = w_2 \sin \beta_2' \quad (4\text{-}17)$$

Axial velocity in relation to the area of passage through the blading sets the gas-flow capacity of the turbine, and changes in axial velocity through a moving blade row also contribute to axial thrust although usually to a minor extent.

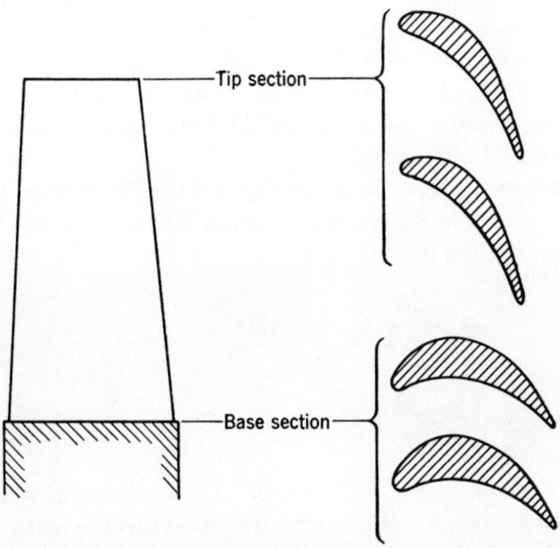

FIG. 4-10. Reaction-type moving blade profiles. (*After Emmert, Ref.* 1.)

4-4. Kinetic Form of Blade Equation. The general expression derived in Eq. (4-8) for the force on a blade (bucket) is applicable to any type of blading or impeller whether in a turbine or compressor. It is the force associated with the interchange of energy between the flowing gas and the rotor elements (blades, impeller, etc.). Because of value in certain types of analyses, it is desirable to express the relations of Eqs. (4-8) and (4-9) in a different form. Apply the law of cosines to Fig. 4-7.

$$w_1^2 = C_1^2 + u_1^2 - 2u_1C_1 \cos \theta_1 \quad (4\text{-}18)$$
$$w_1^2 = C_1^2 + u_1^2 - 2u_1C_{1u} \quad (4\text{-}19)$$
$$w_2^2 = C_2^2 + u_2^2 - 2u_2C_2 \cos \theta_2 \quad (4\text{-}20)$$
$$w_2^2 = C_2^2 + u_2^2 - 2u_2C_{2u} \quad (4\text{-}21)$$

Subtract Eq. (4-21) from Eq. (4-19), giving

$$\begin{aligned} w_1^2 - w_2^2 &= (C_1^2 - C_2^2) + (u_1^2 - u_2^2) - 2(u_1C_{1u} - u_2C_{2u}) \\ u_1C_{1u} - u_2C_{2u} &= \tfrac{1}{2}[(C_1^2 - C_2^2) + (u_1^2 - u_2^2) + (w_2^2 - w_1^2)] \end{aligned} \tag{4-22}$$

Multiply both sides of this expression by G/g_c, and compare with Eq. (4-8).

$$P = \frac{G}{g_c}(u_1C_{1u} - u_2C_{2u})$$

$$P = \frac{G}{2g_c}[(C_1^2 - C_2^2) + (u_1^2 - u_2^2) + (w_2^2 - w_1^2)] \tag{4-23}$$

This as before is an expression for the power in foot-pounds per second interchanged between the rotor blade elements and the flowing gas. It will be noted that it appears in the same form as though kinetic-energy interchanges measured the energy transferred. This expression is perfectly general and in terms of the end velocities given is a measure of the power interchange. However, tip and clearance leakage gas will not follow the same velocity pattern, and Eq. (4-23) is not too useful for predicting actual performance.

4-5. Blade Speed Ratio (Impulse Turbine). The power developed by a gas flow of G lb per sec in passing through blading at a given radius was shown by Eq. (4-9) to be

$$P = \frac{G}{g_c}(C_{1u} - C_{2u})u \tag{4-9}$$

and

$$P = \frac{G}{g_c}u(C_1 \cos\theta_1 + C_2 \cos\theta_2') \tag{4-24}$$

with reference made to Fig. 4-7 or 4-9. Let us define the term u/C_1 as the blade speed ratio σ. In a pure impulse turbine with no reaction (expansion) the relative velocity w_1 remains constant or because of frictional and turbulence effects decreases to a lower value w_2 in flowing through the blade passage.

$$w_2 = k_b w_1 \tag{4-25}$$

The value of the k_b varies with turning angle in the blading, shape and length of blade passage, blade roughness, and shock effect at the inlet and outlet edges of the blade and becomes progressively lower as velocities increase, particularly into the supersonic ranges. As a rough guide, take $k_b = 0.92$ for $w_1 = 500$ fps, 0.9 for 1000 fps, 0.88 for 1500 fps, 0.85 for 2000 fps, and 0.8 for 3000 fps. For average conditions a value of $k_b = 0.85$ is suggested. Gas friction losses vary approximately as the velocity raised to the second power.

In Figs. 4-7 and 4-9 call $u_2 = u_1 = u$, and note that

$$C_2 \cos \theta_2' = w_2 \cos \beta_2' - u \tag{4-26}$$
$$C_2 \cos \theta_2' = k_b w_1 \cos \beta_2' - u \tag{4-27}$$

Also,

$$w_1 \cos \beta_1 = C_1 \cos \theta_1 - u$$
$$w_1 = \frac{C_1 \cos \theta_1 - u}{\cos \beta_1} \tag{4-28}$$

Substitute (4-28) and (4-25) in (4-11).

$$\begin{aligned} P &= \frac{G}{g_c} (C_1 \cos \theta_1 + k_b w_1 \cos \beta_2' - u)u \\ &= \frac{G}{g_c} \left(C_1 \cos \theta_1 + k_b \frac{C_1 \cos \theta_1 - u}{\cos \beta_1} \cos \beta_2' - u \right) u \\ &= \frac{G}{g_c} \left[(C_1 \cos \theta_1 - u) \left(1 + k_b \frac{\cos \beta_2'}{\cos \beta_1} \right) \right] u \end{aligned} \tag{4-29}$$

The efficiency of impulse blading can be defined as the ratio of the delivered power or energy to the power or energy supplied in kinetic f orm, *viz.*

$$\eta_b = \frac{P}{G(C_1^2/2g_c)} \tag{4-30}$$

$$\eta_b = \frac{\dfrac{Gu}{g_c} \left[(C_1 \cos \theta_1 - u) \left(1 + k_b \dfrac{\cos \beta_2'}{\cos \beta_1} \right) \right]}{G(C_1^2/2g_c)}$$

$$\eta_b = \frac{2u}{C_1^2} \left[(C_1 \cos \theta_1 - u) \left(1 + k_b \frac{\cos \beta_2'}{\cos \beta_1} \right) \right] \tag{4-31}$$

Use the symbol σ for the blade-gas velocity ratio u/C_1, and substitute in Eq. (4-31).

$$\eta_b = 2\sigma^2 \left[\left(\frac{\cos \theta_1}{\sigma} - 1 \right) \left(1 + k_b \frac{\cos \beta_2'}{\cos \beta_1} \right) \right]$$

$$\eta_b = 2 \left[(\sigma \cos \theta_1 - \sigma^2) \left(1 + k_b \frac{\cos \beta_2'}{\cos \beta_1} \right) \right] \tag{4-32}$$

To find the blade-gas speed ratio for which η_b has its maximum value, differentiate Eq. (4-32) with respect to σ, and equate to zero.

$$\frac{d\eta_b}{d\sigma} = 2 \left[(\cos \theta_1 - 2\sigma) \left(1 + k_b \frac{\cos \beta_2'}{\cos \beta_1} \right) \right] = 0 \tag{4-33}$$

It is obvious that the factor $\cos \theta_1 - 2\sigma$ equals zero, and the efficiency is a maximum when

$$\sigma = \frac{\cos \theta_1}{2} \tag{4-34}$$

Substituting this in Eq. (4-32) gives

$$\eta_{b\,\max} = 2\left[\left(\frac{\cos^2\theta_1}{2} - \frac{\cos^2\theta_1}{4}\right)\left(1 + k_b\frac{\cos\beta_2'}{\cos\beta_1}\right)\right]$$

$$\eta_{b\,\max} = \frac{\cos^2\theta_1}{2}\left(1 + k_b\frac{\cos\beta_2'}{\cos\beta_1}\right) \tag{4-35}$$

Observation of Eq. (4-35) shows that diagram blade efficiency depends directly on the nozzle angle and increases as θ_1 approaches zero. In particular β_2' is not usually greatly different from β_1, and when $\beta_2' = \beta_1$, Eq. (4-35) reduces to the form

$$\eta_{b\,\max} = (1 + k_b)\frac{\cos^2\theta_1}{2} \tag{4-36}$$

Example 1. An impulse gas-turbine stage operates with a blade speed at the mean position of 646 fps, and gases enter the blades at $C_1 = 1556$ fps at a nozzle angle of 20°. The exit blade angle β_2' is 33.1°, and the blade friction coefficient k_b is 0.88. (*a*) Compute values for and draw the velocity triangles. (*b*) Compute the velocity of whirl. (*c*) Find the power produced by each pound per second of gas passing through the blade at the mean position. (*d*) Find the approximate rotor power for the blade row if 20 lb per sec flows through the whole blade passage. (*e*) Check the answer to (*c*) by the kinetic-energy form of blade equation. (*f*) What is the blade speed ratio for this turbine?

Solution. (*a*) Figure 4-9 can be used to illustrate this example, and it can be seen that

(1) $$C_1 \cos\theta_1 = w_1 \cos\beta_1 + u$$
(2) $$C_1 \sin\theta_1 = w_1 \sin\beta_1$$

Substituting in (1) and (2) with cos 20° = 0.9397 and sin 20° = 0.3420,

(1) $$(1556)(0.9397) = w_1\cos\beta_1 + 646 \qquad 1462 - 646 = w_1\cos\beta_1$$
(2) $$(1556)(0.342) = w_1\sin\beta_1 \qquad 532 = w_1\sin\beta_1$$

Dividing (2) by (1),

$$\tan\beta_1 = 0.6519 \qquad \beta_1 = 33.1°$$
$$w_1 = 974 \text{ fps}$$
$$w_2 = (974)(k_b) = (974)(0.88) = 857 \text{ fps}$$
$$w_2\cos\beta_2' = (857)(0.8377) = 718 \text{ fps}$$

C_2 is found by further calculation as 474 fps. Use of pertinent data from the preceding will produce velocity triangles closely resembling Fig. 4-9.

(*b*) The velocity of whirl ΔV can most readily be found from the preceding data by use of Eq. (4-3), which is, of course, the exact equivalent of Eq. (4-1).

$$\Delta V = V_w = w_1\cos\beta_1 - w_2\cos\beta_2 = w_1\cos\beta_1 + w_2\cos\beta_2'$$
$$= (974)(0.8377) + (857)(0.8377) = 1534 \text{ fps}$$

(*c*) By Eq. (4-11)

$$P = \frac{1}{32.17}(1534)(646) = 30{,}800 \text{ ft-lb per sec}$$

(*d*) If approximately the same power is produced at all positions on the blade from root to tip,

$$P = (20)(30{,}800) \text{ ft-lb per sec}$$

$$\text{hp} = \frac{(20)(30{,}800)}{550} = 1120 \text{ hp}$$

(*e*) Use Eq. (4-23) for 1 lb per sec flow.

$$P = \frac{1}{(2)(32.17)}[(1556^2 - 474^2) + (646^2 - 646^2) + (857^2 - 974^2)]$$
$$= 30{,}800 \text{ ft-lb per sec}$$

(*f*) Blade speed ratio $\sigma = \frac{u}{C_1} = {}^{646}\!/_{1556} = 0.415$

4-6. Blade Speed Ratio of Velocity-compounded Turbine. The velocity-compounded (Curtis) turbine was described earlier in this chap-

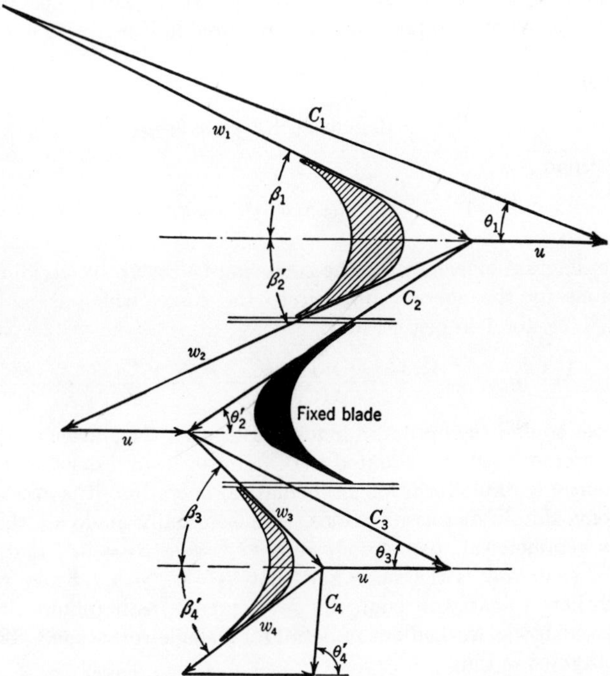

FIG. 4-11. Velocity diagram for a two-row velocity-compounded wheel considering friction (Curtis turbine).

ter. Here an analysis will be made of the velocity relationships associated with a turbine of this type. Reference to Fig. 4-11 shows the gas with absolute velocity C_1 entering the first moving row of blades. As this is an impulse turbine, theoretically there is no pressure drop in the blading and the energy transfer to the rotor is associated solely with velocity decrease. Because of gas friction, turbulence, and shock the entry relative velocity w_1 in passing through the blades is dissipated to some extent, and w_2, the leaving velocity, is less than w_1. Following the notation previously used, $w_2 = k_{b1}w_1$. In the next blade row, which is fixed to the casing and not in motion, a similar dissipation of velocity occurs because of frictional effects, and $C_3 = k_{b2}C_2$. On leaving the fixed blade row the gas is redirected into the second row of moving blades, and for this row $w_4 = k_{b3}w_3$. The representative values of k_b, given in the previous section, are also applicable here. However, for general use with a two-moving-row stage, the following values are suggested: $k_{b1} = 0.88$, $k_{b2} = 0.90$, and $k_{b3} = 0.92$. Power is developed in both of the moving rows, and it is possible to write the power equation (for 1 lb per sec flow) following the equational relationships presented in Eqs. (4-10) and (4-29), giving:

For first row,

$$P = \frac{1}{g_c}(C_1 \cos\theta_1 + C_2 \cos\theta_2')u_1 \tag{4-37}$$

For second row,

$$P = \frac{1}{g_c}(C_3 \cos\theta_3 + C_4 \cos\theta_4')u_2 \tag{4-38}$$

Blade-diagram efficiency for the stage can be set up by dividing these expressions by the energy supplied to the stage, which is, of course, $(1 \times C_1^2)/2g_c$ for 1 lb per sec flow.

$$\eta_b = \frac{(C_1 \cos\theta_1 + C_2 \cos\theta_2')u_1 + (C_3 \cos\theta_3 + C_4 \cos\theta_4')u_2}{g_c(C_1^2/2g_c)} \tag{4-39}$$

In order to find the optimum blade speed ratio, the efficiency equation can be differentiated and equated to 0. However, in order to get this in conveniently suitable form for mathematical operation, it is necessary to make some simplifying assumptions. Those usually made are that each blade is symmetrical, namely: $\beta_1 = \beta_2'$, $\theta_2' = \theta_3$, $\beta_3 = \beta_4'$; that the k_b value for each row is the same and that $u_2 = u_1 = u$. Space will not be taken here to carry through the development, as it follows the same method which was worked out in detail for a single-row stage. The final result obtained is that

$$\sigma = \frac{K \cos\theta_1}{4} \tag{4-40}$$

where K is related to the appropriate values of k_b and in the limit has a value of 1, even for poorly designed blading seldom dropping below 0.95. Thus

$$u = KC_1 \frac{\cos\theta_1}{4} \tag{4-41}$$

Considering K to be unity and comparing with Eq. (4-34), it is apparent that the blade speed of a velocity-compounded wheel thus need be only half as great as that of a single-stage impulse wheel. This makes it possible to absorb large amounts of kinetic energy in a small, compact unit which can operate at a reasonable speed.

One velocity-compounded stage can easily replace two or more pressure (Rateau) stages and three to five reaction stages of a turbine. Unfortunately, the fluid frictional effects in a velocity-compounded unit are high because of the high velocities, and shock losses, when these occur, are of much greater magnitude. Consequently, the Curtis (or the velocity-compounded) unit is not inherently as efficient as a multistage impulse or

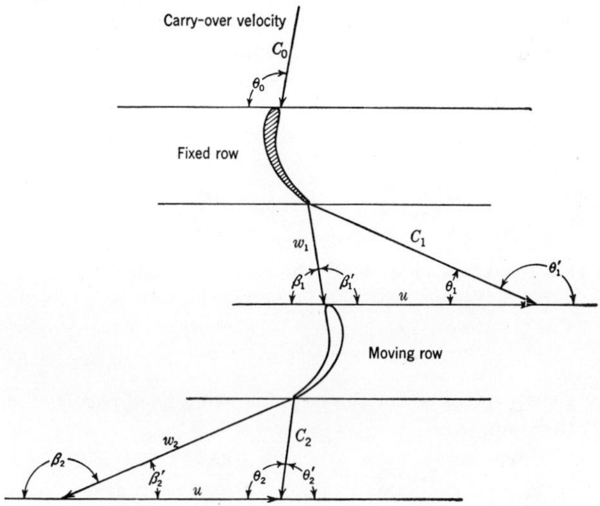

FIG. 4-12. Velocity triangles for a reaction turbine.

a multistage reaction unit. Thought will show that with the increasingly lower velocities as the gas moves through the three blade rows the cross-sectional area of the blade passages must increase. This is usually accomplished by making the second moving row larger than the first moving row either by greater opening between blades or by giving greater length to the blades or by both methods. In steam-turbine practice it is very common to find turbines which use a Curtis stage in the high-pressure section, followed by a number of Rateau stages or by a number of reaction stages.

4-7. Reaction-turbine Blade Speed Ratio and Characteristics. Consider the flow path through the blading of a reaction-type turbine. Figure 4-12 shows the first fixed row of blades, which act as guides to direct the gas into the moving row and also serve as expansion nozzles. In the moving row the gas (air) increases from an inlet velocity of w_1, relative to the moving blade, to its final value of w_2. In entering the blade some shock loss, even though small, is unavoidable, and rubbing friction also is present as the gas moves through the passage. In the impulse-turbine analysis, use was made of the factor $1 - k_b$ to indicate a fraction of inlet relative velocity dissipated by friction and turbulence in passage of the gas through the blading. Here use will also be made of an expression involving k_b, and

$$\frac{(1 - k_b)^2 w_1^2}{2g_c J} \tag{4-42}$$

then represents the fraction of outlet kinetic energy from the previous row dissipated by turbulence and friction. The moving blades also act as nozzles, and their efficiency as expansion devices usually ranges between $\eta_n = 0.88$ to 0.92, say 0.9, for a representative value. The isentropic enthalpy change converted into useful kinetic energy thus appears as

$$\eta_n(\Delta h_s)_m = \frac{w_2^2}{2g_c J} - \frac{(k_b w_1)^2}{2g_c J} \tag{4-43}$$

Similarly in a fixed row (Fig. 4-12) the relation would be

$$\eta_n(\Delta h_s)_f = \frac{C_1^2}{2g_c J} - \frac{(k_b C_o)^2}{2g_c J} \tag{4-44}$$

and writing the expression for the next following fixed row (not shown in Fig. 4-12) there appears

$$\eta_n(\Delta h_s)_f = \frac{C_3^2}{2g_c J} - \frac{(k_b C_2)^2}{2g_c J} \tag{4-45}$$

Values of k_b for reaction blading range between 0.7 and 0.8.

It is convenient to express blade speed ratio in two forms, *viz.*,

$$\sigma = \frac{u}{C_1} \text{ or } \frac{u}{C_3}, \text{ etc.} \tag{4-46}$$

and

$$\rho = \frac{u}{w_2} \text{ or } \frac{u}{w_4}, \text{ etc.} \tag{4-47}$$

Refer to Fig. 4-12, and solve for w_1 by use of Eq. (4-18).

$$w_1^2 = C_1^2 + u^2 - 2C_1 u \cos\theta_1$$

$$w_1^2 = C_1^2\left(1 + \frac{u^2}{C_1^2} - 2\frac{u}{C_1}\cos\theta_1\right)$$

$$w_1^2 = C_1^2(1 + \sigma^2 - 2\sigma\cos\theta_1) \tag{4-48}$$

Also for the moving-row exit triangle

$$C_2^2 = w_2^2 + u^2 - 2w_2u \cos \beta_2'$$

$$C_2^2 = w_2^2 \left(1 + \frac{u^2}{w_2^2} - 2\frac{u}{w_2} \cos \beta_2'\right)$$

$$C_2^2 = w_2^2(1 + \rho^2 - 2\rho \cos \beta_2') \tag{4-49}$$

Substituting Eq. (4-48) in Eq. (4-43) and Eq. (4-49) in Eq. (4-45), there results

$$\eta_n(\Delta h_s)_m = \frac{w_2^2}{2g_cJ} - \frac{k_b^2C_1^2(1 + \sigma^2 - 2\sigma \cos \theta_1)}{2g_cJ}$$

$$(\Delta h_s)_m = \frac{w_2^2 - (k_bC_1)^2(1 + \sigma^2 - 2\sigma \cos \theta_1)}{\eta_n 2g_cJ} \tag{4-50}$$

$$(\Delta h_s)_f = \frac{C_3^2 - (k_bw_2)^2(1 + \rho^2 - 2\rho \cos \beta_2')}{\eta_n 2g_cJ} \tag{4-51}$$

The total isentropic change in enthalpy throughout a stage consisting of one fixed and one moving row would be

$$(\Delta h_s)_m + (\Delta h_s)_f = \Delta h_s$$

In symmetric stages the enthalpy drop is made equal in the fixed and moving rows; $\theta_1 = \beta_2'$, $w_2 = C_1 = C_3$, $w_1 = C_o = C_2$, etc., and consequently $\sigma = \rho$. For symmetric staging utilizing the above information and adding Eqs. (4-50) and (4-51), there results

$$(\Delta h_s)_m + (\Delta h_s)_f = 2\frac{C_1^2 - (k_bC_1)^2(1 + \sigma^2 - 2\sigma \cos \theta_1)}{\eta_n 2g_cJ}$$

$$\Delta h_s = \frac{2C_1^2[1 - k_b^2(1 + \sigma^2 - 2\sigma \cos \theta_1)]}{\eta_n 2g_cJ} \tag{4-52}$$

Degree of reaction is defined as the ratio of the enthalpy drop in the moving row to the enthalpy drop across the whole stage, and a symmetric stage is thus a 50 per cent reaction stage.

Article 3-5 discussed turbine stage efficiency and indicated that it was equal to the work delivered to the rotor per unit of gas flowing divided by the isentropic enthalpy drop available in each unit of gas flowing. Technically the denominator of this expression should also receive credit for the kinetic energy supplied to the stage from a preceding stage. However, for the center stages in a multistage turbine each stage passes on as much kinetic energy as it receives, and the kinetic-energy term in the denominator if disregarded will not critically change the expressed stage-efficiency value.

Blade-diagram calculations are based on the gas which actually passes through the blades and contributes to useful power. However, some gas, because of tip (or clearance) leakage, does not contribute its share of power; also disk friction, fanning loss at partial admission, and unaccounted-for turbulence conditions necessarily reduce the rotor power developed for transmission through the glands to the external shafting. When blade-diagram calculations are used to find developed power and compared with the isentropic enthalpy drop of the stage, the resulting diagram stage efficiency η_{SD} is somewhat higher than stage efficiency.

Set up an expression for diagram stage efficiency by making use of Eq. (4-10) and Eq. (4-52) each for 1 lb per sec flow; thus

$$\eta_{SD} = \frac{(C_1 \cos \theta_1 + C_2 \cos \theta_2')u(2g_cJ)\eta_n}{Jg_c2C_1^2[1 - k_b^2(1 + \sigma^2 - 2\sigma \cos \theta_1)]}$$

For the 50 per cent reaction stage, this becomes

$$\eta_{SD} = \frac{(C_1 \cos \theta_1 + C_1 \cos \theta_1 - u)u\eta_n}{C_1^2[1 - k_b^2(1 + \sigma^2 - 2\sigma \cos \theta_1)]}$$

Substitute $u = \sigma C_1$, and simplify to

$$\eta_{SD} = \frac{(2\sigma \cos \theta_1 - \sigma^2)\eta_n}{1 - k_b^2(1 + \sigma^2 - 2\sigma \cos \theta_1)} \tag{4-53}$$

The conditions under which maximum diagram stage efficiency holds would normally be found by differentiating η_{SD} with respect to σ, considering η_n and k_b as constants, and equating the result to zero. However, rearrangement of Eq. (4-53) as follows shows the same variable term appearing in the numerator and denominator so that when the numerator is a maximum it follows that the whole term will be a maximum:

$$\eta_{SD} = \frac{\eta_n(2\sigma \cos \theta_1 - \sigma^2)}{(1 - k_b^2) + k_b^2(2\sigma \cos \theta_1 - \sigma^2)} \tag{4-53a}$$

Differentiating the numerator and equating to zero,

$$2 \cos \theta_1 - 2\sigma = 0$$

$$\cos \theta_1 = \sigma \tag{4-54}$$

Equation (4-54) and its equivalent form

$$u = C_1 \cos \theta_1 \tag{4-55}$$

show the optimum blade speed ratio for the 50 per cent reaction turbine. For θ_1 equal to 18° the ratio is 0.95. Reference to Fig. 4-13 further indicates that the efficiency curve is relatively flat and satisfactory operation can be had with σ ranging from 0.6 to 1.2 or more. It will later be shown that as a result of this, relatively small enthalpy drops per stage are employed in reaction-turbine design. Enthalpy-change values might be expected to range from 4 to 15 Btu per stage row per lb of gas flowing.

Mention has been made of the energy supplied by carry-over velocity into a stage. In the last stage no use can be made for its leaving velocity, and so diagram stage efficiency for this stage should charge its entering carry-over velocity showing on this basis (see Fig. 4-12):

$$\eta_{SD} = \frac{(C_1 \cos \theta_1 + C_2 \cos \theta_2')u}{g_c[J \, \Delta h_s + (C_o^2/2g_c)]} \tag{4-56}$$

Or if the carry-over velocity supply is disregarded, the equation corresponding to Eq. (4-53) would appear as

$$\eta_{SD} = \frac{(C_1 \cos \theta_1 + C_2 \cos \theta_2')u}{g_c J \, \Delta h_s} \tag{4-57}$$

The isentropic enthalpy drop Δh_s which appears in some of the preceding equations can be found as a first approximation by dividing the total enthalpy drop in the turbine by the number of stages. This value is then prorated between the fixed and moving rows as might be desired. For the symmetric stage, the distribution of enthalpy between the fixed and the moving rows is made equal. Because the approach velocity to the first stage of the reaction turbine is frequently less than the carry-over velocity from stage to stage in the turbine, it may be necessary to allow a greater enthalpy drop for the fixed row of the first stage.

An additional correction should be made in all stages, except the first, for the reheat carry-over. Technically, it is not correct to multiply the enthalpy drop of succeeding stages by the reheat factor of the whole turbine. However, if a correction factor for reheat is employed in all but the first stage, reasonable agreement with experience can be expected. It is suggested that a reheat correction factor of $R' = 1.015$ be employed. This would then be used in equations of the type of (4-43) and (4-45), for all stages except the first, yielding

$$\eta_n R'(\Delta h_s)_m = \frac{w_2^2 - (k_b w_1)^2}{2g_c J} \tag{4-58}$$

$$\eta_n R'(\Delta h_s)_f = \frac{C_3^2 - (k_b C_2)^2}{2g_c J} \tag{4-59}$$

On this basis, Eq. (4-57) would appear as

$$\eta_{SD} = \frac{(C_1 \cos \theta_1 + C_2 \cos \theta_2')u}{g_c J(R' \, \Delta h_s)} \tag{4-60}$$

For the reaction blading equations of Art. 4-7,

C_o = absolute velocity carry-over into a fixed stage row, fps
C_1, C_3, etc. = absolute inlet velocities to moving rows, fps
C_2, C_4, etc. = absolute velocities from moving rows, fps
u = blade speed, fps

R' = reheat correction factor for the turbine stage; 1.015 may be used as a representative value
$(\Delta h)_s$ = isentropic enthalpy drop for the stage, Btu per lb
$(\Delta h_s)_m$ = same for the moving row only
$(\Delta h_s)_f$ = same for the fixed row only
g_c = 32.17
J = 778 ft-lb per Btu
θ_1, β_1, etc. = angles as indicated in Fig. 4-12
$\sigma = u/C_1$
$\rho = u/w_2$
w_1, w_2 = relative gas velocities, respectively, into and out of moving blades, fps
η_{SD} = stage diagram efficiency

Example 2. A four-stage reaction turbine receives gases, considered to be air, at 200 fps, 45 psia, 1600 R and exhausts at 15 psia. Consider that one-fourth of the isentropic enthalpy drop is allocated to the first stage and divided equally between the fixed and moving row. Consider blade-nozzle efficiency as 0.9, and take $k_b = 0.8$. Set the fixed-blade outlet angle θ_1 as 22°, and make $\beta_2' = 22°$. Set $\sigma = 0.87$ for the first-stage row. (*a*) Compute necessary data to lay out the velocity triangles. (*b*) Find the velocity of whirl and the stage power for 1 lb per sec flow. (*c*) Find stage efficiency. (*d*) Find the conditions entering the next stage.

Solution. (*a*) $r_e = 45/15 = 3$. From Table 2-4, at 1600 R read $h_1 = 286.100$ and $p_{r1} = 90.990$.

$$p_{re} = \frac{p_{r1}}{r_e} = \frac{90.990}{3} = 30.330$$

For this value of p_r,

$$h_{es} = 181.280$$

Total $(\Delta h)_s = 286.100 - 181.280 = 104.820$
For the first stage

$$(\Delta h_s)_f + (\Delta h_s)_m = \frac{104.820}{4} = 26.205$$

$$(\Delta h_s)_f = 13.103 = (\Delta h_s)_m$$

Using Eq. (4-44),

$$(0.9)(13.103) = \frac{C_1^2}{50{,}070} - \frac{[(0.8)(200)]^2}{50{,}070}$$

$$C_1 = 768 \text{ fps}$$

$$\sigma = \frac{u}{C_1} = 0.87 = \frac{u}{768} \qquad u = 668 \text{ fps}$$

The inlet velocity triangle is similar to Fig. 4-12, and by computation or graphically w_1 and β_1 can be found.

$$C_1 \cos \theta_1 = u + w_1 \cos \beta_1$$
$$C_1 \sin \theta_1 = w_1 \sin \beta_1$$
$$(768)(0.9272) = 668 + w_1 \cos \beta_1$$
$$(768)(0.3746) = w_1 \sin \beta_1$$
$$\beta_1 = 81.3° \qquad w_1 = 291 \text{ fps}$$

By Eq. (4-43)

$$(0.9)(13.103) = \frac{w_2^2}{50{,}070} - \frac{[(0.8)(291)]^2}{50{,}070}$$
$$w_2 = 803 \text{ fps}$$

By construction or by calculation for the second velocity triangle it can be found that $C_2 = 310.4$ fps. The triangles for this problem closely resemble Fig. 4-12 but are not exactly symmetric as the velocity C_o into the fixed row differs from the velocity w_1 into the moving row. Consequently the triangles are different even with the same enthalpy releases for both rows.

(*b*)
$$\Delta V = w_1 \cos \beta_1 + w_2 \cos \beta_2'$$
$$= (291)(0.1513) + (803)(0.9272) = 788.5 \text{ fps}$$

By Eq. (4-11)

$$P_1 = \frac{1}{32.17}(788.5)(668) = 16{,}400 \text{ ft-lb/(sec)(lb)/(sec) flow}$$

$$\text{Work} = 16{,}400 \text{ ft-lb or } \frac{16{,}400}{778} = 21.08 \text{ Btu per lb of gas}$$

(*c*) Stage efficiency based on velocity-diagram values and actual useful output.

$$\eta_s' = \frac{\text{work output}}{\text{input}} = \frac{21.08}{26.205 + (200^2/2g_cJ)}$$
$$= 0.78 = 78.0\,\%$$

Although this is a true efficiency as far as the actual work output of the stage is concerned, it does not present a true picture as the leaving kinetic energy from this stage carries to the next and represents a mechanical energy term associated with the stage output even though it is not work in a true sense. When this energy also is considered,

$$\eta_s = \frac{21.08 + (C_2^2/50{,}070)}{26.205 + (200^2/50{,}070)} = \frac{21.08 + 1.924}{26.205 + 0.799} = 85.2\,\%$$

Note that Eq. (4-53) is not applicable for this problem as $\sigma \neq \rho$; $u/C_1 \neq u/w_2$.

(*d*)
$$h_1 + \frac{C_o^2}{50{,}070} - \text{work out} = h_{\text{leaving}} + \frac{C_2^2}{50{,}070}$$
$$286.100 + 0.799 - 21.08 = h_{\text{leaving}} + 1.924$$
$$h_{\text{leaving}} = 263.895 \text{ Btu per lb}$$

The corresponding temperature is 1516.6 R. The exit pressure from the stage can closely be found by use of Eq. (3-20) and the energy stage efficiency.

$$\eta_s = \frac{1 - (h_2/h_o)}{1 - (h_{2s}/h_o)} = \frac{1 - 263.895/286.100}{1 - (h_{2s}/286.100)} \cong 0.852$$
$$h_{2s} = 260.039$$

Corresponding to h_{2s} read $p_{r2} = 71.218$ in Table 2-4 and as $p_{r1} = 90.990$ at inlet to the stage

$$\frac{p_1}{p_2} = \frac{p_{r1}}{p_{r2}} \qquad \frac{45}{p_2} = \frac{90.990}{71.218}$$
$$p_2 = 35.2 \text{ psia exit pressure from stage}$$

4-8. Comparison of Turbine Types. Figure 4-13 shows the speed-ratio curves for the three types of turbines which have been considered in this chapter. At the left, employing the lowest blade speed ratio, appears the velocity-compounded turbine (Curtis stage) with its peak efficiency value occurring at σ = 29 per cent for the particular design. Next appears the graph for a Rateau or pressure-compounded stage peaking at σ equal to 49 to 50 per cent. The efficiency curve for the reaction stage is appreciably flatter in its maximum range and peaks between 0.85 and 0.95. These graphs represent test data on regular turbine components and include the losses which are associated with a particular type of turbine. The peak ranges for these types are as represented in Fig. 4-13.

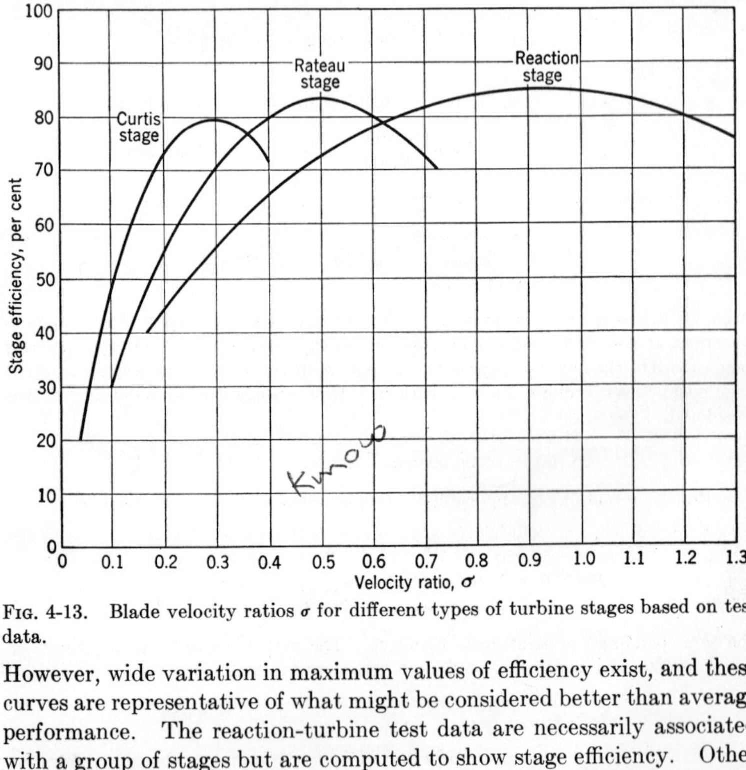

FIG. 4-13. Blade velocity ratios σ for different types of turbine stages based on test data.

However, wide variation in maximum values of efficiency exist, and these curves are representative of what might be considered better than average performance. The reaction-turbine test data are necessarily associated with a group of stages but are computed to show stage efficiency. Other items than the theoretical value of blade speed ratio enter the turbine designer's choice so that the actual blade speed ratio may deviate slightly from the previously indicated best theoretical ratios.

A study of Fig. 4-14 indicates for impulse blading the effect of operating at velocity ratios other than the ratio for which the design was made. Design speed conditions are indicated for an inlet absolute velocity C_1, nozzle angle θ_1, blade inlet angle β_1, and relative velocity w_1. Leaving design conditions here show the relative outlet velocity w_2 (taken as

$0.9w_1$) with the absolute velocity C_2, resulting from this relative velocity, indicated. If the geometry of the blade and the blade speed both remain unchanged while the jet inlet velocity is reduced to 70 per cent of C_1, it can be seen that the inlet jet velocity $w_{0.7}$ impinges on the back of the blade, causing frictional shock and unduly turbulent flow through the passage. If the jet enters at 130 per cent of design velocity, some choking of inlet flow occurs and the jet is impacted on the face of the blade as shown by the direction of $w_{1.3}$. Leaving conditions are indicated in the diagram on the assumption that the relative velocity on outlet follows the blade passage contour and angle under all conditions. Lowering the

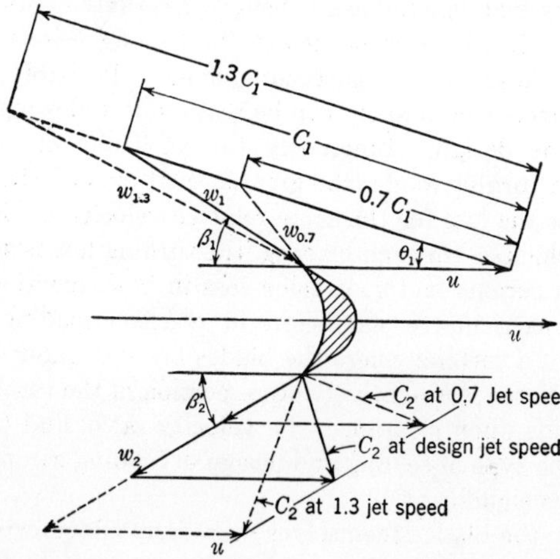

FIG. 4-14. Effect on blade performance of operation at jet velocities varying from design conditions.

velocity ratio σ (as indicated with the jet at $1.3C_1$) does not cause as high a percentage leaving loss as when the value of σ increases (jet at $0.7C_1$). Relatively the same unsatisfactory conditions would result if the jet speed C_1 were constant and the blade speed u changed.

Diagrams of a similar nature apply for reaction blading; however, the effects cannot as readily be shown on a diagram such as Fig. 4-14 because of expansion and the resulting increase of relative velocity in the moving row. The rounded nose of the reaction blading reduces somewhat the turbulence effects occurring when the jet enters at other than design velocity conditions.

The efficiency of reaction-type blading is less sensitive to variations in design velocity ratio than the impulse-type blading. However, the efficiency of all designs decreases if operation occurs at a point radically different from the design-ratio value. Lowering or increasing the velocity ratio causes a flow disturbance at inlet, in some cases causing the jet to impinge upon the back of the blade, affecting the diffusion characteristics in the blade passage, changing the swirl conditions from the blade row, and nearly always contributing to an increase in blade leaving loss in addition to increased frictional effects. It should be realized that a design could be made with the angles and passages so arranged that the blade speed ratio differed greatly from that associated with normal maximum efficiencies.

Leaving velocity loss may or may not be a serious factor, as it is possible in multistage turbines to recover a portion of this energy in succeeding stages, and very high leaving loss is usually associated only with the last, or final, stage. In the reaction design the leaving velocity is integrally woven into the design of each succeeding stage. Probably 75 per cent or better of the carry-over velocity can be realized in following stages in any effective turbine design. Inherently the efficiency of any blading is associated with turning angle; the greater the turn or swirl for any blading, the greater the loss for the same relative velocities. However, with low relative velocities through blading, the turning loss is small.

Leakage is a serious factor, causing loss in both reaction and Rateau stages. It is particularly significant in reaction blading in the high-pressure part of a turbine where the blades are short and the percentage of tip clearance area is a relatively large portion of the total flow passage. Leakage depends upon clearance, the velocity ratio, and the stage pressure ratio. The type of sealing and design of blading can reduce the loss, but it is always significant.

The losses in the blades themselves arise from effects which have been named secondary flow, profile friction, and the leakage previously discussed. Secondary flow occurs when a flowing fluid in a duct or blade passage undergoes a turning action along a wall section. Centrifugal effects tend to build up to a maximum value near the center of the outermost wall, while a lower pressure exists at the confining sides of the duct or passage where boundary-layer friction retards the flow in that vicinity. The greater pressure at the center of the curved wall causes flow outward to the confining side sections (at top and bottom of Fig. 4-15). The turbulence and eddies which result from such flow patterns dissipate some of the directed kinetic energy. The greater the width of the duct passage or, specifically, the length of a blade, the less is the secondary flow a significant factor.

Profile loss is associated with velocity ratio, pitch-width ratio, section geometry, and the Mach number. Many data have been obtained on isolated airfoil (blade) sections, and it has been found desirable to form blading so that it resembles an airfoil section. However, blading does not operate as a group of single airfoils, and consequently airfoil test data cannot well be used to predict or indicate the best turbine-blade form. Cascade tests in which a group of blades are tested are more significant in predicting the performance of a blade group. Airfoil tests usually determine the coefficient of lift C_L and the coefficient of drag C_D, and although these cannot be used directly in blade-group testing, they are of significance in determining some performance characteristics. A blade form with high drag, for example, would be less satisfactory than a similar blade with the same turning angles which had a lower drag characteristic. Most turbine-blade design analyses are carried out on the basis of what has been called channel analysis. That is, it is assumed that the blade

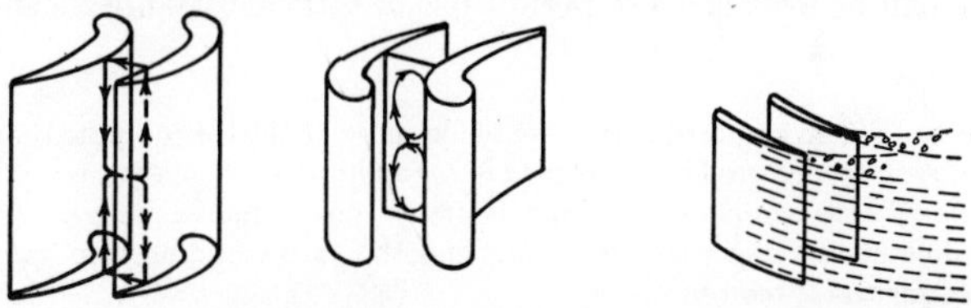

FIG. 4-15. Secondary and profile losses in blading.

passages are relatively so shallow that the stream follows the contour of the passage essentially independent of aerodynamic effects such as might occur if the blades were isolated airfoils.

The enthalpy drop per stage when transformed to velocity appears as

$$\Delta h = \frac{C^2 - C_1^2}{2g_cJ} \tag{4-61}$$

If carry-over velocity from a preceding stage is considered small ($C_1 \rightarrow 0$), Eq. (4-61) becomes

$$C^2 = 2g_cJ\,\Delta h = 50{,}070\,\Delta h \tag{4-62}$$

The blade speed ratio σ for any type of turbine gives the wheel speed u which should be employed to absorb effectively the velocity C. Thus

$$\Delta h = \frac{C^2}{50{,}070} = \left(\frac{u}{\sigma}\right)^2 \frac{1}{50{,}070} \tag{4-63}$$

$$\Delta h = \frac{1}{50{,}070}\left(\frac{\pi DN}{60 \times 12\sigma}\right)^2 = \left[\frac{\pi DN}{(223.8)(720)\sigma}\right]^2 \tag{4-64}$$

where D = turbine wheel diameter, in.
N = wheel (rotor) rpm
σ = ratio of u/C with u and C usually expressed in fps

A study of Eq. (4-64) shows that the enthalpy drop absorbed in each stage per pound of gas flowing varies directly as the square of the wheel diameter and as the square of the rpm and inversely as the square of blade speed ratio. The larger the product $\pi DN/720 = u$, the higher can be the enthalpy absorption per stage. However, the value of u is set by stress considerations and is definitely limited in maximum value. The gas passage through the blading is also related to diameter, as the flow annulus at the periphery of a wheel for a fixed blade length increases in area directly with diameter. Thus on a large-diameter wheel for full gas flow the passage area may be excessively large unless ill-proportioned, short, stubby blades are used. In the case of impulse turbines, particularly in the high-pressure stages, admission around only a portion of the blade ring can be used so as to permit use of satisfactory nozzle and blade heights.

Example 3. For an assumed maximum blade speed of 600 fps compute the approximate enthalpy drop that could be absorbed in (*a*) an impulse stage (Rateau); (*b*) a two-moving-row velocity-compounded stage (Curtis); (*c*) a reaction stage. (*d*) If the maximum diameter of the wheel is 30 in., find the corresponding rpm value which would apply for the three types.

Solution. (*a*) By Fig. 4-13 the maximum value of σ for the Rateau stage is 0.505, and by Eq. (4-63)

$$\Delta h = \left(\frac{600}{0.505}\right)^2 \frac{1}{50{,}070} = 28.2 \text{ Btu per lb gas}$$

(*b*) By Fig. 4-13 for the Curtis stage, $\sigma_{max} = 0.29$ and

$$\Delta h = \left(\frac{600}{0.29}\right)^2 \frac{1}{50{,}070} = 85.5 \text{ Btu per lb gas}$$

(*c*) By Fig. 4-13 for the reaction stage, $\sigma_{max} = 0.85$ and

$$\Delta h = \left(\frac{600}{0.85}\right)^2 \frac{1}{50{,}070} = 9.96 \text{ Btu per lb gas for each row of the stage.}$$

Disregard variations between mean and maximum wheel-diameter values to obtain a representative rpm. Using Eq. (4-64) and the Rateau-stage data,

$$28.2 = \left[\frac{\pi(30)N}{(223.8)(720)\sigma}\right]^2 = \left(\frac{1}{1710}\,\frac{N}{0.505}\right)^2$$

$$N = 4580 \text{ rpm}$$

With a fixed wheel diameter and the same maximum wheel speed, the rpm is the same for all three types.

An analysis based on Eq. (4-64) readily shows that the Curtis stage can effectively utilize the largest enthalpy drop, followed in turn by the Rateau (pressure-compounded) stage and finally by the reaction stage. Expressed in another way, one Curtis stage can do the work of two or three Rateau stages and of five to eight reaction stages. Gas velocities in the blading are higher in the Curtis stage, so that fluid friction losses usually are greater in such a stage. The lowest fluid friction loss is associated with the reaction stage. However, reaction stages have pressure drops across both the fixed and moving rows, and there is gas leakage through the clearance spaces at the tips of both the fixed and moving blades. Where the blades are short as in the high-pressure section of the turbine, this loss can be excessive and it may be desirable to employ a Curtis stage or several pressure stages to absorb energy in the high-pressure region of the machine. Even in a multistage pressure-compounded turbine, some leakage occurs where the diaphragms between stages meet the shaft or rotor element. However, by use of a labyrinth design at this point the leakage between stages can be reduced to a small amount.

For aircraft propulsion one or two Rateau stages have come into common use for generating the total power. For stationary, locomotive, and marine work both Rateau- and reaction-type designs have been employed. Curtis stages have not been used extensively in the gas-turbine field

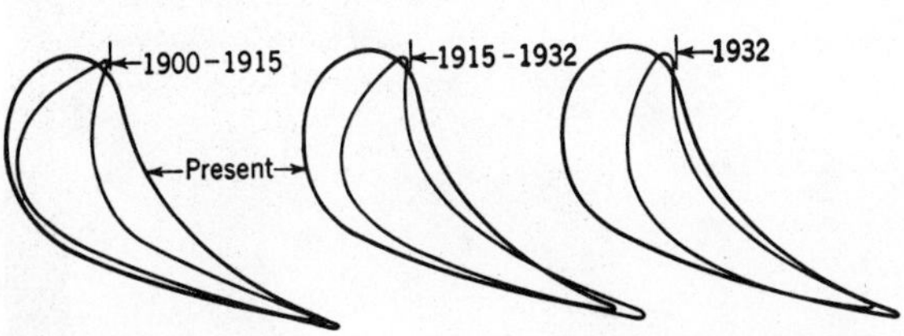

Reaction blade shapes indicating changes which have taken place, primarily with steam. (From *Westinghouse Engr., after Morgan, Ref.* 12.)

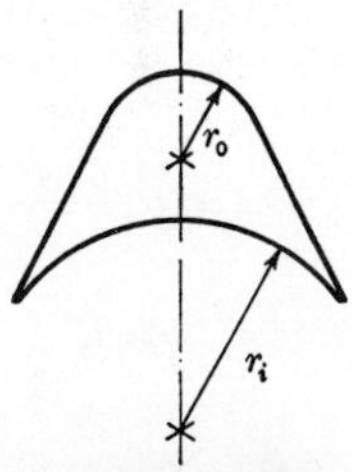

Typical impulse blading form

FIG. 4-16. Turbine blade shapes.

largely because the available enthalpy drops are not nearly so large as in steam-turbine practice. Both Rateau and Curtis stages have one supposed advantage in that with the high-temperature gas expanding through a large pressure ratio, there is an appreciable drop in temperature of the gas stream, which means that on entry to the moving row the gas-stream temperature is not as high as at inlet. However, there is always a certain temperature recovery, as the moving stream in immediate contact with the blade slows down to a low velocity so that the blade temperature is at an intermediate value between the original gas-stream temperature and the actual temperature of the leaving stream of gas. In working with the ultimate design of the turbine, the factors of allowable temperature, available pressure drop, simplicity, cost of construction, and weight requirements all enter into the design picture so that the resultant type is a compromise between these and other objectives.

4-9. Blade Construction. The form of the blade varies through wide limits, depending upon the designer's objectives for the channel section which passes the fluid. Impulse blades, although theoretically without reaction, seldom are made in this manner, as it has been found that a slight choking of the exit passage to create some reaction may improve performance. Impulse blades are relatively deep at the center section from front to back of the blade, and a construction formerly quite commonly used was to form the front and back sections of the blade in the shape of circular arcs, the inner face of the blade being of greater radius than the back face of the blade. The form was completed by drawing tangent lines from the back circle toward the front face of the blade as is shown in Fig. 4-16.

Reaction blading formerly had a sharp leading edge similar to the leading edges still used in many types of impulse blades (Fig. 4-16). However, this design for reaction blading has been superseded to give a

FIG. 4-17. Reaction-turbine blades of Allis-Chalmers 12,240-gross-hp gas turbine (Fig. 8-7).

rounded nose to the leading edge of the blade, which improves the aerodynamic effectiveness of the section and reduces the shock loss which exists when the jet approach angle deviates from design conditions. Actually, this blunt nose may cause a minor drop in efficiency at exact design conditions, but the over-all performance is so much better that its use is almost universal. Note the inlet edges in Fig. 4-17.

Most blades are tapered from hub to tip. This reduces the weight of the peripheral portions of the blade and consequently the hub stresses which must be carried under the high rotative speeds. The center line of the blade section should be located on a radial line to avoid bending stresses in the blading except in cases where the blades are bent forward in the direction of motion to permit the centrifugal bending moment to offset the gas thrust moment. In long blades, it is usually necessary also to twist the blade from hub to tip to compensate for blade-velocity variations and in some cases as well to satisfy radial-pressure-equilibrium conditions. Blade vibration has, in some cases, been the cause of failure, and it is important that the natural frequency of the blade be such that it is outside of the operating frequency range of the turbine and does not coincide with harmonic speeds. In some cases groups of blades or even the whole ring of blades are either shrouded or laced together to give a

FIG. 4-18. Eight-stage 6000-hp Westinghouse reaction turbine for a locomotive.

rigid unit less subject to vibration. Careful planning, based on experience, is required in blade design to reduce the possibility of fatigue failure.

The flow passage between adjacent blades is set by the contour of the blades, by the blade height, and by blade pitch. When the pitch is reduced, surface friction increases because of increased rubbing surface. On the other hand, increased pitch tends to increase the magnitude of secondary flows and may also increase the drag loss at the trailing edge of the blade. It is impossible to set absolute rules as to how to form the passage section. Large turning angles per stage give lower efficiency in general than is true when the turning angles per stage are small. However, if the turning angles are made too small, resulting in less work per stage, the effect of the extra rubbing surface, leakage, etc., in the additional stages more than counteracts the effect of high turning angles per stage. It might again be mentioned that one Curtis stage in the high-pressure section of a turbine can effectively replace several reaction stages and give better over-all performance.

In connection with reaction blading, the term *gaging correction* is used. This means an adjustment of the leaving blade edge by twisting the blade with a tool made for that purpose. After the blades are assembled in the rotor, with ductile blade materials it is possible for the blader to adjust each blade to have the proper flow passage and position in the blade row. *Gaging* is defined as the ratio of the net area for gas flow to the total free

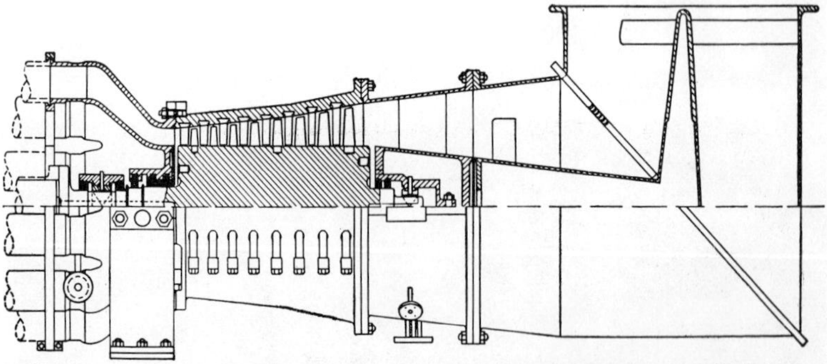

FIG. 4-19. Longitudinal section through the 6000-hp Westinghouse turbine of Fig. 4-18.

annular area in the blade ring. In changing the gaging from 30 to 20 per cent, for example, the blader would twist each blade inward toward the next blade so that the free passage for gas flow would be reduced. It is possible for the blader to increase or decrease the gaging by twisting the blades. This practice, although common with steam-turbine blading, has not been fully established with gas-turbine designs.

Many methods are used in the manufacture of turbine blades. Steam-turbine blades are very frequently made by forging the stock to approximately finished size and then performing a small amount of machining on them. However, with many of the alloys required to withstand the high temperature of the gas turbine, it has been found desirable to develop other methods of fabrication. A discussion of the methods employed is presented in Chap. 12.

The life of blades ranges from a few hundred to thousands of hours, depending on the type of duty and the temperature range employed. For stationary power units which must have long life and reliability, the operating temperatures are sufficiently low (under 1350 F) so that a reasonable life expectancy is possible. With gas turbines and jet units, which necessarily operate at high temperatures for shorter periods of time, the life expectancy is much shorter. Various methods of cooling blades have been proposed and used. One of these which has some promise envisages making the blade hollow and passing cooling air outward through a channel inside the blade.

Corrosion and erosion of blades occur under certain conditions. The greatest difficulty from corrosion apparently takes place when fuel oils high in vanadium are burned in the combustion chamber. Much blade corrosion has been attributed to deleterious effects from this cause. The leading tip of long blades is more prone to erosion than other parts of the blade.

4-10. 6000-horsepower Gas Turbine. Figure 4-18 is a photograph and Fig. 4-19 is a longitudinal section through the 6000-hp reaction tur-

Fig. 4-20. Rotor of four-stage impulse turbine for locomotive drive. (*Elliott Co.*)

bine built by the Westinghouse Corporation for driving the generators of a locomotive. The gas turbine has eight stages, each designed for equal enthalpy drops over the stationary and rotating blades at mean diameter. Shaft seals are of the labyrinth type, and pressure-lubricated sleeve bearings are employed. The rotating blades are tapered and twisted and have a serrated root machined in the base. The stationary blades, which are also tapered and twisted, have a T-type fastening. The blading is unshrouded and is profiled at the tip. The particular blades for this turbine were made of a cobalt-chromium-tungsten alloy, and precision casting was used to form them. The rotor was machined from a solid forging of stabilized 19-9 stainless steel, 14¼ in. in diameter and 24 in. long. The turbine was designed to operate satisfactorily under rapid temperature changes of from 700 F to 1350 F between no load and full load. It will be noted from Fig. 4-19 that the exhaust from the turbine leads into a diffuser designed to recover part of the leaving velocity energy and thereby build up the reduced leaving pressure from the last row to discharge pressure. By use of the diffuser it is thus possible to have the turbine expansion occur through a greater pressure ratio.

The exhaust velocity from the turbine is approximately 500 fps. The design rotative speed of the turbine and its driven compressor is 9200 rpm. The complete unit, which is illustrated in Figs. 8-11 and 8-12, includes a gear reduction to drive a dc generator at 1200 rpm. Of the 6000 hp developed by the turbine, approximately 4000 is used to drive

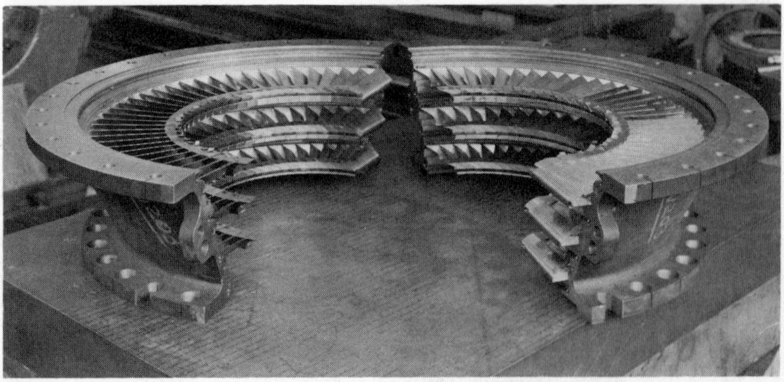

FIG. 4-21. Stator of four-stage locomotive turbine pictured in Fig. 4-20. (*Elliott Co.*)

the compressor, with the remaining 2000 hp constituting the useful output. The pressure supply to the unit is approximately 75 psia. It was thought that this unit could serve as a railroad-locomotive drive or might be suitable as a stationary unit. The compressor is designed to handle 25,000 cfm of free air through a pressure ratio of 5:1. The complete unit is 26 ft long, 3½ ft wide, and 6 ft high and weighs 38,000 lb (19 lb per hp, including auxiliaries).

4-11. Elliott Locomotive Gas Turbine. Figures 4-20 and 4-21 are photographs of the rotor and turbine stator of a gas turbine designed for locomotive drive. The locomotive itself is pictured in Figs. 1-7 and 1-8. As can be seen from the photographs, this particular turbine is a four-stage unit. The first-stage nozzles are attached to the turbine casing adjacent to the combustion chamber and consequently do not appear in the stator section of Fig. 4-21.

This turbine unit is primarily an impulse-type (Rateau-staged) unit with an enthalpy drop of approximately 33 Btu per stage (132 Btu total for the four stages) under design conditions. The maximum diameter of the last moving row is closely 40 in., and at design rpm of 3910, the tip speed is 683 fps. This particular unit under full-load conditions receives air at 68 psia, 1275 F and with an air flow of 67 lb per sec develops 3910 net output hp (approximately 10,500 hp gross).

4-12. Elliott Marine Turbine. The Elliott turbine pictured in Figs. 4-22 and 4-23 is the high-pressure-turbine unit of a marine power plant.

FIG. 4-22. Rotor of 10-stage impulse turbine of marine unit. (*Elliott Co.*)

The maximum diameter of the tenth (last) stage is approximately 30 in. The stator of the turbine shows the nine nozzle rows and diaphragms for the second and succeeding stages of the turbine. The nozzles for the first stage are attached to the combustion-chamber section of the turbine and do not appear in this picture.

Note should be taken in Fig. 4-23 of the diaphragms which, on their inner circumference, are adjacent to the outer periphery of the rotor shaft. The thin metal strips on the diaphragms reduce leakage across each stage without causing danger should the turbine rotor rub against the very light strips. Operating data for this machine at two loads are tabulated below.

Inlet temp, °R	1665	1732
Air flow, lb per sec	16.26	28.60
Rpm	2297	3792
Inlet pressure, psia	46.89	81.00
Pressure drop, psia	17.62	34.60
Hp	990	2170

This particular unit drives the low-pressure compressor of the gas-turbine plant illustrated in Fig. 6-12 and operates at varying speed and load so that the air flow to the other parts of the system can be varied as may appear desirable. A temperature drop of approximately 300° occurs in this particular turbine.

FIG. 4-23. Stator of 10-stage turbine pictured in Fig. 4-22. (*Elliott Co.*)

4-13. De Havilland Aircraft Turbine. Figures 4-24 and 4-25 are photographs illustrating the impulse blading used in the turbine of the Goblin jet-propulsion unit manufactured by the De Havilland Engine Company, Ltd. It can be seen that these blades are both tapered and twisted and use a fir-tree type of mounting in the turbine disk. They are also short relative to the diameter of the turbine disk. This is an advantage, as an increase in disk diameter permits a lower number of rpm for a limiting peripheral blade speed. Also, for a given blade height, as disk diameter increases, the annulus increases in area with increased capacity for gas flow and power. From a stress viewpoint short blades give a lesser loading on their base section than would correspondingly longer blades. The nozzles for this turbine are mounted integrally with the 16 collector tubes from the combustion chambers and are not seen in the illustration. In general layout, the Goblin engine is similar to the Ghost engine illustrated in Fig. 7-13. The turbine disk is cooled by compressed air, which is arranged so as to flow outward along both faces of the disk.

FIG. 4-24. Turbine blades of Goblin jet engine showing fir-tree mounting and edge of turbine disk. (*De Havilland Engine Co., Ltd.*)

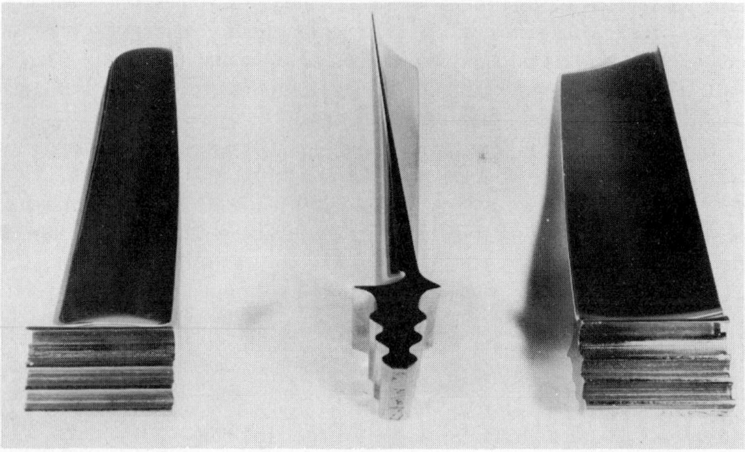

FIG. 4-25. Turbine blades for the Goblin jet engine. (*De Havilland Engine Co., Ltd.*)

This unit produces a maximum static thrust for jet propulsion of 3,500 lb when turning at 10,750 rpm. For cruising, 9,500 rpm is used.

4-14. Turbine Design Procedure. In making the turbine design, the decision must be made as to the type of turbine, whether pressure-compounded, velocity-compounded, or reaction, and a second decision must be made as to whether a single-stage turbine with its probable low efficiency or a more efficient multistage turbine should be considered. In terms of the turbine type it is then possible to select a σ value to assure a suitable blade-diagram efficiency. Figure 4-13 should be consulted in this connection.

It is next necessary to decide on a limiting blade speed. This will normally be set by the maximum allowable stress but may be set to meet a desired rpm of the machine and a fixed diameter or size. Maximum tip speed of the long low-pressure blades cannot greatly exceed 1200 fps, which implies a lower mean blade speed for the high-pressure stage; 600 to 800 fps for a general trial value is suggested.

For the pressure range available, the isentropic enthalpy drop can be found, and when this is modified by a representative turbine efficiency, it is possible to obtain a value of the mass flow rate which the turbine must carry in order to develop the desired power output. With these preliminary decisions made, it is now possible to carry out the design objectives.

The design is usually planned for the mean blade diameter, and then suitable modifications are made to suit flow conditions at the tip and hub.

1. For the enthalpy drop available for the given stages, select the nozzle and blade angles for effective power production under the required conditions of u and σ, usually with an axial velocity also fixed.

2. The mass flow rate should then be computed to a first approximation. This is required for

3. Determining the passage and blade length to pass the required flow rate at the velocities which have been computed.

4. The blade angles at the tip and hub should then be checked to ensure proper flow distribution along the blade section with minimum flow and disturbance.

5. The blade section, particularly at the hub, must be checked to see that it operates under sufficiently low stress. The blade form must be selected to give satisfactory aerodynamic properties and not have undesirable vibration characteristics.

The design pattern for the blade itself is open to wide variation. In short blades much design freedom exists. In long blades, it may be desirable to consider (1) vortex design or (2) constant-reaction design. The following comments apply to any type of blading but are particularly significant for reaction blading, as it is only in reaction blading that appreciable pressure drop occurs in the moving row. In the impulse turbine thermal-energy release occurs exclusively in the nozzles.

*Free vortex design.** For this, the following conditions hold:

1. The value of $\Delta C_u \times r$ = a constant. That is, the product of velocity of whirl at any radius times blade radius is a constant throughout the blade length.

* Refer to Art. 5-6 for demonstration.

2. This condition satisfies radial equilibrium so that there is no radial pressure unbalance at various blade positions.
3. The axial velocity is constant.
4. Work done is the same at each radial location.
5. The velocity triangles are not symmetrical.
6. Inlet Mach numbers are not constant throughout the blade length.

Constant reaction. For this blading design it follows that:

1. The pressure drop across the blade is the same at every radius.
2. Radial equilibrium is not theoretically satisfied.
3. The velocity triangles are symmetrical.
4. The energy released and work done are the same at each radial location.
5. The inlet Mach numbers are essentially constant.

The design of the turbine to fit into the cycle is quite critical, as contrary to conditions in a steam power plant the performance of the turbine must be related directly to the operation of the compressor and wide variations from design conditions are not usually possible. In general, the entire compressor output must pass through the turbine, even though external power requirements may greatly change and make this action undesirable. Pressure ratios of the turbine and compressor are also intimately related, and pressure losses in the system can be very critical in affecting performance. Of the pressure delivered by the compressor, a portion of this is consumed in the ducting to the combustor, in the combustor itself, and finally in the passage to the turbine, so that turbine inlet pressure is 1 to 5 psi lower than compressor discharge pressure. The turbine discharge pressure in addition has to exceed normal exhaust pressure in order to overcome the resistance offered by the exhaust-pipe ducting and, if a regenerator is attached, to overcome the flow resistance of this unit.

Factors reducing turbine expansion ratio thus act both at the inlet and exhaust ends of the turbine. As the power developed by the turbine per unit of gas flow is directly related to expansion ratio, it is thus necessary that the highest efficiency, or best performance possible, for the limited expansion ratio be obtained. Particularly in land and marine power units, the turbine design requires the highest performance and efficiency possible. In the case of an aircraft turbine unit, the residual velocity from the turbine is not wasted, as it can contribute to the jet thrust of the unit. However, in the case of other power plants, the residual velocity should be kept as low as possible consistent with space requirements. Unless the cross sections of the exhaust-gas passages assume large dimensions, the leaving velocity is necessarily high.

In the aircraft, also, the problems of weight and size are of such paramount importance that efficiency may not be the main controlling factor in the turbine design.

Cycle performance is related directly to turbine supply temperature, and this in turn is limited by metallurgical considerations so that any design is necessarily a balance between cycle efficiency and metallurgical life of the operating parts exposed to the high temperatures. Aircraft turbines are not expected to have the long life desired for a stationary unit and so frequently operate at very high temperatures and employ a single-stage impulse-turbine unit. On the other hand, stationary, locomotive, and marine units would employ more conservative designs and use more efficient turbines, with most designs consisting of either multistage reaction or Rateau-type-staged impulse units.

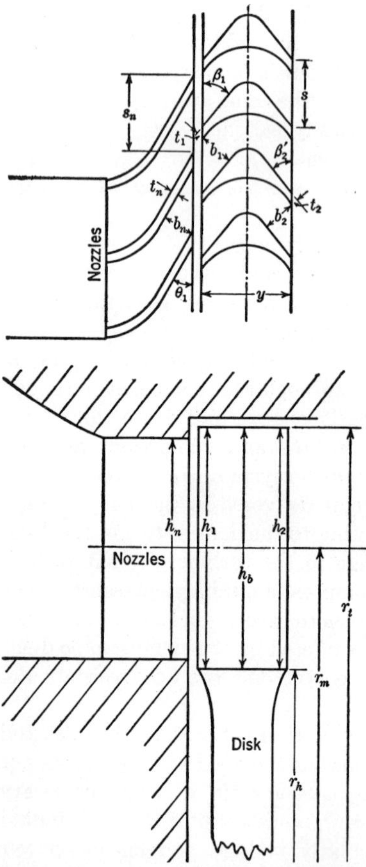

FIG. 4-26. Representation of entry nozzles and blading (buckets) of impulse turbine.

4-15. Turbine Flow Passages. Although influenced by other factors, the power developed by a turbine is directly proportional to the mass rate of gas flow (usually expressed in pounds per second). The continuity-of-mass flow equation [see Eq. (2-53)] must hold both in the nozzles and blading, while at the same time the velocity patterns which directly contribute to the power production have to be maintained throughout the blading.

Full admission is the terminology employed when the nozzles subtend the whole blade circumference (annulus area). However, particularly with impulse turbines, *partial-admission* designs can be employed, with the nozzles covering only a fraction of the inlet blade circumferential annulus. Partial admission is seldom used after the first stage in multistage turbines.

Figure 4-26 indicates the layout and nomenclature applicable to a representative stage of an impulse (Rateau-type) turbine. The nozzle blocks are customarily cast or welded, with the nozzle having a rectangular-shaped passage. Thus for a velocity of efflux of C_1 fps from a nozzle and a specific volume of the gas at the nozzle of v_1 ft³ per lb, the mass rate of flow in pound per second per nozzle appears as

$$G_n = \frac{A_n C_1}{v_1} = \frac{h_n b_n C_1}{144 v_1} \tag{4-65}$$

Here h_n and b_n are expressed in inches and A_n is in square feet. In terms of pitch s_n or the spacing between successive nozzles measured on the mean radius line of the nozzles, Eq. (4-65) becomes

$$G_n = \frac{h_n(s_n \sin \theta_1 - t_n)C_1}{144 v_1} \tag{4-66}$$

It is convenient to define a coefficient m_n to account for reduction in flow area because of wall thickness, where

$$m_n = \frac{s_n \sin \theta_1 - t_n}{s_n \sin \theta_1} \tag{4-67}$$

Thus

$$G_n = \frac{h_n m_n s_n \sin \theta_1 C_1}{144 v_1} \tag{4-68}$$

If the axial-velocity component is employed, namely, $C_1 \sin \theta_1$, Eq. (4-66) becomes

$$G_n = \frac{h_n \left(s_n - \dfrac{t_n}{\sin \theta_1} \right) C_1 \sin \theta_1}{144 v_1} \tag{4-69}$$

This reduces to Eq. (4-68) when m_n is introduced. The maximum number of nozzles pitched at s_n in. measured along the mean radius is

$$n_n = \frac{2\pi r_m}{s_n} = \frac{\pi D_m}{s_n} \tag{4-70}$$

The designer can always modify the pitch s_n to make n_n a whole number of nozzles. It has been mentioned that fewer nozzles can be used than the number required for the total circumference, under which condition partial admission exists.

The total gas flow G in pound per second will be that passing through the total number of nozzles, n_n,

$$G = n_n \frac{h_n(s_n \sin \theta_1 - t_n)C_1}{144 v_1} \tag{4-71}$$

Reference to Figs. 4-26 and 4-7 shows for the gas entering each blade-row passage that

$$G_b = \frac{h_1 b_1 w_1}{144 v_1} \tag{4-72}$$

$$b_1 = s \sin \beta_1 - t_1 \tag{4-73}$$

$$G_b = \frac{h_1(s \sin \beta_1 - t_1)w_1}{144 v_1} \tag{4-74}$$

Introduce blade edge-thickness factors defined as

$$m_1 = \frac{s \sin \beta_1 - t_1}{s \sin \beta_1} \tag{4-75}$$

$$m_2 = \frac{s \sin \beta_2' - t_2}{s \sin \beta_2'} \tag{4-75a}$$

Substituting Eq. (4-75) in (4-74), there results

$$G_b = \frac{m_1 h_1 s w_1 \sin \beta_1}{144 v_1} \tag{4-76}$$

Reference to Fig. 4-7 shows that the axial velocity is

$$w_1 \sin \beta_1 = C_1 \sin \theta_1$$

Thus Eq. (4-76) can also be written as

$$G_b = \frac{m_1 h_1 s C_1 \sin \theta_1}{144 v_1} \tag{4-77}$$

For exit conditions from the blade row

$$G_b = \frac{m_2 h_2 s w_2 \sin \beta_2'}{144 v_1} \tag{4-78}$$

$$G_b = \frac{m_2 h_2 s C_2 \sin \theta_2'}{144 v_1} \tag{4-79}$$

In an impulse blading row the specific volume v_1 is closely constant, as the reaction is small and frictional reheat is not excessive. However, it is often desirable to increase slightly the exit height of the blade to accommodate for the lower relative velocity w_2, particularly when β_2' is less than β_1, as is true with a nonsymmetric blade.

The total number of blades, n_b, in a row of the turbine disk or rotor can be found when the pitch, s in., measured along the circumference at any radius is known. In particular for s at the mean blade radius r_m,

$$n_b = \frac{2\pi r_m}{s} = \frac{\pi D_m}{s} \tag{4-80}$$

The total gas flow, G lb per sec, passes through all the blade passages in a row when full admission is employed; otherwise n_b must be modified to include only those blades adjacent to nozzle banks. For the blade row

$$G = n_b \frac{m_1 h_1 s w_1 \sin \beta_1}{144 v_1} \tag{4-81}$$

$$G = n_b \frac{m_1 h_1 s C_1 \sin \theta_1}{144 v_1} \tag{4-82}$$

$$G = n_b \frac{m_2 h_2 s w_2 \sin \beta_2'}{144 v_1} \tag{4-83}$$

$$G = n_b \frac{m_2 h_2 s C_2 \sin \theta_2'}{144 v_1} \tag{4-84}$$

Impulse Blading. Although flexibility is allowable to the designer of impulse turbines, a background of experience is essential to the building of successful turbines. For making preliminary impulse-turbine design layouts use can be made of the representative data which follow. Refer also to Fig. 4-26.

Blade widths y are not less than one-tenth to one-eighth of the height and range as high as $h/3$ or $h/2$. Widths also should range from 1.0 to 1.7 times the pitch.

Edge thickness t can be thinned to sharp edge, but this is undesirable, and edges should be rounded off to prevent erosion and incipient cracking and also to provide rigidity. Projected edge thickness can range from 0.015 to 0.05 in. or to greater than 0.1 in. for very heavy duty service.

Thickness of nozzle partitions t_n is variable and ranges from 0.03 to 0.125 in.

Pitch s is variable, depending on relative size of turbine, nozzle (or blade) height, and angle; it ranges from 0.5 to 4.0 in.

Nozzle angle θ_1 is from 12 to 18° in early stage or stages, opening to as much as 26 to 28° in later stages to allow for flow.

Inlet blade angles β_1 are set by the velocity diagrams at mean radius, hub and tip; however, it is desirable to open the blade angle some 2 to 5° greater than diagram values to reduce the possibility of the gases striking the back of the blades.

Outlet blade angles β_2' are selected from the velocity diagrams. They can be closed by amounts ranging from 0 to 10° less than β_1 to give a greater velocity of whirl or even to produce reaction (nozzle) action. For jet-propulsion turbines where excess turbine work may not be important β_2' can be opened to greater than β_1 if this produces a greater efflux velocity from the turbine.

Blade overlap (spillage allowance). It is desirable for the blades to overlap the nozzle by 0.25 to 0.5 in. or more to allow for end spillage or oblique entry of the jet. It is important that the whole gas flow pass through the blading; *i.e.*, make h_1 of the blade greater than h_n of the nozzle, not less than $\frac{1}{8}$ in. on each end.

Turbine blade temperatures, if uncooled, range from possibly 125 to 200° colder than the gas stagnation temperature at the blade tips and some 175 to 300° colder at the roots (hub). The disk (rotor) must dissipate the heat flowing into it from the blades by supplementary cooling.

Allowable blade speeds are set by stress considerations and operating temperature (see Chap. 10). For rough estimates hot blading at greater than 1000 F should run at tip speeds not in excess of 550 to 750 fps.

Reaction Blading. With reaction blading the inlet edge is rounded, and it is not possible clearly to indicate the inlet blade angle β_1 (see Fig. 4-12). This rounded inlet is particularly important as a means of reducing entry shock under the varying load or speed conditions between full and part load on a turbine. An edge-thickness coefficient, for the leading edge in particular, as expressed in Eq. (4-75) can be transformed as follows:

$$m = \frac{s \sin \beta_1 - t_1}{s \sin \beta_1} = \frac{s - (t_1/\sin \beta_1)}{s} = \frac{\text{pitch} - \text{thickness of blade edge}}{\text{pitch}} \tag{4-85}$$

Although $t_1/\sin \beta_1$ is indefinite, it is possible to find and use representative values for m.

For the exit edge of the blades, the gaging is the important factor. Gaging is concerned with the closing or opening of the blades to control the exit passage. Gaging is defined as the ratio of available exit area for flow to the annulus area in the blade ring. Referring to Fig. 4-27, this is closely approximated by b_2/s measured at the mean height. However, this is not exact, as the trailing thickness t_2 is not zero and blades are usually twisted from root to tip.

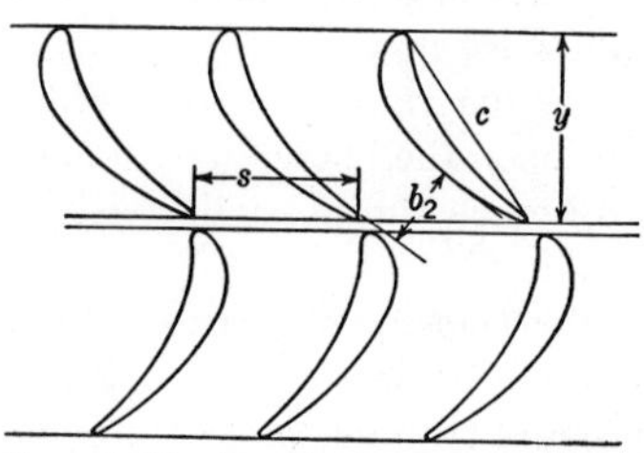

FIG. 4-27. Representation of fixed and moving blade rows of a reaction stage.

In reaction blades, the profile changes with radius. Near the hub, the shape approaches and resembles that of an impulse blade. At the mean radius the profile is that of a conventional reaction blade. Near the tip, the profile resembles an airfoil more than it does a blade section. This is illustrated in Fig. 4-10. In impulse blades, the blade form tends to approach that of a reaction blade toward the tip of the blade.

Representative reaction-blade data follow:

Blade width y. Make this not less than one-tenth or one-eighth of the blade height.

Pitch s ranges usually from 0.6 to 0.7 of the blade chord.

Gaging is variable, from some 22 to 40 per cent; at 25 per cent β_2' is approximately 14°.

Solidity is the ratio of blade chord to pitch c/s and ranges from about 1.0 to 2.

Axial-velocity values range from 300 to 600 fps for power units and 700 to 1000 fps for jet units.

Example 4. A single-stage impulse gas turbine with gas (air) supply at 1340.3 F (1800 R) at 40 psia is to exhaust to atmosphere at 15 psia. The unit is to turn at 6000 rpm, and the approximate maximum diameter to fit space requirements cannot exceed 30 in., nor should it be less than 22 in. The turbine has to develop 3600 shaft hp so that, considering bearing and gland losses, essentially 3675 rotor hp must be designed for. Allow for 95 per cent nozzle efficiency, and take $k_b = 0.84$. Make the fundamental calculations required for the thermodynamic design of this unit, and determine a set of basic dimensions which could be used as a starting point for developing a set of final dimensions. The final design must consider blade stresses, vibration aspects, fatigue, clearances, etc. This is a high-temperature unit, and it is suggested that a blade speed of 800 fps not be exceeded. The blades need not be symmetric and can vary in length on leading and exit edges; in fact β_2' could well be made 3 to 10° less than β_1. A nozzle angle of 17° is suggested. The velocity of approach from the combustors to the nozzles can be taken as 180 fps. Work out blade design at mean radius, and check at hub and tip.

Solution. The expansion pressure ratio is $^{40}/_{15} = 2.667$. Select values from Table 2-4 to find the isentropic enthalpy drop in the nozzles.

$$T_A = 1800 \text{ R} \quad h_A = 340.155 \quad p_{rA} = 144.761 \quad c_A = 180 \text{ fps}$$

$$p_{rB} = \frac{144.761}{2.667} = 54.274$$

and

$$\text{Corresponding } h_{Bs} = 233.139 \text{ Btu per lb}$$

In the nozzle, 0.95 of the isentropic enthalpy change is converted into kinetic energy,

$$(0.95)(340.155 - 233.139) = \frac{C_1^2 - 180^2}{2g_cJ}$$

$$C_1^2 = 5{,}121{,}000$$

$$C_1 = 2263 \text{ fps velocity of efflux from nozzle}$$

$$h_A + \frac{C_A^2}{2g_cJ} = h_1 + \frac{C_1^2}{2g_cJ}$$

$$340.155 + \frac{180^2}{50{,}070} = h_1 + \frac{2263^2}{50{,}070}$$

$$h_1 = 238.49 \text{ Btu per lb}$$

and

$$T_1 = 1420.2 \text{ R temperature of gases leaving nozzle at 15 psia}$$

By Eq. (2-3) find the specific volume v_1.

$$(144)(15)(v_1) = (1)(53.34)(1420.2)$$

$$v_1 = 35.1 \text{ ft}^3 \text{ per lb}$$

The specific volume will be considered essentially constant from the nozzle exit and throughout the blade row although friction and turbulence losses in the blade row cause some increase in enthalpy (temperature) and specific volume.

Select a mean blade speed for use in laying out velocity triangles. If a mean radius of 26 in. is chosen,

$$u = \frac{\pi D N}{(12)(60)} = \frac{(\pi)(26)(6000)}{720} = 681 \text{ fps}$$

Lay out the blade velocity diagram either graphically or by trigonometric functions. Referring to Fig. 4-28,

$$\begin{aligned}
(2263)(\cos 17°) - 681 &= w_1 \cos \beta_1 \\
(2263)(\sin 17°) &= w_1 \sin \beta_1 \\
\tan \beta_1 &= \frac{2263 \sin 17°}{2263 \cos 17° - 681} \\
\tan \beta_1 &= \frac{662}{2165 - 681} = 0.4458 \\
\beta_1 &= 24.04°, \text{ closely } 24° \\
w_1 &= \frac{1484}{\cos \beta_1} = 1624 \text{ fps} \\
w_2 &= k_b w_1 = (0.84)(1624) = 1364 \text{ fps}
\end{aligned}$$

Make a nonsymmetric blade to produce more power from this turbine; say reduce exit by 3°, making $\beta_2' = 24 - 3 = 21°$.

$$\begin{aligned}
w_2 \cos \beta_2' &= (1364)(0.9336) = 1273 \text{ fps} \\
w_2 \sin \beta_2' &= (1364)(0.3584) = 489 \text{ fps} \\
\tan \theta_2' &= 489/592 = 0.8260 \qquad \theta_2' = 39.6° \\
C_2 &= \frac{592}{\cos \theta_2'} = \frac{592}{0.7710} = 767 \text{ fps} \\
\text{Velocity of whirl} &= V_w = w_1 \cos \beta_1 + w_2 \cos \beta_2' \\
&= 1484 + 1273 = 2757 \text{ fps}
\end{aligned}$$

Consider that 95 per cent of mean velocity of whirl is converted to disk (rotor) power. This allows for less effective operation near the blade root and tip and the small, though possible, end leakage with an impulse blade.

Useful power per 1 lb per sec flow by Eq. (4-12) is

$$\text{hp} = \frac{(0.95 \times 2757)(681)}{(32.17)(550)} = 100.8$$

$$\text{Required gas flow for design rotor hp} = \frac{3675}{100.8} = 36.5 \text{ lb per sec}$$

Nozzles. To approximate a number of nozzles, use Eq. (4-70), and assume a pitch, for example, $s_n = 3.0$ in.

$$n_n = \frac{\pi(26)}{3} = 27.2$$

Make n_n a whole number, say 27 nozzles, and recompute s_n.

$$\begin{aligned}
27 s_n &= \pi(26) \\
s_n &= 3.02 \text{ in.}
\end{aligned}$$

Arbitrarily set t_n at some value, say 0.05 in., and use Eq. (4-71) to determine h_n.

$$\begin{aligned}
G = 36.5 &= 27 \frac{h_n[(3.02)(\sin 17°) - (0.05)](2263)}{(144)(35.1)} \\
h_n &= 3.62 \text{ in.}
\end{aligned}$$

This nozzle height appears reasonable and for the width of passage, namely, $(3.02)(\sin 17°) = 0.882$ in. measured at the mean radius, indicates an aspect ratio of $3.62/0.882 = 4.1$. This is somewhat high but acceptable. For this nozzle row the mean diameter is 26 in., the extreme diameter $26 + 3.62 = 29.62$ in., the inner diameter $26 - 3.62 = 22.38$ in. Although the nozzle passages in this annulus are not perfect rectangles, the deviation can usually be disregarded except for very high nozzle banks.

Blades. The inlet blade edge height h_1 should be made somewhat greater than the h_n of the nozzle to accommodate for spread and spillage of the jet (Fig. 4-26). Add about ¼ in., say, at the bottom and top of the blade, making the diameter to the tip of the blades 29.87 in. and at the hub 22.13 in. This makes

$$h_1 = 3.62 + 0.25 + 0.25 = 4.12 \text{ in.}$$

However, at the immediate inlet to the blade it is probable that the nozzle jet will not have fully spread radially, and the blade passage width should be made adequate to accommodate the flow in the circumferential path. Consequently, in computing the blade spacing s, use a value of h_1 only slightly larger than h_n, perhaps ¹⁄₁₆ in. or so for the top and bottom of the blade, *i.e.*, use $h_1 = 3.62 + 0.06 + 0.06 = 3.74$ in. in computation.

Impulse-blade-tip edge thicknesses are frequently thin, of the order of magnitude 0.02 to 0.10 in. For this case call $t_1 = 0.04$ in., and first employ Eqs. (4-75) and (4-80).

$$m_1 = \frac{s \sin \beta_1 - t_1}{s \sin \beta_1} = \frac{0.4074s - 0.04}{0.4074s}$$

$$n_b = \frac{\pi D_m}{s} = \frac{26\pi}{s}$$

Substitute the above values in Eq. (4-81).

$$G = 36.5 = \frac{26\pi}{s} \frac{(0.4074s - 0.04)(3.74)(s)(1624)(0.4074)}{0.4074s\ (144)(35.1)}$$

$$\frac{(36.5)(144)(35.1)}{26\pi(1624)(3.74)} = \frac{0.4074s - 0.04}{s} = 0.4074 - \frac{0.04}{s}$$

$$0.372 - 0.4074 = -\frac{0.04}{s}$$

$$0.0354s = 0.04$$

$$s = 1.13$$

$$n_b = \frac{26\pi}{1.13} = 72.3$$

Call this 72 blades, and readjust the pitch.

$$72s = 26\pi$$

$$s = 1.134 \text{ in. pitch at mean circumference on the blade row}$$

For the exit side of the blade, use Eqs. (4-75*a*) and (4-83). It is also in order to make the exit edge quite thin, say $t = 0.02$ in.

$$m_2 = \frac{(1.134)(0.3584) - 0.02}{(1.134)(0.3584)} = 0.9509$$

$$36.5 = 72 \frac{(0.9509)(h_2)(1.134)(1364)(0.3584)}{(144)(35.1)}$$

$$h_2 = 4.85 \text{ in. leaving-edge length}$$

For this blade with heights of 4.12 and 4.85 in. on the entering and leaving edges, respectively, a blade width must be selected. The width of the blade should be related both to the pitch and height, more particularly the pitch. Values of blade width ranging from 1.4 to 1.7 times the pitch have been employed in several successful designs. For example, using a mean ratio of 1.55, the blade width is

$$y = (1.55)(s) = (1.55)(1.134) = 1.76 \text{ in.}$$

Call y 1¾ in. wide.

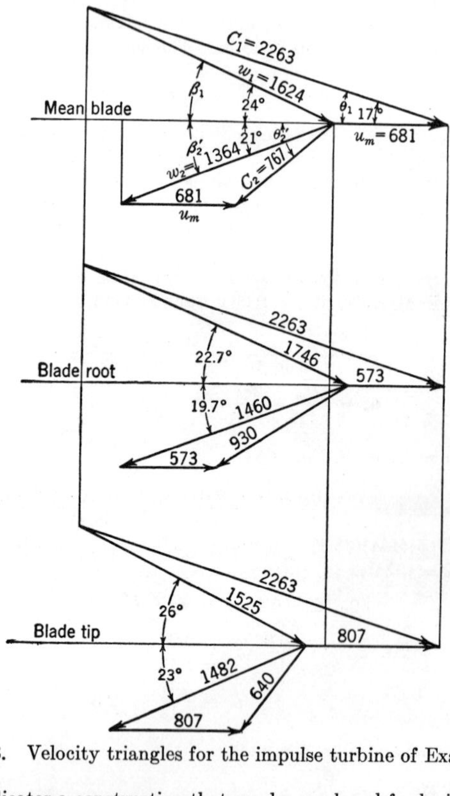

FIG. 4-28. Velocity triangles for the impulse turbine of Example 4.

Figure 4-29 indicates a construction that can be employed for laying out the blade profile. The inside curvature of the blade is drawn to a radius R which can be found in terms of the inlet and outlet blade angles β_1 and β_2'. To determine R, note that the blade width y can be found from the equation

$$y = R \cos \beta_1 + R \cos \beta_2'$$

The blade thicknesses on the leading and trailing edges are next laid off perpendicular to the inlet and outlet blade-angle lines, and the sides of the blade are drawn in by extending parallels through the thickness measurement points. A circular arc to meet these sides is then swung and located to give a depth between the back of one blade and the face of the other equal to that which exists in the entry passage. The blades are positioned with respect to each other by means of the pitch length s.

Conditions at the tip and hub of the blade should be investigated by drawing additional velocity triangles. These can be seen in Fig. 4-28. Notice that the gas velocity from the nozzle is considered constant at 2263 fps, but because of the blade speed variations at the hub and tip, the inlet blade angle changes greatly at the three positions considered.

$$u_h = \frac{\pi D_h N}{(12)(60)} = \frac{\pi(26 - 4.12)(6000)}{720} = 573 \text{ fps}$$

$$u_t = \frac{\pi D_t N}{(12)(60)} = \frac{\pi(26 + 4.85)(6000)}{720} = 807 \text{ fps}$$

Notice that the inlet blade angle β_1 becomes greater as one moves from the root (hub) to the tip of the blade. It is also obvious that the work per pound of gas is different at the three positions with more work being produced as the flow-path position moves outward on the blade and the blade speed ratio moves toward the optimum. In this design, the leaving blade angle was arbitrarily made 3° less than the inlet blade angle.

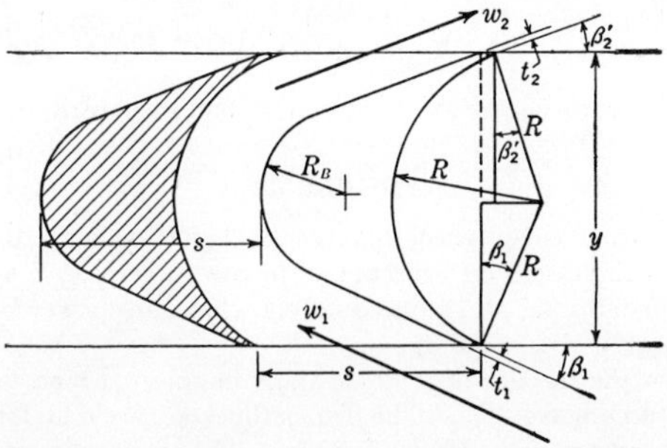

FIG. 4-29. Construction of impulse blade profile based on data of Example 4.

If these blades had been longer, it might have been necessary to check carefully the flow distribution throughout the blade, as pressure variations along the length of the blade could upset the normal flow patterns. It has already been mentioned that one method of accomplishing this result would be to employ so-called "vortex flow," *viz.*, to keep the product, (velocity of whirl) (radius from center of rotation), equal to a constant, for each position along the blade.

This problem shows a simple method of roughing out a set of preliminary nozzle and blade calculations. Various design refinements can be made using this basic design as a starting point. However, it should be realized that this design is inefficient, as it is not usually desirable to employ a single-stage turbine for so large an energy release, namely, $(340.155 - 233.139) + (180^2/2g_cJ)$, or 107.663 Btu per lb. Because of centrifugal stresses the blade speed is limited and cannot reach the optimum value required (Fig. 4-13) for effective absorption of the resulting large kinetic energy in the nozzle jet (2263 fps). Moreover, the high velocity of the gases moving through the blade passages contributes to greater than usual frictional and turbulence losses. For these gases at 1420 R the acoustic velocity is 1820 fps so that the Mach number for relative velocity in the blades is $^{1624}/_{1812} = 0.89$. More efficient designs would envisage a two- or three-stage pressure-compounded turbine or a three- or four-stage reaction turbine. However for a jet-propulsion unit which can use the leaving jet energy C_2 the inefficiency of this single-stage turbine is not so important.

Efficiency of the turbine referred to disk (rotor) power, η_{tr}, can be computed (based on the disk horsepower produced by 1 lb per sec flow) as follows:

$$\eta_{tr} = \frac{(\text{disk hp})(0.707)}{(1)(\text{energy release})} = \frac{(100.8)(0.707)}{(1)(107.663)}$$
$$= 0.662 = 66.2\%$$

Loss to residual (leaving) velocity expressed as a fraction of the energy release available, based on mean blade conditions, appears as

$$\text{Residual loss} = \frac{C_2^2/50{,}070}{107.663} = \frac{767^2/50{,}070}{107.663} = \frac{11.7}{107.663} = 0.108$$
$$= 10.8\%$$

Loss from friction and turbulence for the blade row at the mean blade position appears as

$$\text{Frictional loss} = \frac{\dfrac{w_1^2 - (kw_1)^2}{50{,}070}}{107.663} = \frac{\dfrac{1624^2 - 1364^2}{50{,}070}}{107.663}$$
$$= \frac{15.5}{107.663} = 0.144 = 14.4\%$$

The two preceding items constitute the greatest factors contributing to loss in this design.

PROBLEMS

4-1. A turbine with a mean wheel diameter of 25 in. turns at 10,000 rpm. Blade triangles show a mean velocity of whirl amounting to 2200 fps. When the machine is passing 60 lb of medium per sec, approximately what horsepower is being developed at the turbine wheel? *Ans.* 8130 hp.

4-2. Gases leave the nozzles of a single-stage impulse turbine to enter the blade row at 1500 fps at an angle of 25°. The blades (buckets) are 6 in. long, and the wheel diameter to the outer tip of the blades is 30 in. The rotor turns at 7000 rpm. The blade outlet angle is fixed at 38°, but the inlet blade angle is varied to permit entry of the gases with minimum shock. Consider that there is no loss in relative velocity in flowing through the blade row. (*a*) Find the blade speed at tip, mean radius, and hub. (*b*) Draw velocity triangles for these three positions, and determine the proper inlet blade angle for each location. (*c*) Compute the velocity-of-whirl values at tip, mean-radius, and hub positions. (*d*) For a total gas flow of 30 lb per sec, compute the diagram horsepower produced if one-third of the total flow can be considered to pass, respectively, at tip, mean radius, and hub positions. *Ans.* (*a*) 916, 733, and 550 fps; (*b*) 55°, 45.5°, and 38°; (*c*) 1060, 1336, 1622 fps; (*d*) 549, 553, 504 hp.

4-3. Work Prob. 4-2 with the data the same throughout except for a change in nozzle angle to 20° and a change in outlet blade angle to 31°. *Ans.* (*a*) 916, 733, and 550 fps; (*b*) 44.5°, 36.8°, and 31°; (*c*) 1101, 1400, 1712 fps; (*d*) 570, 577, 533 hp.

4-4. Rework Prob. 4-3 with allowance for a blade friction factor of $k_b = 0.9$. Make no other change in the data of the problem. *Ans.* (*c*) 1038, 1328, 1626 fps at hub; (*d*) 538, 550, 506 hp at hub.

4-5. Rework Prob. 4-2 with allowance for a blade friction factor of $k_b = 0.9$. Make no other change in the data of the problem.

4-6. A single-stage 11,500-rpm simple impulse turbine is supplied with hot gas (air) entering the turbine nozzles at 1492.3 F and a velocity of 300 fps. The pressure drop from 40 to 10 psia takes place exclusively in the nozzles. (*a*) Find velocity and temperature leaving nozzles. *Ans.* 2816 fps, 1370.1 R. (*b*) Draw velocity triangles for impulse blades if mean diameter = 22 in. and nozzle angle is 22½° with exit angle of the blades 60°. *Ans.* $w_1 = 1845$, $w_2 = 1845$, $C_2 = 1610$, $u = 1104$ fps. (*c*) Find mean velocity of whirl and horsepower developed by 79 lb per sec flow, first assuming no loss in nozzle or blades. *Ans.* $V_w = 2425$ fps, horsepower = 11,970. (*d*) Next, if there is an 80 per cent realization of this power to the shaft, compute the exit conditions from the turbine, *i.e.*, subtract useful power (in Btu/(sec)(lb)/sec flow) and leaving kinetic energy from original kenthalpy. The leaving velocity may be considered closely equal to the absolute velocity leaving the blade row as given by the blade diagram. *Ans.* $h = 246.7$; $T = 1451.5$ R. (*e*) This hot gas next drops in pressure (and temperature) in the jet nozzle as it expands to discharge pressure from the nozzle at the altitude pressure of 9 psia. Compute the exit velocity relative to the nozzle. *Ans.* 1762 fps.

4-7. An impulse turbine stage has a mean blade speed of 600 fps, and hot air enters the blade at $C_1 = 1230$ fps at a nozzle angle of 18°. The blade exit angle is the same as the inlet blade angle, and k_b can be considered to be 0.89. (*a*) Compute values for and draw the velocity triangles. (*b*) Compute the velocity of whirl. (*c*) Find the power produced by each pound per second of air passing through the blade at the mean position. (*d*) Check the preceding answer by the kinetic-energy form of blade equation. (*e*) Estimate the rotor horsepower produced by this turbine stage for an air flow of 50 lb per sec. (*f*) What is the blade speed ratio for this turbine stage?

4-8. For the four-stage reaction turbine discussed in Example 2, consider the second stage, which is supplied with gases, taken to be air, at 310.4 fps, 35.2 psia, $h = 263.895$ Btu per lb, and with exhaust from the turbine at 15 psia. Consider that one-third of the isentropic enthalpy drop available to the second, third, and fourth stages is allocated to the second stage and is divided equally between the fixed and moving row. Consider blade nozzle efficiency to be 0.9, and take $k_b = 0.8$. Set the fixed blade outlet angle θ_1 as 22°, and make $\beta_2' = 22°$. Set $\sigma = 0.86$ for second-stage blade row. (*a*) Compute the necessary data to lay out the velocity triangles. (*b*) Find the velocity of whirl and the stage power for 1 lb per sec flow. (*c*) Find the stage efficiency. (*d*) Find the conditions entering the next stage. *Ans.* (*a*) $C_1 = 811.7$ fps, $u = 698$ fps, $w_1 = 308.9$, $w_2 = 811.4$, $C_2 = 309$; (*b*) $\Delta V = 807$ fps, power = 22.50 Btu/(sec)(lb)/sec; (*c*) 85.8%; (*d*) 309 fps, $h = 241.43$ Btu, 27.2 psia.

4-9. For the four-stage reaction turbine discussed in Example 2, consider the third stage, which is supplied with heated air at 309 fps, 27.2 psia, 1431.33 R and with exhaust from the turbine at 15 psia. Consider that one-half of the isentropic enthalpy drop for the third and fourth stages is allocated to the third stage and is divided equally between the fixed and moving row. Consider blade nozzle efficiency to be 0.9, and take $k_b = 0.8$. Set the fixed blade outlet angle θ_1 as 22°, and make $\beta_2' = 22°$. Set $\sigma = 0.86$ for the third-stage blade row. (*a*) Compute the necessary data to lay out the velocity triangles. (*b*) Find the velocity of whirl and the stage power for 1 lb per sec flow. (*c*) Find the stage efficiency. (*d*) Find the conditions entering the next stage. *Ans.* (*a*) $C_1 = 818.6$ fps, $u = 704$ fps, $w_1 = 311.5$, $w_2 = 819.1$, $C_2 = 311.8$; (*b*) $\Delta V = 814.5$ fps, power = 22.91 Btu/(sec)(lb)/sec; (*c*) 86.0%; (*d*) 311.8 fps, 1343.3 R, 20.6 psia.

4-10. The fourth and last stage of the reaction turbine discussed in Example 2 is supplied with heated air at 311.8 fps, 20.6 psia, 1343.3 R and with exhaust at 15 psia. Consider that the isentropic enthalpy drop for this stage is divided equally between the fixed and moving row. Consider blade nozzle efficiency to be 0.9, and take $k_b = 0.8$. Set the fixed blade outlet angle (θ_1) as 22°, and make $\beta_2' = 22°$. Set $\sigma = 0.86$ for the last-stage blade row. (*a*) Compute the necessary data to lay out the velocity triangles. (*b*) Find the velocity of whirl and the stage power for 1 lb per sec flow. (c) Find the true last-stage efficiency. (d) Find the conditions entering the exhaust. *Ans.* (a) $C_1 = 832$ fps, $u = 716$ fps.

4-11. Hot gas (air) at 45 psia and 1600 R flowing at 70 fps is supplied to a turbine which exhausts at 15 psia. (a) Compute the ideal enthalpy drop available for work, and allocate one-sixth of this drop to the first stage of a six-stage reaction turbine. (b) For this enthalpy drop, which is divided equally between the fixed and moving row, what is the pressure drop across both rows of the stage? (c) Set θ_1 at 15°, and using a mean blade speed of 480 fps while allowing for the 70 fps entering the fixed blade, compute the velocity C_1 entering the moving row, and draw the inlet velocity triangle. (d) Using the carry-over velocity w_1 from the diagram and the remainder of the enthalpy drop, compute w_2 and draw the exit triangle with $\beta_2' = 15°$. (e) Compute the velocity of whirl and the diagram horsepower for this stage for a flow of 30 lb of gas per sec. (f) If the horsepower realized at the shaft is 84 per cent of the horsepower from the diagram, find the enthalpy leaving this stage. Assume the leaving velocity to be that indicated on the velocity diagram. *Ans.* (a) 104.82 and 17.470 Btu per stage. (b) 38.3 psia leaving and $\Delta p = 6.7$ psi. (c) $C_1 = 665$ fps, $w_1 = 236.5$ fps. (d) $w_2 = 702.6$ fps, $C_2 = 269.5$ fps. (e) $642.3 + 199 = 841.3$ fps and 684 hp. (f) 574.5 hp and $h = 271.21$ Btu per lb.

4-12. Air is available for a gas turbine at 60 psia and 1400 R. Discharge pressure from the turbine is 15 psia; maximum blade speed is 850 fps. The mean blade diameter is 24 in., and relatively short blades (not over 4 in.) are called for. It is desired to make preliminary calculations to determine the suitability of a multistage impulse or a multistage reaction turbine. Making use of desirable blade speed ratios, calculate the most desirable number of stages for both a pressure-compounded impulse and a reaction turbine.

4-13. Work Prob. 4-12 for an expansion from 45 psia to 15 psia with no other change in the data.

4-14. A two-stage impulse gas turbine with gas supply at 1340.3 F (1800 R) at 40 psia is to exhaust at 15 psia. The unit is to turn at 6000 rpm, and the approximate maximum diameter to fit space requirements should not exceed 30 in. or be less than 22 in. The turbine has to develop 3600 shaft hp so that considering bearing and gland losses the rotor must be designed to supply essentially 3675 hp. Allow for 95 per cent nozzle efficiency in each stage based on stage enthalpy drop, and use k_b equal to 0.86. Make the fundamental calculations required for the thermodynamic design of this unit, and determine a set of basic dimensions for the first stage which could be used as a starting point for developing a set of final dimensions. For this unit it is suggested that a blade speed of 750 fps not be exceeded. The blades need not be symmetric and can vary in length on leading and exit edges. It is suggested that β_2' be made 2° less than β_1. A nozzle angle of 17° is suggested. The velocity of approach from the combustors to the nozzle can be taken as 180 fps. As carry-over and reheat from the first stage will be available in the second, a greater portion than one-half of the isentropic enthalpy drop should be allocated to the first stage; 54 per cent is suggested, and on this enthalpy allocation the first-stage pressure after the nozzles can be found. Work out the design at mean radius, and check at blade hub and tip. An additional assumption which should be made is that equal work might be expected in the two stages, and on this premise flow in pounds per second can be determined. Answers given are based on a mean diameter of 25 in. furnished as a convenient starting dimension.

Answers to a possible design solution for the first stage

$u_m = 655$ fps

$p_B = 24.28$ psia after nozzle row

C_1 = 1668 fps leaving nozzles, $w_1 = 1059$, $w_2 = 911$, $C_2 = 424$
$\beta_1 = 27.4°$, $\beta_2' = 25.4°$
Hp per lb per sec flow = 65.3 using fixed velocity of whirl throughout blade
28.15 lb per sec flow for equal work per stage
24.5 ft^3 per lb for air at outlet from nozzle
Nozzle pitch of 3.02 and assumed $t_n = 0.1$ in. shows 26 nozzles 2.93 in. high.
Making the inlet edge of blade ⅛ in. greater on top and bottom shows $h_1 = 3.18$ in.
With $t_1 = 0.06$ in. the blade pitch appears 0.707 in. and $n_b = 111$ blades.
With $t_2 = 0.02$ in. $h_2 = 3.30$ in.
Blade row diameters 25 in. mean, 28.18 in. outer on inlet, 28.30 in. outer on outlet

REFERENCES

1. Emmert, H. D.: Current Design Practices for Gas-turbine Power Elements, *ASME Paper* 48-A-69, *Trans. ASME*, vol. 72, pp. 189–200, 1950.
2. Yellott and Lype: Some Effects of Pressure Loss on the Open Cycle Gas Turbine Plant, *Trans. ASME*, vol. 69, pp. 903–911, 1947.
3. Fentress, W. K.: Design of the Turbine Element for the Gas Turbine, lecture notes presented at Northwestern University, ASME-ASEE Summer School, June, 1947.
4. Turunen, W. A.: Gas Turbines in Automobiles, *Trans. SAE*, vol. 4, pp. 102–115, January, 1950.
5. Jacklin, H. M.: Boeings' Model 502 Gas Turbine, *J. SAE*, vol. 58, pp. 53–57, 72, August, 1950.
6. Woodhouse, H.: Turbine Blade Fastenings, *Machine Design*, vol. 23, pp. 148–153, February, 1951.
7. Eckert and Vietinghoff-Scheel: Experimental Study of Flow Past Turbine Blades, *NACA Tech. Mem.*, June, 1949.
8. Knowles, D. W.: Development of the Avro Orenda Jet Engine, *J. Eng. Inst. Canada*, vol. 34, pp. 1183–1190, December, 1951.
9. Putz, T. J.: A Compact Locomotive-type Gas Turbine, *Westinghouse Engr.*, vol. 7, pp. 35–39, March, 1947.
10. Redding, A. H.: Future Forms of Aviation Gas Turbines, *Westinghouse Engr.*, vol. 7, pp. 110–114, July, 1947.
11. Brown, T. W. F.: Marine Gas Turbine Research in Britain, *Mech. Eng.*, vol. 72, pp. 379–388, May, 1950.
12. Morgan, D. W. R.: Central Station Steam Power Generation, *Westinghouse Engr.*, vol. 10, pp. 7–17, January, 1950.
13. Tucker, W. B.: Construction of Gas Turbine for Locomotive Power Plant, *Mech. Eng.*, vol. 70, pp. 877–882, November, 1948.
14. Church, E. F.: "Steam Turbines," 3d ed., McGraw-Hill Book Company, Inc., New York, 1950.
15. Godsey and Young: "Gas Turbines for Aircraft," McGraw-Hill Book Company, Inc., New York, 1949.
16. Kearton, W. J.: "Steam Turbine Theory and Practice." 5th ed., Pitman and Sons, Ltd., London, 1948.
17. Sorenson, H. A.: "Gas Turbines," Ronald Press Co.. New York, 1951.
18. Vincent, E. T.: "The Theory and Design of Gas Turbines and Jet Engines," McGraw-Hill Book Company, Inc., New York, 1950.
19. Zucrow, M. J.: "Principles of Jet Propulsion and Gas Turbines," John Wiley & Sons, Inc., New York, 1948.

CHAPTER 5

AXIAL-FLOW COMPRESSORS

5-1. Types of Compressors. Because of the enormous quantities of air required in gas-turbine systems, conventional positive-displacement reciprocating compressors are not employed. Units of the lobe type, such as the Roots blower and the Lysholm compressor, have been used to a certain extent. Although Roots and Lysholm compressors operate differently, their fundamental action is that of trapping air between the rotating elements of the compressor and squeezing it into a narrowing space from which the compressed air is discharged. The Lysholm compressor (Fig. 6-10) is described in detail in Chap. 6. Displacement blowers, in general, are noisy, subject to wear, and also of limited capacity compared with the high-speed centrifugal and axial-flow types.

Centrifugal compressors for handling air and industrial gases have been used for many years, as blowing engines in steel mills, as compressors for low-pressure refrigerants and industrial gases, as superchargers for aircraft and other reciprocating engines (Figs. 1-6 and 1-7), and more recently as compressors for certain types of gas-turbine-driven aircraft (Figs. 1-13 and 7-13). Centrifugal compressors can handle large quantities of gas, are stable over a relatively broad operating range, and can be developed to have efficiencies in excess of 80 per cent. For aircraft work their frontal area is somewhat greater than that of an axial type of the same capacity. Satisfactory axial-flow compressors are a more recent development than the other types, but because of higher efficiency, breadth of design possibilities, and relative ease of mass manufacture they are assuming an increasingly important position in the field of gas-turbine design and layout.

5-2. Axial-flow Compressors, General. Axial-flow compressors resemble in appearance the reaction steam turbine with the direction of flow reversed, the inlet to the compressor being associated with the large blades and the outlet with smaller blading to accommodate the decreased volume as the gas is compressed. Axial-flow compressors are illustrated in many points throughout this text; Figs. 5-1, 5-2, 1-9, and 1-12 show several designs.

Referring to Fig. 5-1, it can be seen how the gas flows through a rounded inlet passage into the first moving blade row of the compressor. In this row, the velocity configuration of the gas is changed, and the gas is directed into a fixed row of blading attached to the compressor casing. By varying the shape and direction of the gas passage in the blading, it is possible to alter the pressure rise which takes place in the moving row and

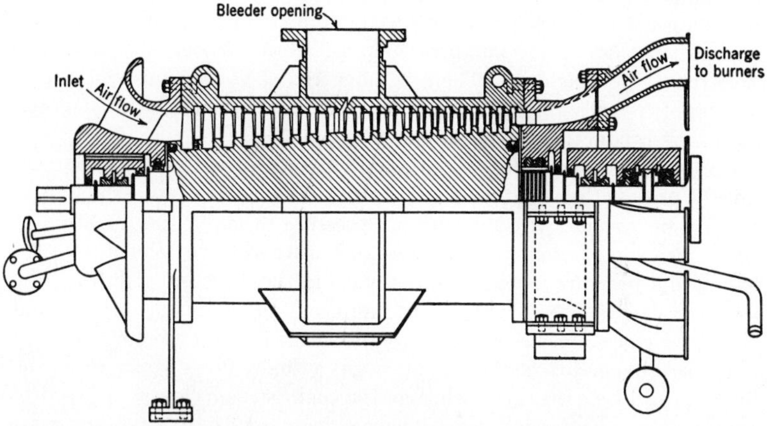

FIG. 5-1. Cross section through a 20-stage axial-flow compressor. (*Westinghouse Electric Corp.*)

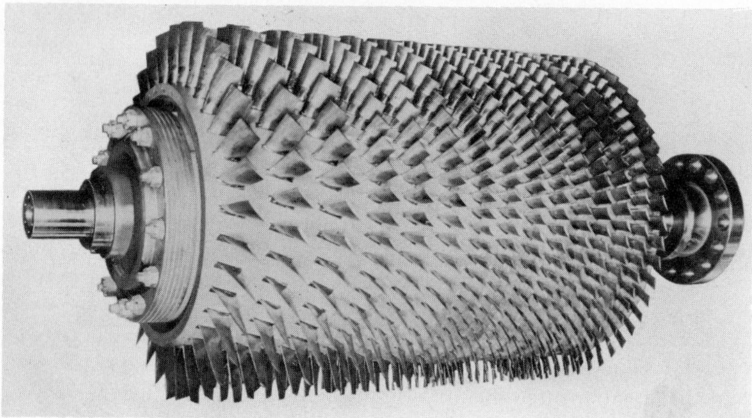

FIG. 5-2. Fifteen-stage compressor rotor for a 4800-hp locomotive gas-turbine power plant; inlet at left (Fig. 1-9). (*General Electric Co.*)

in the fixed row of the compressor stage. The compressor stage is normally considered to consist of a moving row and a fixed row. However, in some cases the inlet casing has a guide blade to constitute a fixed row at inlet and the outlet discharge scroll can serve as a fixed row at the compressor outlet. The compressor in Fig. 5-1 is classed as a 20-stage compressor consisting of 20 moving rows and 20 fixed rows of blading. Figures 1-9 and 5-2 both show the compressor of an axial-flow machine. This compressor has 15 moving rows with a fixed blade (or guide) row preceding the first moving row. It is obvious that the term "axial flow" in a compressor of this type means that compression takes place and flow proceeds throughout the compressor in a direction essentially parallel to the axis of the machine. There is some radial flow in most machines because of changes in length of blading, but the effect of this in the design and in the production of pressure is negligible.

The compression process in an axial compressor is inherently a more difficult process than that associated with the expansion process in a reaction steam turbine. In the turbine, pressure drops in the direction of flow. In a compressor, the diffusion process is necessarily one of retarded flow with a pressure increase taking place in the direction of flow. The boundary layer of fluid which exists adjacent to the blade surfaces and clings to them is not associated directly with the main stream flow, and under certain conditions the boundary gas actually flows backward in the direction of lower pressure, whereas the main stream flow is forward in the direction of pressure rise. This tendency leads to potential instabilities in a machine. At design conditions, instability is relatively unimportant, amounting to little more than localized eddies and turbulence, but with machines operating far from design or optimum conditions, it is possible for the flow to reverse completely and set up a surge. It happens that the range of operation of an axial-flow compressor is somewhat limited. The problem of surging is not unique to the axial-flow compressor but also exists in centrifugal compressors. Surging will be discussed in more detail later in this chapter.

Figure 5-3 is a generalized plot of the flow characteristics of an axial-flow compressor, showing pressure ratio and efficiency plotted against a weight-flow parameter. This shows that at increasing speeds (rpm) both the pressure ratio and the capacity increase. It will be noticed that at the varying speeds, the pressure-ratio–capacity lines are relatively steep and stop abruptly at an upper limit which is called the *surge line* or *limit of stability*. If an attempt is made to operate at a higher discharge pressure than that for the surge line, it is impossible for the machine to develop this pressure and back flow, or surging, will start, with resulting unstable operation. The possible pressure ratio increases as the speed of the machine rises. It will be noticed from the efficiency curves that maximum efficiency occurs at a point somewhat below the surge-limit point and holds over a relatively narrow range so that if operating conditions are far from those for which the machine is designed a decrease in operating efficiency results.

Underlying true surge in a compressor is the stalling of the individual blade rows of which the compressor is composed. Stalling is sensitive not only to blade profile and angle of attack but also to cascade spacing and solidity of the blades (see Fig. 5-9). Increasing the camber, stagger, or solidity increases the pressure rise of a stage and tends to make the stall more severe when it does occur. Stall, in general, is characterized by reverse flow near the region of the blade tip. This recirculation upsets velocity distribution and adversely affects the performance of succeeding

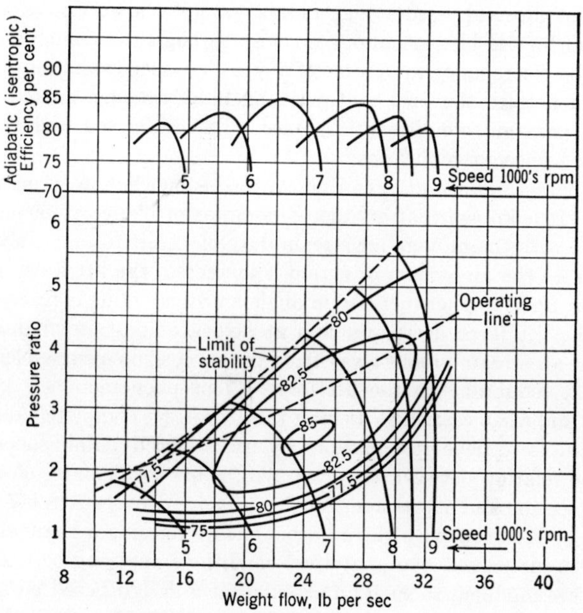

FIG. 5-3. Typical performance-characteristic curves of an axial-flow compressor, indicating the limits of stable operation. (*After Ponomareff, Ref.* 3.)

stages. Careful investigations have shown that when surge really develops the compressor surges as a unit, with the flow pulsating at the same frequency throughout the machine.

A study of Fig. 5-1, which indicates the representative annulus area of a compressor, clearly shows the great difference in area existing between inlet and outlet. The reduction in area exists largely because the compressor operating under normal conditions will deliver gases at much lower specific volume than they have on entry to the machine. Under starting conditions, the specific volume of the gas throughout the compressor is the same. Thus, at starting, the area at outlet and inlet might desirably be the same. However, as the machine rises in speed and compression begins, the maladjustment of areas at starting is gradually eliminated.

There is no absolute cure for compressor surge, although several approaches have contributed to successful operation. One suggestion has been to limit the pressure ratio in a compressor so as not to exceed 7 to 9 in a single machine. This can be done by breaking a total compression ratio into two (or more) steps. By compounding axial-flow compressors so that they are used in series, pressure ratios in the range of 12 to 16 are not illogical. Interstage bleeds, whereby a portion of the compressor airflow is bled off about midway through the compressor, helps reduce the discrepancy in volume of flow at low speeds and improves performance despite the fact that working fluid is being wasted. Under normal operating conditions, the bleeds can be closed and the stages matched to design conditions.

Stall, or *limit of stability*, can be clarified somewhat in terms of airfoil theory. It is known that an airfoil (compressor blade) at certain angles of attack with the moving air stream develops a lift force. As the angle of attack to the air-stream direction is increased, the lift force increases. However, upon further increase in angle a critical value is reached above which the lift force decreases, and at attack angles above this critical value the air stream breaks away from the surface, creating violent eddies and great reductions in the lift force. This phenomenon is known as stalling, and when carried to its limit the lift force is completely dissipated. It is possible to develop equations by use of aerodynamic theory which show the relation between the lift force (or lift coefficients) of the blade (airfoil) section and expressions for the pressure rise (see Art. 5-7). However, in a blade row with the adjacent blades forming a channel through which the air passes, the analytical conditions applying to a single foil (blade) are modified to such an extent that a design based on single-foil aerodynamic theory is inadequate. However, the aerodynamic aspects of flow through compressor blading are extremely important in blade-development theory, and one of the most effective ways of analyzing the performance of blade designs is by means of cascade testing. In cascade testing a grid, or model blade pattern, is constructed. This is placed in an experimental flow chamber, and at various attack angles and at varying air velocities measurements are made from which conclusions can be drawn as to how that blade design would perform in an actual compressor. Cascade tests all utilize aerodynamic theory in their analysis.

The form of the blade and the amount of turning are closely related to the velocity diagrams for the compressor blade. The resultant blade shape must meet the desired velocity requirements, aerodynamic effectiveness, and stress considerations. Typical blading has a heavy base and tapered section to satisfy stress considerations and a twisted form to allow for the variation in tangential velocity along the blade length as well as to reduce the possibility of widely variant pressure changes at different radial positions along the blade length. The turning angles which can be employed in reaction steam turbines are greater than those which can effectively be used in the case of compressors. Before discussing this matter further, typical compressor blading diagrams will be developed.

5-3. Compressor Blade Diagrams. In a typical axial-flow compressor as illustrated in Fig. 5-2 the blading attached to the turbine rotor forms a series of passages through which the gas (air) passes. In flowing through these passages, various velocity and pressure changes take place. The space between adjacent blades constitutes a channel for directing the gas flow, and it is customary to design on the basis of this channel. In this approach, vectors are drawn to indicate by their position the general direction of stream flow and to indicate by their length the magnitude of velocity in the same manner that was followed in the case of the turbine. This approach, in general, disregards the aerodynamic character of flow around foils.

Figure 5-4* shows a typical arrangement of a stage in an axial-flow compressor with the gas leaving a previous fixed row (or the inlet guides of a compressor) at velocity C_1 and angle θ_1. When the absolute blade speed u_1 is combined with the gas velocity, the relative velocity at inlet to the blade is w_1 at angle β_1. In passing through the moving row, the relative velocity decreases to a design value w_2, and the outlet velocity triangle can be drawn in similar manner. In the following fixed row, the velocity C_2 is changed to a leaving value C_3, and the velocity diagram for the fixed row consists merely of the two lines C_2 and C_3, with C_3 placed to suit the outlet angle of the fixed blade row, here shown as being θ_3 in magnitude. If the gas can be considered essentially incompressible, the idealized pressure rise in the moving row can be written as

$$\Delta p_m = \frac{\rho}{2g_c}(w_1^2 - w_2^2) \tag{5-1}$$

and in a fixed row the pressure rise is measured by

$$\Delta p_f = \frac{\rho}{2g_c}(C_2^2 - C_3^2) \qquad (5\text{-}2)^*$$

* Note that compressor blade angle notation differs from that used with the turbine, *i.e.*, θ_1 replaces θ_1' etc.

* Dimensionally these equations appear as

$$\Delta p = \frac{\text{lb}_m}{g_c\,\text{ft}^3}\left[\frac{\text{ft}^2}{\text{sec}^2}\right] = \frac{(\text{slugs})(\text{ft/sec}^2)}{\text{ft}^2} = \frac{\text{lb}_f}{\text{ft}^2}$$

In these equations Δp_m and Δp_f in pounds per square foot represent the pressure rise, respectively, in the moving and the fixed row; w_1 and w_2 are relative velocities in feet per second on inlet to and exit from the moving blade; C_2 and C_3 in feet per second are absolute velocities on inlet to and exit from a fixed row. The basic theory upon which this statement is based was developed for the turbine element in Eq. (4-23) although (4-23) as written is a power equation for a whole stage. Whenever it happens that the tangential velocity components u_1 and u_2 are different for a streamline of gas passing through the blade, an additional pressure effect would exist for these terms. However, for a turbine or axial-compressor element, the effect of variation in u_1 and u_2 is relatively small and usually neglected.

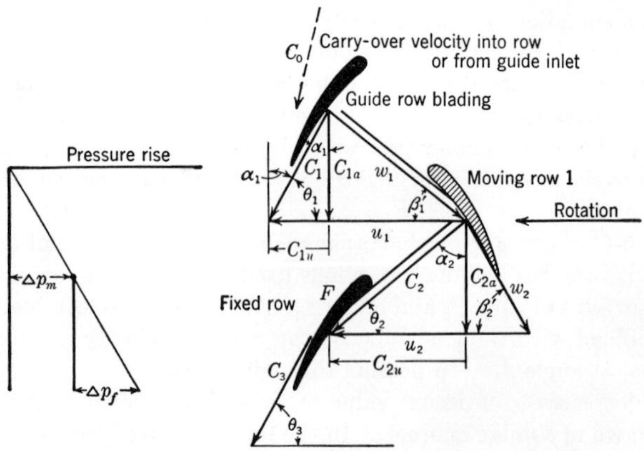

FIG. 5-4. Velocity diagram of an axial-flow compressor stage.

Symmetric Compressor. The diagram of Fig. 5-4 as drawn applies to a symmetric blade arrangement. In symmetric blading, the angles and flow pattern are so arranged that $w_1 = C_2$ and C_3 or $C_1 = w_2$, and for any streamline at fixed radius from the center of rotation (shaft) $u_1 = u_2$. In a symmetric arrangement such as this, the pressure rise in the moving row and the pressure rise in the fixed row are equal. This is not an essential feature of a blade design and can be varied. In general, symmetric compressors are characterized by high axial velocities and high blade tip speeds, while the velocities through the moving blades are relatively low. Where high frictional losses are associated with high relative velocities, the low magnitude of the w terms represents an advantage. It is possible for blade tip speeds with this type of compressor to exceed the velocity of sound, and axial velocities in excess of 600 fps are common. The pressure rise per stage can be relatively large. Compressors of this type have been used with aviation gas turbines, as for a given number of stages it is possible to obtain maximum pressure rise. The residual velocity loss, however, may be somewhat high unless an efficient diffusion passage follows the compressor.

Nonsymmetric Stage Compressor. In the nonsymmetric arrangement, lower blade speeds are employed than in the symmetric type. Inlet to the first moving row is essentially axial, and consequently it is frequently possible to omit the inlet guide vanes. A typical diagram is shown in Fig. 5-5 along with a representative pressure rise in both the moving and the fixed row. With axial inlet to the first row, the absolute velocity on inlet is lower than in the preceding type. Pressure rise in the moving row is appreciable, whereas a much smaller pressure rise is planned for the fixed row. Blade-tip speeds are usually kept under 800 fps, and axial velocities of some 400 fps are employed. Because about 90 per cent of the pressure rise takes place in the moving row and some 10 per cent in the fixed row, a compressor of this type is sometimes referred to as a 10 per cent reaction machine. With a ratio of 85 to 15 per cent, the machine would be called a 15 per cent reaction unit. In the case of the symmetric machine, the term 50 per cent reaction is employed.

Vortex Stage Compressor. In this design it is customary to make the velocity entering the fixed row completely axial, *i.e.*, have $\theta_2 = 90°$. A swirling condition, or vortex, is set up, as C_{1u} is opposite to the direction of blade motion. In the figure the outlet swirl is eliminated, as C_{2u} is zero. Low blade-tip speeds, of some 500 fps, and low axial gas velocities of some 200 fps characterize this type of machine. Small pressure rise per stage and low velocities indicate a larger size machine for a given capacity compared with other types. This design may be preferable for closed-cycle units, where the specific volume of the gas is smaller. Efficiencies are comparable with those of the nonsymmetric type.

To summarize, Fig. 5-5, which is drawn to relative scale, indicates for conventional design conditions the varying amounts of pressure build-up which might be expected. In the symmetric arrangement, Δp_m, the pressure increase in the moving row, is equaled by Δp_f, the pressure increase in the following fixed row. In the nonsymmetric arrangement with axial inlet, the total stage pressure increase is lower because of the much smaller Δp_f term. The vortex arrangement with the inlet swirl opposite to the direction of rotation (note component C_{1u}) shows the smallest pressure rise of all, partly because of the pressure drop in the fixed row, where C_2 increases to C_3 and expansion takes place. The question may arise as to why u for all types cannot be made the same, but a study of the diagrams readily shows that, at the same u, the nonsymmetric and vortex arrangements would give excessively high relative velocities w_1 and w_2. This can lead to high friction and turbulence losses and even to local supersonic gas velocities. To meet continuity-of-mass flow conditions in moving and fixed rows of essentially the same annulus areas, the axial velocities have been made equal for each type, *i.e.*,

$$C_{a1} = C_1 \sin \theta_1 = C_2 \sin \theta_2 = C_{1u} \frac{1}{\tan \alpha_1} = C_{a2}$$

or

$$C_{a1} = w_1 \sin \beta_1' = w_2 \sin \beta_2' = C_{a2}$$

For a given pressure increase in a multistage compressor, it is obvious that a symmetric design would require the fewest stages. However, non-symmetric and vortex arrangements show higher efficiencies and are

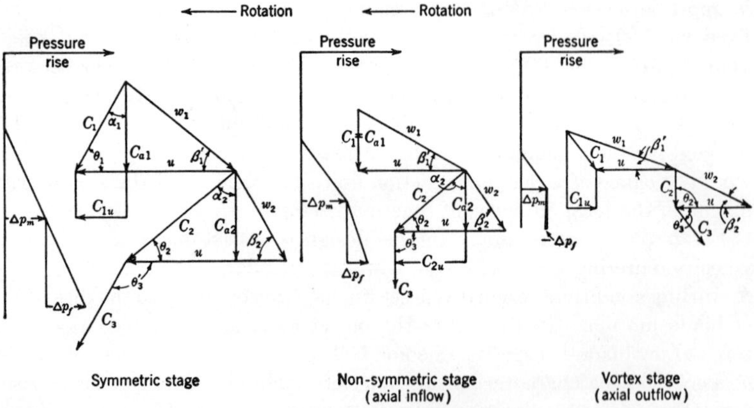

FIG. 5-5. Velocity diagrams to indicate blade positioning for various axial-flow arrangements. $C_{a1} = C_{a2}$ for each respective type.

employed where over-all performance is important and additional compressor weight is not too objectionable.

The total pressure rise in an axial-flow stage for any flow line at fixed radius from the center of rotation appears by Eqs. (5-1) and (5-2) as:

$$\Delta p = \frac{\rho}{2g_c} (w_1^2 - w_2^2 + C_2^2 - C_3^2) \tag{5-3}$$

By the methods of Art. 4-4 (when $C_3 \equiv C_1$) this can be reduced to the form

$$\Delta p = \frac{\rho}{g_c} (uC_{2u} - uC_{1u}) \tag{5-4}$$

$$\Delta p = \frac{\rho}{g_c} u(C_{2u} - C_{1u}) = \frac{\rho}{g_c} u \, \Delta C_u \tag{5-5}$$

Equations (5-3) to (5-5) are approximations except in the case of an incompressible fluid. However, as the pressure rise in each row is small, ρ does not change greatly and the pressure rise per stage can be approximated by Eqs. (5-3) or (5-5). The velocity of whirl ΔC_u has units of feet per second, and as before Δp has units of pounds force per square foot. By removing ρ from Eqs. (5-1) to (5-5) the units become *feet head of fluid flowing* (air, gas, etc.). A cubic foot of a fluid weighs $|\rho|$ lb per ft^3, and a column 1 ft high exerts a pressure of $|\rho|$ lb$_f$ per ft^2; thus for Eq. (5-5)

$$H = \frac{\Delta p}{\rho} = \frac{u}{g_c} \Delta C_u \tag{5-6}$$

where H is pressure rise in the stage measured in feet head of the fluid flowing.

5-4. Compressor Stage Performance. Apply Eq. (2-2) to a gas compressor stage. Under adiabatic conditions the work done *on* G lb of gas per sec in the moving row, or the power input, appears as

$$P_T = G \frac{{}_1W_2}{J} = G \left(h_2 - h_1 + \frac{C_2^2 - C_1^2}{2g_cJ} \right) \tag{5-7}$$

If we now adapt the notation of Fig. 5-4 for this equation, considering a moving- and fixed-row stage, there results

$$P_T = G \frac{{}_1W_3}{J} = G \left(h_3 - h_1 + \frac{C_3^2 - C_1^2}{2g_cJ} \right) \tag{5-8}$$

With G expressed as a flow rate, pound per second, both Eqs. (5-7) and (5-8) are power equations, and P_T is the total power input to the stage in Btu per second. Although the whole stage is considered, it must be realized that power interchange occurs only in the moving row. The fixed blades serve either as diffuser passages or as expanding passages; the only energy interchange possible is that between kinetic energy and enthalpy or between enthalpy and kinetic energy, *i.e.*, no work is done or power absorbed in the fixed row.

Let us rewrite Eq. (5-8) to show pressure explicitly, by expanding the enthalpy terms into internal energy and pv, thus:

$$\begin{aligned} P_T &= G \left[\left(\frac{p_3v_3}{J} + u_3 \right) - \left(\frac{p_1v_1}{J} + u_1 \right) + \frac{C_3^2 - C_1^2}{2g_cJ} \right] \\ &= G \left[\frac{p_3v_3 - p_1v_1}{J} + u_3 - u_1 + \frac{C_3^2 - C_1^2}{2g_cJ} \right] \end{aligned} \tag{5-9}$$

Equation (5-9) indicates the pressure terms p_3 and p_1, but their magnitudes cannot be found for a given power input without knowing the value of the other terms in the equation as well. Power input for a flow rate of G lb per sec can increase the pv term, the internal energy term $u_3 - u_1$, or the kinetic-energy term.

For a compressible fluid (gas or air) u increases with adiabatic pressure rise, even under reversible conditions. By use of Eq. (2-12), Eq. (5-9) can be written as

$$P_T = G\left[\frac{p_3v_3 - p_1v_1}{J} + c_{vm}(T_3 - T_1) + \frac{C_3^2 - C_1^2}{2g_cJ}\right] \tag{5-10}$$

By similar analysis, Eq. (5-8) can be written as

$$P_T = G\frac{{}_1W_3}{J} = G\left[h_3 - h_1 + \frac{C_3^2 - C_1^2}{2g_cJ}\right]$$
$$= G\left[c_{pm}(T_3 - T_1) + \frac{C_3^2 - C_1^2}{2g_cJ}\right] \tag{5-11}$$

To arrive at a determination of pressure rise, let us take the right-hand side of Eq. (5-8) and consider it in connection with Fig. 5-6. Equation

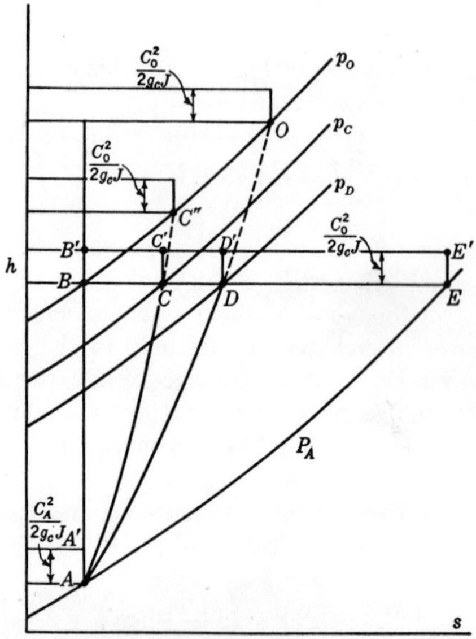

FIG. 5-6. *hs* diagram showing pressure rise in compressor or compressor stage in terms of work added. Inlet condition at A, exit condition at B, C, D, or E for same work added.

(5-8) is true not only for reversible conditions AB but also for the irreversible conditions illustrated as AC and AD. If the process were such that the gas were compressed adiabatically and reversibly, the isentropic line to B would indicate the final pressure to which the gas arrived in the compressor stage. However, the pressure realized is never this high, but arrives on the lower pressure line passing through D or, if the compressor is more efficient, arrives on an intermediate pressure line passing through C. Test performance alone will indicate what pressure can be realized in a stage of a compressor. It should be noted that the work supplied, which is represented by $h_B - h_A$, is the same for any of the processes, even though the pressure rise differs in each case. The work, when kinetic energies are included, is precisely represented by $h_{B'} - h_{A'}$ in the diagram. Here, C_A represents the velocity at inlet to the stage and C_O the velocity on outlet.

Example 1. A compressor employing a symmetric design shows in one of its stages a velocity triangle (see Fig. 5-4) at mean radius with values as follows:

$$C_1 = w_2 = C_3 = 636 \text{ fps},$$

$w_1 = C_2 = 819$ fps, $u = 1100$ fps, and $\beta_1' = 35°$. (*a*) Estimate the pressure rise in the moving and fixed rows of the compressor on the basis of compressing air entering the stage at 520 R and 14.7 psia and considering the air as essentially incompressible throughout the stage. (*b*) Compute the velocity of whirl for the stage, and compute the work absorbed from the rotor per pound of air compressed. (*c*) Compute the enthalpy increase in the stage and the temperature rise. (*d*) For the conditions of part (*c*), assuming reversible-adiabatic compression, find the stage pressure rise.

Solution. Symmetric stage. (*a*) By Eq. (2-3) air at 520 R and 14.7 psia has a density of 0.0763 lb per ft^3 $= \rho$. Use Eq. (5-1) to find the pressure rise in the moving row

$$\Delta p_m = \frac{0.0763}{64.34}(819^2 - 636^2) = 315.6 \text{ psf}$$

By Eq. (5-2) the pressure rise in the fixed row is

$$\Delta p_f = \frac{0.0763}{64.34}(819^2 - 636^2) = 315.6 \text{ psf}$$

For a symmetric stage the pressure rise is the same in both rows and totals 631.2 psf, or $631.2/144 = 4.38$ psi.

(*b*)

$$w_1 \cos \beta_1' = (819)(0.8192) = 671 = C_2 \cos \theta_2 = C_{2u}$$
$$C_{1u} = C_1 \cos \theta_1 = 1100 - 671 = 429 = w_2 \cos \beta_2'$$

The velocity of whirl is

$$\Delta C_u = C_{2u} - C_{1u} = 671 - 429 = 242 \text{ fps}$$

$$\text{Work absorbed} = \frac{1}{g_c}\Delta C_u\, u = \frac{(242)(1100)}{32.17} = 8270 \text{ ft-lb per lb of air}$$
$$= 10.63 \text{ Btu per lb}$$

(*c*) The work absorbed in an adiabatic stage all reappears as enthalpy increase when there is no over-all change in kinetic energy, and Eq. (5-8) or (5-11) shows

$$(1)(10.63) = (1)(h_3 - h_1) = (1)\,(c_p)(T_3 - 520)$$

From Table 2-4, $h_1 = 14.393$ at 520 R.

$$h_3 = 25.023 \text{ Btu per lb} \qquad \text{and} \qquad T_3 = 564.3 \text{ R}$$

(*d*) For this stage the enthalpy increased from 14.393 Btu per lb to 25.023 Btu per lb, and when this is considered to occur reversibly (isentropically), the corresponding pressure rise can easily be found by use of Table 2-4. Read for the h_3 and h_1 values $p_{r3} = 2.043$ and $p_{r1} = 1.535$, respectively.

$$\frac{p_{r3}}{p_{r1}} = \frac{2.043}{1.535} = 1.332 = \frac{p_3}{p_1}$$
$$p_3 = (1.332)(14.7) = 19.56 \text{ psia}$$
$$\Delta p = 19.56 - 14.7 = 4.86 \text{ psi}$$

The stage pressure rise is greater than indicated in part (*a*), where the assumption was made that the gas was incompressible, which is of course an approximation. However, the general method of part (*a*) is useful in indicating the relative magnitudes of the pressure changes which occur in the moving and in the fixed rows.

Consider the problem of how to evaluate stage performance. Figure 5-6 along the line *BCDE* indicates discharge conditions under various performance conditions for a fixed amount of work supplied, namely, $h_B - h_A$ or $h_{B'} - h_{A'}$. However, a compressor is called upon to deliver its air at a definite outlet pressure p_O, and the work input has to be increased by an amount adequate to deliver against this outlet pressure. For the irreversible compression *ADO*, the work supplied is equivalent to the enthalpy change, $h_O - h_A$, and, as a result of the losses in the machine, the delivery temperature T_O is higher than the temperature which would exist at *B* after isentropic compression. If compression had proceeded from temperature T_A to temperature T_O isentropically, the pressure would have been appreciably higher than that existing in following the irreversible path *ADO*. Internal compressor efficiency η_C (or compressor stage efficiency η_{CS}) is defined as the ratio of the minimum work required to compress a gas from inlet pressure p_A to outlet pressure p_O compared with the actual work of compression through the same pressure ratio. It can be seen for an adiabatic compressor that the minimum work is the isentropic and for a particular compressor with a characteristic compression pattern such as that for *A* to *O* the efficiency appears

$$\eta_C \text{ or } \eta_{CS} = \frac{(H_{oB} - H_{oA})_s}{H_{oO} - H_{oA}} = \frac{\left[\left(h_B + \frac{C_B^2}{2g_cJ}\right) - \left(h_A + \frac{C_A^2}{2g_cJ}\right)\right]_s}{\left(h_O + \frac{C_O^2}{2g_cJ}\right) - \left(h_A + \frac{C_A^2}{2g_cJ}\right)} \tag{5-12}$$

where H_o = kenthalpy, Btu per lb
h = enthalpy, Btu per lb
C = velocity, fps
η_C = internal compressor efficiency
η_{CS} = compressor stage efficiency

Fortunately Eq. (5-12) can be simplified because the velocities (kinetic energies) in the entering and leaving streams are not greatly different, and when this is the case,

$$\eta_C \text{ or } \eta_{CS} = \frac{(h_B - h_A)_s}{h_O - h_A} = \frac{(h_B/h_A)_s - 1}{(h_O/h_A) - 1} \tag{5-13}$$

$$\eta_C \text{ or } \eta_{CS} = \frac{(T_B/T_A)_s - 1}{(T_O/T_A) - 1} = \frac{(p_O/p_A)^{(k-1)/k} - 1}{(T_O/T_A) - 1} \tag{5-14}$$

where T_A = inlet compressor (or stage) temperature, °R
T_O = outlet compressor (or stage) temperature, °R
T_B = temperature resulting after isentropic compression from T_A through pressure ratio p_O/p_A
p_O = outlet static pressure from compressor (or stage), usually psia
p_A = inlet static pressure to compressor (or stage), usually psia

For stage analyses as previously considered in this chapter, the subscripts 1 and 3 have been used for inlet and outlet conditions, respectively. These correspond to A and to B or O, respectively, in Eqs. 5-12 to (5-14).

Example 2. An axial-flow compressor has an intake capacity of 20 lb per sec when supplied with air at 14.7 psia and 59.3 F (519 R). When tested at these inlet conditions and flow rate, it was found that the air temperature at outlet was 720 R and the internal compressor efficiency was 87 per cent. Velocities in the inlet and outlet compressor passages were equal and amounted to 120 fps. Find (*a*) the discharge pressure; (*b*) the internal compressor (rotor) horsepower; (*c*) the shaft horsepower input if the mechanical efficiency considering gland and bearing losses was 96.5 per cent.

Solution. (*a*) Both Eqs. (5-13) and (5-14) are applicable. From Table 2-4 and first using Eq. (5-13), $h_A = 14.153$, $p_{rA} = 1.524$ at 519 R, and $h_O = 62.560$ at 720 R.

$$0.87 = \frac{(h_B/14.153)_s - 1}{(62.560/14.153) - 1}$$

$h_B = 56.233$ Btu per lb, $T_B = 693.9$ R, and $p_{rB} = 4.224$.

$$\frac{p_B}{p_A} = \frac{p_{rB}}{p_{rA}} \qquad p_B = (14.7)\frac{4.224}{1.524} = 40.7 \text{ psia}$$

$$p_B = p_O = 40.7 \text{ psia}$$

To use Eq. (5-14), read $k = 1.398$ in the range of 519 to 720 R from Fig. 2-1.

$$0.87 = \frac{(p_O/14.7)^{(1.398-1)/1.398} - 1}{^{720}/_{519} - 1} = \frac{(p_O/14.7)^{0.2846} - 1}{0.3874}$$

$$\frac{p_O}{14.7} = (1.337)^{1/0.2846} = 2.773$$

$$p_O = 40.7 \text{ psia}$$

(*b*)
$$\text{Rotor hp} = \frac{G\left(h_O - h_A + \dfrac{C_O^2 - C_A^2}{2g_cJ}\right)}{0.707} = \frac{20(62.560 - 14.153 + 0)}{0.707} = 1371$$

(*c*)
$$\text{Shaft hp} = \frac{1371}{0.965} = 1422$$

A study of Eq. (5-11) shows that for a stage in which the entering and leaving kinetic energies are approximately equal ($C_1 \equiv C_3$) all the power delivered to the rotor appears in the gas undergoing compression as increased enthalpy. This increase for a gas is evidenced by a rise in temperature. When $C_1 \equiv C_3$, there results

$$P_T = G[c_{pm}(T_3 - T_1)] = G(h_3 - h_1)$$

In the event that there are no losses, the isentropic enthalpy increase can be found by determinations from the velocity diagrams and

$$P_T = \frac{P_m}{J} = G[c_{pm}(T_3 - T_1)]_s = G(h_3 - h_1)_s \tag{5-15}$$

where P_m is *diagram power, viz.*, power input determined in terms of velocity-diagram values and usually measured in foot-lb per second.

The diagram power absorbed in a compressor stage is the reverse of the power delivered in a turbine stage so that Eq. (4-9) would become for a compressor (Fig. 5-4)

$$P_m = \frac{Gu}{g_c}(C_{2u} - C_{1u}) \tag{5-16}$$

$$P_m = \frac{G}{g_c}u(C_2 \cos \theta_2 - C_1 \cos \theta_1) \tag{5-17}$$

$$P_m = \frac{G}{g_c}u(C_{a2} \tan \alpha_2 - C_{a1} \tan \alpha_1) \tag{5-18}$$

As the axial velocities at inlet to and outlet from the moving row are frequently equal ($C_{a1} = C_{a2}$),

$$P_m = \frac{G}{g_c}uC_a(\tan \alpha_2 - \tan \alpha_1) \tag{5-19}$$

Substitute Eq. (5-19) in Eq. (5-15), giving

$$\frac{G}{g_c}uC_a(\tan \alpha_2 - \tan \alpha_1) = GJc_{pm}(T_3 - T_1)_s = GJc_{pm}(\Delta T)_s$$

$$(\Delta T)_s = \frac{uC_a}{g_c c_{pm} J}(\tan \alpha_2 - \tan \alpha_1)$$

$$(\Delta T)_s = \frac{u}{g_c c_{pm} J}(C_{2u} - C_{1u}) \tag{5-20}$$

The temperature rise indicated by Eq. (5-20) is higher than would be realized in the actual compressor because of variations in diagram performance along the blade, blade end (annulus) effects, and flow deviations from the channel passage between the blades. The temperature rise is also affected by the ratio of c/s, namely, the blade chord c and the pitch s between blades (see Fig. 5-9). Howell[2] suggests the use of a *work-done factor* Ω for modifying the diagram temperature rise of Eq. (5-20) to meet actual performance and test conditions; thus

$$\Delta T = \Omega \frac{uC_a}{g_c c_{pm} J} (\tan \alpha_2 - \tan \alpha_1) \tag{5-21}$$

$$\Delta T = \Omega \frac{u}{g_c c_{pm} J} (C_{2u} - C_{1u}) \tag{5-22}$$

Also

$$h_3 - h_1 = \Delta T \, c_{pm} = \Omega \frac{u}{g_c J} (C_{2u} - C_{1u}) \tag{5-23}$$

A value of $\Omega = 0.86$ can be considered representative for general use. The fact that Ω is less than unity means not that less work is required for the inefficient compressor but merely that churning turbulence is less effective in power absorption than productive compression leading to pressure rise.

Comparison of Eqs. (5-22) and (5-20) shows that

$$\Delta T = \Omega(\Delta T)_s \tag{5-24}$$

Stage pressure factor or simply *pressure factor* ψ is defined as

$$\psi = \frac{(\Delta h)_s}{u^2/2g_c J} = \frac{2g_c J (\Delta h)_s}{u^2} \tag{5-25}$$

where $(\Delta h)_s$ is the isentropic enthalpy change per stage. The numerator of this expression is in velocity form by Eq. (2-44), and as the earlier discussions of this chapter in turn have shown how velocity is related to pressure, Eq. (5-25) is really a pressure parameter.

The term *flow coefficient* ϕ is defined as

$$\phi = \frac{C_a}{u} \tag{5-26}$$

where C_a is axial velocity and u is blade speed.

The blade speed u which appears in both these factors is exceedingly important in compressor design. For minimum weight of machine it is desirable to keep it as high as practicable. The ultimate value of u is set by stress considerations for the blading and rotor material available. However, u is also dependent on the desired rotor rotational speed and at the same time on compressor dimensions required, primarily over-all diameter and the annulus depth which sets the free area at the blade ring.

The available area at the blade ring must pass the required flow of air, and the area and flow together set the axial velocity C_a.

Figures 5-7 and 5-8 (after Woodworth[1]) show values of pressure coefficient and stage efficiency plotted against ϕ for varying Mach numbers. These are indicative of what might be expected in a conventional design,

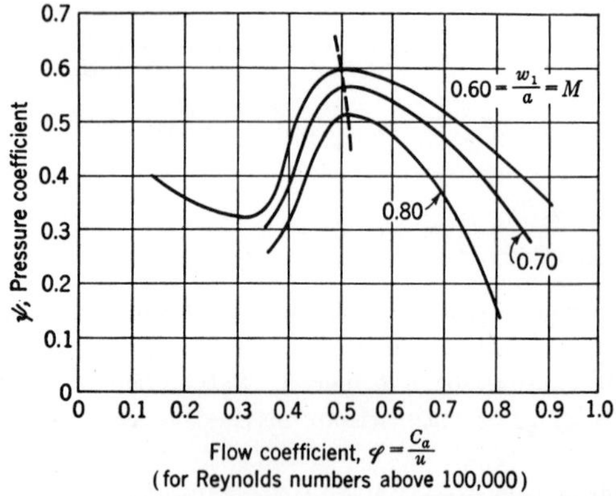

FIG. 5-7. Axial-flow pressure coefficients. Pressure coefficients require a multiplying factor of 2 when used in certain equations. (*After Woodworth, Ref.* 1.)

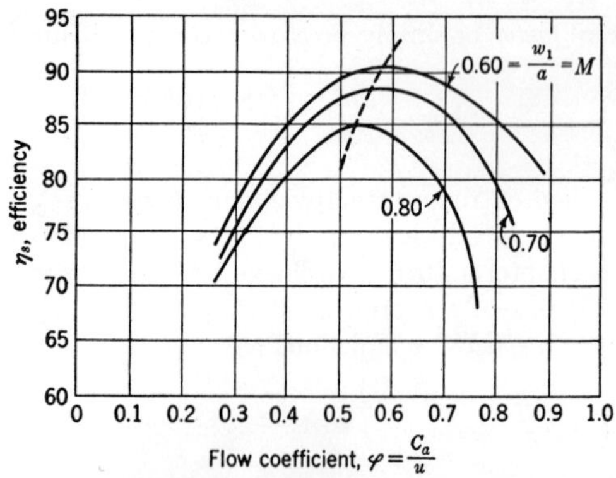

FIG. 5-8. Single-stage efficiency for axial-flow compressors showing relation to varying Mach number. (*After Woodworth, Ref.* 1.)

but because of design variables other than Mach number, they cannot be used to predict performance over broad operating or design conditions.

Reference to Eqs. (4-13), (4-14), and (4-15) gives values of turbine blade speeds at the root or hub, r_h, at mean blade height and at the tip. These can be used in exactly the same form for axial compressor blades. The horsepower input to each compressor stage can be found starting with

Eq. (5-23) or (5-21); thus

$$\mathrm{hp} = \frac{Gc_{pm}\,\Delta T\,J}{550} \tag{5-27}$$

$$\mathrm{hp} = \frac{G\Omega u(C_{2u} - C_{1u})}{550g_c} \tag{5-28}$$

$$\mathrm{hp} = \frac{G\Omega u(C_{2a}\tan\alpha_2 - C_{1a}\tan\alpha_1)}{550g_c} \tag{5-29}$$

where G = gas flow through the stage or portion of blade length, lb per sec
hp = horsepower required for a stage (or portion of a stage dependent on G and u) when variations in kinetic energy to and from the stage are negligible.
u = blade speed, fps

In case of a blade with widely varying power requirements at hub, mean radius, and tip, more than one calculation must be made for stage horsepower with the value of G prorated for each segment of the blade. Other symbols are as used in Eqs. (5-14) to (5-23).

In testing of complete compressors where it is not possible to measure the power input through a torque-measuring device or dynamometer, performance characteristics can yet be found by making precision measurements of the temperature and pressure at inlet to and from the compressor. Then in terms of adiabatic- (isentropic) efficiency characteristics as given by Eq. (5-12), it is possible to estimate the performance of the machine. If the mass flow is known, use of Eq. (5-8) in expanded form can yield the horsepower absorbed in the machine exclusive of bearing and radiation losses.

$$\mathrm{hp} = \frac{JG\left[h_o - h_1 + \dfrac{C_o^2 - C_1^2}{2g_cJ}\right]}{550} = \frac{G\left[Jc_{pm}(T_o - T_1) + \dfrac{C_o^2 - C_1^2}{2g_c}\right]}{550} \tag{5-30}$$

where G = air (gas) flow, lb per sec
T_o, T_1 = static temperature of air (gas) at outlet from and inlet to compressor, °F or °R
h_o, h_1 = corresponding enthalpy values, Btu per lb
c_{pm} = mean sp heat over range T_1 to T_o for air (gas)
C_o and C_1 = air (gas) velocities at outlet and inlet, respectively, fps

Example 3. An axial-inflow nonsymmetric compressor stage has a mean blade speed of 750 fps, and the velocity entering axially at inlet to the stage is 400 fps. Referred to the nomenclature of Fig. 5-5, β_2' is 60°. The axial velocity component on outlet is 400 fps. The inlet pressure is 13.4 psia, inlet temperature 516.4 R, and $\rho = 0.07$ lb per ft.^{3} The air flow through the compressor is 30 lb per sec. Test data show that stages of this type have an efficiency of 87 per cent. (*a*) Based on mean blade conditions and velocity-diagram data, estimate a probable maximum pressure rise in this stage if the air is considered essentially incompressible. (*b*) What portion of this pressure rise occurs in the moving row? (*c*) Find the ideal and actual temperature rise associated with the mean blade velocity-diagram values. (*d*) Compute the delivery static pressure from the stage. (*e*) Find the horsepower required for this stage based on modified velocity-diagram values at mean blade location.

Solution. (*a*) From a graphical plot of the velocity triangles or by trigonometry it can be found that $w_1 = 850$ fps, $w_2 = 462$ fps, $C_2 = 657$ fps, $C_{1u} = 0$, and $C_{2u} = 519$ fps. By Eq. (5-5), which is applicable for total pressure rise when $C_1 = C_3$,

$$\Delta p = \frac{0.07}{32.17}(750)(519 - 0) = 848 \text{ psf}$$

$$\Delta p = 848/144 = 5.89 \text{ psi}$$

or by Eq. (5-3),

$$\Delta p = \frac{0.07}{(2)(32.17)}(850^2 - 462^2 + 657^2 - 400^2)$$

$$= \frac{0.07}{64.34}[(1312)(388) + (1057)(257)] = 848 \text{ psf} = 5.89 \text{ psi}$$

(*b*) By Eq. (5-1)

$$\Delta p_m = \frac{0.07}{64.3}(850^2 - 462^2) = 554 \text{ psf} = 3.85 \text{ psi pressure rise in moving blade row}$$

By Eq. (5-20) and for low-temperature air with c_{pm} as 0.240 from Table 2-4

$$(\Delta T)_s = \frac{750}{(32.17)(0.24)(778)}(519 - 0) = 64.8°$$

$T_s = 516.4 + 64.8 = 581.2$ R ideal temperature after compression if isentropic

Use $\Omega = 0.86$.

$\Delta T = \Omega(\Delta T)_s = (0.86)(64.8) = 55.7°$

$T = 516.4 + 55.7 = 572.1$ R probable actual temperature after compression

(*d*) By Eq. (5-14)

$$\eta_s = 0.87 = \frac{(p_0/13.4)^{1.399-1/1.399} - 1}{(572.1/516.4) - 1} = \frac{(p_0/13.4)^{0.285} - 1}{1.108 - 1}$$

$$p_0 = 18.35 \text{ psia}$$

$$\Delta p = 18.35 - 13.4 = 4.95 \text{ psi}$$

(*e*) By Eq. (5-27)

$$\text{hp} = \frac{(30)(0.24)(55.7)(778)}{550} = 567$$

5-5. Blading Characteristics. Figure 5-9 shows the representative nomenclature and outline of typical axial-compressor (or turbine) blading. The notation used corresponds to that which has been previously employed in Figs. 5-4 and 5-5. However, here a necessary distinction is made between blade angles and fluid-flow angles. In the previous discussions, it was tacitly assumed that design conditions obtained and that the channel passages made by adjacent blades were sufficiently narrow so that the flow directions of the gas were tangent to the respective blade angles. This condition is not actually realized over a range of operating conditions.

Referring to Fig. 5-9, the inlet blade angle β_1, which is drawn to the mean camber line, and the outlet blade angle β_2, drawn in similar manner, are shown. The actual fluid angle on inlet β_1' may be less than or greater than β_1, and the same is true of the exit flow angle β_2'. The actual deflection through which the fluid turns in passing through the blading is known

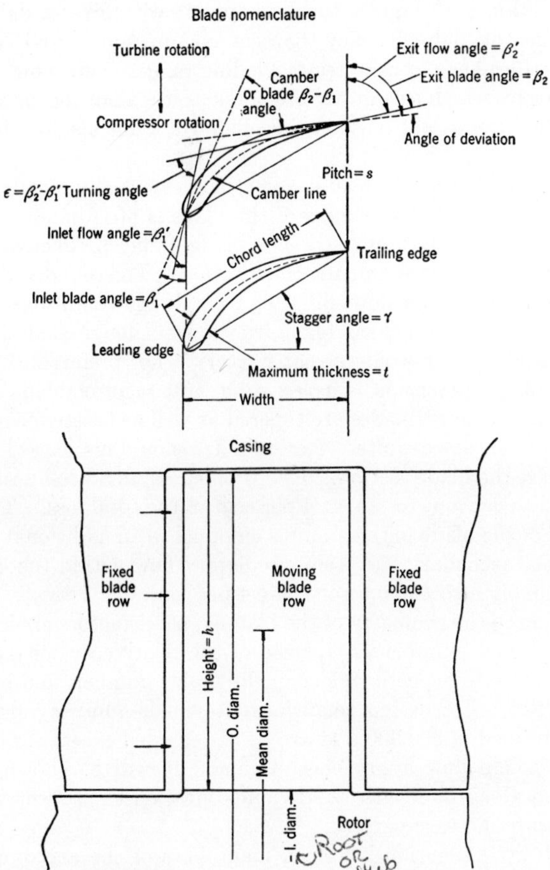

FIG. 5-9. Axial-flow compressor (turbine) blading nomenclature.

as the turning angle and is obviously $\beta_2' - \beta_1'$ in magnitude. The mean camber, or blade, angle is obviously $\beta_2 - \beta_1$. It is conceivable that these turning angles $\beta_2' - \beta_1'$ and $\beta_2 - \beta_1$ could be the same, but in most cases there is some difference, even at design conditions, between the turning angle and the camber angle. The angle of incidence i, also known as the angle of attack, is the angle at which the blading meets the air stream and is measured by $\beta_1 - \beta_1'$. It is obvious that this can be negative or positive, depending on the direction of the stream. With β_1' and β_1 as drawn, the angle of incidence i is positive. For lowest coefficients of drag (loss) of a blade section, it is desirable to keep the incidence angle in a range of -10 to $+10°$.

Both reaction and impulse turbines operate with greater deflection or turning angles at high efficiency than can compressors. Fluid deflection for good performance in compressor blading ranges from some 15 to 30°. The amount by which the air turning angle is less than the camber angle of the blade depends not only on the solidity c/s but also on the stagger angle.

The mean camber line of blades, sometimes essentially circular, is more usually parabolic. The thickness of the blade is often made to suit the dimensions of a standard airfoil. The blading may be overcambered in order to give the required amount of deflection. The rounded nose of the blade, over a range of air flows differing from design conditions, gives less shock loss than would a sharp blade front, even though at design conditions the sharper front would be satisfactory if not preferable. Another important blade dimension is *aspect ratio, viz.*, ratio of blade height to blade chord. As most blades are tapered as well as twisted, aspect-ratio values are not always definite. Ratios of h/c should not exceed 10. The drag loss past the blade section, called profile loss, in a good design might be expected to account for about 4 per cent of the total loss. The loss at the tip end of the blade in the annulus amounts to an additional 2 to 3 per cent, and the secondary loss from circulatory flow within the interblade section probably also amounts to some 4 per cent.

In addition to the geometry of the blading, other factors are important, such as the Mach number (w/a, ratio of the relative velocity of the air stream to the local acoustic velocity), Reynolds number, and blade base characteristics. For design conditions, Reynolds-number values are in the neighborhood of 500,000 and can be disregarded except for extremely low flows, where flow approaches stalling. It will be shown how the blade design along its length requires modification to prevent variations in the pressure-increase pattern.

Although little mention has been made here of the aerodynamic approach to compressor stage design, a relation exists between the lift force of a blade section and the pressure rise which can be produced in the blading. A development of this relation in Art. 5-7 shows that pressure rise varies with the angle of attack (incidence). When the flow decreases, the angle of attack increases, with a resultant higher pressure rise. This is indicated by the steep flow lines of Fig. 5-3 and confirms the well-known fact that when the angle of attack reaches the critical value, flow separation and surge occur, with extremely unsatisfactory performance as a result. Figure 4-14 showed how the blade flow angles changed in the case of a turbine blade and indicated relatively the same condition that would apply in the case of a compressor, although in the case of a compressor this is brought about in a different manner. When solidity (c/s) values are in excess of 1 to 1.3, the velocity-diagram theory can be employed and give a reasonable basis of analysis. However, in the range below this more dependence must be placed on the results of cascade tests of prototype grids.[2,4]

5-6. Velocity Distribution in Blading. In Art. 4-9, in discussing turbine blading, mention was made of various velocity distributions in the blading passage. In either turbine or compressor blading, radial variations in gas pressure beyond those associated with the centrifugal pressure field can cause radial gas flows which, if excessive, could upset design patterns and lower expected performance. The centrifugal field to which the fluid is subjected is not associated with the blade velocity u, as the fluid itself is moving with a tangential velocity of only C_u as it passes through the blade. In the case of the nonsymmetric blade of Fig. 5-5, for example, it should be noted that with axial inlet there is no tangential component and consequently no swirl or centrifugal field associated with the gas on inlet. As the gas passes through the blade, however, the tangential velocity increases to C_{2u} with its associated centrifugal field.

The power for a flow of G lb per sec through any unit height of blading passage by Eq. (5-16) can be expressed

$$P_m \propto u\,\Delta C_u \propto 2\pi N r\,\Delta C_u$$

or for any given blade turning at N rpm

$$P_m \propto r\,\Delta C_u \tag{5-31}$$

This states that, for the same flow rate throughout the blade length, to produce or absorb equal power at each radius the product (radius length times the velocity of whirl at that radius) must be constant throughout the length of blade.

The following discussion is of interest only to those who wish to follow an analytical development of the conditions of radial equilibrium.

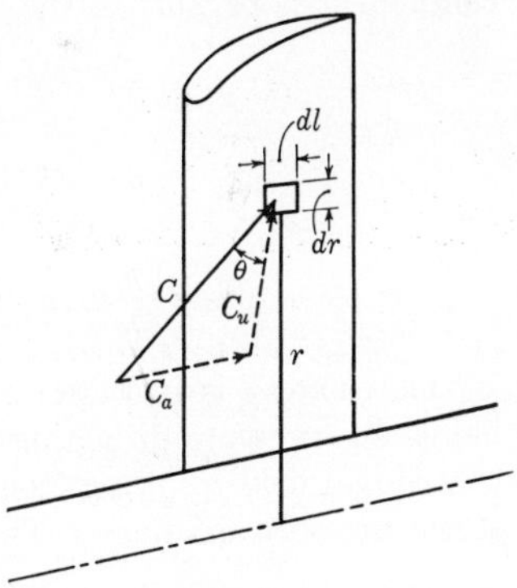

FIG. 5-10. Velocity components in a blade passage.

Let us analyze in detail the effect of velocity variation at progressive outward positions in a moving blade. Consider the diagrammatic arrangement indicated in Fig. 5-10. Here the blade at a given radius r is moving with a velocity u, and the gas is moving in the blade passage with a directed absolute velocity C. The component of this velocity in the direction of u is indicated as C_u, and if θ is the angle between C and C_u, then $C_u = C \cos \theta$. However, C_a is $C \sin \theta$ only when C lies in a plane parallel to the axis of the machine.

The energy of this moving gas at a point is

$$\frac{C^2}{2g_cJ} + h = H_o$$

where H_o is the kenthalpy (total energy) of the moving stream.

For the case of C lying in a plane parallel to the axis of the machine, we can write the previous equation in terms of its components, C_u, tangentially, and C_a, axially:

$$\frac{C_u^2}{2g_cJ} + \frac{C_a^2}{2g_cJ} + h = H_o \tag{5-32}$$

Differentiate (5-32) to show the change in kenthalpy in a radial direction (*i.e.*, out from the axis).

$$\frac{1}{g_cJ}\left(\frac{C_u\,dC_u}{dr} + \frac{C_a\,dC_a}{dr}\right) + \frac{dh}{dr} = \frac{dH_o}{dr} \tag{5-33}$$

Consider an infinitesimal element of gas of height dr, breadth dl, and unit width moving with tangential velocity C_u. The velocity C_u alone contributes to the centrifugal force on this element as the axial-velocity component is parallel to the axis of rotation.

$$F_c = \frac{mC_u^2}{r} \tag{5-34}$$

$$m = \frac{\rho\,dl\,dr\,(1)}{g_c} \qquad \text{where } \rho \text{ is the density}$$

$$F_c = \frac{\rho}{g_c}\,dl\,dr\,(1)\,\frac{C_u^2}{r} \tag{5-35}$$

If this element is in radial equilibrium, F_c must be counteracted by a balancing pressure dp acting over the area dl (1). In other words, the centrifugal field resulting from the swirling gas is balanced by the radial static pressure gradient. Thus

$$F_c = \frac{\rho}{g_c}\,dl\,(1)\,\frac{dr}{r}\,C_u^2 = dl\,(1)\,dp \tag{5-36}$$

$$\rho = \frac{1}{v} \qquad \text{where } v = \text{sp vol of the gas}$$

$$\frac{C_u^2}{vg_c}\frac{dr}{r} = dp$$

$$\frac{C_u^2}{g_c}\frac{dr}{r} = v\,dp \tag{5-37}$$

Thermodynamic theory can show that in a reversible-adiabatic process a change in enthalpy is equivalent to $v\,dP$ for the process, and thus

$$v\,dp = J\,dh \tag{5-38}$$

Substituting in Eq. (5-37),

$$\frac{C_u^2}{Jg_c}\frac{1}{r} = \frac{dh}{dr} \tag{5-39}$$

Put this expression in the differentiated form of the energy equation (5-33).

$$\frac{1}{g_c J}\left(\frac{C_u\,dC_u}{dr} + \frac{C_a\,dC_a}{dr} + \frac{C_u^2}{r}\right) = \frac{dH_o}{dr} \tag{5-40}$$

Consider the case where no change in the kenthalpy of the gas flowing in the blade passage takes place in a radial direction, that is, $dH_0/dr = 0$. For this desired condition Eq. (5-40) becomes

$$\frac{C_u\,dC_u}{dr} + \frac{C_a\,dC_a}{dr} + \frac{C_u^2}{r} = 0 \tag{5-41}$$

For C lying in a plane parallel to the axis of the machine

$$\tan\theta = \frac{C_a}{C_u} \tag{5-42}$$

If in addition we make the blade angle θ the same at every radius (*i.e.*, uncorrected with constant angle from hub to tip), it follows that

$$\frac{1}{\tan\theta} = \frac{C_u}{C_a} = K \qquad C_u = KC_a$$

$$dC_u = K\,dC_a \tag{5-43}$$

Substituting in Eq. (5-41),

$$\frac{K^2C_a\,dC_a}{dr} + \frac{C_a\,dC_a}{dr} + \frac{K^2C_a^2}{r} = 0 \tag{5-44}$$

$$\frac{dC_a}{dr}[K^2 + 1] = -\frac{K^2C_a}{r}$$

$$\frac{dC_a}{C_a} + \frac{K^2\,dr}{(K^2 + 1)r} = 0 \tag{5-45}$$

This differential equation has a solution,

$$\log C_a + \log r^n = \log \text{constant} \tag{5-46}$$

$$C_a r^n = \text{constant} \tag{5-47}$$

where

$$n = \frac{K^2}{K^2 + 1} = \frac{1/\tan^2\theta}{(1/\tan^2\theta) + 1} = \cos^2\theta$$

Thus

$$C_a r^{\cos^2\theta} = \text{constant} \tag{5-48}$$

A flow satisfying this equation would be in radial equilibrium, but the velocity distribution is not satisfactory, as C_a would vary greatly from hub to tip, the velocity triangles would be very much distorted from hub to tip, and the greater flow near the hub would give much greater work there than at the blade tip.

A preferable equilibrium arrangement is to make use of what is known as the free-vortex-type distribution. A free vortex when unconfined is in radial equilibrium. As before, let us assume no radial change in kenthalpy, and in addition let us set the axial velocity as constant at every radius. With dC_a necessarily zero, Eq. (5-41) becomes

$$\frac{C_u\,dC_u}{dr} + \frac{C_u^2}{r} = 0 \tag{5-49}$$

$$\frac{dC_u}{C_u} + \frac{dr}{r} = 0$$

$$\log C_u + \log r = \text{constant}$$

$$C_u r = \text{constant} \tag{5-50}$$

This equation, which is a satisfactory basis for design, gives radial equilibrium and equal work per pound of gas radially from hub to tip. To bring about this result, the blade angles must necessarily change from the hub to the tip.

Of the two designs presented, the second is preferable, but both are purely bases for theoretical design. Viscosity, boundary effects, and radial motion of the supply stream tubes can alter the true flow, and merely twisting the blade angles does not ensure perfect performance. Experimental confirmation of a design should be made in all cases. As we are concerned with the pressure change along the blade chord at any radius in relation to the centrifugal field, it may be necessary to use Eq. (5-50) as

$$\Delta C_u\, r = \text{constant} \tag{5-51}$$

even if $C_{1u}r$ or $C_{2u}r$ is not exactly constant.

5-7. Aerodynamic Considerations in Blading Design. In Fig. 5-11 appear representative curves of the lift and drag coefficients of an airfoil of a type which in modified form might be used as a basis for developing a blade section. A medium of constant density, ρ lb per ft³, flowing around an airfoil causes a lift force, per unit length of blade, dh, which can be expressed as

$$\Delta L = C_L\rho\,\Delta A\,\frac{w^2}{2g_c} = C_L\rho(c\,dh)\,\frac{w^2}{2g_c} \tag{5-52}$$

and the drag force is similarly

$$\Delta D = C_D\rho(c\,dh)\,\frac{w^2}{2g_c} \tag{5-53}$$

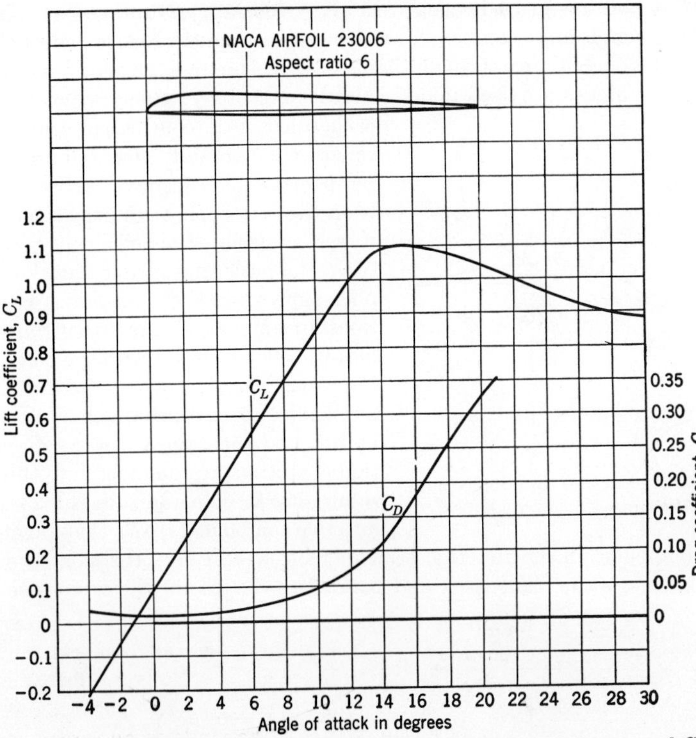

FIG. 5-11. Lift-drag coefficients of NACA airfoil. (*Jacobs, Pinkerton, and Greenberg, Ref.* 10.)

where ΔL = lift, lb force
ΔD = drag, lb force
ΔA = an element of area, ft^2
C_L = lift coefficient, dimensionless
C_D = drag coefficient, dimensionless
ρ = density, lb per ft^3
c = chord length, ft
dh = increment of airfoil (or blade) length, ft
w = velocity of air over the airfoil (blade) section, fps, usually taken as a mean of velocities w_1 and w_2 as previously defined in the blade diagrams

A great variety of foil (blade) forms have been tested in wind tunnels, and test data appear throughout the literature and in standard aerodynamics texts. In general, the blade form must differ appreciably from standard foils although the foil represents a good starting point for design as the high lift and low drag developed in many foils are advantageous. The blade section is more cambered, *i.e.*, bent more than the foil, to deflect the stream through a greater angle than would be true in a conventional foil. Moreover blades are used under conditions whereby, acting adjacent to each other in a continuous grid arrangement around the compressor periphery, they behave differently from the way each blade would act if it were an isolated airfoil.

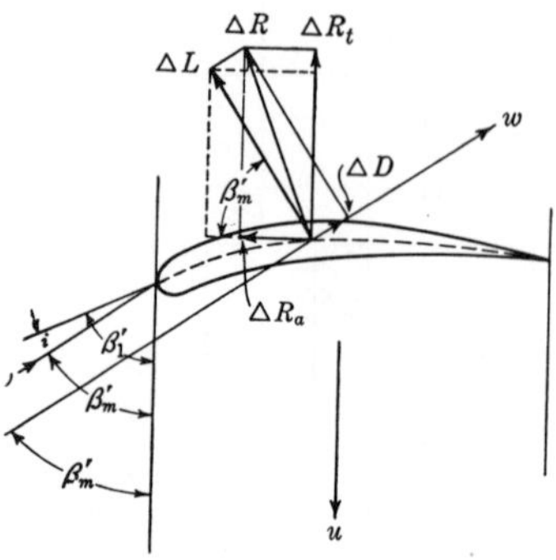

FIG. 5-12. Lift and drag forces on a blade element.

On the basis that airfoil theory applies approximately to blading, a relationship among the important variables will be developed with the help of Fig. 5-12. In this figure, the lift ΔL perpendicular to the direction of air motion and the drag ΔD in the direction of motion are combined into their resultant ΔR. This in turn is broken into two components, ΔR_t, the tangential component in the direction of blade motion, and ΔR_a, the axial component across the blade itself and parallel to the axis of the compressor.

$$\begin{aligned}\Delta R_t &= \Delta L \sin \beta_m' + \Delta D \cos \beta_m' \\ &= \frac{C_L \rho c\, dh\, w^2}{2g_c} \sin \beta_m' + \frac{C_D \rho c\, dh\, w^2}{2g_c} \cos \beta_m' \\ &= \frac{C_L}{2g_c} \rho c\, dh\, w^2 \sin \beta_m' \left(1 + \frac{C_D}{C_L} \frac{\cos \beta_m'}{\sin \beta_m'}\right)\end{aligned}$$

The value of C_D/C_L is small, $0.1\pm$, and as $\cos \beta_m'/\sin \beta_m'$ approximates unity under usual conditions, the final term can be dropped with little error, giving

$$\Delta R_t = \frac{C_L}{2g_c} \rho c\, dh\, w^2 \sin \beta_m' \tag{5-54}$$

where β_m' is the angle indicating the direction of the mean air flow measured with respect to the direction of blade motion. Note that i in Fig. 5-12 is a representative angle of incidence (angle of attack) where as drawn it is positive in direction.

Equation (5-16) represents the power delivered for a flow of G lb per sec. If for a blade element of height dh we call the corresponding flow ΔG, Eq. (5-16) could appear as

$$\Delta P_m = \frac{\Delta G}{g_c} u(C_{2u} - C_{1u}) = \frac{\Delta G\, u}{g_c} \Delta C_u \tag{5-55}$$

but

$$\Delta G = \rho s\, dh\, C_a \tag{5-56}$$

where s = blade pitch, ft
dh = a blade-height elemental length, ft
C_a = axial velocity, fps
ρ = density, lb per ft^3

Thus Eq. (5-55) with the substitution becomes

$$\Delta P_m = \frac{\rho s\, dh\, C_a}{g_c} u\, \Delta C_u \tag{5-57}$$

If u is removed from Eq. (5-57), it changes from a power equation to an expression for the tangential force on the blade element,

$$\Delta R_t = \frac{\rho s\, dh\, C_a}{g_c} \Delta C_u \tag{5-58}$$

Equate (5-58) to (5-54).

$$\frac{\rho s\, dh\, C_a}{g_c} \Delta C_u = \frac{C_L}{2g_c} \rho c\, dh\, w^2 \sin \beta_m'$$

$$C_L \frac{c}{s} = 2 \frac{C_a}{w^2} \frac{\Delta C_u}{\sin \beta_m'} \tag{5-59}$$

Reference to Fig. 5-4 or 5-5 will show that

$$\sin \beta_1' = \frac{C_a}{w_1} \qquad \sin \beta_2' = \frac{C_a}{w_2}$$

Therefore

$$\sin \beta_m' = \frac{C_a}{w}$$

and substituting in Eq. (5-59) there results

$$C_L \frac{c}{s} = \frac{2\, \Delta C_u}{w} \tag{5-60}$$

where w is the mean relative velocity through the blading. This relation derived for a blade element is equally applicable anywhere along the blade. It should be noticed that the velocity of whirl ΔC_u is directly associated with the lift coefficient (and consequently angle of attack), the solidity c/s, and the mean relative velocity in the blade passage

$$\Delta C_u = C_L \frac{c}{s} \frac{w}{2} \tag{5-61}$$

This expression is but one of a large number of relationships used in relating blade performance to foils or blade cascades. Additional forms will not be derived here but can be found in the literature (Refs. 1,2,3).

The coefficients of lift and drag when considered in regard to cascades (grids of test blades) are not so important as they are in the case of isolated airfoils. In a grid the lift coefficient alone does not express the stalling limit, nor is efficiency defined by drag characteristics. The results of cascade tests must be modified when stability conditions are being investigated, and the blade shape is a compromise between that indicated by aerodynamic theory and experience with previous compressor blade designs.

Example 4. An axial-flow air compressor in one of its stages passes air at $\rho = 0.053$ lb per ft.[3] It has blades which closely resemble the airfoil of Fig. 5-11. The blades are 6.25 in. long and have a chord of 1.25 in. The pitch of the blades is 1.25 in. During a certain test it was found that the mean relative velocity of the air was 608 fps, the blade speed 750 fps, and the air entered the passage at an equivalent angle of attack of $+6°$ (*i.e.*, in Fig. 5-9, $i = \beta_1 - \beta_1' = +6°$). Compute the probable velocity of whirl, and estimate a pressure rise in the stage which employed the previously described moving-row blading.

Solution. The lift coefficient from Fig. 5-11 is 0.55, and by Eq. (5-61)

$$\Delta C_u = C_L \frac{c}{s} \frac{w}{2} = 0.55 \frac{1.25}{1.25} \frac{608}{2} = 167 \text{ fps}$$

By Eq. (5-5), the stage pressure rise

$$\Delta p = \frac{0.053}{32.17} (750)(167) = 206 \text{ psf}$$

It should be realized that an application of single-foil theory to a cascade of moving blades (as in a compressor) is necessarily open to some question. However, a representative pressure rise for the stage is definitely indicated. For the stage in this case, it need only be considered that a stationary guide vane or diffuser row is available to cause the leaving absolute velocity to be the same as the entering absolute velocity. The particular airfoil of Fig. 5-11 would have a camber line of much greater bend in a real blade to produce a greater deflection of the fluid in passage through the row.

5-8. Compressor Design. The stage-by-stage design of an axial-flow compressor is not complicated after a decision has been made as to the form of the velocity diagrams, the radial-flow pattern, and the manner in which $\phi = C_a/u$, ψ, the stage efficiency, the solidity c/s, and blade lift characteristics vary throughout the compressor. From compressors already constructed, an idea of expected stage efficiency is available. The decision as to whether the compressor be built with constant over-all diameter or with gradually decreasing diameter is a designer's choice. The shape and twist of the blade, in turn, must fit with previously made considerations and stress requirements. In order to reduce the number and style of the different blades required in a multistage compressor, blade types are sometimes standardized into two or three patterns. The number of blades, stagger angles, and height are changed to suit desired conditions. As the air passes through the compressor, it decreases in specific volume, and the over-all flow annulus must decrease to account for the change in specific volume, or the axial velocities must be greatly changed.

Because of the lower temperatures employed in a compressor in contrast to a turbine, a greater range of materials is available for the compressor. These can be light-alloy materials, although it is probably desirable to have the blading made of stainless steel with good vibration-damping characteristics. The fixed blading, because it is not subjected to rotational stress, may be made of different material from the moving blades. Under conventional conditions, the whole compressor can operate as designed with regard to the flow pattern and pressure rise of each stage. At variable capacities, however, some of the rows of the compressor may be operating at much less desirable conditions than other rows and even beyond the point of stall. Particularly at extremely light flows, axial-compressor performance is unsatisfactory and it may be desirable to set up a return by-pass around the compressor so that a proper flow balance is maintained in the compressor even though the delivery to the remainder of the system has to be reduced.

Figure 5-13 represents a set of velocity triangles drawn to scale for the tip, mean position, and hub of a relatively long blade. The triangles, as drawn, represent a free-vortex velocity pattern [Eq. (5-51)]. The blade throughout its length receives gas both at constant inlet velocity C_1 and fixed inlet angle from guide vanes or from a previous blade row. It can be seen that the velocity of whirl is least at the outer tip and maximum near the hub. Notice also that in this distribution the tangential velocity component in the direction of blade motion is appreciably greater near the hub than at the tip.

In Fig. 5-13, the blade diagrams are set up in two forms. The form used at the right of the figure is the conventional one, while the polar form at the left of the figure is convenient for showing graphically variations in velocity of whirl and in related velocity vectors. In this polar form, the relative velocities appear at the right of each diagram, and an average value between the entering and leaving relative velocity gives the mean relative value which can be employed when aerodynamic analyses are being made of the blading. The angle between w_1 and w_2 represents the deflection of the stream in the moving blade. At the left of the polar diagrams, the corresponding values of absolute velocities appear. It should also be observed that the velocity of whirl can be expressed in terms either of the absolute (C_1 and C_2) or of relative (w_1 and w_2) velocities.

In this free-vortex distribution the pressure rise in the moving row is greatest at the tip and least at the hub, but with the higher absolute velocity leaving the moving row at the hub it is easily possible to offset this effect in the fixed row and have essentially the same pressure rise throughout the stage at every position. Notice also that the relative velocities near the hub are lower, implying lower Mach numbers for the flow than at the tip. However, near the hub the blade turning angles are

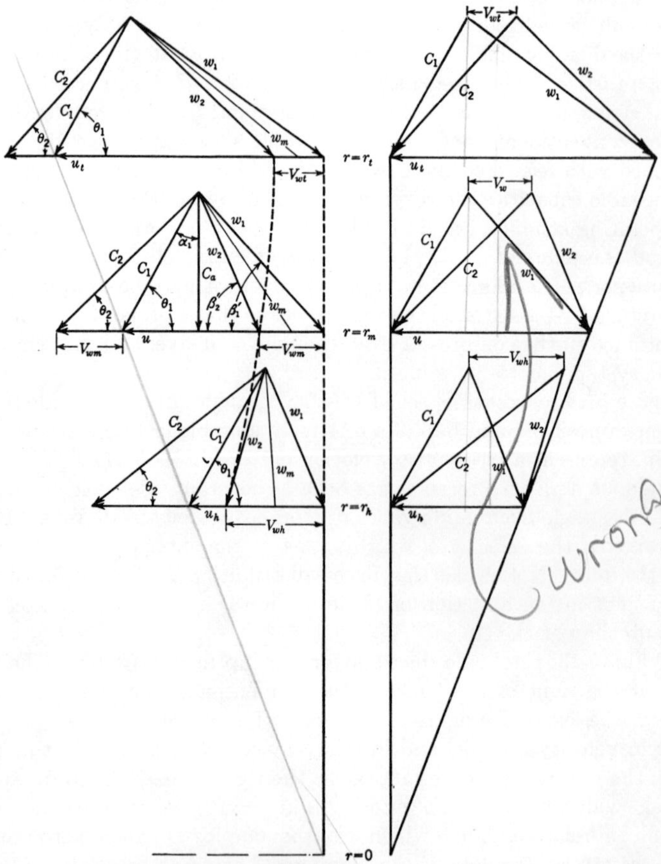

FIG. 5-13. Representative blade velocity triangles for tip, mean, and hub positions showing free-vortex distribution. $r_t = 2r_h$).

greatest. Wide variation in blade angles exists from hub to tip. At the tip, the velocity triangles are almost symmetrical, whereas this is far from true at the hub.

A similar set of diagrams can be constructed for any velocity distribution which might be selected for a stage. For example, one distribution which has been employed is the so-called "50 per cent reaction," which means that the enthalpy increase in both moving and fixed rows is the same at all positions throughout the blade. With short blades, variations in the pattern along the blade are not too important. With long blades, difficulties can arise if a suitable distribution is not chosen.

Throughout the compressor, the continuity-of-mass equation must be satisfied, and the relations developed for the turbine in Eqs. (4-65) to (4-79) are applicable to the compressor as well. However, again it should be mentioned that the gas-flow-direction vectors seldom exactly agree with blade-angle directions, and this is particularly true when the solidity c/s is lower than 1.5.

Example 5. Air at 640 R enters the moving row of the fourth stage of an axial-flow compressor from the previous fixed row at an angle of $\theta_1 = 60°$ ($\alpha_1 = 30°$) with an axial velocity of 500 fps. The moving blades are 6 in. long from hub to tip and have a mean diameter of 18 in. Compute the blade speeds at hub, mean position, and tip when the compressor turns at 9560 rpm. The axial velocity is to remain constant at 500 fps across this blade row, and also a free-vortex distribution is to be maintained in the row. For conditions such that the leaving relative velocity at the mean blade position is 550 fps, draw the mean-position velocity triangles, and compute velocity of whirl, ΔC_u; also find $rV_w = r\,\Delta C_u$ in square feet per second. Use the mean position value of rV_w as determined to find the velocity-of-whirl magnitude at the hub and tip of the blades. With these known, draw the velocity triangles for the hub and tip. If the mean density of the air in the stage is 0.053 lb per ft³, estimate the maximum pressure rise in the moving row at the three positions. Use velocity-of-whirl values to find the diagram power per 1 lb per sec of air flow at the three positions. Find the probable stage enthalpy increase for 86 per cent ($\Omega = 0.86$) realization of diagram power to enthalpy form. If the stage efficiency is 87 per cent, find the pressure ratio across the stage.

Solution. For this blade row the mean diameter is 18 in., the hub diameter 12 in., and the tip diameter 24 in.

$$u_h = \frac{\pi(12)(9560)}{(12)(60)} = 500 \text{ fps}$$

$$u = \frac{\pi(18)(9560)}{720} = 750 \text{ fps}$$

$$u_t = \frac{\pi(24)(9560)}{720} = 1000 \text{ fps}$$

The velocity triangle is drawn in Fig. 5-13 as the middle diagram, using either row by making $\theta_1 = 60°$, $u = 750$ fps, $C_a = 500$ fps, $w_2 = 550$ fps. From the diagram the velocity of whirl $V_w = \Delta C_u = \Delta w_u$ scales to be 240 fps, and

$$rV_w = \frac{18}{(2)(12)}(240) = 180 \text{ ft}^2 \text{ per sec}$$

For the tip and hub positions, the V_w values are

$$180 = \frac{24}{(2)(12)} V_{wt} \qquad V_{wt} = 180 \text{ fps at tip}$$

$$180 = \frac{12}{(2 \times 12)} V_{wh} \qquad V_{wh} = 360 \text{ fps at hub}$$

Use these V_w values to complete the construction of the tip and hub triangles. This can easily be carried out with the construction used on the left of Fig. 5-13 by superposing V_w on the u line at its right end and then drawing in the w_2 line. Then lay out V_w at the end of the u line to locate the end position for C_2. Note that C_1 is fixed in direction and magnitude from the end of the u line and should first be located to form the basic triangle pattern for each position. It is easy to transform polar velocity diagrams into the more conventional forms shown on the right. Scaled values from the triangles show:

At the tip:

$$w_1 = 874 \text{ fps} \quad w_2 = 733 \quad w_m = 797 \quad C_2 = 685 \quad V_{wt} = 180$$

At mean radius:

$$w_1 = 681 \text{ fps} \quad w_2 = 550 \quad w_m = 608 \quad C_2 = 732 \quad V_w = 240$$

At the hub:

$$w_1 = 540 \text{ fps} \quad w_2 = 523 \quad w_m = 533 \quad C_2 = 823 \quad V_{wh} = 360$$

The maximum pressure-rise values in the moving blade row on the assumption of an incompressible fluid are by Eq. (5-1)

$$\Delta p = \frac{0.053}{64.34}(874^2 - 733^2) = 187 \text{ psf pressure rise at tip}$$

$$\Delta p = \frac{0.053}{64.34}(681^2 - 550^2) = 133 \text{ psf pressure rise at mean}$$

$$\Delta p = \frac{0.053}{64.34}(540^2 - 523^2) = 15 \text{ psf pressure rise at hub}$$

In the fixed row of the stage, let us assume that the velocity is diffused to a lower value C_3 which will be made equal to the C_1 value entering the stage of 577 fps. Maximum pressure rise in the fixed row by Eq. (5-2) is thus

$$\Delta p = \frac{0.053}{64.34}(685^2 - 577^2) = 112.3 \text{ psf at tip}$$

$$\Delta p = \frac{0.053}{64.34}(732^2 - 577^2) = 167.0 \text{ psf at mean position}$$

$$\Delta p = \frac{0.053}{64.34}(823^2 - 577^2) = 284.0 \text{ psf at hub}$$

The maximum pressure rise across the stage is the sum of the pressure rise created in both the moving and fixed rows. For these cases it can be seen as constant at some 300 psf at the respective tip, mean, and hub positions on the blade.

By Eq. (5-16) the stage diagram power can be found for 1 lb per sec air flow.

$$P_m = \frac{G}{g_c} u(C_{2u} - C_{1u}) = \frac{1}{g_c} u V_w$$

At tip:

$$P_m = \frac{1}{32.17}(1000)(180) = 5{,}600 \text{ ft-lb per sec}$$

At mean position:

$$P_m = \frac{1}{32.17}(750)(240) = 5{,}600 \text{ ft-lb per sec}$$

At hub:

$$P_m = \frac{1}{32.17}(500)(360) = 5{,}600 \text{ ft-lb per sec}$$

By use of Eq. (5-23) the stage temperature rise and enthalpy increase can be found.

$$h_3 - h_1 = \Delta T\, c_{pm} = \Omega \frac{u}{g_c J} (C_{2u} - C_{1u})$$

$$h_3 - h_1 = \Delta T\, c_{pm} = (0.86) \left(\frac{5{,}600}{778}\right) = 6.19 \text{ Btu per lb increase in stage}$$

With the inlet temperature to the stage, 640 R, c_{pm} is about 0.241 so that

$$\Delta T = \frac{6.19}{0.241} = 25.7 \qquad \text{and} \qquad T_3 = 25.7 + 640 = 665.7 \text{ R}$$

Use Eq. (5-14) to find the pressure ratio.

$$\eta_c = \frac{(p_3/p_1)^{(k-1)/k} - 1}{(T_3/T_1) - 1} = \frac{r^{(1.397-1)/1.397} - 1}{(665.7/640) - 1} = 0.87$$

$$r = 1.129, \text{ the pressure ratio increase across the stage}$$

5-9. Westinghouse J-34 Compressor. Figure 5-14 is a photograph of the rotor of the axial compressor employed on the J-34 turbojet engine. This unit is illustrated in Fig. 1-12. As can be seen, the compressor consists of 11 stages. The first 10 stages are mounted on an aluminum

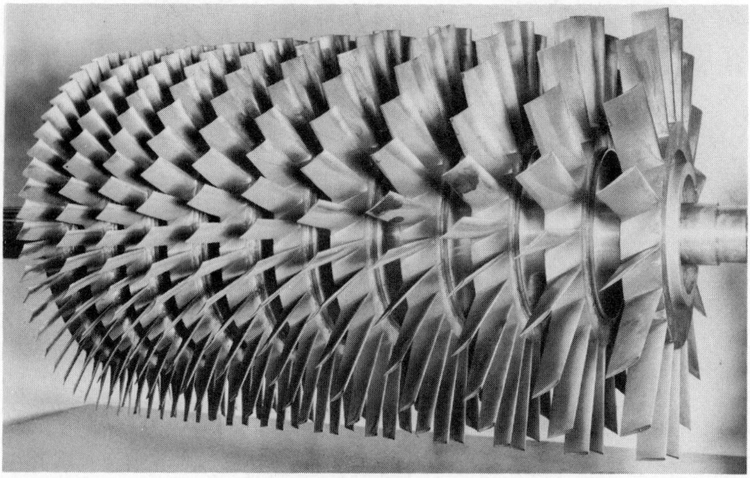

FIG. 5-14. Rotor of 11-stage axial-flow compressor of Westinghouse J-34 jet engine.

spindle, and the eleventh, or final, stage is a steel disk bolted to the spindle. The stator of the compressor consists of fabricated steel diaphragms made in halves. They are slid into position and assembled into machined grooves in the housing of the cast-aluminum stator.

5-10. General Electric Rotor. The compressor rotor for the General Electric TG-180 (J-35) turbojet unit is pictured in Figs. 5-15 and 5-16. In Fig. 5-17 the stator for this rotor is also shown. This compressor consists of 11 bladed disks which are shrunk onto a steel shaft. The disks for the first 10 stages are machined from aluminum forgings. The final disk for the last stage of the compressor is a heat-treated steel forging directly linked by a splined fit to the turbine shaft. The steel compressor rotor blades, which are forged to size, are attached to the rims of the disk

FIG. 5-15. Rotor for 11-stage axial-flow compressor for type TG-180 General Electric aircraft gas turbine.

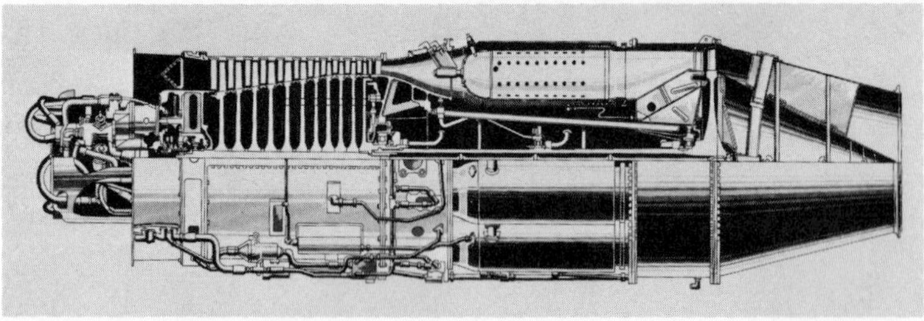

FIG. 5-16. Longitudinal semisection of TG-180 General Electric gas-turbine jet unit.

with trapezoidal-shaped dovetails. The compressor rotor disks with their blades have a constant outside diameter to the tips of the blades of slightly less than 30 in. The blades themselves decrease in length from inlet to outlet. A modified vortex design with pressure rise in both moving and fixed rows is employed for the air flow. The disks themselves are connected by cylindrical aluminum spacer rings which are shrunk under the rim shoulders, and each disk is secured to its adjacent spacer ring by steel pins to permit even distribution of the driving torque. The casing of the compressor is horizontally split into halves which are bolted together. The casing is also bolted around the circumference at the ends to the forward frame and to the mid-frame. The frames themselves, which constitute the main mounting structure of the unit, are not split. The stator blades are also forged and finished to size. They also dovetail into split rings, which, in turn, are assembled into the stator halves. Eleven rows of stationary blades are used, followed by two rows of straightening vanes.

FIG. 5-17. Stator of TG-180 General Electric axial-flow compressor.

The compressor and its turbine in this unit turn at 7700 rpm. The whole unit weighs 2450 lb and can produce a take-off thrust of 4125 lb with a guaranteed minimum thrust of 3750 lb. The air leaving the compressor passes into eight unit combustors before passing through the turbine. Operating temperatures as high as 1500 F can be used with the turbine. This particular unit is 166 in. in over-all length, with a maximum diameter of 37 in. It has been used on a number of military airplanes.

PROBLEMS

5-1. An axial-flow-stage blade diagram at mean radius has values of $u = 518$, $C_1 = 437$, $w_1 = 800$ fps and $u = 518$, $C_2 = 400$, $w_2 = 634$. C_3 leaving the next fixed row is 437. Find the probable maximum pressure change in the moving and in the fixed row if the medium flowing has a density of 0.1 lb per ft³ which can be considered essentially constant throughout the stage. What type of compressor blading is this? *Ans.* 2.56 psi and −0.34 psi.

5-2. An axial-flow compressor in its first stage has blades which are 0.9 in. long attached to a rotor which is 8.0 in. at the blade root. The compressor turns at 20,000 rpm. The guide vanes lead air into the first stage in such manner that the absolute-velocity vector of the inlet air makes an angle of 60° with the direction of motion of the blade. The blade diagrams are symmetric, and the absolute velocity of the air on inlet to the blade is 500 fps. (*a*) For the mean blade position draw the velocity triangles to scale. (*b*) Compute the velocity of whirl. (*c*) Find the probable work absorbed from the rotor per pound of air passing through the stage. (*d*) Find the temperature rise in the moving row of the stage; (*e*) In a following fixed row the absolute velocity is reduced to 500 fps; estimate a possible pressure rise in this stage if the air under compression can be considered to have an essentially constant density of 0.08 lb per ft.^{3}

5-3. Work Prob. 5-2 if it is applied to conditions at the tip of the blade.

5-4. A compressor stage operates with a mean blade speed of 730 fps. The velocity at inlet to the stage is 400 fps and axial. The blade leaving angle is 45°, and the axial component of velocity is 400 fps. Inlet pressure is 13.4 psia, inlet temperature 516.4 R, and $\rho = 0.07$ lb per ft^3. Stage efficiency is 86 per cent, and 10 lb per sec of air flows through the compressor. (*a*) Based on mean blade conditions, draw the velocity diagram, and estimate the probable maximum pressure rise in a moving row if the air is considered essentially incompressible. (*b*) What probable maximum pressure rise occurs in the whole stage? (*c*) Find an ideal and probable actual temperature rise based on blade-diagram values (use Ω as 0.86). (*d*) Making use of actual leaving temperature and stage efficiency, compute the discharge pressure from the stage. (*e*) Find the probable horsepower absorbed by the stage. *Ans.* (*a*) $w_1 = 834$, $w_2 = 564$, $\Delta p_m = 411$ psf; (*b*) 528 psf; (*c*) 40.3° and 34.7°; (*d*) 16.3 psia; (*e*) 117 hp.

5-5. Air enters a compressor at 14.7 psi and 530 R at low velocity. It is compressed through a pressure ratio of 11. Find the actual work of compression in Btu per pound and the final temperature at outlet from the compressor. The compressor efficiency is to be taken as 84 per cent.

5-6. A 20-stage axial-flow compressor has a maximum tip diameter of 18.375 in. which is constant throughout the length of the machine. The design is arranged so that absolute axial velocities are constant throughout the compressor and equal on inlet and outlet from each stage. The inlet moving blade is 3 in. high, and the last moving blade row is 1.54 in. high. Assume the axial velocity is 525 fps at mean blade position, and use this as representative of average conditions for each blade row. The pressure and temperature at inlet to the first row are 12.6 psia and 507.4 R. (*a*) The discharge pressure is 70 psia leaving the last moving row, and the axial velocity is 525 fps. If the internal efficiency of the compressor to this point is 85 per cent, compute the probable air temperature at this row if the machine can be considered adiabatic. (*b*) If 90 per cent of the annulus area of the first row can be considered air-passage area, find the pounds per second flowing through the machine. (*c*) Find the internal horsepower absorbed by the rotor of the compressor for the air flow found above. (*d*) For the conditions leaving the last moving row, check to find what fraction of the annulus area is free air-flow area. *Ans.* (*a*) 885 R, (*b*) 31.9 lb per sec, (*c*) 4130 hp, (*d*) 0.5.

5-7. The total velocity of whirl in a stage of an axial-flow compressor is 300 fps, and the wheel speed is 500 fps. The air flow is 20 lb per sec. Compute the stage input horsepower from these data. If 86 per cent of this horsepower is actually absorbed in the stage, find the horsepower received from the shaft.

5-8. For the compressor of Prob. 5-2, draw the velocity diagrams for the conditions indicated at the mean position. Compute the velocity of whirl. Next compute the velocity-of-whirl values which must hold at the hub and tip to produce an over-all free-vortex distribution in the blade row. Keeping the same axial velocity throughout the blade row and maintaining the absolute inlet velocity to the row at 60°, draw the velocity triangles at hub and tip to satisfy free-vortex distribution.

5-9. A 25,000 cfm compressor with supply at 14.7 psia, 70.3 F has a constant tip diameter of 18.375 in. The inlet blade is 3 in. high, and the outlet fixed (guide) blade is 1.5 in. high. No fixed or guide blade is used at entry to the compressor, but a gently tapering annulus funnel leads the air to entry of the first moving row. There are 20 stages in the rotor, which is 35.5 in. long (axially). The rpm is 9200, and a pressure ratio of 5 can be obtained. The annulus area of the first stage (moving) is 1.006 ft^2, of which 90 per cent can be called available or free area and the flow to the stage can be considered axial. Under certain conditions of full flow it was found that the temperature and pressure at entry to the first stage (a moving row) were 12.6 psia at 507.4 R and the axial velocity $C_1 = 525$ fps $= C_{a1}$.

a. Computation will show that the mean blade speed of the first moving row is $u = 617$ fps. Draw the velocity diagram for this moving row; make $C_1 = 525$ fps and axial; fix $\beta_2' = 55°$ and $C_{a2} = 525$ fps. *Ans.* $w_1 = 810$ fps, $w_2 = 641$ fps, $\theta_1 = 90$, $\theta_2 = 64.6$, $\beta_1' = 40.4°$, $C_2 = 581$ fps.

b. Compute the velocity of whirl, ΔC_u. *Ans.* 249.

c. Find the diagram power for the first stage, based on 1 lb per sec of air flowing, in foot-pounds per second. *Ans.* 4780 ft-lb per lb per sec.

d. If 86 per cent of the diagram power appears in the form of increased enthalpy for the air, find the temperature of the air leaving the stage. The fixed row following delivers its air axially at the same C_a as the moving row (525 fps). *Ans.* 529.4 R.

e. Compute the real pressure rise for the first stage based on the actual temperature rise and a stage efficiency of 87 per cent. *Ans.* 1.76 psi.

f. For the 25,000 cfm entering the compressor, how much horsepower is supplied to the first stage from the shaft? *Ans.* 236 hp.

5-10. Continue with the data of Prob. 5-9.

a. Draw velocity diagrams for the second, third, and fourth stages. Keep C_a constant at 525 fps. The absolute velocity is always axial on entry to the moving row. The angle similar to β_2' for the first stage is to be kept at 55°. *Ans.* $u_2 = 620$ fps, $u_3 = 623$, $u_4 = 626.5$, $\Delta C_u = 252$ fps for second, 255 for third, 258 for fourth.

b. Compute the enthalpy and temperature rise in each stage on the basis of 86 per cent of the diagram work appearing as increased enthalpy of the air ($\Omega = 0.86$). Using a stage efficiency of 87 per cent and the computed enthalpies or temperatures, find the pressure leaving each stage. *Ans.* Temperature leaving first = 529.4 R, second = 551.8 R, third = 574.5 R, fourth = 597.4 R; pressure leaving first = 14.36 psia, second = 16.29, third = 18.43, fourth = 20.80 psia.

c. In terms of the over-all pressure increase and the over-all enthalpy (or temperature) change, compute the internal compressor efficiency for the four complete stages considered in Probs. 5-9 and 5-10. *Ans.* 86.5%.

REFERENCES

1. Woodworth, L. R.: Axial Flow Compressor Design Considerations, lecture notes presented at Northwestern University ASME-ASEE Summer School, June, 1947.
2. Howell, R. A.: Fluid Dynamics of Axial Compressors, in "Development of the British Gas Turbine Jet Unit," pp. 441–462, Institution of Mechanical Engineers, London, reprinted for American Society of Mechanical Engineers, New York, 1947 (8 authors).

3. Ponomareff, A. I.: Axial Flow Compressors and Gas Turbines, *Trans. ASME* vol. 70, pp. 295–306, 1948.
4. Spurr and Allen: A Theory of Unstaggered Airfoil Cascades in Compressible Flow, *NACA Rept.* 888, Washington, D.C., 1947.
5. Katzoff, Finn, and Laurence: Interference Method for Obtaining the Potential Flow Past an Arbitrary Cascade of Airfoils, *NACA Rept.* 879, Washington, D.C., 1947.
6. Lieblein and Sandercock: Compressibility Correction for Turning Angles of Axial-flow Inlet Guide Vanes, *NACA Tech. Note* 2215, Washington, D.C., December, 1950.
7. Bullock, R. O., and H. B. Finger: Surging in Centrifugal and Axial Flow Compressors, *J. SAE*, vol. 59, pp. 42–45, September, 1951.
8. Foley, John R.: Compressor Surge, *J. SAE*, vol. 59, pp. 46–50, September, 1951.
9. Wu and Wolfenstein: Application of Radial Equilibrium to Axial-flow Compressor and Turbine Design, *NACA Tech. Note* 1795, January, 1949.
10. Jacobs, Pinkerton, and Greenberg: Tests of Related Forward-camber Airfoils in the Variable Density Wind Tunnel, *NACA Rept.* 610, 1937.
11. NACA Subcommittee on Compressors: Standard Procedures for Rating and Testing Multistage Axial-flow Compressors, *NACA Tech. Note* 1138, September, 1946.
12. Huppert and MacGregor: Comparison between Predicted and Observed Performance of Gas-turbine Stator Blade Designed for Free-vortex Flow, *NACA Tech. Note* 1810, April, 1949.
13. Tucker, W. B.: Construction of Gas Turbine for Locomotive Power Plant, *Trans. ASME*, vol. 70 pp. 877–882, November, 1948.
14. Bowen, Sabersky, and Rannie: Investigations of Axial Flow Compressors, *Trans. ASME*, vol. 73, pp. 1–15, 1951.

CHAPTER 6

CENTRIFUGAL AND DISPLACEMENT COMPRESSORS

6-1. Centrifugal Compressor. The previous chapter considered various aspects of the axial-flow compressor, in which type the gas flows essentially parallel to the axis of the machine, usually through several stages of blading. In the centrifugal compressor, although admission to the impeller necessitates the air having an axial component, the flow through the impeller takes place largely in a plane which is perpendicular to the axis or shaft of the machine. In the axial-flow compressor, centrifugal force was not significant in the development of pressure. This is not the case with the centrifugal machine, as in the impeller the gas moves from a smaller to a greater radius in the machine and thereby flows in a field in which the pressure gradient increases outwardly in terms of the peripheral velocity possessed by the gas at a given radius. The same retardation of flow or diffuser action which causes pressure build-up in the axial-flow machine also occurs in the centrifugal unit.

Equation (4-23), when referred to a compressor with power supplied by the impeller, appears as

$$P_m = \frac{G}{2g_c}[(C_2^2 - C_1^2) + (u_2^2 - u_1^2) + (w_1^2 - w_2^2)] \qquad (6\text{-}1)$$

An equation of the form of (4-8) is also applicable and would appear for a compressor as

$$P_m = \frac{G}{g_c}(C_{2u}u_2 - C_{1u}u_1)$$

It is worth while to study Eq. (6-1) in terms of the idealized pressure rise which occurs in the compressor. The term $(C_2^2 - C_1^2)/2g_c$ represents the increase in kinetic energy contributed to the moving fluid by the impeller. The absolute velocity C_1 at entry to the impeller increases in magnitude to the value of C_2 at leaving (Fig. 6-1). The value of C_2 in the limit approaches the value of impeller peripheral tip speed u_2. The increase of kinetic energy of the fluid stream in the impeller in no way contributes to pressure increase in the impeller itself. However, in the diffusing section of the pump outside the moving impeller, pressure increase can be produced at the expense of the leaving kinetic energy $(C_2^2/2g_c)$. In fans with narrow blading in a radial direction nearly all the pressure increase is produced in this way. In centrifugal compressors (and pumps) with deep (radial-direction) impellers, a significant pressure rise occurs in the impeller in relation to the second and third terms of Eq. (6-1). The term $(u_2^2 - u_1^2)/2g_c$ measures the pressure rise associated with the radial centrifugal field, while the term $(w_1^2 - w_2^2)/2g_c$ is associated with the relative velocity of the gas at inlet to and leaving the impeller.

Figure 6-2 shows diagrammatically a radial impeller turning in a volute section and delivering gas into a diffusing section. In the volute and the

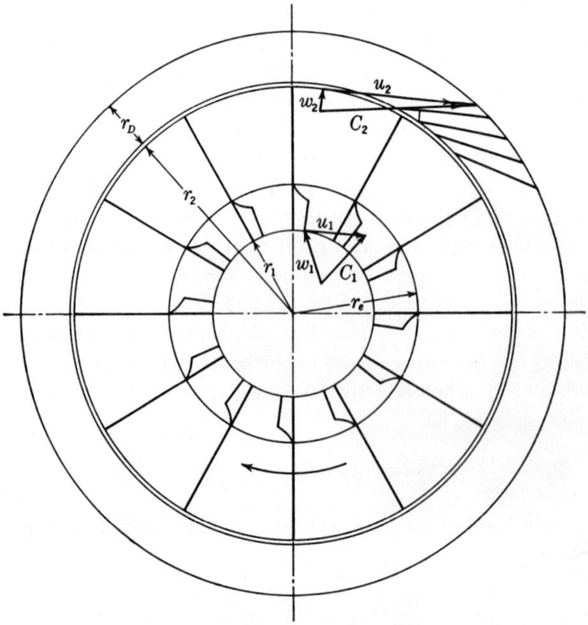

FIG. 6-1. Flow velocity triangles superimposed on radial-vane centrifugal-compressor impeller.

following diffusing section, the pressure builds up at the expense of reduced kinetic energy. Most centrifugal compressors at the present time, instead of using a volute, employ diffusion vanes for the pressure build-up. In an axial-flow compressor, diffuser effect in all rows alone is able to create pressure, and consequently the pressure rise per stage is less in an axial machine than in a centrifugal machine when each machine has the same maximum radius and rotational speed.

Axial-flow machines show improved performance as the quantity of gas flowing increases above a certain minimum amount. With small passages and high friction, the efficiency of an axial-flow machine is markedly reduced. This effect is less noticeable in a centrifugal machine, which can be built with reasonably high efficiency when of small capacity. Inherent characteristics of radial and axial compressors are shown in Fig. 6-3. For any given speed, such as at n_4 rpm, it will be noticed that the pressure characteristic of the centrifugal compressor is much flatter than the corresponding characteristic of the axial-flow compressor. Thus when flow volume is reduced in a centrifugal compressor, a considerably greater reduction can take place before the surge line is reached than is the case for an axial machine. Stated in another way, the centrifugal machine is stable over a greater flow range than is the axial machine.

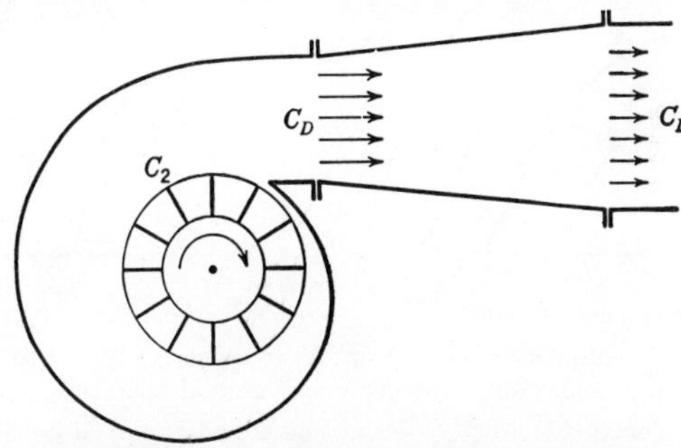

C_2 = velocity of medium at exit from impeller, fps
C_D = velocity at entry to diffuser tube, fps
C_{DO} = velocity at outlet from diffuser tube, fps
$C_2 > C_D > C_{DO}$

FIG. 6-2. Diagrammatic illustration of impeller delivering air into a volute and into a following diffusing section.

The blading of an axial machine acts as a diffuser and is effective only when the flow streams follow the restraining surfaces, in this case the blade surfaces. When flow decreases to a point such that this is not possible, breakaway occurs and this will eventually result in surging. However, when working within its normal design range, the efficiency of an axial-flow machine is greater than the efficiency of a centrifugal machine. This lower efficiency of the centrifugal machine exists because, in the passages of the impeller, guidance is not always as direct as might be hoped, and friction also occurs because of the bends and diffusing passages which must be placed in the machine.

6-2. Centrifugal Pressure Field. It has already been mentioned that the pressure rise in the impeller of the machine is brought about by change in the $w^2/2g_c$ term and in the $u^2/2g_c$ term. The first of these, associated with the change in relative velocity of the fluid in moving through the impeller, is a diffusing action and needs little further explanation at this point. The pressure effect of the centrifugal field is not so obvious; consequently a development, based on elementary principles of mechanics, will be carried through to show how a pressure difference is produced in the centrifugal field of a rotating fluid.

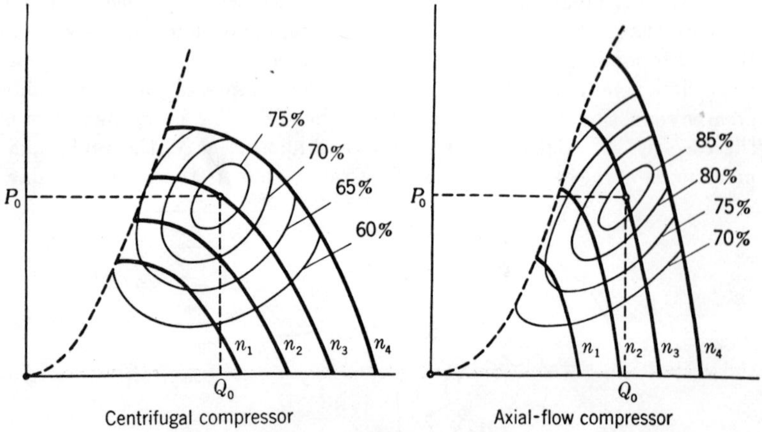

n_1, n_2, n_3, n_4, compressor speed, rpm; Q = capacity, cfm or lb per sec; P_0 and Q_0 = pressure and capacity under normal operating conditions.

FIG. 6-3. Comparative characteristics of centrifugal and axial-flow compressors.

Consider a rotor, or impeller (Fig. 6-4), turning in any plane, and disregard the effect of gravity between the topmost and lowermost point of the rotor. An element of fluid mass dm of density ρ lb per ft³ is acted upon by a centrifugal force of magnitude F_r. Let θ represent the central angle and ω represent the angular velocity in radians per second. The pressure at any radius r has a magnitude p_r. The tangential velocity is u fps at radius r ft.

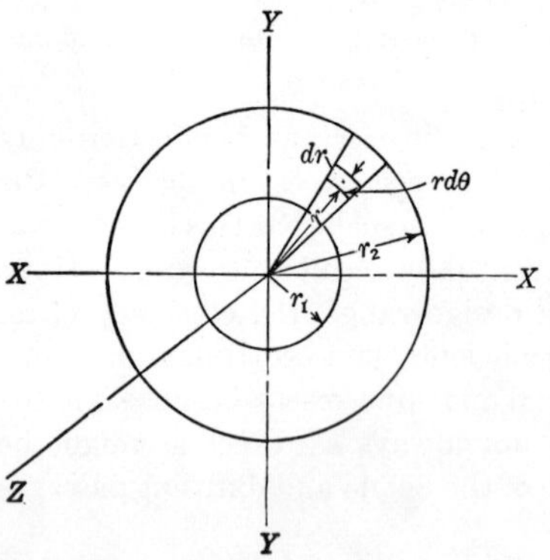

FIG. 6-4. Centrifugal field in a rotating fluid.

$$dm = r\,d\theta\,dr\,dz\,\frac{\rho}{g_c}$$

$$F_r = dm\,\frac{u^2}{r} \tag{6-2}$$

$$F_r = \frac{\rho}{g_c}\,r\,d\theta\,dr\,dz\,\frac{r^2\omega^2}{r} \tag{6-3}$$

In equilibrium the centrifugal force is balanced by the radial pressure on the element, and so

$$F_r = p_{r+dr}[(r + dr)\,d\theta\,dz] - p_r r\,d\theta\,dz$$
$$F_r = p_{r+dr}(r\,d\theta\,dz + dr\,d\theta\,dz) - p_r r\,d\theta\,dz$$

Omitting the third-order differential product, there results

$$\frac{\rho}{g_c} r\, d\theta\, dr\, dz\, \frac{r^2\omega^2}{r} = p_{r+dr}(r\, d\theta\, dz) - p_r r\, d\theta\, dz$$

$$\frac{\rho}{g_c}\, \omega^2 r\, dr = dp \tag{6-4}$$

$$\frac{\omega^2}{g_c}\int_{r1}^{r2} \rho r\, dr = p_{r2} - p_{r1} \tag{6-5}$$

if ρ is essentially constant in the path of length $r_2 - r_1$,

$$p_{r2} - p_{r1} = \frac{\rho\omega^2}{g_c}\left[\frac{r^2}{2}\right]_{r_1}^{r_2} = \frac{\rho\omega^2}{2g_c}(r_2^2 - r_1^2) \tag{6-6}$$

$$\Delta p = \frac{\rho}{2g_c} 4\pi^2 n_s^2 (r_2^2 - r_1^2) = \frac{\rho}{2g_c}(u_2^2 - u_1^2) \tag{6-7}$$

$\omega = 2\pi n_s$ rad per sec, where $n_s = rps$ and Δp is the pressure difference, which, for conventional units, is expressed in pounds force per square foot. This equation represents the pressure difference developed because of centrifugal effects by a fluid rotating about a central axis.

Thus, pressure can be created by means of a centrifugal field and is particularly significant in the case of deep impellers. By using the centrifugal field for creating pressure rise, the difficulties associated with diffusion are reduced. However, all centrifugal machines in addition utilize diffusion effects within the impeller. Outside of the impeller, pressure can of course be created only by absolute-velocity diffusion.

Example 1. A centrifugal compressor with a 3600-rpm radial-vaned impeller, 39 in. in outer diameter, handles air supplied at 60.3 F and 14.7 psia. The outer diameter of the eye is 8 in., and the hub is 4.5 in. The inlet velocity triangle near the outer edge of the eye lies in a plane parallel to the axis and has sides with magnitudes of $u_1 = u_e = 125.6$ fps, $C_1 = 377.5$ fps, and $w_1 = 370$ fps with the angle between u_1 and $C_1 = 77°$. The velocity triangle at the impeller tip in a plane perpendicular to the axis shows $u_2 = u_t = 612$ fps, $w_2 = 75.9$ fps (perpendicular to u_t), and $C_2 = 616.8$ fps. The air leaving the impeller enters the diffuser, where it reduces in velocity to $C_3 = 205.6$ fps. Assume that the air in passing through the compressor is essentially incompressible and keeps the same specific volume it had on entry to the compressor. (*a*) Find the ideal pressure rise associated with the centrifugal field from eye tip to impeller tip. (*b*) Find the ideal pressure rise associated with change in relative velocity. (*c*) Find the ideal pressure rise in the diffuser. (*d*) Compute the diagram power absorbed by each pound per second of air flowing through the rotor.

Solution. (*a*) Density of air undergoing compression at inlet from Eq. (2-3) is $(144)(14.7)(1) = \rho(53.34)(530)$, $\rho = 0.0763$ lb per cu ft. By Eq. (6-7)

$$\Delta p = \frac{\rho}{2g_c}(u_2^2 - u_1^2) = \frac{0.0763}{(2)(32.17)}(612^2 - 125.6^2)$$

$$= 425 \text{ psf pressure increase from the centrifugal field}$$

b. Ideal relative velocity contribution to pressure increase

$$\Delta p = \frac{\rho}{2g_c}(w_1^2 - w_2^2) = \frac{0.0763}{(2)(32.17)}(370^2 - 75.9^2) = 156 \text{ psf}$$

c. Ideal pressure rise in diffuser

$$\Delta p = \frac{\rho}{2g_c}(C_2^2 - C_3^2) = \frac{0.0763}{(2)(32.17)}(616.8^2 - 205.6^2) = 400 \text{ psf}$$

d. The velocity-of-whirl component C_{2u} at the tip is equal to 612 fps, that is, u_2, as w_2 is perpendicular to u_2. At inlet

$$C_{1u} = C_1 \cos 77° = (377.5)(0.225) = 84.9 \text{ fps}$$

By Eq. (6-8)

$$P_m = \frac{1}{g_c}(C_{2u}u_2 - C_{1u}u_1) = \frac{1}{32.17}[(612)(612) - (84.9)(125.6)]$$
$$= 11{,}270 \text{ ft-lb per sec} = 20.5 \text{ hp for each lb per sec of air compressed}$$

While these results cannot be realized in the actual compressor, they indicate the order of magnitude of the pressure rise contributed by the centrifugal field, by the relative velocity change, and by the diffusion process.

FIG. 6-5. Impeller and pivot shaft of centrifugal compressor for Ghost engine. (*De Havilland Co., Ltd.*)

6-3. Centrifugal-machine Elements. Figure 6-1 shows in diagrammatic form the impeller of a centrifugal compressor. At the outer part of the figure is indicated the stationary ring which holds the diffusion vanes. Figure 6-5 shows the impeller of a De Havilland compressor built for the Ghost gas turbine. Inlet to the impeller takes place essentially axially as the air flows into the inducer section at the eye of the compressor. The air then turns and moves outward through the passages between the impeller blades with the relative-velocity direction essentially radial.

Figure 6-7 shows the two impellers of an Elliott gas-turbine centrifugal compressor. The left impeller clearly shows the eye section with the inducer edges visible. It is into these that the gas enters on its way to the passages between the radial impeller blades. It will be noticed that the Elliott compressors have shrouded impeller blades. That is, the front

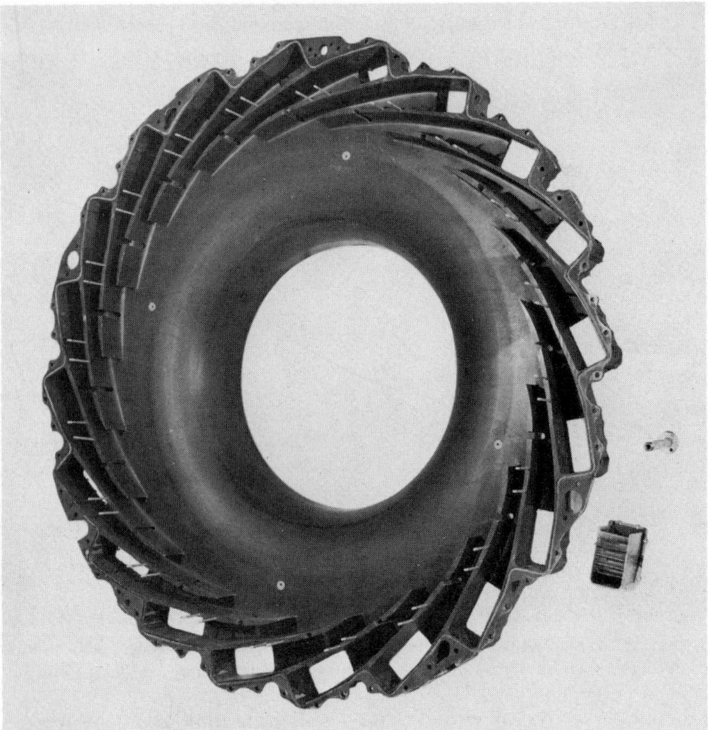

FIG. 6-6. Casing for Ghost engine compressor with diffusers in place; one cascade box for turning outlet air from compressor in axial direction shown in reversed position at lower right. (*De Havilland Co., Ltd.*)

as well as the rear side of the impeller blades is covered, and the air passages lie between the covers.

The phantom view of Fig. 1-8 clearly shows the operation of this compressor. Here air enters at the left and flows through the inducer section (eye) of the impeller, and thence outward through the impeller. The diffuser section receives air from the impeller, and in its passages a pressure increase takes place. From the diffuser, the air flows into a connecting passage which crosses over and then leads to the inlet of the second impeller. This performs in exactly the same manner as the first.

A centrifugal compressor is also shown in Fig. 1-5. Here air can be seen entering and leaving the impeller, which occupies a central position in the photograph. This particular compressor is of interest in that its outer casing is of the volute type, and in it much of the diffusion (pressure rise) takes place.

The De Havilland impeller of Fig. 6-5 is open, and the air is constrained to flow through proper passages in the impeller because the outer casing of

FIG. 6-7. Shrouded impellers of two-stage centrifugal compressor. The inlet eyes of each impeller are shown as well as the tips of the radial vanes. (*Elliott Co.*)

the compressor fits so closely that small clearance exists between the impeller and the casing. Figure 6-6 shows the matched casing which constitutes the static element of this compressor with the vaned diffusion passages. The air leaving the impeller at its periphery enters these diffusing passages. These increase in cross section as the flow progresses outward, and the reduction in velocity results in a pressure increase. This particular unit uses a gas turbine which is placed directly behind and in line with the compressor. Consequently, it is necessary to turn the air through 90° rearward from each diffuser, and the small insert of Fig. 6-6 shows the cascade (airfoil) element which turns the compressed air rearward.

In Fig. 1-11 the compressor employs a double-inlet impeller. Here two similar impellers are placed back to back. Air feeds into both the front and rear impeller elements, and the diffusion-ring vanes must have sufficient flow area to pass the air delivered by both halves of the double-inlet impeller. One advantage of a double-inlet unit is that axial thrust is effectively balanced. However, the ducting arrangements for inlet air to the rear impeller are less direct than in a single-inlet unit and may call for considerable design ingenuity to meet space limitations and keep inlet pressure loss to a minimum.

It will be noticed that the inlet eye for each of these compressors (Figs. 6-6 to 6-8) constitutes a relatively large proportion of the radial height of the impeller. Although the eye with its inducers is needed to admit the entering air to the impeller, that air which enters at the outer part of the eye section (Fig. 6-8) has a shorter path in which to build up pressure by means of the centrifugal field. In Fig. 6-1, the eye is represented by the annulus between r_1, the hub radius, and r_e, the maximum eye radius. In Fig. 6-1, the inlet to the eye is shown as having a relative velocity w_1, an absolute velocity C_1, and a hub velocity u_1. At the outer radius of the eye, u_1 would be greater in magnitude, and this, in turn, would change the values of w_1 and C_1 (increasing one or both of them if the axial velocity is constant over the eye area). It will also be noted that the plane of action of the inlet velocity triangle, w_1, u_1, and C_1, is not clearly indicated. At the inlet to the eye of the impeller, it lies necessarily in planes parallel to the axis of the machine, but as the air moves into the impeller, the planes in which w and C lie greatly change until finally they are perpendicular to the axis of the machine. Design for inlet conditions to the impeller is thus complex because of variation in impeller speed as the radius changes and because of resulting air-velocity variations.

In particular, mention should be made of the use of inlet swirls for the fluid entering the eye of the compressor. Reference to Fig. 6-8 will show that these consist of guide vanes attached to the casing of the machine and so arranged as to direct the fluid into the eye of the rotating impeller at an angle. At the right of Fig. 6-8 is shown the effect of prewhirl on reducing the relative velocity at inlet to the impeller compared with the diagrams at the left of Fig. 6-8, where inlet swirl (prewhirl) vanes are not employed. Both these sets of diagrams are drawn on the basis of constant inlet axial velocity of the fluid entering the impeller, which is considered to be operating at the same speed, whether or not prewhirl is used. Notice that velocity triangles are drawn for two positions, *viz.*, at the hub and at the tip of the eye. Prewhirl reduces the relative velocity at entry to the impeller and permits lower Mach numbers (w_1/a) for a constant axial inlet velocity C_{a1}. This in turn should lead to somewhat higher efficiency and better performance in the impeller. Prewhirl is usually in the direction of rotation of the impeller, but some pump designs have been made with prewhirl opposed to the direction of motion.

The energy analysis for fluid flow through blading in a turbine or through an axial-flow compressor expressed in Eqs. (4-9) and (5-16) is equally applicable to a centrifugal compressor, provided consideration is

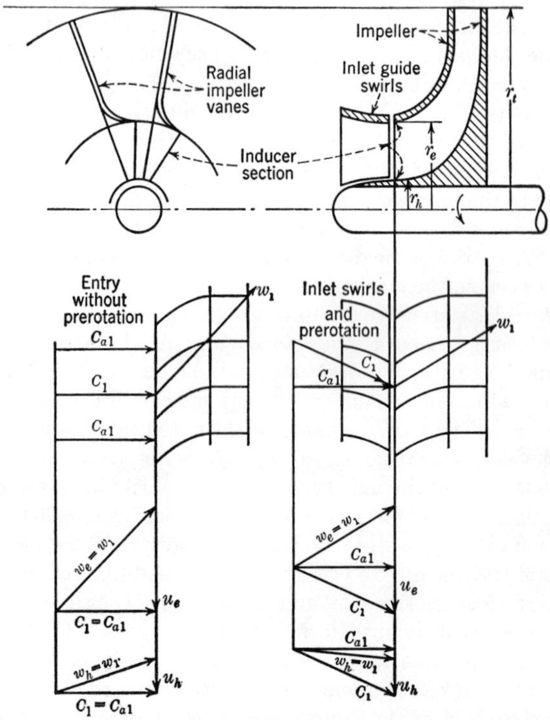

FIG. 6-8. Effect of inlet swirl (prewhirl) vanes on inlet conditions to centrifugal compressor.

given to the fact that the blade (impeller) speed is not constant at varying radii. Consequently the power expression must be written

$$P_m = \frac{G}{g_c} (u_2 C_{2u} - u_1 C_{1u}) \tag{6-8}$$

where P_m = power absorbed by impeller, ft-lb per sec
G = fluid flow, lb per sec
u_2 = velocity of impeller at tip, fps
u_1 = velocity of impeller at eye, fps
C_{2u}, C_{1u} = respective velocity of whirl components, fps

For each pound of fluid flowing, the corresponding work expression in foot-pounds appears as

$$ {}_1W_2 = \frac{1}{g_c}\,(u_2C_{2u} - u_1C_{1u}) \qquad \textbf{(6-9)} $$

From either of the preceding equations it can be seen that when prewhirl exists in the direction of rotation the subtractive term u_1C_{1u} reduces the power which can be absorbed by the impeller for each pound of fluid flowing. Note that with no prewhirl $u_1C_{1u} = 0$. With prewhirl opposite to the direction of rotation and u_1C_{1u} additive, more work can be done by the impeller.

Prewhirl (swirl) vanes have the disadvantage of introducing additional parts and additional weight, which can be objectionable on jet airplanes. In addition there is the added danger of possible icing in the vanes under unfavorable operating conditions. However, by reducing the Mach number at the eye inlet, it is possible to design with eyes of greater depth and corresponding higher flow (compressor) capacity. In this connection the ratio of r_e/r_t can move toward values in the range of 0.4 to 0.65, possibly even to 0.7 (see Fig. 6-8).

The impeller blades of most machines are made radial because the stresses induced in the impeller would be increased if these were bent forward or backward. Centrifugal effects on the curved blades create a bending moment and produce increased stresses which reduce the maximum speed at which the impeller can turn. Good performance can be had with radial impeller blades. However, it will be recalled that pumps which operate at lower speeds nearly all have backward-tip blades. Drawing of the velocity triangles will readily show that backward-tip blades create a leaving-velocity component C_2 (Fig. 6-1) which is less in magnitude than u_2. Consequently, a lesser portion of the pressure rise of the pump is brought about by diffusion action and a smaller over-all pressure ratio is produced with this type than with radial or forward-inclined impellers.

Backward-tip (leaning) blades are slightly better in efficiency and are stable over a larger range of flows than either radial-tip or forward-tip blades. The forward-inclined impeller can produce the highest pressure ratio for a given blade-tip speed. It is inherently less stable and possesses a narrow operating range, and efficiencies are lower than are possible with the backward tip or radial tip. The potentially high pressure ratios available with this type of compressor for given blade-tip speeds justifies further efforts toward developing really efficient types. In the field of fans, many successful forward-tip designs are in use. In so far as gas-turbine centrifugal compressors are concerned, practically all the designs employ radial-tip impellers.

The number of blades in the impeller is set by the necessity of providing proper flow passages in the impeller with the blades acting as both side restraining walls for the fluid and for transmitting power to the fluid. With a finite number of blades (vanes), as must necessarily be the case, there always exist conditions which lead to secondary flows. Referring to Fig. 6-9, it is possible to envisage what happens to a fluid between the vanes of an impeller on starting. As the front, or driving, face of each impeller vane moves against the fluid, inertia effects cause the fluid first to move inward from 3 to 4 and 4 to 5 (as shown at A of Fig. 6-9), and the

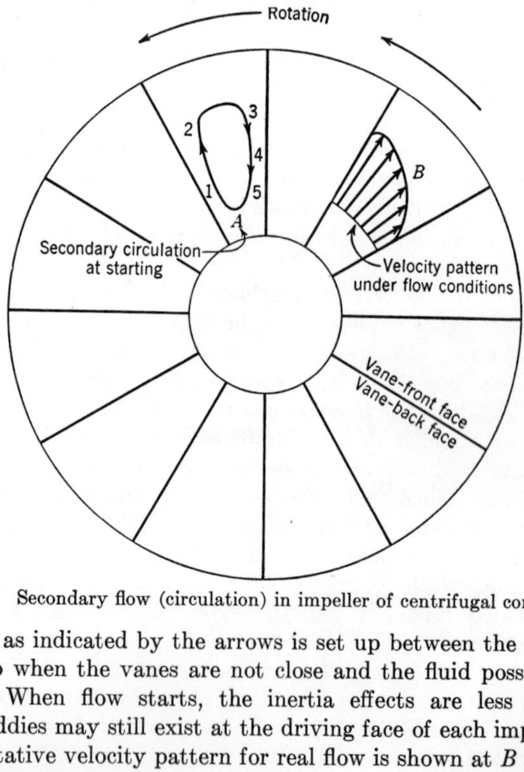

FIG. 6-9. Secondary flow (circulation) in impeller of centrifugal compressor.

circulation as indicated by the arrows is set up between the vanes, particularly so when the vanes are not close and the fluid possesses a low viscosity. When flow starts, the inertia effects are less significant, although eddies may still exist at the driving face of each impeller vane. A representative velocity pattern for real flow is shown at B in Fig. 6-9. This indicates that the highest velocities exist close to the back face of an impeller vane. With high values of flow coefficient, however, the velocity distribution across the vaned passage is relatively uniform. The flow pattern is also improved when impeller-tip speeds operate with lower Mach numbers relative to the gaseous fluid in question. The presence of secondary flows or eddies of any type in an impeller leads to the production of nonuniform flow patterns and reduced effectiveness.

6-4. Design Considerations and Power Requirements. The tip diameter of a centrifugal compressor is set by space requirements and by the allowable tip velocity u_t, where

$$u_t = \frac{2\pi r_t N}{12 \times 60} \quad \text{fps} \tag{6-10}$$

where r_t = impeller-tip radius, in.
N = rpm

Knowing the desired flow and with a decision made concerning rpm and tip diameter, it is then possible to determine the proportions of the inlet or eye of the compressor. The eye must be capable of passing the desired flow after proper allowance is made for the shaft and hub thickness, which reduces the effective flow area. The angles on the inducer section, which lead the fluid into the eye, are set so as to permit entry of the fluid without shock and without an excessive relative velocity at the tip of the inducer (r_e in Fig. 6-8). It is most desirable to avoid flow separation in the inducer section, as this can lead to high inefficiencies.

An important but idealized relation can be developed from Eq. (6-9) if it is considered, first, that the fluid enters the impeller axially ($C_1 = C_{a1}$) so that C_{1u} is necessarily zero and, second, that the fluid flow w_2, is radial at the outlet from the impeller, causing C_{2u} to be equal to u_2.

$$\begin{aligned} {}_1W_2 &= \frac{1}{g_c}(u_2C_{2u} - u_1C_{1u}) \\ &= \frac{1}{g_c}[u_2u_2 - u_1(0)] = \frac{u_2^2}{g_c} \end{aligned}$$

As $u_2 = u_t$, the impeller-tip velocity in feet per second, it follows

$${}_1W_2 = \frac{u_t^2}{g_c} = 2\left(\frac{u_t^2}{2g_c}\right) \tag{6-11}$$

This result shows that, for an impeller with the aforementioned flow conditions and further assuming no losses in the impeller, the work done (energy added) by the impeller for each pound of fluid flowing is exactly equal to twice the kinetic energy possessed by the fluid moving at impeller tip velocity.

The pressure rise resulting from this energy addition appears partly in the rotor, with the remainder occurring in the diffuser. For a fluid under adiabatic conditions

$${}_1W_3 = \left(h_3 + \frac{C_3^2}{2g_cJ}\right)J - \left(h_1 + \frac{C_1^2}{2g_cJ}\right)J$$

$${}_1W_3 = \frac{u_t^2}{g_c} = \left(h_3 - h_1 + \frac{C_3^2 - C_1^2}{2g_cJ}\right)_s J \tag{6-12}$$

$$ {}_1W_3 = \frac{u_t^2}{g_c} = \left((c_{pm})(T_3 - T_1) + \frac{C_3^2 - C_1^2}{2g_cJ} \right)_s J \tag{6-13} $$

where h_3, T_3, and C_3 refer to enthalpy, temperature, and velocity at the outlet from the diffuser. C_3 and C_1 are reasonably close in magnitude, and the difference of their squares is consequently trivial. For a gaseous medium the reversible pressure rise can now be developed.

$$ \frac{u_t^2}{g_cJ} = (h_3 - h_1)_s = c_{pm}(T_3 - T_1)_s = c_{pm}T_1 \left(\frac{T_3}{T_1} - 1 \right)_s \tag{6-14} $$

$$ \frac{u_t^2}{g_cJ} = c_{pm}T_1 \left[\left(\frac{p_3}{p_1} \right)^{(k-1)/k} - 1 \right] \tag{6-15} $$

$$ \left(\frac{p_3}{p_1} \right)_s = \left(\frac{u_t^2}{c_{pm}g_cJT_1} + 1 \right)^{k/(k-1)} \tag{6-16} $$

If a solution by use of the gas tables is desired (as with Table 2-4 for air), write Eq. (6-14) in either of the following forms:

$$ \left(\frac{h_3}{h_1} \right)_s = \frac{u_t^2}{g_cJh_1} + 1 \tag{6-17} $$

or

$$ \left(\frac{T_3}{T_1} \right)_s = \frac{u_t^2}{g_cJc_{pm}T_1} + 1 \tag{6-18} $$

With u_t and either h_1 or T_1 known, h_3 or T_3 can readily be found. The p_{r3} value for h_3 or T_3 and the p_{r1} value for h_1 or T_1 can then be read from Table 2-4.

$$ \frac{p_3}{p_1} = \frac{p_{r3}}{p_{r1}} \tag{6-19} $$

Equations (6-16) and (6-19) are valuable in showing that the maximum (isentropic) pressure ratio which can be attained is directly related to the tip velocity of the impeller and inlet temperature. However, tip velocity is important also in actual compressors, and a parameter ψ known as the pressure coefficient is used in this connection. If both sides of Eq. (6-15) are divided through by u_t^2/g_cJ, there results

$$ 1 = \frac{g_cJc_{pm}T_1[(p_3/p_1)^{(k-1)/k} - 1]}{u_t^2} = \psi_s \tag{6-20} $$

Unity (or 100 per cent) thus represents the pressure performance factor for a reversible-adiabatic compressor with no losses. Let us however employ in an expression of this type the actual pressure ratio produced by a compressor, and although the preceding expression is satisfactory, it is more usual to define ψ as it appears in Eq. (5-25), where it will be noticed that a factor 2 appears in the numerator.

$$\psi = 2\frac{g_c J(\Delta h)_s}{u_t^2} = 2\frac{g_c J c_{pm} T_1[(p_3/p_1)^{(k-1)/k} - 1]}{u_t^2} \tag{6-21}$$

For compressors having impellers 28 in. and larger in diameter it can be expected that the value of ψ will range between 1.30 and 1.45, with somewhat lower values holding for smaller impellers. Where values of ψ in the literature are based on Eq. (6-20), their magnitudes are half as large as indicated above and range from 0.60 to 0.73.

The compressor air-flow rate or capacity, G lb per sec, is set by the intake capacity of the compressor eye and the specific volume of the entering medium. For conditions under which the axial velocity, C_{a1} fps, is constant over the eye annulus and the density is ρ_1 lb per ft^3 for air at temperature T_1 and pressure p_1 psi,

$$G = C_{a1}\rho_1 A_e = C_{a1}\rho_1 \frac{\pi(r_e^2 - r_h^2)}{144}$$

$$G = \frac{C_{a1}144p_1(28.9)}{1545T_1}\frac{\pi(r_e^2 - r_h^2)}{144}$$

$$G = \frac{C_{a1}p_1}{53.34T_1}\pi(r_e^2 - r_h^2) \tag{6-22}$$

Introduce a Mach number based on entering air temperature and the axial velocity.

$$C_{a1} = M_1 a_1 = M_1\sqrt{g_c k \frac{1545}{28.9} T_1}$$

$$G = \frac{M_1\sqrt{g_c k 1545 T_1}\,(28.9)p_1\,\pi(r_e^2 - r_h^2)}{1545T_1}$$

$$G = p_1 M_1\sqrt{\frac{g_c k(28.9)}{1545T_1}}\,[\pi(r_e^2 - r_h^2)] \tag{6-23}$$

where G = air flow into eye, lb per sec
p_1 = pressure at eye inlet, psia
M_1 = Mach number based on C_{a1} at inlet
T_1 = air temperature, °R
g_c = 32.17, k = 1.4 for air at ambient temperature
r_e = outer radius of eye, in. (Fig. 6-8)
r_h = outer radius of hub, in. (Fig. 6-8)

Equations (6-22) and (6-23) both assume full uniform flow over the eye area, which is not exactly true but represents a limit condition. It should also be realized that the particular M_1 used in the equation is lower than the actual Mach number occurring in the eye itself, as this would be the ratio

$$M_{1w} = \frac{w_1}{a_1} \tag{6-24}$$

and would of course vary in magnitude across the eye annulus (see Fig. 6-8). It is important in laying out the design that the value of M_{1w} be kept below unity to reduce shock losses. It is at the eye that high Mach-number values will be met.

The flow coefficient ϕ defined in Eq. (5-26) is also applicable to centrifugal compressors.

$$\phi = \frac{C_a}{u} \tag{5-26}$$

It will vary across the eye as u varies from the inner to the outer eye radius. A variation of the flow coefficient is sometimes seen in the form

$$\phi_c = \frac{144Q}{\pi r_t^2 u} \tag{6-25}$$

where Q = flow, cfs
r_t = tip diameter, in.
u_t = peripheral velocity at impeller tip, fps

In the form of Eq. (6-25) values of ϕ_c can range from approximately 0.06 to 0.12 and give good performance.

Reference to Fig. 5-6 will show that for an adiabatic compressor raising the fluid from pressure p_A to pressure p_O the minimum (isentropic work) is $(h_B - h_A)_s$. With an actual compressor a greater work value such as $h_{C''} - h_A$ is required, or for a less efficient compressor the work required is $h_{o''} - h_A$. Internal compressor efficiency thus appears as

$$\eta_{ic} = \frac{(h_B - h_A)_s}{h_{C''} - h_A} \text{ or } \frac{(h_B - h_A)_s}{h_{o''} - h_A} \tag{6-26}$$

Apply to this the nomenclature previously used in this chapter, where compressor leaving conditions have been designated with subscript 3 and entering conditions with subscript 1, thus

$$\eta_{ic} = \frac{h_{3s} - h_1}{h_3 - h_1} \tag{6-27}$$

Equations (6-26) and (6-27) both tacitly assume that the kinetic-energy values of the fluid at exit from and on entry to the compressor are essentially equal. Rearrangement of Eq. (6-27) will show the actual enthalpy rise in the compressor as

$$h_3 - h_1 = \frac{h_{3s} - h_1}{\eta_{ic}} \tag{6-28}$$

Let us use the numerator of Eq. (6-28) to represent the isentropic enthalpy increase required to produce the actual pressure ratio of the compressor.

$$h_{3s} - h_1 = c_{pm}T_1\left[\left(\frac{p_3}{p_1}\right)^{(k-1)/k} - 1\right]$$

Then

$$h_3 - h_1 = \frac{c_{pm}T_1[(p_3/p_1)^{(k-1)/k} - 1]}{\eta_{ic}} \tag{6-29}$$

The rotor power absorbed for G lb of gas compressed per sec can be expressed as

$$\text{Total rotor hp} = \frac{G(h_3 - h_1)778}{550} \tag{6-30}$$

$$\text{Rotor hp} = \frac{Gc_{pm}T_1[(p_3/p_1)^{(k-1)/k} - 1]}{{}^{550}\!/_{778}\eta_{ic}} = \frac{Gc_{pm}T_1[(p_3/p_1)^{(k-1)/k} - 1]}{0.707\eta_{ic}} \tag{6-31}$$

where η_{ic} = internal compressor efficiency (often called adiabatic efficiency) expressed as a decimal

p_3/p_1 = actual pressure ratio produced

h_3 = enthalpy, Btu per lb, at leaving conditions from compressor

h_1 = enthalpy, Btu per lb, at entering conditions to compressor

C_3 is considered $\equiv C_1$ for velocity at compressor outlet and inlet, fps

When the actual temperatures and velocities are known at inlet to and exit from an essentially adiabatic compressor,

$$\begin{aligned}\text{Total rotor hp} &= \frac{G\left[c_{pm}(T_3 - T_1) + \dfrac{C_3^2 - C_1^2}{2g_cJ}\right]}{0.707} \\ &= G\left[\frac{h_3 - h_1}{0.707} + \frac{C_3^2 - C_1^2}{(550)(2g_c)}\right]\end{aligned} \tag{6-32}$$

The power required from the drive shaft (shaft horsepower) must exceed the rotor horsepower by a sufficient amount to provide for bearing and gland losses

$$\text{Shaft hp} = \frac{\text{rotor hp}}{\eta_m} \tag{6-33}$$

where η_m = mechanical efficiency expressed as a decimal and ranging from 0.94 to 0.98 for well-designed compressors operating near rated load

Over-all centrifugal-compressor performance values η_{oc} range from 0.78 to 0.82 for better designed compressors operating at rated conditions. For small and less carefully designed compressors 0.7 to 0.75 might be thought of as most representative.

$$\eta_{oc} = \eta_{ic}\eta_m \tag{6-34}$$

Equations (6-16) to (6-19) showed methods suitable for determining the ideal pressure ratio which could be produced by an impeller operating at a given tip speed. For the real compressor, modifications have to be made to permit prediction of the actual pressure ratio. First it will be realized that even with radial vanes located close to each other in the impeller the tangential fluid velocity C_{2u} does not reach the tip velocity u_t. Numerical values of the ratio

$$\mu = \frac{C_{2u}}{u_2} = \frac{C_{2u}}{u_t} \tag{6-35}$$

are a measure of the effectiveness in a given impeller of bringing the fluid to tip velocity. The term slip is used in naming

$$1 - \mu = 1 - \frac{C_{2u}}{u_t} = \frac{u_t - C_{2u}}{u_t} \tag{6-36}$$

Slip approaches zero with a large number of vanes and may rise to about 0.25 with 12 vanes. Compressors in common use employ 10 to 30 vanes. For general design analyses, values of μ should be selected ranging from 0.86 to 0.92 with 0.9 being often used as a round figure. For nonradial vanes select C_{2u} from the constructed outlet velocity triangle, and use 90 per cent of this value in computations.

Now employ μ in a derivation similar to that employed for Eq. (6-11), for this case,

$${}_1W_2 = \frac{1}{g_c}(u_2C_{2u} - u_1C_{1u}) = \frac{1}{g_c}[u_2\mu u_2 - u_1(0)]$$

$${}_1W_2 = \frac{\mu u_2^2}{g_c} = \frac{\mu u_t^2}{g_c} \tag{6-37}$$

and Eq. (6-37) represents the probable work added by the impeller. This work with $C_3 \equiv C_1$ must all appear as enthalpy increase, and by Eq. (6-28)

$$\mu\frac{u_t^2}{g_cJ} = h_3 - h_1 = \frac{h_{3s} - h_1}{\eta_{ic}}$$

$$h_{3s} - h_1 = \eta_{ic}\mu\frac{u_t^2}{g_cJ} \tag{6-38}$$

$$\frac{h_{3s}}{h_1} = \frac{\eta_{ic}\mu u_t^2}{h_1g_cJ} + 1 \tag{6-39}$$

Solution of this case by use of the isentropic p_{r3}/p_{r1} ratio will give the *actual pressure ratio* which can be produced. If gas tables such as Table 2-4 for air are not available, use can be made of a form corresponding to Eq. (6-16),

$$\frac{p_3}{p_1} = \left(\frac{\eta_{ic}\mu u_t^2}{c_{pm}g_cJT_1} + 1\right)^{k/(k-1)} \tag{6-40}$$

Example 2. The first stage of a centrifugal compressor is single inlet with an eye 22 in. in diameter, hub 8 in., and an over-all impeller diameter of 39¾ in. When supplied with 70.3 F air at 14.7 psia and turning at 5910 rpm, it compresses 67 lb of air per sec through a pressure compression ratio of 2.122. (*a*) Compute the ideal pressure ratio for this compressor when operating under the above conditions. (*b*) Find the internal rotor efficiency based on impeller tip velocity with μ considered as 0.9. (*c*) Use the efficiency found in (*b*) to determine the leaving-air enthalpy and temperature. (*d*) Find the rotor horsepower and the shaft horsepower.

Solution. (*a*) Impeller tip velocity, $u_t = \pi\,(39.75)(5910)/(12)(60) = 1024$ fps. Using Eq. (6-17) with $h_1 = 16.80$, for $T_1 = 530$ R,

$$\frac{h_3}{16.80} = \frac{(1024)^2}{(32.17)(778)(16.80)} + 1$$

$$h_3 = 58.76 \text{ Btu per lb}$$

for which $p_{r3} = 4.454$ (Table 2-4); for $h_1 = 16.800$, $p_{r1} = 1.640$.

$$\left(\frac{p_3}{p_1}\right)_s = \left(\frac{p_{r3}}{p_{r1}}\right)_s = \frac{4.454}{1.640} = 2.71$$

(*b*) Employ the real pressure ratio to find the working value of p_{r3}.

$$2.122 = \frac{p_3}{p_1} = \frac{p_{r3}}{p_{r1}} = \frac{p_{r3}}{1.640}$$

$$p_{r3} = 3.480$$

For this value of p_{r3} from Table 2-4 read $h_{3s} = 47.120$ and $T_{3s} = 656.75$ R. Equation (6-39) can now be used to find η_{ic}.

$$\frac{47.120}{16.80} = \frac{\eta_{ic}(0.9)(1024)^2}{(16.80)(32.17)(778)} + 1$$

$$\eta_{ic} = 0.802 = 80.2\%$$

Equation (6-40) can be used for an alternate solution if desired.

$$2.122 = \left\{\frac{\eta_{ic}(0.9)(1024)^2}{(0.241)(25{,}030)(530)} + 1\right\}^{1.4/1.4-1}$$

$$1.2402 - 1 = \frac{\eta_{ic}(0.9)(1024)^2}{(0.241)(25{,}030)(530)}$$

$$\eta_{ic} = 0.802 = 80.2\%$$

(*c*) By Eq. (6-28)

$$h_3 - h_1 = \frac{47.120 - 16.80}{0.802} = 37.83$$

$$h_3 = 37.83 + 16.80 = 54.63 \text{ Btu per lb}$$

$$T_3 = 687.2 \text{ R}$$

(*d*) By Eq. (6-30) the rotor horsepower is

$$\text{Rotor hp} = \frac{67(54.63 - 16.80)}{0.707} = 3585$$

For the shaft horsepower the mechanical efficiency must be known or, as here assumed, say 0.97.

$$\text{Shaft hp} = \frac{\text{rotor hp}}{\eta_m} = \frac{3585}{0.97} = 3700$$

6-5. Lysholm Compressor. A Lysholm compressor is illustrated in Fig. 6-10. This is a positive-displacement compressor using two meshing rotors. One of the rotors is directly driven from the turbine shaft, while the other rotor is turned by gearing from the first rotor so that it always rotates in proper synchronism. The rotors are machined and ground to have close clearances without coming into mechanical contact. As the lobes move into position at the suction port, a fixed volume of air is trapped between adjacent meshing segments of the lobes. This volume of air, sealed off by the turning lobes, is progressively compressed as it

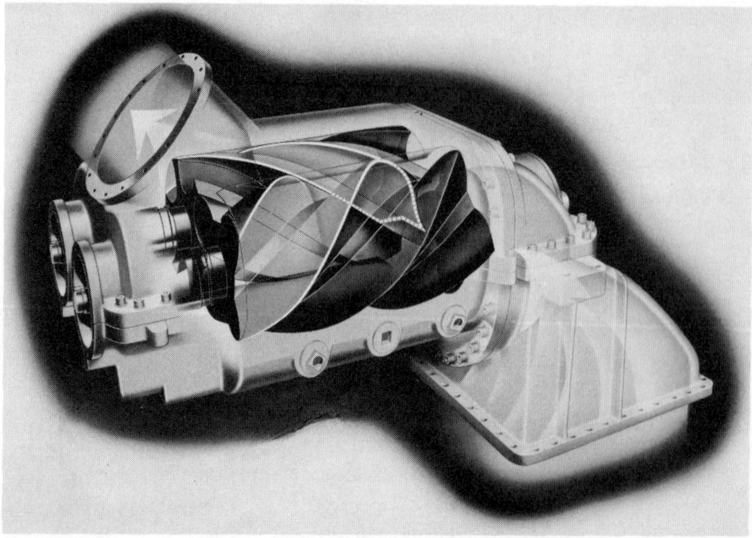

FIG. 6-10. Cutaway drawing of Lysholm compressor showing lobes. (*Elliott Co.*)

moves through the narrowing passageway of the lobes. As progressively lower volume exists between the lobes, when the air is brought to the discharge port a definite ratio of inlet volume to discharge volume exists in the compressor. This could be called the fixed-volume ratio of the compressor, and using the isentropic pressure relationships it is possible to convert this into a pressure ratio which is known as the *built-in* pressure ratio. The particular unit shown in the illustration has a capacity of 10,000 cfm. It can operate at varying speeds up to a maximum of 4000 rpm, with a pressure ratio of 2.8.

In a displacement compressor of this type, instability does not exist, as is true with centrifugal- and axial-type machines, so that it is possible to operate it over an extensive capacity pressure range. Adiabatic compression efficiencies of some 84 per cent have been obtained with this machine. Unfortunately, such machines are of limited capacity in contrast to centrifugal and axial-flow compressors.

Figure 6-11 shows performance tests of a typical Lysholm unit. It will be noticed that the volumetric efficiency is high over a wide range of pressure-ratio values.

One of the marine plants designed for the United States Navy and illustrated in Fig. 6-12 shows the use of Lysholm compressors. In this particular plant air drawn from outside enters the first Lysholm compressor indicated by the letter *A*. It undergoes a pressure ratio of 2.75:1 and then enters an intercooler which drops the air temperature to within

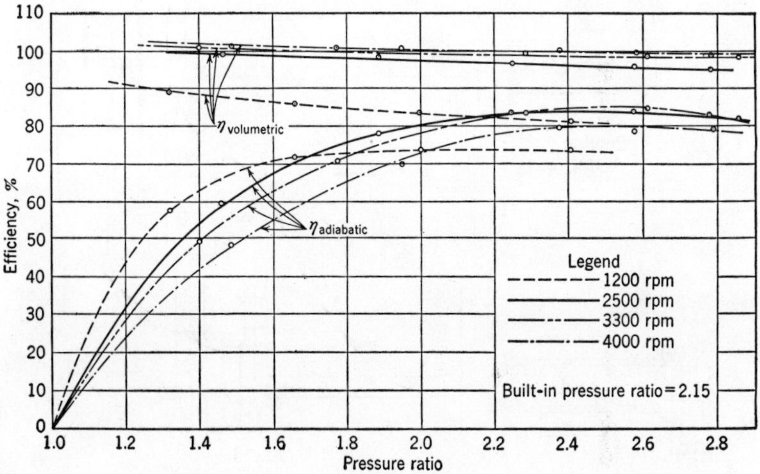

FIG. 6-11. Typical performance curves, Lysholm compressor. (*Elliott Co.*)

20° of ambient air. It then enters the second Lysholm compressor labeled *C*, where it undergoes an additional 2.1 pressure ratio. The compressed air at a temperature in the neighborhood of 225 F enters the regenerator labeled *D*, where the air is warmed to about 815 F. In the high-pressure combustion chamber *E* the gas can be heated up to 1400 F, but the temperature rise is limited to satisfying the energy required to drive the low-pressure compressor *A* which the turbine *F* drives. From this turbine, the gases pass to the low-pressure combustor *G*, where they are warmed up to a temperature of 1400 F before entry into the second turbine, which operates at lower pressure. This turbine drives the high-pressure compressor and in addition provides the useful output of the system.

6-6. The Free-piston Engine and Compressor. For some twenty years there has been active interest in the idea of a free-piston engine, but only during the Second World War and the postwar period has outstanding development work been done with this type of unit. The engines also go under the name of floating-piston units. A Frenchman named Pescara is credited with having invented this type of engine. In addition to work

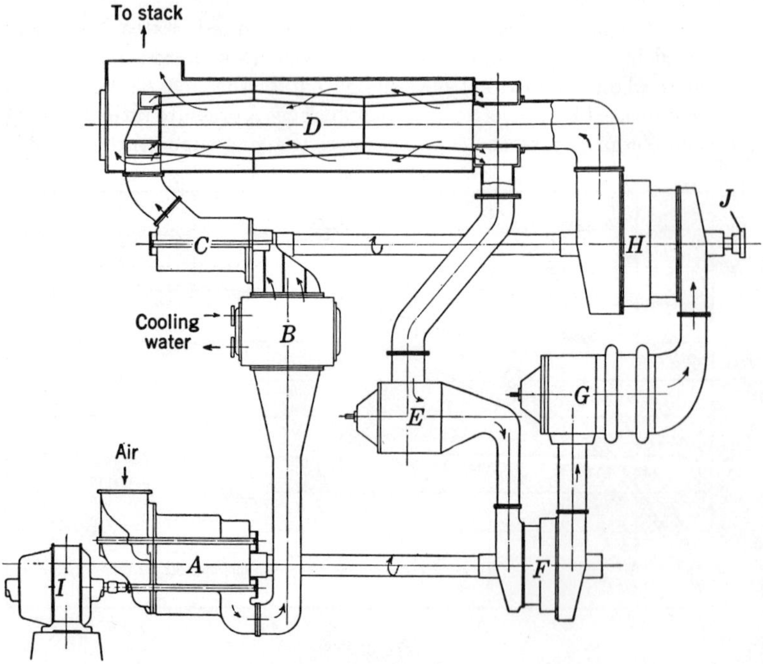

Legend

A Low-pressure compressor
B Intercooler
C High-pressure compressor
D Regenerator
E High-pressure combustion chamber
F High-pressure turbine
G Low-pressure combustion chamber
H Low-pressure turbine
I Starting motor
J Output shaft

FIG. 6-12. Diagram of marine turbine plant with Lysholm compressors. (*Elliott Co.*)

by Pescara development has also been actively conducted by Junkers in Germany and by Sulzer Brothers in Switzerland. In this country some research has been successfully carried out, and one company is endeavoring to develop a commercial unit.

Figure 6-13 shows the fundamental elements of a free-piston engine and compressor. The power unit consists of a two-stroke-cycle opposed-piston diesel engine. When the power pistons are brought together and fuel is injected, combustion occurs and the pistons are forced outward. This power stroke is the delivery stroke for the compressed air and the

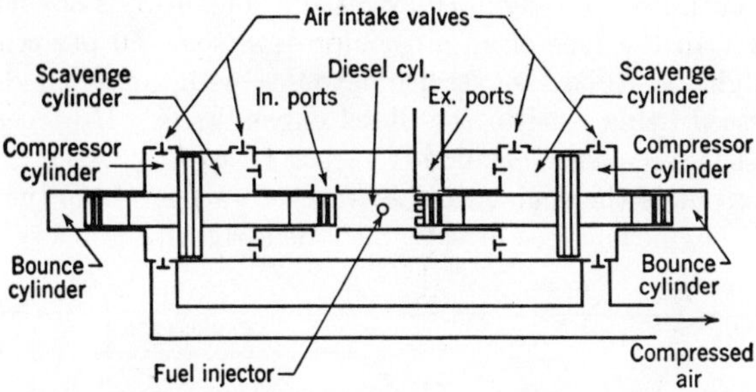

FIG. 6-13. Diagrammatic layout of free-piston engine and compressor. (*After Oppenheim and London, Ref.* 4.)

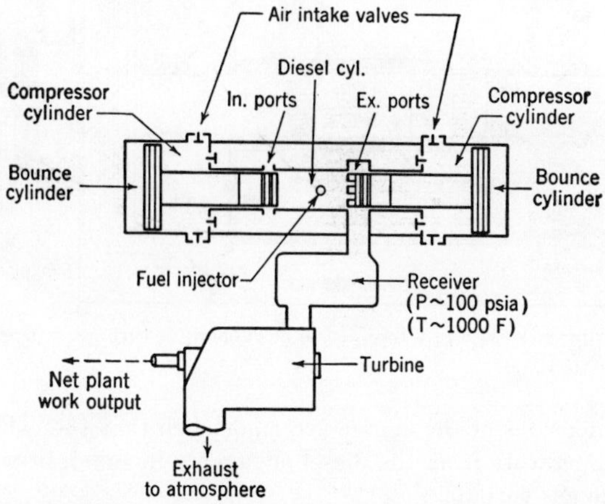

FIG. 6-14. Combination of diesel-engine-operated free-piston compressor with exhaust serving a gas turbine. (*After Oppenheim and London, Ref.* 4.)

suction stroke for the scavenging air. The bounce cylinders, which are shown in the figure, are critical elements of the free-piston compressor. During the outward power stroke, air is compressed in the bounce cylinders, and the spring action of this compressed air is adequate to bring the diesel pistons toward each other, where, at the end of the stroke, fuel is again injected and a second power stroke takes place. Near the end of each power stroke the exhaust ports open, and later the charge of scavenging, supercharged air enters the diesel engine.

The time cycle of the free-piston compressor is determined by a resonant frequency which is established by the spring action of the compressed air and gases acting in concert with the masses of the reciprocating system. The fuel-pump injection is timed by the returning opposed pistons in such way as to work harmoniously in connection with the cycle of operations. The useful output of the engine, illustrated in Fig. 6-13, is of course compressed air. In the case of an air compressor, some 50 per cent of the total air might be utilized for service purposes, with the remainder of the air compressed being used in the diesel engine itself. However, in the event that the free-piston engine is to serve a gas-turbine power system, the total output of the engine can be arranged to come from the exhaust of the diesel engine and feed directly into the turbine (Fig. 6-14). In this

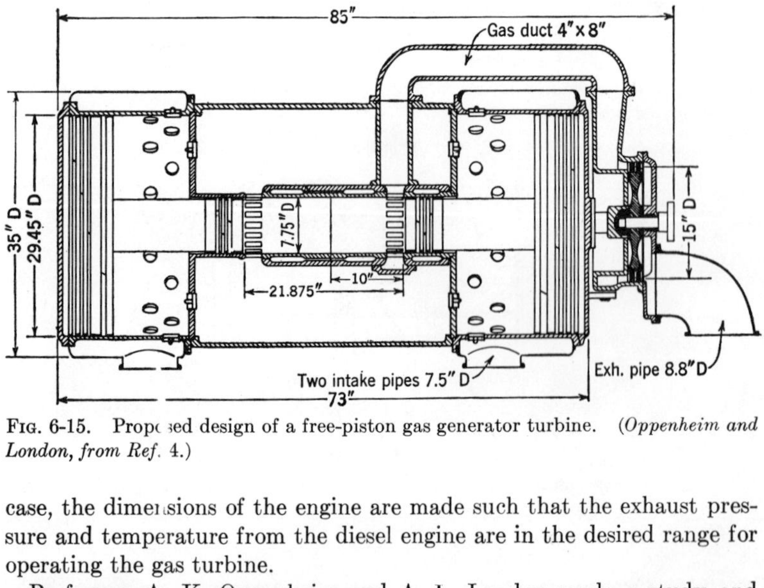

FIG. 6-15. Propc sed design of a free-piston gas generator turbine. (*Oppenheim and London, from Ref.* 4.)

case, the dimensions of the engine are made such that the exhaust pressure and temperature from the diesel engine are in the desired range for operating the gas turbine.

Professors A. K. Oppenheim and A. L. London made a study and analysis of the free-piston engine, and much of the material in the following discussion is taken from their paper covering this subject.[4] The writers laid out a complete theoretical design of a free-piston engine constructed for the purpose of supplying high-pressure high-temperature gas in adequate amounts to power a gas-turbine unit. Figure 6-15 shows the proposed design computed by Oppenheim and London, and this also shows the dimensions of the unit. It will be noted that the inner side of the compressor is used exclusively for supplying air to the diesel unit, while the outer side of the large piston works in the bounce cylinder.

This unit is designed to deliver its whole output as the high-pressure high-temperature energy supply for a gas-turbine power unit. The design shows that it would be possible for the diesel exhaust of the free-piston engine to deliver gases at 92.6 psia at a temperature of 1523 F, and with a computed gas-flow rate of 19,800 lb per hr the gas turbine could deliver 1000 hp. The diesel-engine gas generator has a stroke of 10 in., with a piston area of 0.3273 ft^2 and a bounce piston area of 4.73 ft^2. Under these conditions, considering the resonating mass of the system (1385 lb per piston) a frequency of 847 cycles per min could be expected. The computed air delivery is 0.1925 lb per cycle. The pressure ratio of the compressor is 6.98:1.

The air-fuel ratio is 20:1, and the computed fuel rate of this unit is 0.333 lb per shaft hp-hr delivered from the turbine. This represents a

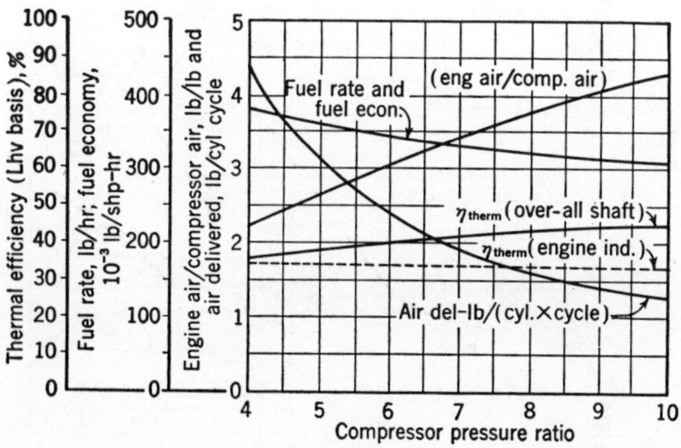

FIG. 6-16. Thermodynamic characteristics of free-piston-type gas generator turbine. (*Oppenheim and London, from Ref.* 4.)

thermal efficiency of 42 per cent based on the lower heating value of the 18,200 Btu per lb fuel or 39.6 per cent on the higher heating value. This high efficiency can be explained in terms of the effectiveness of the diesel engine, which serves as the gas-generation stage of the process. In a normal gas-turbine system, approximately two-thirds of the turbine output is required to compress the gases used in the turbine, leaving one-third of the power available for useful external output. The free-piston design eliminates this compressor tax and supplies the gases at high pressure and high temperature almost as a by-product. In the design analysis, the diesel unit showed an indicated thermal efficiency itself of 34.2 per cent based on the lower heating value. Figure 6-16 taken from the aforesaid paper indicates the expected performance of this unit at varying pressure compression ratios based on 1000 shaft hp delivered from the turbine.

Figure 6-17 illustrates a design developed by Sulzer in Switzerland for a

7000-hp gas turbine which receives its gas supply from three floating-piston engines. The bounce cylinders, as before, are shown at the extreme ends of each of the floating-piston engines, and the exhaust from each engine feeds into a common header which serves the exhaust-gas power turbine. A charging compressor also driven from the hot gas supplied by the free-piston engines is used to supercharge the gas supply to the free-

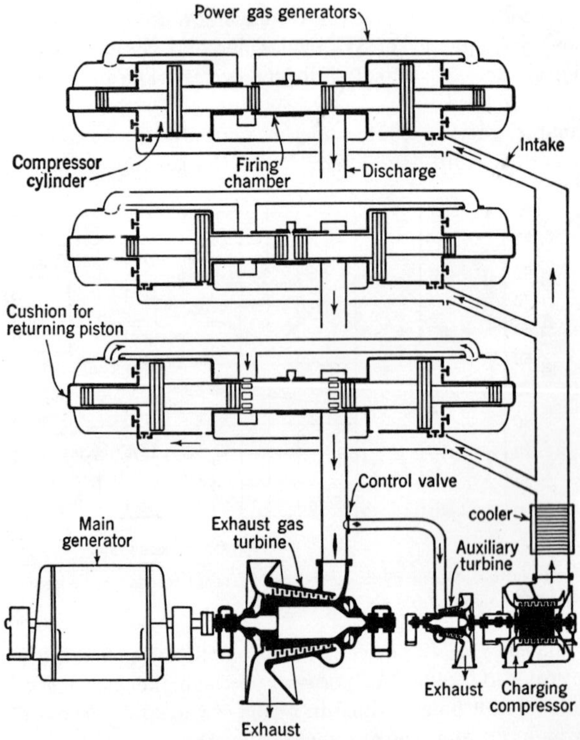

FIG. 6-17. Combination design of three floating-piston engines serving a gas turbine. (*Sulzer Co.*)

piston engines. A synchronization system, which can be controlled by fuel-injection timing and the resonance conditions of the compressors, can provide a uniform flow of gases to the exhaust-gas turbine. It is thought that the Sulzer design has an over-all efficiency in excess of 40 per cent based on lower heating value of the fuel.

The starting of any free-piston unit can be accomplished by supplying compressed air to the far face of the compressor cylinder to provide the initial compression stroke for the opposed piston diesel or even by supplying air to the bounce cylinder for the same purpose. It is, of course, necessary to see that the bounce cylinder under operating conditions is provided with an adequate amount of air to give the spring action required for effective operation of the unit.

Data on operation of a Pescara unit are given by Oppenheim and London[4] as follows: 1138 shaft hp developed at a gas-flow rate of 29,000 lb per hr with the gases at 64.5 psia and 1405 R entering the turbine. The Pescara unit has a compressor pressure ratio of 4.85 with an engine compression ratio of 8.5. The air-fuel ratio employed is 32 and the stroke of the compressor 17.5 in. The engine piston area is 0.979 ft^2 with a cylinder length of 38 in., and the ratio of compressor piston area to engine piston area is 5.86. The frequency of firing is 613 cycles per min. The piston mass is 2080 lb. The flow rate of fuel oil to the system is 460 lb per hr, and the thermal efficiency of the unit based on the lower heating value is 34.6 per cent.

A free-piston gas-turbine unit employed for operating a 10- to 15-ton truck has been designed by Robert Huber in France.[11] This design, which is quite unique, consists of two 120-hp free-piston engines placed on each side of the gas turbine in the truck. Each of the free-piston engines weighs 550 lb, and operating together or singly they will supply the hot gases for the turbine. Each free-piston engine operates at 870 pulsations per minute to deliver 0.84 lb per sec of gases at 56.9 psig and 1060 F to the turbine. The two-stage turbine, which turns at 30,000 rpm, should develop better than 220 hp and should propel the truck at 50 mph. A reducing gear and a transmission having two speeds forward are used in the design.

It is obvious that, because of the potentially high thermal efficiency, the free-piston-engine–gas-turbine combination merits serious consideration from designers. Such a combination brings the gas turbine into a competitive position for economy in use of fuel with other types of prime movers without the necessity of excessively high temperatures at the turbine inlet. The turbine can be of conventional design. The free-piston engines, having no cranks or external shafting and operating in a horizontal plane with no side thrust, can keep wear to a minimum.

6-7. Water Injection in Compressors. The net power of any gas-turbine system is merely the excess of turbine power over that required by the compressor. Consequently, any provision whereby the power requirement of the compressor can be reduced will increase the net power of the system and, in general, the efficiency of the cycle. Mention has been made of breaking the compression phase into a series of steps with use of intercooling between the steps, and previous numerical examples have shown how this reduces the compressor power required. With a large number of intercoolers the compression phase could be made to reach the conditions of isothermal compression. For a perfect gas, isothermal compression represents a minimum work-input condition.

Water injection into the moving air stream undergoing compression can also reduce the work required to compress the air. The injection of a water mist in the air stream reduces the temperature of the air leaving the compressor as well as the amount of work required to compress each pound of air. The cooling action occurs as a result of utilizing an enthalpy reduction (temperature decrease) in the air stream to supply the latent heat for vaporizing the water droplets carried in suspension in the stream.

When water is atomized into the stream and vaporized as the stream passes through the compressor, the resulting water vapor remains in the air stream and takes part in the remainder of the cycle. The amount of water that can be added varies from a trivial amount to 5 per cent or more by weight in the mixture. Theoretically, with infinitely small droplets and no irreversibilities from mixing and turbulence, work reductions reaching 20 per cent or more can take place in compressing a pound of air from a given inlet to a given outlet pressure when wet compression is compared with dry compression. With wet compression the temperature of the compressed air is lower, and each pound of air passing through the system carries an additional amount of water vapor. Because of both these conditions, a greater fuel input is required with water injection. However, more work is produced in the turbine by each pound of air with its associated moisture. Under this condition, if the hot products are discharged at high temperature, the cycle efficiency is definitely lower than with dry air. However, when a regenerator is used to warm the incoming compressed air and water vapor at the expense of the waste gases, the efficiency of a wet-compression cycle can be appreciably higher than that of a dry cycle.

Several experimental programs have been carried out using water as an injection means,[7,9,10] and it seems possible that further investigations will be made to determine the feasibility of water injection for land and marine units. In the case of aircraft units, an extra amount of water would have to be carried on the airplane, thereby offsetting useful payload. This makes water injection somewhat impracticable for aircraft.

Among the difficulties which might be experienced with water injection are potential deposition on and erosion of the compressor blading or of the impeller, possible interference with combustion and flame stability, and the possibility that under certain conditions surges of unevaporated water might move through the system and cause serious damage. Experimentation has indicated, however, that none of these conditions is serious, and water injection may come into more extensive use in gas-turbine systems.

PROBLEMS

6-1. A low-speed centrifugal compressor turning at 3600 rpm has an impeller with a 39-in. outer diameter. The extreme diameter of the eye is 8 in., and the hub measures 4.5 in. The fluid at entry into the impeller flows in planes essentially parallel to the axis of the compressor and has an absolute velocity C_1 which makes an angle θ_1 of 77° with u_1 (the blade speed at the extreme eye diameter). The impeller blade tips are curved backward and make an angle β_2' of 62°, measured with respect to a tangent drawn to the outer impeller circumference. Assume an idealized compressor running full and with the relative velocity w_1 of the fluid at the outermost diameter of the eye lying in the plane of u_1 and C_1 and having a magnitude of 370 fps.

a. Make use of full-flow characteristics to find the approximate outlet relative velocity $w_2 = w_1 (D_1/D_2)$. *Ans.* 75.9 fps.

b. Draw a representative diagram to scale for this impeller. Sketch in the shape of the vanes, and show the outlet angle to scale.

c. Draw the inlet and outlet velocity triangles to scale, each in its appropriate plane.

d. Assuming an incompressible fluid of 0.08 lb per ft,^{3} compute the ideal pressure rise in the impeller based on changes in u and w as found from the velocity triangles. *Ans.* 611 psf.

e. If the absolute velocity C_3 leaving the diffuser reduces to $\frac{1}{2}C_2$, where C_2 is the absolute velocity of the fluid on outlet from the impeller, find the ideal pressure rise in the diffuser. *Ans.* 316 psf.

f. With the help of velocity triangles and the use of momentum-energy relationships, compute the energy imparted to each pound of fluid by the impeller. *Ans.* 10,700 ft-lb per lb.

g. Assuming that approximately the same flow and momentum conditions hold over the whole eye, find the pounds per second of gas flow and the impeller horsepower. *Ans.* 7 lb per sec, 137 hp.

6-2. Compute the data for and plot a graph of ideal pressure ratio as ordinate and impeller tip speed as abscissa. Use tip speeds of 800 to 1600 fps, and assume air at 70.3 F (530 R) as the fluid compressed. *Ans.* R for 800 fps = 1.94, R for 1200 fps = 3.7.

6-3. Compute the data for and plot a graph of pressure ratio as ordinate and impeller tip speed as abscissa for a radial compressor with 24 vanes, having axial flow at inlet, for tip speeds varying from 800 to 1600 fps, using the air as the fluid compressed at $T_1 = 530$ R. Assume $\mu = 0.9$ and $\eta_{ic} = 0.75$, with both constant over the range in question. *Ans.* R for 800 fps = 1.56, R for 1200 fps = 2.55.

6-4. For Prob. 6-2, compute the ideal temperature ratio.

6-5. A single-inlet centrifugal compressor supplies air for a gas turbine which is employed to power a pump. The compressor, which is driven at 40,300 rpm, has an impeller with a tip diameter of 6.97 in., with 4.56 in. for the outer diameter of the eye and 2.09 in. at the hub. The compressor pressure ratio is 2.44, and the air flow is 2.35 lb per sec when supplied with air at 14.7 psia and at 81.3 F. (*a*) Compute the internal compressor efficiency on the basis of Eq. (6-40) or (6-39) if μ is considered as 0.87. (*b*) Find the rotor horsepower and the probable shaft horsepower. *Ans.* (*a*) 0.721, (*b*) 173.5 and 180.7 shaft hp.

6-6. A jet engine employs a double-inlet centrifugal compressor for which the following data are pertinent: rpm 11,500, impeller tip diameter 30 in., impeller inlet (eye) diameter 18.25 in., impeller hub diameter 8 in., realized compression ratio 4.13, compressor discharge temperature 413 F from inlet at 14.7 psia at 519 R (59.3 F), air flow 79 lb per sec. For purposes of analysis it can be considered that there are two compressors, each compressing 39.5 lb of air per sec. (*a*) Compute velocities at impeller tip, eye, and hub. (*b*) Compute the ideal pressure ratio for a compressor of this size and rpm while compressing air. (*c*) Use Eq. (6-39) or (6-40) in connection with the realized pressure ratio, and with μ assumed as 0.87 find the internal compressor efficiency. (*d*) Use this efficiency to find the final air temperature leaving the compressor, and compare with the test value. (*e*) Use the test value to estimate the probable rotor horsepower for the double impeller. *Ans.* (*a*) 1504 fps, 915 fps, 401 fps; (*b*) 6.77; (*c*) 78.9%; (*d*) 385 F; (*e*) 9570 hp.

6-7. If in Prob. 6-6 the air entering the eye of the compressor can be considered essentially axial in flow direction and constant across the whole annulus, compute the inlet velocity for the 79 lb per sec flow divided equally between each side of the impeller. Using this axial velocity and the velocities at the hub and extreme eye radius, find the relative velocity at the inlet to the eye at the two extreme positions. *Ans.* 1.469 ft^2 area, $C_{al} = 351.6$ fps, $u_h = 401$, $u_e = 915$, $w_h = 534$, $w_e = 980$ fps.

6-8. The jet engine described in Prob. 6-6 when operating in a more efficient range turns at 9800 rpm. On test, when supplied with air at 14.7 psia and 519 R, it was found that the internal compressor efficiency was 0.78. Consider that μ for this operating condition is 0.90. The measured air flow is 67 lb per sec. (*a*) Compute the ideal pressure ratio for the compressor while operating at 9800 rpm. (*b*) Make use of the internal efficiency and value of μ to find the actual pressure ratio. (*c*) Find the approximate actual temperature of the air leaving the compressor. (*d*) Make use of the temperature found in (*c*), and estimate the probable rotor horsepower for the double impeller. (*e*) If the mechanical efficiency can be considered as 0.965, estimate the shaft horsepower for the compressor drive.

REFERENCES

1. Moss, S. A.: Gas Turbines and Turbosuperchargers, *Trans. ASME*, vol. 66, pp. 351–363, 1944.
2. Cheshire, L. J.: The Design and Development of Centrifugal Compressors for Aircraft Gas Turbines, in "Development of the British Gas Turbine Jet Unit," Institution of Mechanical Engineers, London, reprinted by American Society of Mechanical Engineers, New York, 1947 (8 authors).
3. Crocker, Wilson: Fundamentals of the Elliott-Lysholm Compressor, *Mech. Eng.*, vol. 68, pp. 514–518, June, 1946.
4. Oppenheim and London: Design Analysis of Free-piston Engines, *Automotive Ind.*, vol. 103, pp. 46–50, 76, 78, July, 1950.
5. Traupel, W.: Sulzer Turbo-compressors, *Sulzer Tech. Rev.*, no. 2, pp. 12–23, 1950, Winterthur, Switzerland.
6. Trout, A. M.: Theoretical Turbojet Thrust Augmentation by Evaporation of Water during Compression as Determined by Use of a Mollier Diagram, *NACA Tech. Note* 2104, June, 1950.
7. Wetzel and Jennings: Water Spray Injection of an Axial-flow Compressor, *Proc. Midwest Power Conf.*, vol. 11, pp. 376–380, April, 1949, Chicago, Ill.
8. Samuels and Gale: Effect of Humidity on Performance of Turbojet Engines, *NACA Tech. Note* 2119, June, 1950.
9. Wilcox, E. C.: Turbojet Thrust Augmentation by Evaporation of Water Prior to Mechanical Compression as Determined by Use of Psychrometric Chart, *NACA Tech. Note* 2105, June, 1950.
10. Kleinschmidt, R. V.: Value of Wet Compression in Gas-turbine Cycles, *Mech. Eng.*, vol. 69, pp. 115–116, February, 1947.
11. Bradley, W. F.: French Gas Turbine Truck with Free-piston Engines, *Automotive Ind.*, vol. 106, pp. 52, 106, Feb. 15, 1952.
12. Stanitz and Ellis: Two-dimensional Compressible Flow in Centrifugal Compressors with Straight Blades, *NACA Rept.* 954, 1950.

CHAPTER 7

AIRCRAFT PROPULSION AND JET ENGINES

7-1. General Considerations. The various schemes for aircraft propulsion, whether by propeller, jet, or rocket, have one principle in common, namely, that the reaction resulting from the acceleration rearward of a large weight of air (gases) from the unit creates forward thrust. This principle is based on Newton's third law, that to every action there is an equal and opposite reaction. The rocket carries within itself both fuel and oxygen, and thrust is created by the products of combustion as they discharge at high velocity from a rest condition relative to the shell of the rocket.

To produce thrust in a propeller-type airplane, the propeller driven by the engine displaces rearward a large weight of air in the high-velocity slip stream leaving the airplane. In jet-propulsion engines, the compressor supplies compressed air to the combustors. There energy added by the burning fuel brings the gases to high temperature so that they possess sufficient energy to power the turbine and, in addition, to expand from the tail cone as a high-velocity jet. In other words, the gas-turbine jet engine inducts its air, compresses it, adds heat to the compressed gases, and then expands the high-temperature combustion products. Only enough energy is taken from the combustion products by the turbine to drive the compressor. The remainder of the energy continues in kinetic form in the jet and as enthalpy associated with the high-temperature products leaving in the jet. In the propeller-jet engine, frequently called the prop jet, the turbine develops power in excess of that required to drive the compressor and employs this excess power to drive a propeller through reduction gearing. In the propeller-jet engine the leaving jet also contributes to the thrust power. In some ways the propeller-jet arrangement is approximated by a reciprocating engine drive, as the reciprocating engine also delivers exhaust products rearward and can contribute to jet-power effect. However, two fundamental differences should be noted. The reciprocating engine uses an almost minimum weight of air per pound of fuel burned, and practically all its useful power is absorbed by the propeller shaft. The propeller-jet engine, on the other hand, probably utilizes only 15 to 30 per cent of the gas-turbine power in driving the propeller, with the remainder being absorbed by the compressor; in addition the turbine is so designed that the gases leaving the final moving row of the turbine blading possess a high residual kinetic energy.

Figures 1-11 and 1-12 illustrate two representative gas-turbine jet units, the former of these employing a centrifugal compressor and the latter an axial-flow compressor. The axial-flow compressor has the smaller cross-sectional frontal area and has a slightly higher compressor efficiency at design operating conditions. The centrifugal compressor, with its greater frontal area, offers more drag and is slightly lower in compressor efficiency. However, it is stable over a broader range of air-flow limits. Both types possess simplicity and light weight for a given power. They require a minimum of external cooling and use a relatively simple lubrication system. As will be seen later, it is also possible to operate them at much higher airplane speeds than are attainable with propellers. However, the fuel rate of jet engines (pounds per hour per pound of thrust) is greater than the fuel rate of reciprocating engines.

7-2. Jet Theory. The major part of the thrust produced by a jet engine is equal to the product of the mass (weight) of gas accelerated per

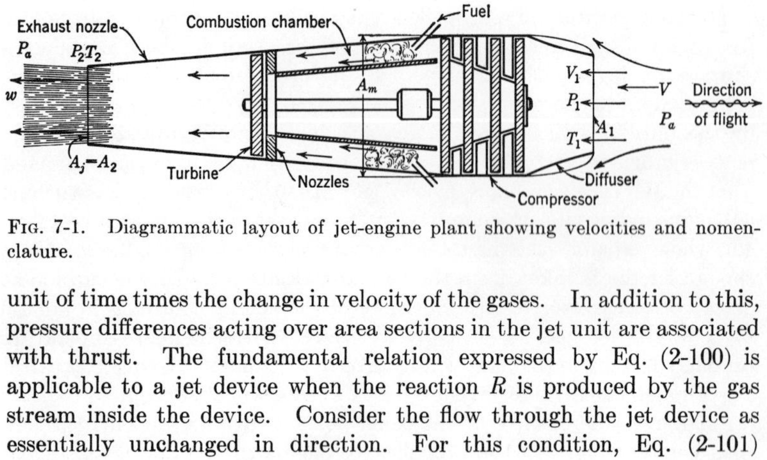

FIG. 7-1. Diagrammatic layout of jet-engine plant showing velocities and nomenclature.

unit of time times the change in velocity of the gases. In addition to this, pressure differences acting over area sections in the jet unit are associated with thrust. The fundamental relation expressed by Eq. (2-100) is applicable to a jet device when the reaction R is produced by the gas stream inside the device. Consider the flow through the jet device as essentially unchanged in direction. For this condition, Eq. (2-101) appears as

$$R + P_1A_1 - P_2A_2 = \frac{G}{g_c}(C_2 - C_1) \tag{7-1}$$

For the absolute velocities C_2 and C_1 (relative to the earth), it is permissible to substitute velocities w and V_1 of the stream relative to the jet unit. From the viewpoint of thrust, the result is the same. Figure 7-1 shows the stream velocities relative to the jet unit, and other nomenclature. Using relative velocities, Eq. (7-1) becomes

$$R + P_1A_1 - P_2A_2 = \frac{G}{g_c}(w - V_1) \tag{7-2}$$

The ambient pressure also acts on the exterior of the jet unit and may contribute to gross thrust. For an ambient fluid *of uniform pressure*, the summation of pressure-area forces on the jet unit in the direction of flight is

$$P_a(A_m - A_2) - P_a(A_m - A_1) \tag{7-3}$$

where A_m is the exterior area of the jet unit projected in a plane at right angles to the line of flight.

The total gross thrust T produced by the jet unit is the summation of R from Eq. (7-2) and the thrust from Eq. (7-3), or

$$T = \frac{G}{g_c}(w - V_1) + P_2A_2 - P_1A_1 + P_a(A_m - A_2) - P_a(A_m - A_1)$$

This can be simplified to

$$T = \frac{G}{g_c}(w - V_1) + P_2A_2 - P_1A_1 - P_aA_2 + P_aA_1 \tag{7-4}$$

It should be realized at this point that the amount of air G_a entering the device is less than the amount of products G leaving the device because fuel is burned inside the unit. To account for the momentum change of the additional mass, Eq. (7-4) is modified to

$$T = \frac{Gw}{g_c} - \frac{G_aV_1}{g_c} + A_2(P_2 - P_a) - A_1(P_1 - P_a) \tag{7-5}$$

where T = gross thrust, lb_f
w = jet efflux velocity relative to the airplane, fps
V_1 = air inlet velocity relative to the airplane, fps
P_1, P_2 = pressure, psf; A_1 and A_2 = ft²
G, G_a = lb per sec flowing, $g_c = 32.17$

The gross thrust T is used to overcome

1. External body friction forces resulting from the motion of the jet unit and its associated (airplane) structure, classified as part of drag with symbol D_1.

2. Summation of external pressure-area forces on the structure (airplane) associated with the jet unit, but not on the jet unit itself. This summation is part of drag and will be given the symbol D_2.

3. The gravity component in the direction of motion of the weight of the jet engine and its associated airplane structure *during climbing*, represented by symbol L_c.

4. The reversed effective force during acceleration a of the jet engine and its associated airplane structure of total weight S lb and represented by $(S/g_c)a$.

It is interesting to note that the exterior surface area of the jet unit itself may be altered when installed in an airplane, and thus the external pressure-area forces on the jet unit may be different for the installed jet than for the same jet if tested separate and apart from the airplane. The conclusion is that the installed thrust may differ to some extent from the thrust of the bare unit. Equation (7-5) is gross thrust for the bare unit.

If the thrust is now set equal to the resistances which it must overcome, another equation for thrust may be written as

$$T = D_1 + D_2 + L_c + \frac{S}{g_c} a$$

$$= \text{total drag } D + L_c + \frac{S}{g_c} a \tag{7-6}$$

Combining the two equations for thrust,

$$\frac{Gw}{g_c} - \frac{G_a V_1}{g_c} + A_2(P_2 - P_a) - A_1(P_1 - P_a) = D + L_c + \frac{S}{g_c} a \tag{7-7}$$

Now define *net thrust* T_n as that part of thrust which is available for climb and acceleration.

$$\text{Net thrust } T_n = \text{gross thrust } T - \text{drag } D$$

Also,

$$T_n = L_c + \frac{S}{g_c} a$$

Therefore,

$$T_n = \frac{Gw}{g_c} - \frac{G_a V_1}{g_c} + A_2(P_2 - P_a) - A_1(P_1 - P_a) - D \tag{7-8}$$

For the jet engine and airplane in flight, the relative entering velocity V_1 is equal to V, the flight velocity; $P_1 = P_a$, and as the discharge pressure from the jet is also at atmospheric pressure, the gross thrust T becomes [from Eq. (7-5)],

$$T = \frac{Gw - G_a V}{g_c} \tag{7-9}$$

If we also consider that the air flow G_a and the products flow G are nearly equal, a further simplification is possible, giving

$$T = \frac{G_a}{g_c} (w - V) \tag{7-10}$$

This equation obviously can also be used for the thrust obtained from air moved by a propeller only.

Thrust *horsepower* by use of Eq. (7-9) is

$$\frac{TV}{550} = \frac{Gw - G_a V}{g_c} \frac{V}{550} \tag{7-11}$$

and as $G_a \approx G$,

$$\frac{TV}{550} = \frac{G_a}{g_c}(w - V)\frac{V}{550} \tag{7-12}$$

A more complete expression for thrust horsepower is obtained from the original gross-thrust equation (7-5).

$$\frac{TV}{550} = \left[\frac{Gw}{g_c} - \frac{G_a V_1}{g_c} + A_2(P_2 - P_a) - A_1(P_1 - P_a)\right]\frac{V}{550} \tag{7-13}$$

This equation reduces to Eq. (7-11) when $P_2 = P_a$, $P_1 = P_a$, and $V_1 = V$. Equation (7-13) may also be written as

$$\frac{TV}{550} = \frac{VG_a}{550g_c}[w(1 + f) - V_1] + \frac{V}{550}[A_2(P_2 - P_a) - A_1(P_1 - P_a)]$$

The symbol f is the fuel-air ratio (by weight) for jet combustion, expressed as a decimal. In usual flight, $P_1 = P_a$, and $V_1 = V$, and this expression becomes

$$\frac{TV}{550} = \frac{VG_a}{550g_c}[w(1 + f) - V] + \frac{V}{550}A_2(P_2 - P_a) \tag{7-14}$$

Here, the symbols A_2, P_2, and w refer to the same location in the jet, but not necessarily to the outlet section only. At the outlet, $P_2 = P_a$, and the last term becomes zero.

An expression for propulsion efficiency will now be developed. Propulsion efficiency η_p is defined as

$$\eta_p = \frac{\text{thrust power}}{\text{jet power}} \tag{7-15}$$

For thrust power, the simplified equations (7-11) and (7-12) will be used. The jet power when defined and expressed as horsepower appears,

$$\text{Jet hp} = \begin{pmatrix}\text{change in kinetic energy of entering air} \\ \text{and leaving products per sec, as measured} \\ \text{by an observer on the jet unit}\end{pmatrix} 1/550$$

$$= \frac{Gw^2 - G_a V^2}{2g_c 550} \tag{7-16}$$

$$\text{Jet hp} = \frac{G_a}{2g_c}(w^2 - V^2)\frac{1}{550} \tag{7-17}$$

Both the equation for thrust horsepower [Eq. (7-12)] and the second expression for jet horsepower neglect the weight of fuel burned in the air and leaving with the jet, that is, $G_a = G$. However, air-fuel ratios in jet units are very high, and the error is slight. It should be recognized at this point that energy (and power) is added to the propelling fluid in the form of enthalpy increase as well as in the form of kinetic-energy increase.

However, the *propulsion* efficiency is concerned only with the conversion of the kinetic-energy increase (with associated velocity and momentum increase) to useful thrust power. The conversion of the enthalpy to kinetic energy is the internal function of the jet unit, and this conversion might be evaluated as a thermal efficiency in contrast to propulsion efficiency.

The expression for propulsion efficiency by Eqs. (7-11) and (7-16) appears as

$$\eta_p = \frac{\dfrac{(Gw - G_aV)V}{550g_c}}{\dfrac{Gw^2 - G_aV^2}{2g_c(550)}} = \frac{(Gw - G_aV)(2V)}{Gw^2 - G_aV^2} \tag{7-18}$$

Further, when G is not greatly different from G_a, Eq. (7-18) becomes

$$\eta_p = \frac{2V}{w + V} \tag{7-19}$$

The same equation for propulsion efficiency can be obtained in a different manner. The viewpoint of an observer on the ground will now be taken. The equivalent equation for thrust horsepower is

$$\frac{TV}{550} = \frac{G_a}{g_c}(C_2 - C_1)\frac{V}{550}$$

and usually C_1, the absolute velocity of the air entering the jet unit, is taken to be zero. Therefore,

$$\text{Thrust hp} = \frac{G_a}{g_c}(C_2)\frac{V}{550}$$

But

$$\begin{aligned} C_2 &= \text{absolute exit air velocity} \\ &= w - V \end{aligned}$$

Therefore, thrust horsepower is, as before,

$$\frac{G_a}{g_c}(w - V)\frac{V}{550}$$

The stationary observer notes the air leaving at high absolute velocity in the jet and understands that additional thrust power could have been developed had that kinetic energy been reduced to zero. He assigns as the total available thrust power (this is the same as jet power) the actual thrust power developed plus the exit-loss horsepower. This exit-loss horsepower is

$$\text{Exit hp} = \frac{G_a}{2g_c}C_2{}^2\frac{1}{550} = \frac{G_a}{2g_c}(w - V)^2\frac{1}{550} \tag{7-20}$$

A new expression for propulsion efficiency can now be written.

$$\eta_p = \frac{\text{thrust hp}}{\text{thrust hp} + \text{exit-loss hp}}$$

$$= \frac{(G_a/g_c)(w - V)(V/550)}{(G_a/g_c)(w - V)(V/550) + (G_a/2g_c)(w - V)^2 \, 1/550}$$

When simplified, the equation becomes

$$\eta_p = \frac{2V}{w + V}$$

This is the same expression for propulsion efficiency as was previously derived from the viewpoint of the observer on the airplane.

If the absolute velocity of exit gas is reduced to zero, so that $w - V$ is zero, there is no exit-loss horsepower. For this condition, since $w = V$, propulsion efficiency reaches 100 per cent. However, as the thrust under this condition has also become zero, this limiting condition is of academic interest only. Flight speeds V in excess of w do not produce propulsive thrust in the direction of motion, and such values lie outside the range of normal operation. The thrust which could develop is negative in direction and becomes a *braking* thrust. Although Eq. (7-19) shows positive values greater than 100 per cent, it should be obvious that the equation is not valid for values of $V > w$.

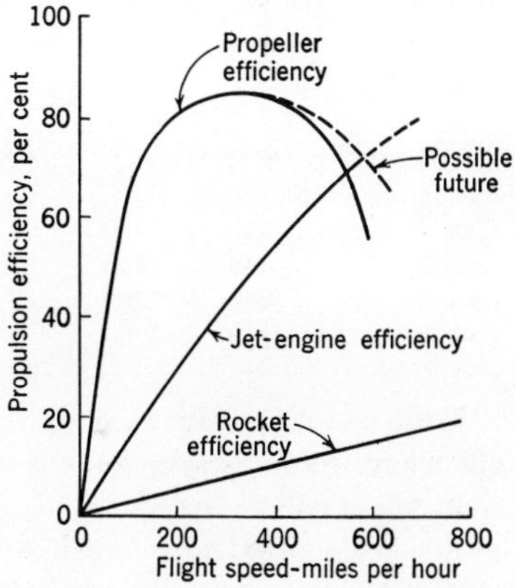

FIG. 7-2. Graph of representative propulsion efficiencies.

A study of Eq. (7-19) shows that for high propulsion efficiency the leaving jet velocity w relative to the airplane must not vary greatly from the velocity of the airplane. To approach 100 per cent propulsion efficiency, it would be necessary for V to approach w with the thrust also approaching zero so that very high propulsion efficiency implies very low thrust per unit of air flowing.

Example 1. Consider a case in which the thrust T and the airplane speed V are kept constant but the weight of air flow is progressively doubled. Consider first $G_a = 60$ lb per sec gas (air) flow, $V = 440$ fps (300 mph), and $w = 1400$ fps.

Solution. By Eq. (7-10)

$$T = \frac{G_a}{g_c}(w - V) = \frac{60}{32.17}(1400 - 440) = 1790 \text{ lb}_f$$

If the air flow is doubled, T and V remaining constant,

$$1790 = \frac{120}{32.17}(w - 440)$$

$$w = \frac{(1790)(32.17)}{120} + 440 = 920 \text{ fps}$$

If the air flow is doubled again, quadrupling the original flow, then

$$w = \frac{(1790)(32.17)}{240} + 440 = 680 \text{ fps}$$

The ideal propulsion efficiency becomes therefore

$$\eta_p = \frac{2V}{w + V} = \frac{(2)(440)}{1400 + 440} = 47.8\% \text{ for case 1}$$

and the other cases are tabulated in the following table:

	V, fps	T, lb	G, lb per sec	w, fps	η_p, eff	Thrust hp	Jet hp
Case 1........	440	1790	60	1400	47.8	1430	2990
Case 2........	440	1790	120	920	64.7	1430	2210
Case 3........	440	1790	240	680	78.5	1430	1822
Case 4........	440	1790	480	560	88.0	1430	1624
Case 5........	440	1790	960	500	93.5	1430	1528

From a study of this example, it is obvious that a reasonable propulsion efficiency would require an exit velocity which is not excessively greater than the airplane speed. This condition is reached very easily with a propeller-operated airplane but is less easily attained with a turbojet machine. In terms of numerical values, w/V is probably in the neighborhood of 1.45 or less for a propeller drive, while for turbojet engines values of w/V less than 3 are infrequent. Graphs of representative propeller, rocket, and turbojet propulsion performance are given in Fig. 7-2.

Figure 7-3 has been plotted to indicate representative temperature, pressure, and velocity variations throughout a typical axial-flow jet engine. A study of the graphs indicates a slight rise in pressure in the inlet diffusing section of the jet, which results from the ramming action produced by the forward motion of the jet. At the same time in the inlet a slight decrease in velocity takes place. In the compressor the maximum pressure rise occurs. In the combustor some pressure is lost although the major pressure utilization occurs in the turbine and finally in the exhaust nozzle as the gases expand to atmospheric pressure.

The corresponding axial-velocity pattern is essentially constant until the turbine is reached, after which the velocity continues to increase, finally reaching the high exit velocity w leaving the exhaust nozzle. The temperature gradually increases throughout the compression process but undergoes a rapid rise in the combustor. A drop in temperature occurs as energy is extracted in the turbine and decreases further as energy interchange to kinetic form occurs in the exhaust nozzle. However, the leav-

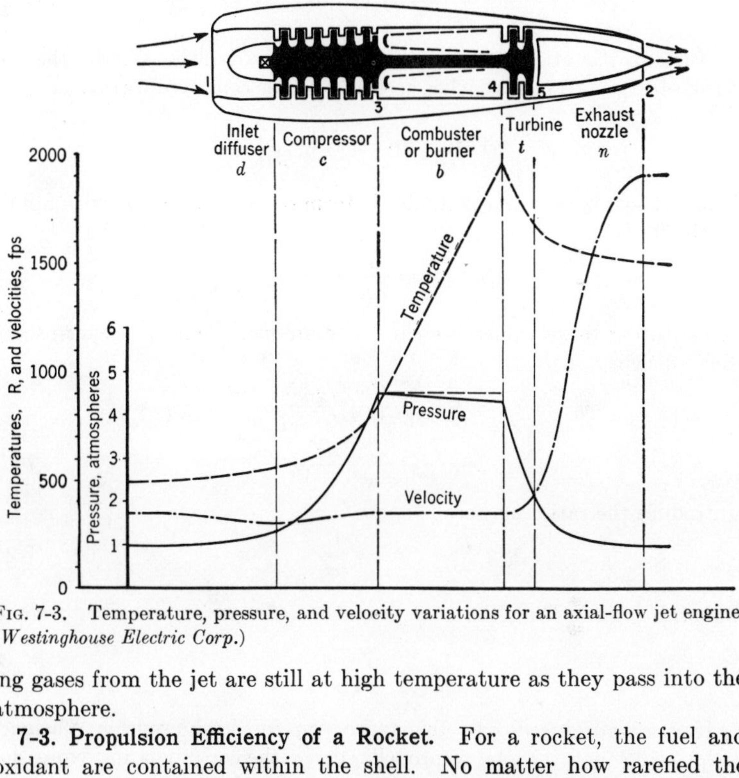

FIG. 7-3. Temperature, pressure, and velocity variations for an axial-flow jet engine. (*Westinghouse Electric Corp.*)

ing gases from the jet are still at high temperature as they pass into the atmosphere.

7-3. Propulsion Efficiency of a Rocket. For a rocket, the fuel and oxidant are contained within the shell. No matter how rarefied the

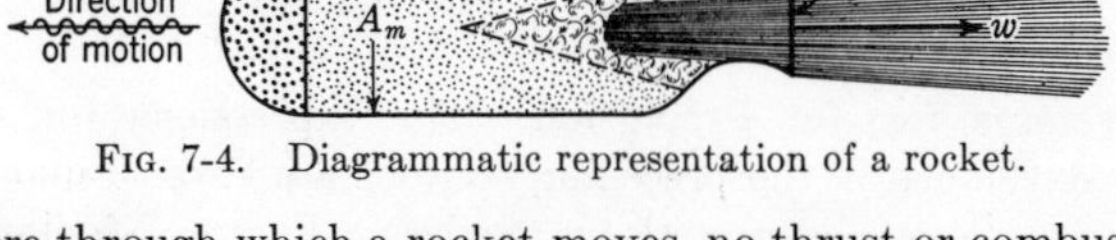

FIG. 7-4. Diagrammatic representation of a rocket.

atmosphere through which a rocket moves, no thrust or combustion difficulty develops, as the energy source is self-contained and there is no dependency on air as an oxidant or as a constituent in propulsion. A rocket would, in fact, be most effective in a vacuum, as the resisting drag of a surrounding medium would be nonexistent.

For the rocket, using previous nomenclature and referring to Fig. 7-4, the expression for thrust becomes

$$\text{Gross thrust } T = \frac{G}{g_c}(w) + P_a(A_m - A_2) - P_a(A_m) + P_2A_2$$

$$= \frac{G}{g_c} w + A_2(P_2 - P_a) \tag{7-21}$$

Further, if static pressure in the jet exhaust is atmospheric, the final term of Eq. (7-21) approaches zero and power can be expressed as

$$\text{Thrust hp} = \frac{G}{g_c} \frac{(w - 0)V}{550} \tag{7-22}$$

The exit-loss horsepower is similar in form to the expression which applies to the jet [Eq. (7-19)]:

$$\text{Exit hp} = \frac{G}{2g_c}(w - V)^2 \frac{1}{550}$$

Substituting these values, we can write an expression for rocket-propulsion efficiency,

$$\eta_p = \frac{GwV/g_c550}{(G/g_c)(wV/550) + G/2g_c)(w - V)^2 1/550}$$

$$\eta_p = \frac{2wV}{w^2 + V^2} \tag{7-23}$$

Introduce the ratio σ, where

$$\sigma = \frac{V}{w} \tag{7-24}$$

Substituting σ in Eq. (7-23), the propulsion efficiency of a rocket becomes

$$\eta_p = \frac{2\sigma}{1 + \sigma^2} \tag{7-25}$$

If σ is introduced into the expression for jet-propulsion efficiency [Eq. (7-19)] for a turbojet or propeller drive, the corresponding expression becomes

$$\eta_p = \frac{2\sigma}{1 + \sigma} \tag{7-26}$$

7-4. Systems for Jet Propulsion. The expressions for propulsion efficiency developed in the preceding sections are keys to understanding the performance of jet-propulsion devices. Figure 7-5 illustrates in diagrammatic form the inherent characteristics of the various propulsion types, starting with the rocket in the upper left corner. Because of the low propulsion efficiency of such a device at low velocities, the over-all efficiency is also necessarily low. In addition to this, the rocket must carry its own oxygen as well as fuel so that it is definitely limited to applications of short duration such as the powering of certain types of projectiles and assisting in the take-off of heavily loaded airplanes.

The athodyd, which takes its name from the expression "Aero-THermO-DYnamic Duct," is really a ram-jet device which can go into automatic operation only at high speeds (above 500 mph). Liquid fuel (usually gasoline) is burned into the air stream of the athodyd, and the hot gases expand rearward with increased velocity. The compression produced by the ramming effect of the athodyd itself is sufficiently high to permit the expansion to take place in a backward direction.

The buzz bomb, which in principle was employed by the Germans in their V-1 and V-2 bombs, operates in a manner similar to that of the athodyd except that a series of flutter valves open under the ramming action of the air. This bomb must be brought up to a critical minimum speed before automatic operation results. Fuel is fed continuously into the air stream, and when ignited a pressure builds up on the far side of the shutters, closing them and preventing further inlet of air until the burned charge has expanded rearward and reduced the pressure inside the chamber. Following this, ram effect opens the shutters, permitting the entry of a fresh charge of air, and combustion again occurs. The combustion cycle occurs with great rapidity, reaching a natural period of resonance dependent upon the size of the tube, in the neighborhood of 40 times per second. By properly tuning the resonance cycle, it has been found that these units can be operated at speeds as low as 250 mph.

The gas-turbine jet, as has been discussed earlier in the text, shows a reasonable propulsion efficiency at speeds above 400 mph and in the range above 600 mph appears to be the answer to aircraft propulsion. At low speeds, the propulsion efficiency is so low that the over-all performance is vastly inferior to the performance which can be reached with a reciprocating engine and propeller or a turbine-propeller jet. The curves of Fig. 7-5 merit careful consideration although they must be interpreted with some measure of judgment, particularly in regard to the conversion of fuel energy to mechanical (shaft) energy or directly to jet energy, as many variables enter.

The representative curves of Fig. 7-2 indicated the vastly greater propulsion efficiency which is possessed by the engine-propeller aircraft unit as contrasted to the turbojet unit at low speeds, with the opposite condition existing at higher speeds. Based upon such data, Fig. 7-6 has been constructed to indicate the amount of thrust horsepower available with a jet engine and with a propeller engine, both systems developing the same thrust horsepower at 375 mph. It can be seen that either type of unit has sufficient power to drive the airplane, except at starting and low velocities, where the particular jet is shown as having inadequate power. This would make it necessary for the ground speed to be higher in magnitude before take-off could be accomplished. The power-required graph rises rapidly with increased speeds, showing the effect of increased drag at greater speeds. The required operating power range of representative airplanes is drawn in as a region on the chart. In the power range indicated, the extreme left line shows the power-flight characteristics of an airplane with conventional drag resistance, whereas the line at the extreme

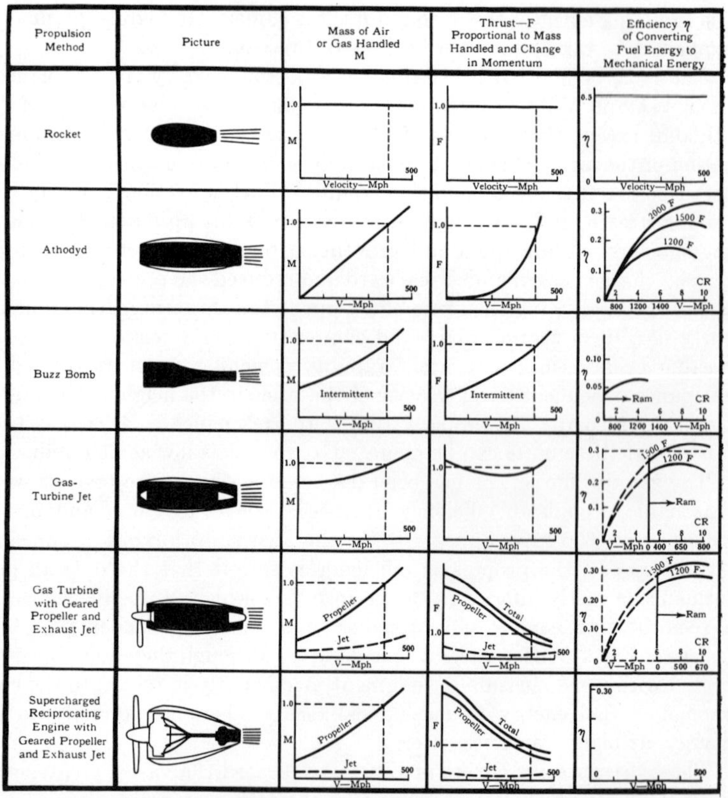

FIG. 7-5. Comparison of various types

right of the range shows the power-flight characteristics of an airplane with minimum drag characteristics. In this connection the jet engine with its better ability to merge into the fuselage presents the possibility of lower drag. No matter which extreme of the power range required is employed, it can be seen that the jet turbine with its rising power characteristic at high speeds can develop the greater velocity. For the extremes of the power-required range it will be noticed that not only is ΔV_J of the jet greater than ΔV_p of the propeller airplane, but the flight speed is also much higher.

7-5. Thermal and Over-all Efficiencies of Airplane Propulsion Systems. The eventual source of the power for propulsion comes from energy in the fuel which is released from its chemical form during the combustion process. In the reciprocating engine, the power from combustion is pro-

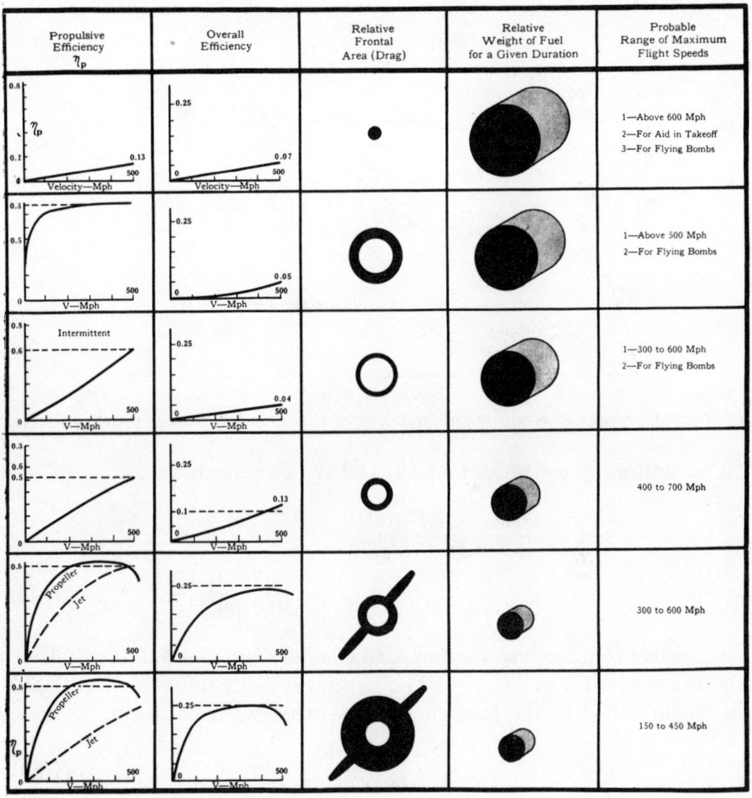

of propulsion devices. (*Ref.* 4.)

duced in the cylinders of the engine and directly or through a gearing system drives the propeller. In a jet system, through the use of the compressor, combustor, and turbine, a portion of the combustion energy is converted directly into an increase in kinetic energy of the jet.

$$\text{Fuel energy input} = G_f H_L \qquad \text{Btu per sec} \qquad (7\text{-}27)$$

$$\text{Fuel energy input} = G_f H_L \,{}^{778}\!/_{550} \qquad \text{horsepower} \qquad (7\text{-}28)$$

where G_f = fuel flow, lb per sec

H_L = lower heating value of the fuel, Btu per lb

Higher heating value can also be employed. Consequently, the appropriate heating value used should be stated in case of doubt.

For the jet engine, thermal efficiency is expressed as the ratio of the jet power produced to the power added by the fuel. The thermal efficiency

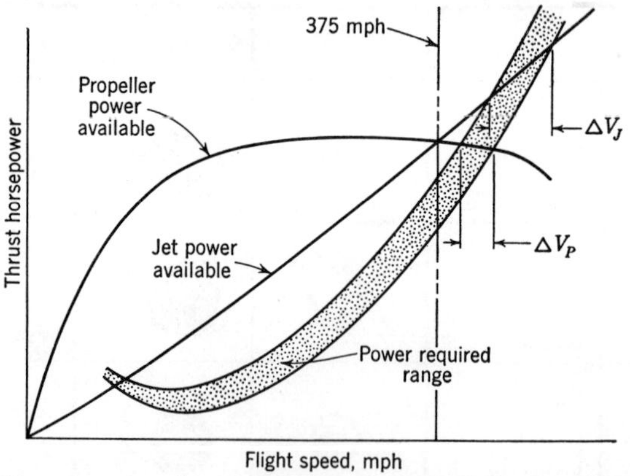

FIG. 7-6. Horsepower requirements and power available for propeller and jet units.

of a jet unit η_t by use of Eqs. (7-16) and (7-28) appears as

$$\eta_t = \frac{(Gw^2 - G_aV^2)(550)}{(550)(2g_c)G_fH_L778}$$

$$= \frac{Gw^2 - G_aV^2}{(2)(778)(32.17)G_fH_L} = \frac{Gw^2 - G_aV^2}{50{,}070G_fH_L} \tag{7-29}$$

It is possible to introduce the fuel-air ratio $f = G_f/G_a$, and if, in addition, it is realized that G and G_a are often not appreciably different, Eq. (7-29) can be simplified to the following approximately correct form:

$$\eta_t = \frac{w^2 - V^2}{50{,}070fH_L} \tag{7-30}$$

Overall-efficiency η_o of a jet-propulsion unit must consider not only thermal efficiency but propulsion efficiency as well.

$$\eta_o = \eta_p\eta_t \tag{7-31}$$

where η_p = propulsion efficiency of jet unit

η_t = thermal efficiency of jet unit

For the simple jet airplane, an expression for over-all efficiency can be obtained by combining Eqs. (7-18) and (7-29),

$$\eta_o = \eta_p \eta_t = \frac{2(Gw - G_a V)V}{Gw^2 - G_a V^2} \frac{Gw^2 - G_a V^2}{50{,}070 G_f H_L}$$

$$= \frac{2(Gw - G_a V)V}{50{,}070 G_f H_L} = \frac{(Gw - G_a V)V}{25{,}030 G_f H_L} \tag{7-32}$$

and this can be expressed approximately as

$$\eta_o = \frac{2V(w - V)}{50{,}070 f H_L} = \frac{(w - V)V}{25{,}030 f H_L} \tag{7-33}$$

The graphs of Fig. 7-7 present representative values of what might be expected with jet-propulsion power-plant systems. It will be noticed that the graphs for propulsion efficiency appear at the top of the chart, for thermal efficiency at the middle of the chart, with over-all efficiency

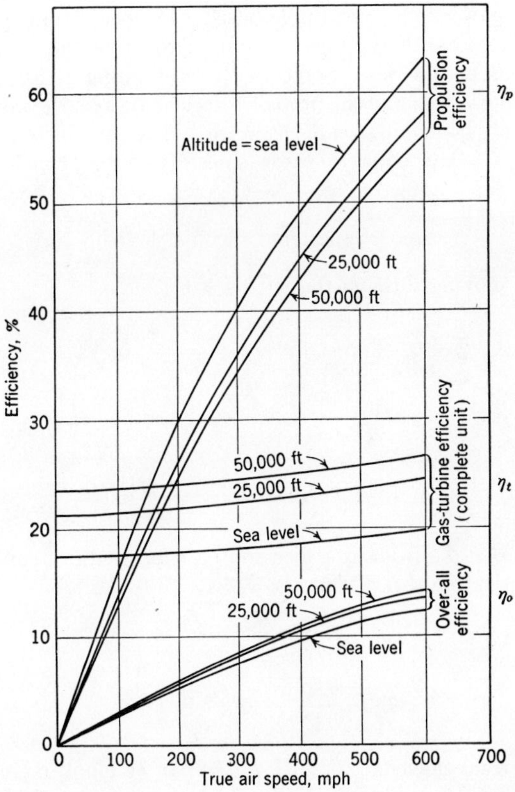

Fig. 7-7. Representative efficiency curves of a jet-propulsion plant. (*After Warner and Auyer, Ref.* 1.)

as the product of these terms appearing as the lowest group of curves. It is of interest to note that the performance of the power unit itself improves with altitude, and in general this is true of over-all efficiency.

To generalize over-all efficiency for a propulsion drive using a propeller, one must introduce a factor to cover losses in the transmission system from the power source to the propeller, η_g, as well as considering the propeller (propulsion) efficiency η_p'. Thus for the general case with a propeller

$$\eta_o = \eta_p'\eta_t'\eta_g \tag{7-34}$$

where η_p' = propeller (propulsion) efficiency
η_t' = thermal efficiency of the engine (power turbine)
η_g = transmission efficiency from prime mover to propeller shaft

For a propeller-jet unit it is necessary to reduce the jet power and the net turbine power to a comparable basis. The net turbine power is that produced in excess of the power required to drive the compressor and miscellaneous accessories. Some of this net power is absorbed in the transmission gearing, and the propeller does not use all the power supplied to it in directing the air stream rearward. The composite factor $\eta_g\eta_p'$ as used in Eq. (7-34) accounts for these losses so that net propeller propulsion horsepower (hp_P) appears as

$$\text{hp}_P = (\eta_g\eta_p')(\text{net turbine shaft hp}) \tag{7-35}$$

For the air moving through the jet, by Eq. (7-12),

$$\text{Thrust hp} = \frac{G_a(w - V)V}{550g_c}$$

Adding the preceding horsepower values, the equivalent total horsepower hp_{ET} appears as

$$\text{hp}_{ET} = \text{hp}_P + \frac{G_a(w - V)V}{550g_c} \tag{7-36}$$

The *total* thrust T_T from propeller and jet expressed in pounds force can follow directly from Eq. (7-36),

$$T_T = \frac{550\text{hp}_{ET}}{V}$$

$$T_T = \frac{550\text{hp}_P}{V} + \frac{G_a(w - V)}{g_c} \tag{7-37}$$

Over-all thermal efficiency for such a unit can be found merely by introducing the input energy furnished by the G_f total lb of fuel per sec and its heating value.

$$\eta_o = \frac{550\text{hp}_{ET}}{778G_fH_L}$$

$$= \frac{550\text{hp}_P + \left(\frac{G_a}{g_c}\right)(w - V)V}{778G_fH_L} \tag{7-38}$$

where hp_{ET} = equivalent total horsepower of the propeller and jet

hp_P = propulsion horsepower delivered by the propeller as useful air-stream power

= $\eta_g\eta_p'$ (net turbine shaft horsepower). In the range of efficient operation for a propeller, for approximate calculation, take $\eta_g\eta_p' = (0.96)(0.81) = 0.78$

G_a = air (gas) flow of jet unit, lb per sec

G_f = fuel used by propeller-jet unit, total lb per sec

w = efflux velocity of jet, relative to the airplane, fps

V = velocity of airplane, fps

H_L = lower heating value of fuel, Btu per lb

It might be well to point out some generalizations concerning jet propulsion. A jet engine moves the maximum weight per second of air flow through the unit, and consequently produces its maximum thrust under sea-level conditions. At higher altitudes, lowered pressure reduces the density of the air to such a point that the air passing through the engine is appreciably diminished. This reduction, however, is partially offset by the lower air temperatures at higher altitude, which tend to increase air density, and by ram effect resulting from faster forward motion of the airplane. The effect of ramming is to increase inlet pressure to the compressor. Compression takes place from this ram inlet pressure to the discharge pressure of the compressor. On the other hand, the turbine and the expansion nozzle expand from essentially compressor discharge pressure on entering the turbine down to ambient atmospheric pressure. As atmospheric pressure is lower than the ram pressure on the inlet to the compressor, the expansion pressure ratio may be greater than compression pressure ratio in the compressor.

The reduction in thrust which occurs in high-level flight is not too serious, as rapid climbing in general is not required and the drag or resistance to motion of the airplane is less in the rarefied high-altitude air. At a given forward velocity of an airplane, the specific fuel consumption in pounds per hour for each pound of thrust is usually slightly lower at a higher altitude than it is at sea level, particularly in the range between 10,000 and 40,000 ft altitude.

If several cycle computations are carried out, it can be shown that increasing the pressure ratio with constant temperature at the turbine inlet of the jet engine leads to decreased fuel consumption until a minimum value is reached. The improvement cannot be carried past the minimum point unless the turbine inlet temperature also is raised. In terms of temperatures, an increase in the temperature of the gases leaving the combustion chamber and entering the turbine causes an increase in thrust even with no change in pressure ratio. In the case of temperature also, with constant pressure ratio, an optimum value for economy exists, and if the temperature is raised above this value, the specific fuel consumption increases instead of decreases. This temperature may even be less than the temperature limit which is imposed by the strength-temperature characteristics of turbojet engine materials. In general, the maximum thrust produced by unit mass rate of flow occurs at a lower compressor pressure ratio than that which produces minimum specific fuel consumption for any given inlet turbine temperature.

7-6. General Considerations for Aircraft Gas Turbines. The reciprocating-piston engine, which was developed to a high degree of efficiency and reliability, approached a practical limit of power output without excessive weight and mechanical complication with horsepower in the range of 5000.

The piston engine, compared with an axial-flow gas turbine, had higher frontal area, higher installed weight per horsepower, and more difficult cooling and lubrication problems. However, its economy was somewhat greater than that of gas-turbine jets and somewhat better than that of gas-turbine propeller-jet engines.

For the highest speed range running past sonic velocity, the gas-turbine jet engine alone is available and in the speed range from 500 to 650 mph should be considered exclusively. Jet engines are now economical in the use of fuel, requiring from 0.5 to 2 lb of fuel per hr per lb of thrust developed, with the fuel consumption particularly high when operation occurs at lower speeds. Developmental work on gas turbines and gas-turbine jets has shown a progressive improvement in these engines.

In terms of rough figures to keep in mind for specific fuel consumption, it should be mentioned that turbojet engines have representative values of specific fuel consumption of 1.4 to slightly less than 0.5 lb per hr per thrust hp. Turbine-propeller engines range from 1 to 0.6 lb per equivalent hp-hr, while reciprocating engines range from 0.65 to 0.4 lb per hp-hr. The term *equivalent horsepower-hour* is used in an endeavor to integrate the combined effects of the horsepower associated with propeller action and the horsepower associated with jet action.

Godsey and Flagle[5] have made a study of the characteristics of different propulsion devices. In Fig. 7-8, which is reproduced from their paper, a fast high-altitude single-seater airplane is compared for propulsion by jet and by reciprocating engines. At any flight speed for this airplane the solid graph shows that greater power is required when the reciprocating engine is used compared with the jet engine (dashed line). This result is true because the jet engine, with its smaller frontal area, blends more readily into the streamlining of the airplane itself, reducing the drag. The power-available curves show that the jet engine does not reach the capacity of the propeller engine until a flight speed in excess of 400 mph has been reached. At low velocities the jet-engine power is extremely low. However, it will be noticed that the maximum flight speed of the propeller engine is but slightly in excess of 400 mph, whereas the jet engine reaches practically 600 mph. The fuel economy of this jet airplane is inferior to that of the engine airplane, except at speeds in excess of 400 mph. Reference to the curve on the right will show that the fuel-rate curve has a minimum value, in this case at a flight speed around 500 mph.

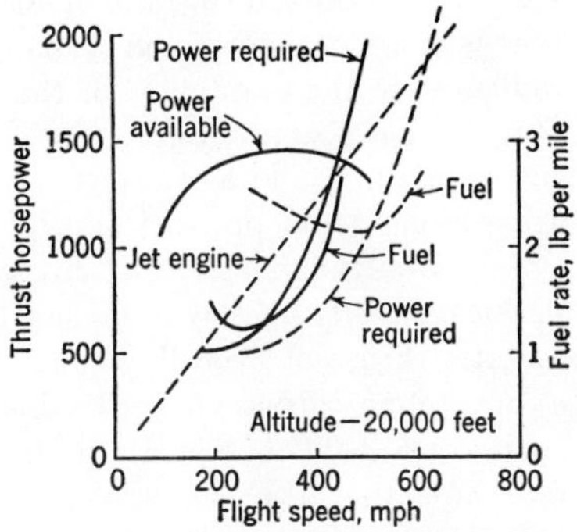

FIG. 7-8. Power-required and power-available curves for single-engine airplane powered by reciprocating engine (solid lines) or jet engine (dashed lines.) (*Godsey and Flagle, Ref.* 5.)

A further study of airplane performance was made by Godsey and Young[15] in connection with a four-engine transport airplane with a gross weight of 120,000 lb. The results of this study are graphed in Fig. 7-8. The power produced by four reciprocating engines lies between 6000 and 7000 hp, and the take-off and low-speed power are adequate for satisfactory performance of the airplane. For the jet arrangement four jet

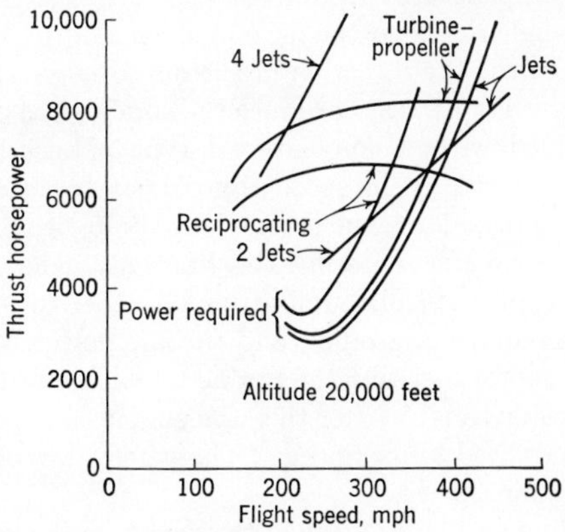

FIG. 7-9. Power available and power required for a large 120,000 lb airplane equipped with four engines. (*Godsey and Flagle, Ref.* 5.)

engines are also installed, as they are needed for satisfactory take-off. However, in the cruising and operating range it is not necessary to operate four jets, and the operation is changed to two jets firing and two idle, as is indicated in the graphs. For the turbine-geared propeller arrangement, four engines also are used. These develop more power than the reciprocating engines, and a study of the graphs shows that two of these operating alone could propel the airplane as effectively as the four engines.

7-7. De Havilland Ghost Engine. The Ghost-50 engine, illustrated in Figs. 7-10 and 7-11, is included to show a former type. It develops a static thrust of 5000 lb at take-off under sea-level conditions and has a maximum continuous operating thrust of 4300 lb. Under cruising conditions, 3300 to 4000 lb of thrust is produced. Both Figs. 7-10 and 6-5 show that the compressor employs a single-inlet front-facing centrifugal impeller which delivers air through a pressure ratio of 4:1. From the compressor the air passes into 10 combustion chambers and thence into the nozzles supplying the single-stage turbine. The turbine wheel and the outlet from the combustion chamber are illustrated in Fig. 9-17.

The whole unit has a length of 121 in., a maximum diameter of 53 in., and a dry weight of 2218 lb. It turns at 9000 to 9500 rpm for cruising, with a take-off rpm of 10,250. In operation it has been found that this engine, for cruising rpm at sea level, had a fuel consumption of 1.01 to 1.04 lb per hr (Fig. 7-12) for each pound of thrust, with the value at take-off rising to a value of 1.08. At 40,000 ft altitude, the fuel rate is slightly higher for usual flight speeds (Fig. 7-13). The single-entry impeller offers the advantage of better utilization of ram effect of the entering air and can operate successfully at tip speeds of the order of 1500 fps. The air leaving the impeller enters diffuser passages cast in the casing surrounding the impeller. Near the far end of each diffuser passage, a blade cascade section turns the direction of the air through approximately 90° so that it can flow axially rearward into the combustion chambers. Each engine is provided with a conventional type of electric starter. The engine reaches correct starting speed approximately 8 sec after the starter button has been pressed, at which time igniters in two of the combustion chambers ignite the fuel entering the burners. The flame started at these two points spreads from combustion chamber to combustion chamber through interconnecting pipes, and turbine power begins to develop, with the starter motor assisting the engine to reach its sustaining speed in the shortest possible time. After this, the engine can accelerate under its own power to a desired idling speed. The whole starting process can be accomplished within 30 sec.

External cooling and ventilation of the engine are accomplished by surrounding the exhaust cone by an outer cone through which air can be

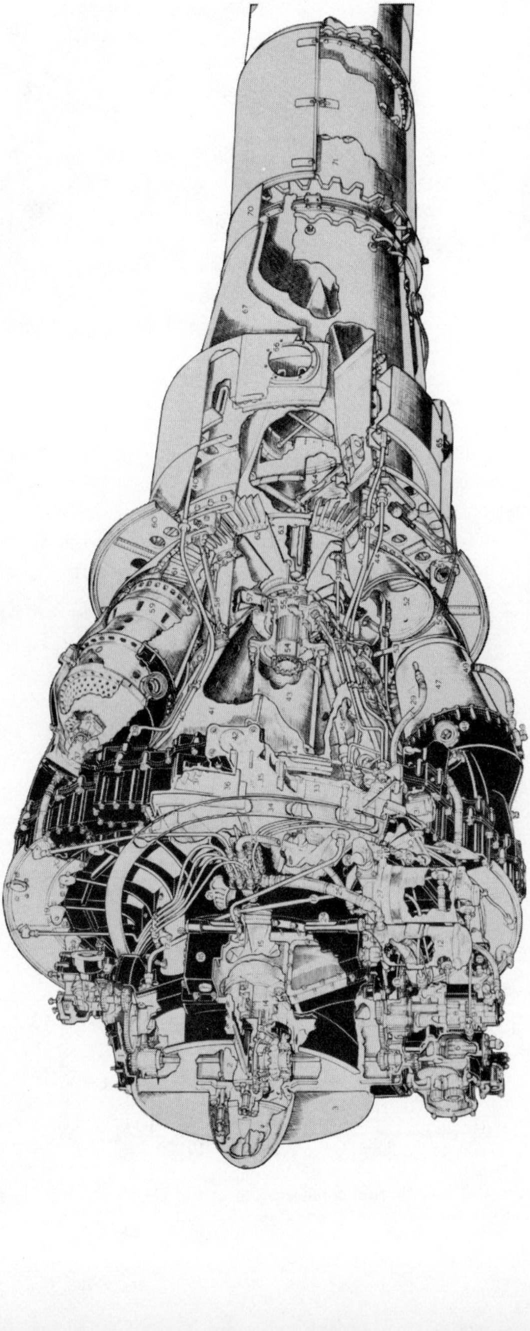

FIG. 7-10. Cutaway view of Ghost jet engine showing centrifugal compressor and other components. (*De Havilland Engine Co., Ltd.*)

FIG. 7-11. Side view of Ghost jet engine showing in particular the combustors (center) and tail cone (at left). (*De Havilland Engine Co., Ltd.*)

passed. By venturi action through a cutoff near the outlet of the main propulsion jet, some cold air is drawn from the space between the lagging and the exhaust cone and maintains a flow of cooling air over the jet pipe and combustion chambers. The exhaust-cone assembly is made of stainless steel and is bolted to the turbine casing. It is possible to add exten-

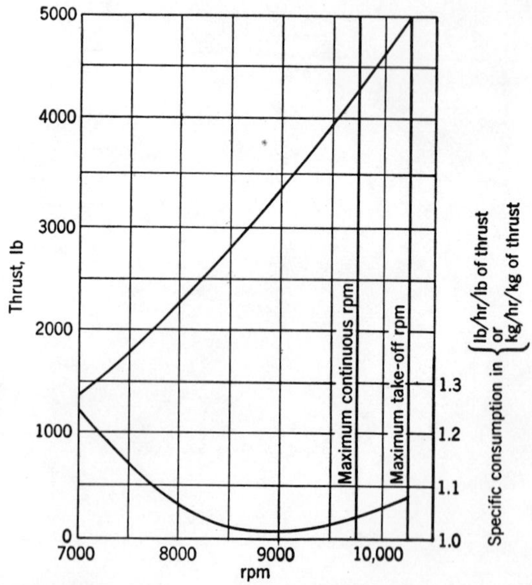

FIG. 7-12. Thrust and specific fuel consumption of the De Havilland Ghost turbojet engine at sea-level conditions.

sions to the exhaust-cone pipe where this is necessary. The effect of an extension pipe is to decrease the performance of the engine, but only slightly.

It is of interest to note that four Ghost engines powered the first jet-operated airplane specifically built for commercial transport use. This airplane, designed and manufactured by the De Havilland Engine Company, Ltd., was named the Comet. It is pictured in Fig. 7-14. It had a wing span of 115 ft, length of 93 ft, height of 28 ft, with a total weight of 105,000 lb and a payload capacity of 14,000 lb. Its cruising speed was 490 mph at an operating altitude of 35,000 to 40,000 ft, with a practical cruising range of some 2200 miles. Cabin pressurization for altitude flying could be accomplished by bleeding air under pressure from compressor diffusers. The bleeding of 10 lb of air per min under cruising conditions at 40,000 ft does not affect the thrust by more than 1 per cent. Air from the same bleeding location could also be used for anti-icing operation.

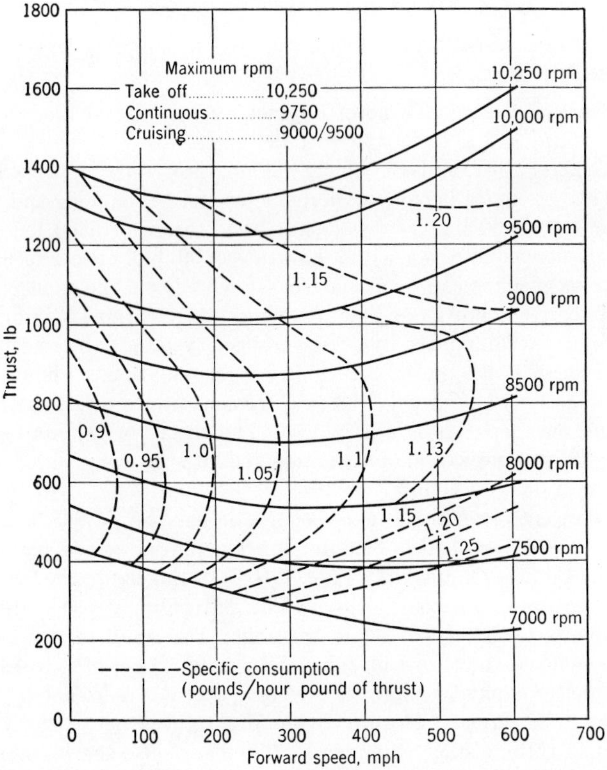

FIG. 7-13. Thrust and specific fuel consumption of the De Havilland Ghost engine at 40,000 ft altitude.

FIG. 7-14. Photograph of the Comet, first commercial airplane powered exclusively by jet engines. (*De Havilland Engine Co., Ltd.*)

7-8. General Electric Turbojet Engines. The General Electric Company has produced a variety of turbojet engines.

The General Electric J-47, which was an early type, appears in Fig. 7-15. This jet engine is 36¾ in. in diameter and 144 in. long and weighs approximately 2500 lb. It develops a basic thrust of 5200 lb. A 12-stage axial-flow compressor with a flow rate of 90 lb of air per sec is used. Kerosene or gasoline can be employed as fuel. From the eight combustion chambers the hot gases flow to a single-stage turbine. Lubrication for the unit is accomplished by a pressure-feed system to the bearings and accessory gears, with the oil returning for recirculation. The electrical system consists of a direct-drive starter-generator unit, two ignition units, and two separate igniter plugs. The thrust of this unit can be augmented by vaporization of water injected into the compression system to give additional mass flow.

A more recent General Electric turbojet design, the CJ 805-3, is shown in Fig. 1-12 and described in Chapter 1. The fan-jet counterpart of that engine, the CJ 805-23, appears as Fig. 7-16. In this figure the fan, placed in an aft position, is clearly visible. The extra airflow accelerated rearward amounts to 251 lb per sec at take-off. The result of this extra air flow is to increase the thrust on take-off by some 35 per cent to 16,100 lb and the fuel economy is improved by some 10 to 14 per cent to 0.56 lb per hr per lb of thrust. To provide the extra power needed for the aft-fan, a fourth turbine stage is required. The figure also shows the blading for this stage with the associated compressor blading protruding past the shroud ring. The aft-fan increases the weight of the unit to 3766 lb total weight, but this is offset by the improved economy, the additional thrust possible, and the resulting lower fuel load needed by the airplane for a given flight distance. The turbofan design is slightly shorter in total length (150 in.) than the straight turbojet, but its diameter at the aft-fan is necessarily greater and reaches some 54 in. Note that the turbofan air jet emerges in parallel with the jet stream which passes completely through the turbojet. For this unit, the maximum continuous rpm is 7460 and a turbine inlet temperature of almost 1800 F can be employed for a short period at take-off.

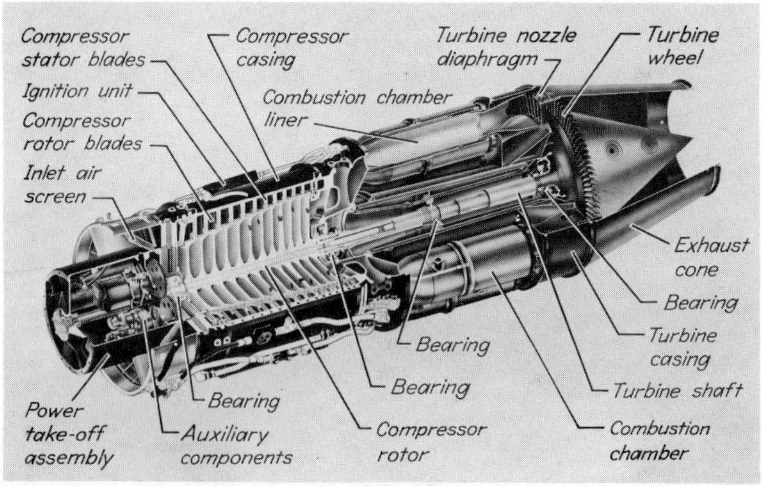

FIG. 7-15. General Electric J-47 (TG-190) turbojet engine in phantom cross section.

7-9. Supersonic Jet Engines. Ever since the capabilities of the gas-turbine engine for aircraft propulsion became fully realized, thought has been given to the possibility of having gas turbines propel aircraft at supersonic speeds. This objective has now been reached for a number of jets and plane structures. One of the very successful supersonic turbojets was produced by General Electric Co. in its J79 series. This engine can drive aircraft at speeds greater than 2M with the thrust supplemented by afterburners. In general 17 stages are used in the axial-flow compressor, with a 3-stage turbine serving as power source. Thrusts in the range of 17,000 to 18,000 lb are produced at take-off and specific fuel consumptions at M = 2 at 35,000 ft are in the range of 2 lb per lb of thrust. This is a very conservative value for such high speeds. Cruising at less than M = 2 gives specific fuel consumption values lower by approximately 50 per cent. Turbine inlet temperatures exceeding 1600 F are employed.

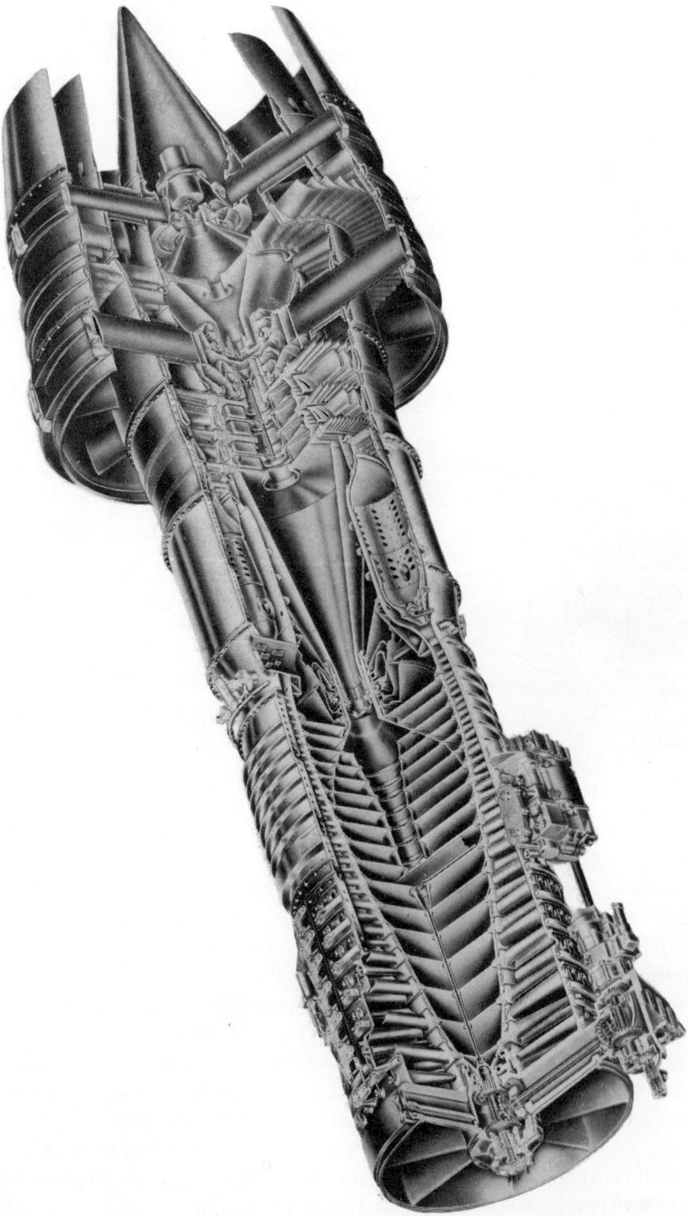

FIG. 7-16. General Electric Co. CJ805-23 turbofan jet engine (17-stage axial-flow compressor with 3-stage turbine plus 4th stage driving aft-position turbofan).

For the large commercial supersonic aircrafts, higher thrusts in the neighborhood of 60,000 lb are needed, and cruise capabilities up to M = 3 are desired. Afterburners and thrust reversers for landing are essential. Fig. 7-17 is a layout of the essential elements of the GE4 supersonic turbojet with augmentor (afterburner).

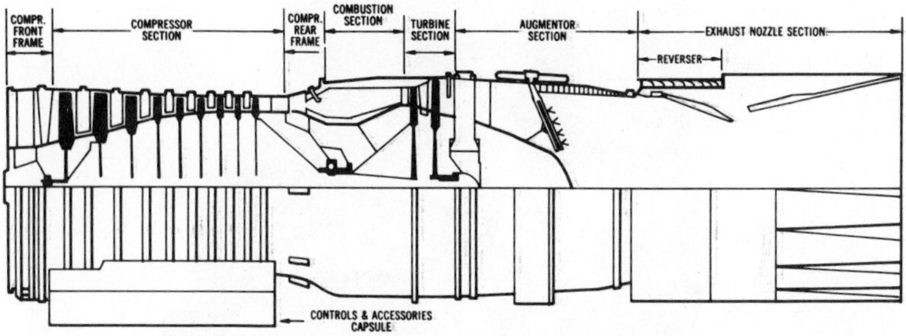

FIG. 7-17. Skeleton outline of the General Electric GE4 turbojet for supersonic service.

7-10. Pratt and Whitney Aircraft Jets. This company produces a variety of gas-turbine engines for commercial and military aircraft as well as for stationary and marine purposes. Fig. 7-18 shows the JT3D-7 turbofan engine produced for commercial aircraft. The turbofan is at the front of this unit and the flow is from left to right. The extreme diameter of the fan is 53 in. and the length of the compressor, combustors, and turbine is 134.4 in. The axial-flow compressor has 8 stages in the low-pressure section of the compressor, including the two in the turbofan section, followed by 7 stages in the high-pressure compressor. The respective maximum rpm's for these rotors are 6800 and 10,200. Air from the compressor enters the combustors and then passes into the 3-stage axial-flow turbine. Turbofan air which does not enter the main stream flow passes rearward around the outside of the engine and joins with the turbine exhaust air at the exit jet nozzles which are of the fixed-area type. This engine with its fan can deliver a take-off thrust of 19,000 lb. At take off, 196 lb per sec flow through the compressor-turbine path while 264 lb per sec flow through the turbo-fan circuit. The weight of the dry engine including its standard equipment is 4260 lb.

The pressure-compression ratio is in the neighborhood of 14 to 1 for this turbojet. Its performance is exceptionally good as can be seen from Figs. 7-19 and 7-20, which give values for the unit under climb and under cruising conditions. The exhaust temperature from the turbine is around 900 F.

The JT9D-1 turbofan jet, made by the same company, is an even larger version which can produce a take-off thrust of 42,000 lb. This unit has

FIG. 7-18. Pratt and Whitney turbofan engine model JT3D-7 (air inlet at left).

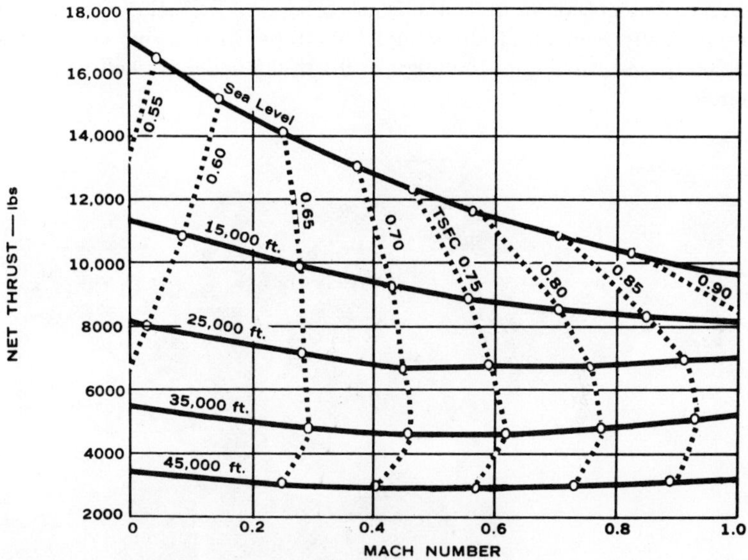

FIG. 7-19. Performance data for the Pratt and Whitney JT3D-7 turbofan under maximum continuous and maximum climb conditions.

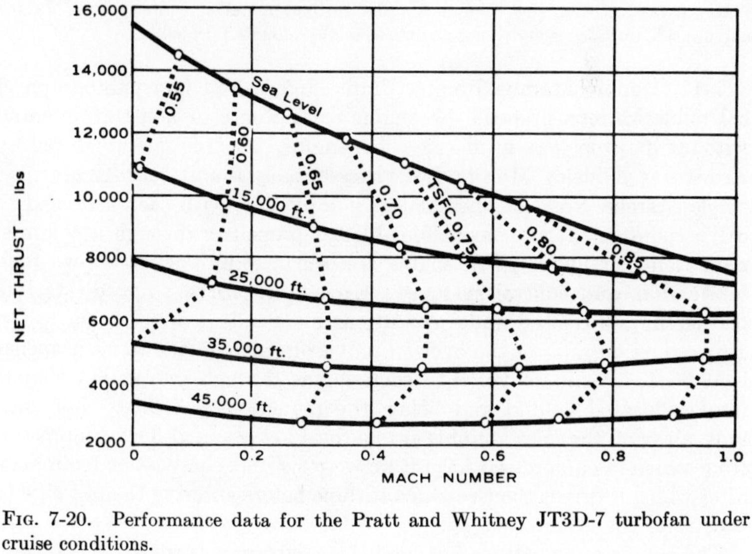

FIG. 7-20. Performance data for the Pratt and Whitney JT3D-7 turbofan under cruise conditions.

an inlet diameter of 96 in. at the front turbofan. Its fuel consumption is exceptionally economical, dropping to 0.6 lb per lb of thrust at 0.8 M at 35,000 ft, and running in the range of 0.4 to 0.5 under take-off and climb conditions.

7-21. Armstrong-Siddeley Double Mamba turbine-propeller engine. Two Mamba engines side by side, each driving its own coaxially mounted propeller.

7-11. Double Mamba Propjet Unit. Figure 7-21 is a photograph of a Double Mamba propeller-jet engine resting on a demonstration stand without its propellers in place. This engine, which is manufactured by Armstrong Siddeley Motors, Ltd., is essentially a unit consisting of two single Mamba No. 2 engines placed side by side, with each arranged to drive counterrotating coaxial four-bladed propellers through a common reduction-gear unit. Each engine is a separate unit with its own fuel, lubrication, and control systems. Each unit can be stopped, started, feathered, and cruised under conditions independent of the other. The gear-reduction ratio of the over-all unit from each engine to its propeller is 10.36:1. Air is delivered to each engine through two kidney-shaped intakes located behind and below the reduction-gear case. For each unit, air from the inlet annulus enters the 10-stage axial-flow compressor, from which in compressed form it flows to six unit combustion chambers, after which it drives the two-stage turbine before entering the tail pipe to constitute the exhaust jet.

The compression ratio is 5:1, and the compressor turns at 15,000 rpm.

Under take-off conditions, each of the Mamba engines develops 1270 shaft hp and a jet thrust of 384 lb. Under conditions of continuous operation, each engine develops 1158 hp along with a jet thrust of 130 lb. The unit has an over-all length of 79.8 in., an over-all width of 52.8 in., and a height of 42.4 in. The maximum take-off horsepower of the unit is 2540 with 768 lb thrust in addition. Kerosene is used as fuel, at a rate of 250 gal per hr under take-off conditions. For emergency conditions at sea level and at 300 mph, the unit develops 3040 equivalent hp at 15,000 rpm. Specific fuel consumption amounting to 0.89 lb of fuel per equivalent shaft hp-hr has been attained by the Double Mamba.

7-12. Other Jet Engines. Almost all of the major gas-turbine manufacturers make smaller units for aircraft-propeller drives, whether these be for commercial passenger airplanes, airplanes for industry use or for helicopters. As examples of these, several units will be described.

The T56-A8 gas-turbine propeller-jet unit manufactured by the Allison Division of General Motors develops 4050 shaft horsepower. It is geared down to drive its propeller at 2450 rpm. A 4-stage axial-flow turbine and a 14-stage axial-flow compressor with a pressure compression ratio of 9.65 are employed. The unit weighs 1872 lb and is 146 in. long by 27 in. wide by 39 in. high. It has 6 combustors, and specific fuel consumptions as low as 0.53 lb per hr per lb of thrust have been reached. The same company supplies its model 250-C18 as a helicopter drive. This develops 270 hp using a 2-stage turbine and a 6-stage axial-flow compressor with a 5.2 pressure compression ratio. It uses a single combustor, weighs 136 lb, and is 41 in. long by 19 in. wide by 23 in. high. Specific fuel consumption is in the neighborhood of 0.69 lb of liquid fuel per hp-hour.

A helicopter turbine engine manufactured by AiResearch Manufacturing Co. produces 400 hp continuously and 500 hp maximum intermittent. This unit rotates at 4950 rpm. It employs a 3-stage axial-flow turbine driving a 2-stage axial-flow compressor to produce a pressure compression ratio of 6.4. Through the two-stage compressor 4.76 lb of air per sec pass. Exhaust from the turbine is at a temperature slightly over 1000 F. The unit weighs 199 lb and is 41 in. long, 19 in. wide and 19 in. high.

The major automotive manufacturing companies have been engaged in the development of gas turbines for driving equipment on the highways. To make such equipment, competitive with reciprocating engines, it was imperative that a successful design of a regenerator (recuperator) be developed. This has been accomplished with the creation of effective heat transfer surfaces which are light-weight and compact. The GT-309 unit made by General Motors produces 280 hp. A single-stage turbine turns at 35,700 rpm, and drives a single-stage centrifugal compressor to produce a pressure-compression ratio of 3.8. It has an air-flow of 4 lb per sec through a single combustor. A second turbine taking gas flow from the first turbine connects to the power shaft through gearing. This turbine uses a rotary heat exchanger which abstracts heat from the exhaust gases leaving the power turbine, lowering the temperature of these gases to 525 F. The unit weighs 950 lb and is 30 in. by 30 in. by 36 in. high. The Ford Motor Company also makes a unit which uses a stationary recuperator instead of a rotary one. The Chrysler Corp. employs a rotary recuperator. At the present time it appears the main application of these automotive gas turbines is for large road units such as trucks and buses or for earth-moving machinery.

7-13. Compound Airplane Engine. The Wright Aeronautical Corporation has successfully developed an engine-turbine unit. This unit has three turbines which use gases from the exhaust of an otherwise essentially conventional 18-cylinder cyclone-type airplane engine. Basic engines of this type were described in Art. 1-3. This particular engine is a radial unit with two rows having nine cylinders each.

Figure 7-22 is a side view of the engine with two of the turbines also apparent. Figure 7-23 shows the drive arrangement of two of the turbines to the crankshaft extension. Special exhaust-pipe connectors lead the exhaust gas from the cylinders to the three turbines. The exhaust pipes are connected in pairs, with three pairs combined into a common manifold for supplying each turbine. Each engine receives the exhaust from six cylinders; the first pair of the six connects to the front row, the second to the rear row, while the third uses one cylinder in the front and one in the rear row. This unusual arrangement of the exhaust piping is made to space most uniformly the exhaust impulses to each turbine and also to cause the minimum disturbance to the exhaust from any cylinder. An aspiration arrangement is made in some of the duct pairs so that the exhaust from one of the cylinders enters an aspiration section in the exhaust pipe from its associated cylinder and thus the back pressure is not so high as to interfere with scavenging during valve-opening overlap of the other cylinder.

Each turbine (Fig. 7-23) is mounted on a flexible shaft which in turn connects with a quill shaft and thence through a pair of bevel gears into a fluid coupling. Through a pair of straight gears the output of the fluid coupling feeds directly to the crankshaft. Vibration forces from either the turbine assembly or the crankshaft system do not readily transmit from one system to the other through the fluid coupling. The oil for the fluid coupling is supplied by the engine itself, and in starting oil is not present in the coupling in sufficient amount to cause drag until the engine reaches operating speed and is developing power. When the normal engine speed is reached, the engine delivers a satisfactory exhaust so that the turbines develop power to add to the output of the engine. This arrangement prevents a high starting drag on the engine which might otherwise be present during starting.

FIG. 7-22. Side view of Turbo-Cyclone engine. (*Wright Aeronautical Corp.*)

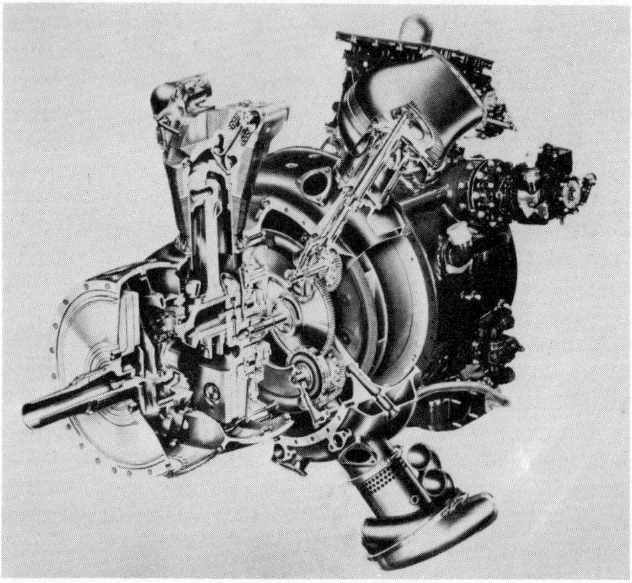

FIG. 7-23. Cutaway view of Wright Turbo-Cyclone engine with two of the three turbines showing. (*Wright Aeronautical Corp.*)

The turbine wheel has a mean diameter of 10½ in. and an outside diameter of 11 in. and turns about eight times crank speed, above 16,000 rpm. The buckets are of stellite and welded to the wheel itself. This construction has been found to give a stronger, lighter structure than would be possible if the blades were slotted or dovetailed into the turbine disk. The energy in the exhaust is sufficiently high to give supersonic velocities through the turbine wheel, and temperatures ranging between 1400 and 1500 F can occur. These high temperatures necessitate some cooling of the disk, which is accomplished by forcing cool air through slots which are cut in the turbine disk just inside the base of the blades. The cooling air then mixes in with the turbine exhaust gases and goes to waste. The three turbines, each of which covers a 120° segment of engine operation naturally receive their energy in the form of intermittent discharges from respective groups of cylinders.

Over an extensive range of testing, the compound engine has been found to perform very satisfactorily, producing about a 20 per cent power increase and using approximately 20 per cent less fuel over the basic operating range of the engine. In addition to this, the specific weight of the engine in pounds per horsepower has been reduced in the neighborhood of 5 per cent and amounts to about 1 lb per brake hp. It appears probable that compound engine arrangements will be further developed, and it is even conceivable that the exhaust from these three turbines may at a later date be combined into a second turbine, which, in turn, could drive a supercharger for the engine itself.

The ratio of the division of power between the propeller and the jet can be modified as desired, but it appears that in general only about 10 to 15 per cent of the useful power will be developed by jet thrust, with the remainder being produced by the propeller.

As better materials and more experience go into the production and operation of jets and turbine-propeller engines, the period of time between overhauls can be increased. Even now, this period between complete overhauls places many of these units in the classification of dependable, reliable power plants.

7-14. Control Systems. Control systems for jet engines are still in a state of transition, as the control problem is extremely complex. For example, the major problems which must be faced are the ability of the operator to produce a definite control thrust for each throttle position, to provide adequate protection of the unit from high temperatures, to make possible operation under conditions of good economy, and finally to avoid the possibility of flame extinction or blowout with resultant stopping of the power unit.

The variables which can produce these end results are directly associated with (1) fuel flow and intake air temperature and pressure as related to allowable rates of acceleration and deceleration; (2) the allowable speed range of the unit, which must be controlled in both directions by a limiting governor; (3) allowable turbine temperature, which must not exceed predetermined limits, or the engine may fail; (4) possible variation of exhaust-nozzle area; (5) operation of supplementary afterburning.

Keeping the gas temperature, on entry to the turbine blades, at a safe maximum limit is essential and protective devices must be installed for this purpose. Because of momentary fluctuations in gas temperature leaving the burners and because high temperature shortens thermocouple life, it is customary to install the control thermocouples in the exit gas stream from the turbine. Here multiple thermocouples are installed and arranged to give an average value of the exhaust temperature across the turbine outlet duct. If this average temperature exceeds a preset safe limit, it energizes an amplifier which sets a control mechanism into play to reduce the fuel flow to the burners. This protective control mechanism overrides any speed or load control on the turbine. Thus, exhaust temperature is used to limit the maximum turbine inlet temperature and provides the means for having the turbine always operate within the design stress limits set by temperature. The exhaust temperature measurements for control must always be taken before augmentors or after burners are brought into use.

Multiple combustion chambers are now used in preference to annular chambers. With these it is necessary to control the air and fuel for each combustion chamber to assure satisfactory life and not overheat any part of the nozzle ring or turbine blades. Since the combustion chambers are constructed of light sheet metal, they cannot serve as air-flow control devices and the air-flow control for the combustion chambers is thus transferred to the turbine nozzle area served by a particular combustion chamber. Since the nozzles are accurately formed, the air flow through each combustion chamber is thus controlled. In spite of this, there is some variation in the temperature of the gases leaving each combustor. This is true because the fuel-flow requirements may change by as much as 25 to 1 for the operating range of a turbine between full and part load. It is difficult to assure perfect performance of fuel nozzles over such a range and balance the fuel flow to each combustion chamber. Although unsatisfactory performance occurred in some early turbine designs, most manufacturers now provide fuel pumping equipment which can meter the fuel to within plus or minus two per cent of that required for an existing turbine load.

PROBLEMS

7-1. The output of jet engines is commonly expressed in pounds thrust. (*a*) Compute the airplane speed in miles per hour at which the thrust in pounds numerically equals the thrust horsepower of the airplane. (*b*) Use as a basis 100 lb of thrust, and at varying speeds plot a graphical relation for horsepower. Use miles per hour as the abscissa and horsepower as the ordinate. *Ans.* (*a*) 375 mph.

7-2. The efflux velocity from a jet unit is 1762 fps for an air flow of 79 lb per sec through the unit. The airplane is flying at 400 mph. Compute the approximate thrust developed, the thrust horsepower, and the propulsion efficiency. *Ans.* 2886 lb_f, 3070 hp, 49%.

7-3. A jet-propelled airplane travels at 550 mph and employs two jet engines, each of which develop 4000 lb of thrust. When flying at an altitude equivalent to a barometric pressure of 7 psia, it was found that the pressure in the tail-cone outlet at the point at which the exit velocity w was measured amounted to 7.31 psia. The tail-cone area at the point of measurement was 1.65 ft^2, and 174 lb of air per sec moved through each compressor. Disregarding the weight of fuel, compute the value of exit velocity from the jet at the measuring point. If 4400 lb of fuel is used per hour by each engine, compute the efflux velocity required to produce the same thrust. *Ans.* 1531 fps and 1521 fps.

7-4. The leaving velocity is 1090 mph from a jet, and the inlet velocity is 550 mph. The specific fuel consumption is 1.2 lb per hr for each pound of thrust. Gasoline of 18,250 Btu per lb lower heating value is used as fuel. For 4200 lb thrust, compute the air flow in pounds per second. Compute the probable propulsion and thermal efficiencies, and finally estimate the over-all efficiency of this jet-propelled unit. *Ans.* 167.8 lb per sec air flow; 67.1%; 25.3%; 17%.

7-5. A jet engine in flight at 580 mph inducted 120 lb of air per sec and used 4650 lb of a gasoline fuel per hr. The tail-jet efflux velocity was 1800 fps. Disregard pressure-area corrections, and compute the gross thrust and the thrust horsepower. *Ans.* 3620 lb_f, 5590 hp.

7-6. An after burner with the preceding jet unit makes it possible to increase the jet efflux velocity to 2100 fps, and when in use fuel is consumed at the rate of 8700 lb per hr. The flight speed also increases to 620 mph. Compute the thrust and thrust horsepower for afterburning conditions. *Ans.* 4506 lb_f, 7440 hp.

7-7. The heating value of the gasoline used for the jet of Prob. 7-5 is 20,390 Btu per lb (higher) and 18,900 Btu per lb (lower). Compute the specific fuel consumption and the thermal efficiency based on thrust horsepower at 580 mph. Make use of the answers given at end of Prob. 7-5. *Ans.* 1.292 lb per hr per lb thrust, 15.1% on higher, 16.2% on lower.

7-8. Compute the specific fuel consumption and the thermal efficiency based on thrust horsepower for the jet unit of Prob. 7-6 when afterburning is employed. Gasoline used has a lower heating value of 18,900 Btu per lb. *Ans.* 1.93 lb per lb_f-hr, 11.4% based on lower heating value.

7-9. A propeller-jet airplane has a net shaft horsepower output to the propeller gearing of 2700. The jet exit velocity is 990 fps from the tail cone, and the airplane speed is 500 mph. Air flow through the jet system is 62 lb per sec. The propeller propulsion efficiency is 78 per cent, and the transmission efficiency through the gearing is 95 per cent. (*a*) Compute the total equivalent horsepower produced by propeller and jet. (*b*) Compute the total thrust in pounds. (*c*) This unit shows a performance of somewhat better than 0.8 lb of gasoline per hr per lb of thrust when using gasoline with heating values of 18,900 (20,390 higher) Btu per lb and specific gravity of 0.73. Compute the probable fuel used per hour and the thermal efficiency based on both lower and higher heating values. *Ans.* (*a*) 2660 hp, (*b*) 1990 lb, (*c*) 273 gal per hr, (*d*) 22.5% and 20.9%.

REFERENCES

1. Warner and Auyer: Contemporary Jet-propulsion Gas Turbines for Aircraft, *Mech. Eng.*, vol. 67, pp. 707–714, November, 1945.
2. Puffer and Alford: The Gas Turbine in Aviation, *Mech. Eng.*, vol. 67, pp. 803–812, December, 1945.
3. Howard and Walker: An Aircraft Gas Turbine and Propeller Drive, *Mech. Eng.*, vol. 69, pp. 827–835, October, 1947.
4. The Day Dawns for Jet Propulsion, *Westinghouse Engr.*, vol. 5, March, 1945.
5. Godsey and Flagle: The Place of the Gas Turbine in Aviation, *Westinghouse Engr.*, vol. 5, pp. 121–127, July, 1945.
6. Burgess, N.: Design Analysis of the General Electric TG-180 Turbojet, *Aviation Week*, vol. 47, pp. 29–33, 36–43, July 7, July 14, 1947.
7. Streid, D.: General Electric Type J-40 Jet Engine, *Aviation*, vol. 45, pp. 51–59, January, 1946.
8. Wells, R. L.: Tale of a Turbojet, *Aero Digest*, vol. 59, pp. 40–41, 101, December, 1949.
9. Redding, A. H.: Future Forms of Aviation Gas Turbines, *Westinghouse Engr.*, vol. 7, pp. 110–114, July, 1947.
10. Kroon, R. P.: The Jet Engine Comes of Age, *Westinghouse Engr.*, vol. 10, pp. 194–200, September, 1950.
11. Tifford, A.: The Application of Gas Turbines to Aircraft, *Aeronaut. Eng. Rev.*, July, 1948.
12. Pinkel and Karp: A Thermodynamic Study of the Turbojet Engine, *NACA Rept.* 891, 1947.
13. Downs, W. H.: Starters for Turbojet Engines, *Tech. Data Digest* (*CADO*), vol. 16, pp. 14–18, February, 1951.
14. Durham, F. P.: "Aircraft Jet Powerplants," Prentice-Hall, Inc., New York, 1951.
15. Godsey and Young: "Gas Turbines for Aircraft," McGraw-Hill Book Company, Inc., New York, 1949.
16. Hall, N. A.: "Thermodynamics of Fluid Flow," Prentice-Hall, Inc., New York, 1951.
17. Vincent, E. T.: "The Theory and Design of Gas Turbines and Jet Engines," McGraw-Hill Book Company, Inc., 1950.
18. Zucrow, M. J.: "Principles of Jet Propulsion and Gas Turbines," John Wiley & Sons, Inc., New York, 1948.

CHAPTER 8

POWER PLANT AND TRANSPORTATION TURBINE UNITS

8-1. General. Mention has been made throughout the book of various power units driven by gas turbines. In this chapter several gas-turbine plants will be discussed from a viewpoint of over-all characteristics rather than from a consideration of the individual components constituting the power plant.

Historically, one of the earliest modern-type gas-turbine installations was made in connection with the refining of gasoline by use of the Houdry process. Figure 8-1 shows a picture of the unit made by the Allis-Chalmers Manufacturing Company and installed at the Tidewater Associated Oil Company Refinery in Bayonne, N. J. In the Houdry process, a deposit of carbon takes place on the catalyst and filler in the reaction chamber. In order to reactivate the catalyst, it is necessary to burn off this carbon, and it was found that this could be accomplished by burning it out with compressed air, blown through the catalyst material. During the combustion of the carbon, the compressed gases can be controlled to reach a temperature slightly under 1000 F. Such a hot gas under pressure can effectively be used in a gas turbine. The output of the turbine in turn is used to compress the gases which go to the catalyst chambers and also for production of useful power. These units have been in use for many years (one since 1936) and have given an extremely creditable record of performance. In Fig. 8-1, the turbine is shown at the left, with the axial-flow compressor in the central part of the picture, followed by the reduction gear, the generator, and the steam turbine used for starting. This particular unit compresses 60,000 cfm of air with delivery at 50 psig.

8-2. Allis-Chalmers High-temperature Experimental Unit. In Fig. 8-2 is pictured the 1500-F gas-turbine unit installed at the U.S. Naval Engineering Experiment Station at Annapolis, Md., for research and test. Information was desired as to the maximum temperature at which it might be possible to operate a gas turbine of this type and also as to other test data on a complete unit. The Allis-Chalmers Manufacturing Company built the unit, which was put into operation in 1944, and the long series of tests which have been carried out have shown good life expectancy even as the temperatures in the turbine have been pushed up toward 1500 F.

Figure 8-3 is a diagrammatic layout of the system. It will be noticed that two separate turbines are used, which makes it possible to vary compressor conditions independent of the output-power requirements. In this figure are shown representative operating data, including pressure losses and flow rates throughout the system.

The axial-flow compressor has 20 stages and handles 40,000 cfm of free air at a discharge pressure of 45 psig. The compressor has an over-all efficiency of approximately 85 per cent. From the compressor the air enters an 8500-ft^2 counterflow regenerator which has an effectiveness of approximately 60 per cent. A pressure loss of low magnitude occurs in

FIG. 8-1. Axial-compressor gas-turbine unit used with Houdry process for gasoline production. (*Allis-Chalmers Manufacturing Co.*)

its passages. Air flows on the outside of the tubes with the hot combustion gases and high-pressure air flowing inside the tubes. Number 2 furnace oil has been used in most of the experiments, although the design was planned for operation with Bunker C oil whenever desired. The oil burners are of the mechanical atomizing type, firing into the combustion chambers.

The combustion air passes through a central flame tube with additional air flow passing between the flame tube and the outer chamber wall. A fuel-heat release of 2½ million Btu/(hr)(ft^3) of combustion chamber has been recorded.

Both the compressor turbine and the power turbine are five-stage units. In the first impulse stage, the initial pressure drop occurs in the nozzle ring and the gases then enter the first-stage blading. The gases from this stage enter the second stationary nozzle ring, which delivers the gases into a second row of impulse blading. The remaining three stages of the turbine are conventional reaction stages. The disk of the first-stage high-temperature wheel is air-cooled. This is accomplished by bringing cool air to it through a double-wall pipe. This air is delivered to a distributing disk ahead of the first wheel. The air, in cooling the wheel, passes radially inward and is then led out through the center of the double-wall pipe, which is located near the shaft of the machine. It is recooled and

FIG. 8-2. Experimental 3500-hp gas-turbine unit installed at U.S. Naval Engineering Experiment Station at Annapolis, Md. (*Allis-Chalmers Manufacturing Co.*)

by use of a positive blower is returned to repeat the cooling process. This arrangement makes it unnecessary to bring the compressed air up to the full pressure used in the first stage. The far side of the first-stage turbine disk is also cooled by air, which, under pressure, is blown against the wheel. This air, however, mixes with the main air stream going through the turbine and completes the cycle.

A study of Fig. 8-2 will indicate the large size of some of the pipes. For this unit, the air intake is 36 in. ID, and the exhaust opening of the compressor turbine is 42 in. ID. The high-temperature piping is double-walled, with about 9 in. of mineral-block insulation between the outer and inner walls. With this arrangement, the outer wall is at low temperature and carries the pressure of the system. The inner wall acts merely as a shield, and there is no pressure differential across it. This arrangement makes it possible to use a metal which can develop its full strength without being endangered by the high-temperature gases.

An electric motor, through a clutch, connects to the compressor and can bring the unit up to its starting speed of 25 to 35 per cent of full-load speed. When this speed is reached, ignition and firing are started. After this, the starting motor can be disengaged, and the unit is then brought up to desired speed by manual fuel control. Indications point to the possibility of efficiencies in the neighborhood of 40 per cent with a unit of this type under operation at 1500 F. As an experimental unit, it has been invaluable in pointing the way toward possible marine designs. In such

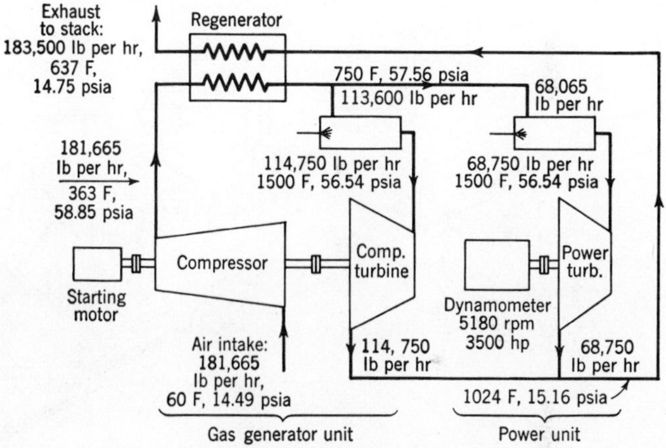

FIG. 8-3. Cycle diagram of Allis-Chalmers 3500-hp high-temperature unit at Annapolis, Md.

a design, additional regenerator surface and operation of the two turbines in series would probably be envisaged. At least one stage of intercooling would also be considered desirable in the compressor. It seems highly probable that the gas turbine may be used extensively in certain types of marine propulsion before many years have passed.

8-3. Locomotive Power Plants. One of the most promising fields of utilization of the gas turbine appears to be as a power unit for railway locomotives. Many of the requirements for locomotive drive are admirably served by the gas turbine. A listing of some of these items will be indicated here and discussed throughout the chapter.

1. *Ability to use low-cost fuel with reasonable economy.* The gas turbine can operate on cheap residual fuel oils such as Bunker C, and it seems possible that at some future time powdered coal can also be adapted for gas-turbine firing.

2. Availability for service. There seems to be little doubt that the gas turbine can be built into a unit of long service life, with the maintenance costs not excessively high.

3. Weight loading. The open-cycle gas turbine and its components are relatively light in weight. Consequently, turbines can be used with a standardized electrical transmission system and not cause additional weight loading on the trucks or other part of the frame.

4. Good efficiency. Previous considerations have already shown that even simple gas-turbine cycles can develop efficiencies in the neighborhood of 20 per cent, and by modification and elaboration of the cycle, efficiencies approaching 30 per cent are possible. These efficiencies are lower

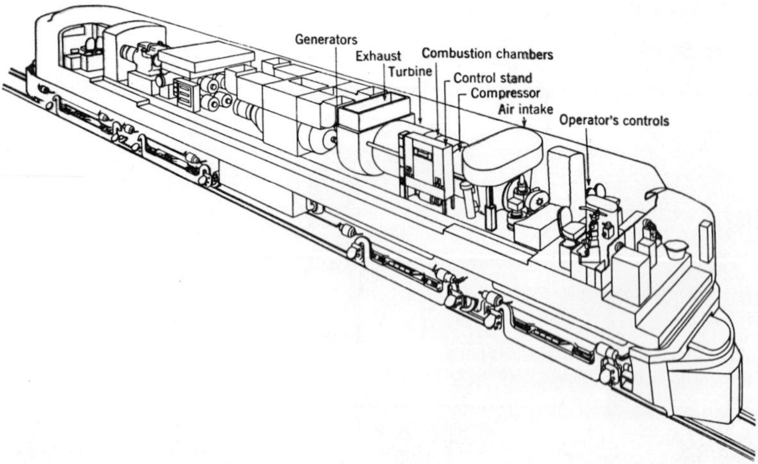

FIG. 8-4. Arrangement of components in General Electric gas-turbine electric locomotive.

than realized with standard diesel-engine equipment, but the lower cost of fuel should offset the lower efficiency.

5. Low maintenance cost. Certainly there is reason to believe that long-life units can be built, and as the units can be standardized, costs of replacements need in no way be excessive. Also, because of relative lightness, it should be easy to replace component parts and not tie up a whole locomotive while repairing elements of the turbine unit.

6. Quantity production of units. The gas turbine lends itself readily to quantity production, and in the field of aircraft propulsion mass production of complete turbine units and elements has already been accomplished.

7. Control arrangement. Although there may be some difficulty in designing a gas-turbine unit to operate with a geared system to the trucks, there should be no difficulty in adapting the gas turbine to work satisfactorily with an electric system. In such a system, the gas turbine would drive direct-current generators, which, in turn, would generate the electrical energy for the traction motors. The additional weight of the electrical equipment is not a complete detriment as it gives increased tractive effort to the locomotive.

8. Water requirements. As gas-turbine cycles need no circulating water, the gas turbine in this respect is comparable in requirements with diesel locomotives. Water may be needed for oil coolers and the like.

8-4. General Electric Gas-turbine Power Plant. The General Electric Company has developed a gas-turbine power unit of versatile design which is being used for locomotive propulsion or with minor modification as the prime mover for power-plant installations. This unit is pictured in Fig. 1-9 in an arrangement as a 4800-hp gas-turbine propulsion unit for a locomotive. In the locomotive, the unit turns at 6700 rpm and is rated to develop 4800 shaft hp when supplied with 1400-F turbine-inlet-temperature gases and when operating at an altitude of 1500 ft with 80-F ambient-air temperature (5000-hp rating at sea level). A thermal efficiency of better than 17 per cent, based on lower heating value of Bunker C fuel oil, has been attained. Figure 8-4 shows the arrangement of the turbine unit in the locomotive.

The 15-stage axial-flow compressor handles approximately 70,000 cfm through a pressure ratio of 6:1. The compressed air at roughly 90 psia passes into six combustion chambers placed around the periphery of the machine, and the hot gases still flowing in essentially a straight line are led into the gas turbine. This turbine, which is a two-stage impulse type, is shown in detail in Fig. 4-5. By making use of impulse construction, it is possible to operate with lower first-stage turbine temperatures than would be used in the first few stages of a multistage reaction turbine. It is estimated that with 1400-F inlet gas the temperature at the pitch line of the turbine blade does not exceed 1160 F. Details on this turbine are given in Chap. 4.

The compressor, as can be seen from Fig. 1-9, is made up of separate wheels which are held firmly together by through bolts and are fitted concentrically on the shaft. The wheels of the first six stages are of aluminum alloy to reduce weight and rotating inertia. Figures 5-15 and 5-17 show a rotor and stator of construction similar to the one used in this unit.

The combustion chambers are of all-metal construction with an outer casing of carbon steel and an inner lining of stainless material. The liners are approximately 35 in. long and 10 in. in diameter. The fuel nozzles are of the air-atomizing type, and ignition is accomplished by means of spark plugs located in two of the six combustors. The other chambers are ignited through cross-ignition tubes, which connect the chambers at the forward end. It is customary to start this unit on diesel fuel and change over to Bunker C fuel. This unit has now undergone an extensive series of tests which have shown that it has lived up to design expectations and even exceeded them. A backlog of orders for these locomotives now exists, and a number of them are in production.

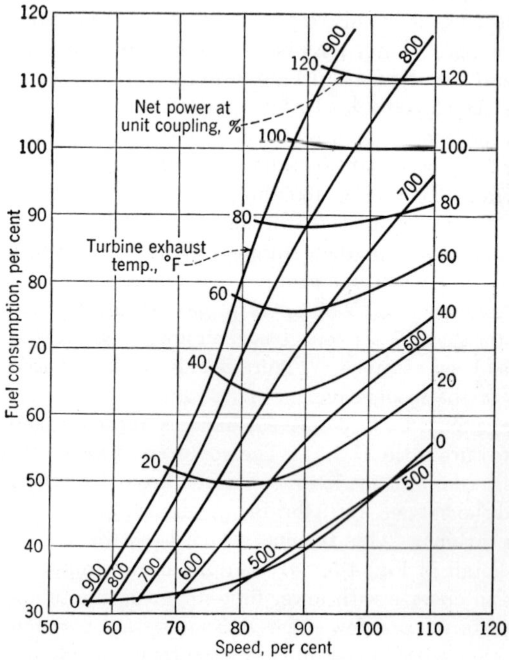

FIG. 8-5. Test-performance values of General Electric gas-turbine power plant. 100 per cent corresponds to 6700 rpm, 5000 HP, at 14.7 psia and 80°F at 75,600,000 Btu per hr, based on LHV of fuel.

Test results on this unit are shown in Fig. 8-5. These curves show the tested fuel consumption versus speed at each of a series of exhaust temperatures, with lines of constant net shaft horsepower output also indicated. The curves are corrected to ambient temperatures of 80 F and to a datum pressure of 14.7 psia. The 100 per cent points correspond to 6700 rpm, 5000 hp, and a heat input of 75,600,000 Btu per hr based on lower heating value (LHV) of the fuel. These curves show that the exhaust temperature at rated power is 780 F. For this temperature it can be shown that the turbine inlet temperature by calculation would be of the order of 1280 F. When the unit is operated at full inlet temperature of 1400 F, the corresponding exhaust temperature would be 850 F, which means that this unit could develop at 6700 rpm a horsepower of some 6000 with a fuel consumption indicating an efficiency of about 18.5 per cent based on LHV.

At high powers, the fuel rate is nearly independent of operating speed, while at low powers there are definite minimums in the fuel-rate curves. For locomotive operation, the control should be set to follow approximately the best fuel rate at idling; for example, 70 per cent speed would give in the neighborhood of approximately one-third of full-load rate. If full speed were maintained at idling, the fuel rate would be in the neighborhood of 47 per cent. Accurate test data are not available on the components, although data can be estimated. It is believed that the compressor efficiency is about 84 per cent at a pressure ratio of 5.9:1. At somewhat reduced speed, it apparently has a higher efficiency. The turbine efficiency is also in the neighborhood of 84 per cent. However, if the high axial leaving velocity is not specifically charged to the turbine, the efficiency is in the neighborhood of 88.5 per cent. The air flow of the unit at rated power is approximately 94 lb per sec.

The unit is started by using one of the main locomotive generators as a motor. The power for this is supplied from a diesel-engine-driven generator. The motor brings the unit to firing speed in some 45 sec, at which time ignition is turned on and diesel fuel is admitted to the nozzles. During the next 155 sec, the unit accelerates to an idling speed of 4700 rpm, where it runs under control of the governor. The combustion system can then be changed over to operate on Bunker C fuel. The driving motor is usually arranged to assist the unit in picking up speed to 3500 rpm. Above this speed, driving-motor power is disconnected by use of reverse-current relays. A control mechanism regulates the fuel input to limit the maximum average exhaust temperature to about 875 F. The starting system is designed to be fully automatic.

Operation with Bunker C fuel shows slightly more visible smoke than with diesel fuel, but the amount of smoke is so slight as to be considered unobjectionable. Full power output can be obtained in 3 to 3.5 min. In the tests run on this unit, the turbine efficiency has held very close to expectation while burning over 200,000 gal of Bunker C oil. A slight deposit from ash in the fuel does, however, appear in the nozzles and on the buckets. This is apparently water-soluble and can be cleaned off easily. The compressor also accumulated deposits, even though a coarse dry metal-wool screen was interposed in the inlet-air passageway. The dirt in the compressor deposits as small accumulations of dry material near the entrance edge of the blades. It is not believed that required cleaning cycles are frequent enough to present a serious problem in connection with operation of the compressor unless the atmospheric air is heavily loaded with oily material or similar substances.

It is readily realized how the capacity of a gas-turbine system is decreased as the absolute pressure of the inlet air decreases. Figure 8-6 shows the sharp decrease in power associated with the moderate pressure changes which a locomotive might meet in traversing mountainous territory. To express these values in other terms, a 5-in.-Hg barometric depression is associated with 4800 ft altitude.

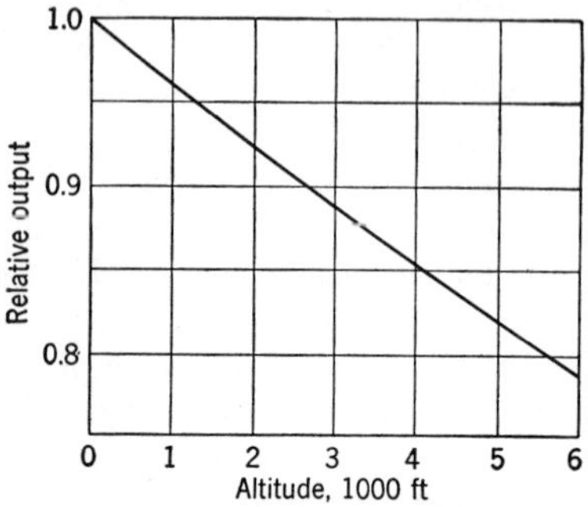

FIG. 8-6. Graph to indicate approximate effect of altitude on output of gas-turbine plant.

A gas-turbine unit of essentially the same design as that for which the aforesaid data were given was installed in the generating station of the Oklahoma City Gas and Electric Company and is pictured in Fig. 1-10. In Art. 1-8 a discussion was presented as to how this unit worked into the power scheme of the station, and by utilization of the high-temperature exhaust gases from the gas turbine it was possible to warm the feedwater for the whole station and thereby reduce the amount of steam which had to be bled from the main steam turbines of the station. Thus a basic design of turbine has shown its utility in two separate fields, locomotive propulsion and straight power generation. It is of interest to note that the power-plant unit uses natural gas as a fuel instead of fuel oil.

Data on the installation cost of this unit released by Oklahoma Gas and Electric Company are tabulated below. For a small plant of this type

INSTALLATION COSTS, 3500-KILOWATT GAS-TURBINE UNIT

Buildings, grounds, and foundations	$ 79,851
Gas-turbine unit erected	400,107
Automatic control	28,393
Recuperator erected	91,140
Stacks, duct work	32,548
Air-intake cooler	9,237
Piping	51,225
Electrical	34,486
Miscellaneous	3,013
	$730,000

the cost of the unit per kilowatt of capacity is reasonable, particularly when some of the items not specifically related to the turbine unit itself are considered in proper perspective. This plant went into service July 29, 1949.

Although the efficiency of the simple-cycle units just described is relatively low (less than 20 per cent), in the case of the locomotive the simplicity of the unit does not particularly justify employing heat-conservation devices. For the power-plant unit the exhaust heat is conserved in the waste-heat boiler, which warms the feedwater for the remainder of the plant, thereby entering into the complete heat balance of the plant itself.

Previous discussions have shown how it is possible to improve the effectiveness of the system by splitting the compression into two steps and

TABLE 8-1. CHARACTERISTIC DATA ON ONE TYPE OF GAS-TURBINE POWER PLANT*

	Simple cycle	Compound cycle with regenerators and intercoolers	Compound cycle with regenerators
Rating at 80 F and 1000 ft altitude	3500 kw	5000 kw	5000 hp
Fuel heat rate (HHV, Btu/kwhr)	22,970	12,950	14,500
Over-all eff (HHV value)	14.9%	26.4%	23.5%
Maximum overloading	5000 kw	6250 kw	7000 hp
Cooling water required	320 or 0 gpm	1100 gpm	350 gpm
Dimensions, length	47 ft	49 ft 7½ in.	33 ft
Width	8 ft 9 in.	27 ft 10¾	15 ft 4 in.
Height	10 ft 4 in.	12 ft 11 in.	22 ft 4 in.
Weight	86,000 lb	310,000 lb	370,000 lb
Fuel	Bunker C oil or natural gas	Bunker C oil or natural gas	Bunker C oil or natural gas
Air requirements	94 lb per sec	72 lb per sec	94 lb per sec
Oversize generator	4000 kw at 0.8 power factor	6000 kw at 0.8 power factor	
Voltages	2400/4160	2400/4160	
Exciter (direct-connected)	25 kw	40 kw	
Starting motors (slip ring)	250 hp	40-hp low-press. shaft 100-hp high-press. shaft	250 hp

* Abstracted from General Electric booklet, Gas-turbine Power Plants (1950).

utilizing an intercooler between stage groups. When intercooling is employed, it is also desirable to use regenerators for conservation of the energy in the hot gases leaving the turbine. By use of heat-exchange devices the efficiency can readily be increased to a value in the neighborhood of 25 per cent, which makes the units very competitive with small steam plants, and even with diesel plants when comparative fuel costs are brought into the picture. Certainly for small outlying power plants, the gas turbine has an extremely promising future. Table 8-1, which is reproduced in modified form from the manufacturer's publication, shows how, starting with a basic unit of the type just described, a set of comparative values can be worked out. This table merits careful study.

8-5. Allis-Chalmers Locomotive. The Allis-Chalmers Manufacturing Company has developed a gas-turbine locomotive unit. This particular unit has been satisfactorily fired with fuel oil, but investigations are under way whereby, under the auspices of the Locomotive Development Committee of Bituminous Coal Research Incorporated, it will be fired experimentally with pulverized coal. The turbine uses a simple open cycle with a regenerator and is geared down to power a direct-current generator. Transmission of power to the trucks is accomplished electrically.

FIG. 8-7. 4100-hp Allis-Chalmers gas-turbine unit for locomotive drive without combustion equipment. (*Allis-Chalmers Manufacturing Co.*)

Figure 8-7 is a drawing of the unit without its combustion equipment, and Fig. 8-8 is a photograph showing the turbine followed by the compressor, which, in turn, connects to the reduction gearing of the generators. Referring again to Fig. 8-7, it can be seen how the compressed gases rise at the center part of the picture and pass to the right through the regenerator on their way to the combustor. The combustor does not appear in this picture. Beneath the regenerator is the gas turbine, for which two inlet connections from the combustors can be seen. The gases, after expanding through the five-stage turbine, rise vertically through the regenerator and pass out to a stack section in the roof of the locomotive.

Figure 8-9 is a flow diagram of this unit. Operation at 1300 F design temperature at turbine inlet with ambient-air conditions of 70 F at 14.7

FIG. 8-8. Locomotive gas turbine and compressor with top casing halves removed. (*Allis-Chalmers Manufacturing Co.*)

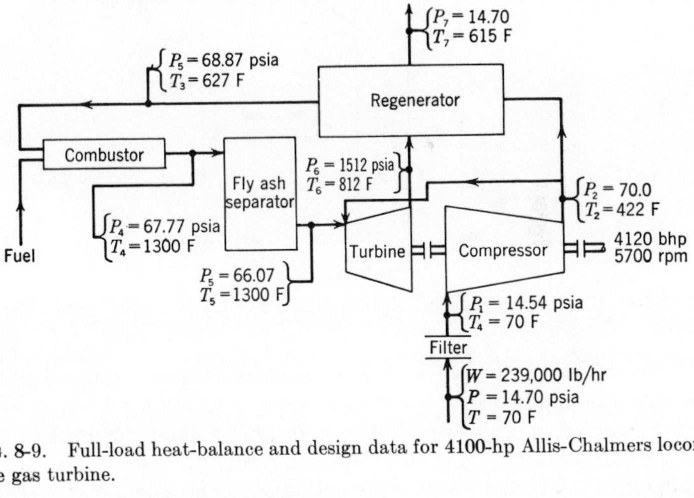

FIG. 8-9. Full-load heat-balance and design data for 4100-hp Allis-Chalmers locomotive gas turbine.

psia produces 4120 hp for delivery to the reduction gear. For pulverized-coal operation, it is necessary to have a fly-ash separator, and this is indicated in the diagram. The figures on the diagram show an allowed pressure loss of 2.8 psi in the separator. In the regenerator, pressure losses of 1.2 psi on the air side and 0.42 psi on the gas side have been used in the design. The initial gas temperature of 1300 F is used in connection with a compressor pressure ratio of 4.8. A thermal efficiency of 24 per cent can be expected when the combustion efficiency is considered as 96 per cent. Starting of the unit will take place on fuel oil with further operation on pulverized coal if desired.

Governing of the unit is accomplished by varying the speed and by varying the temperature as well. Figure 8-10 is a diagrammatic representation of expected performance. At light loads, the turbine inlet temperature is varied, while a constant lower speed is maintained. At higher loads on the unit the temperature is maintained constant at a fixed value, and the speed is varied. The graphs show that in the varying-speed range the decrease in thermal efficiency is not excessive as load decreases.

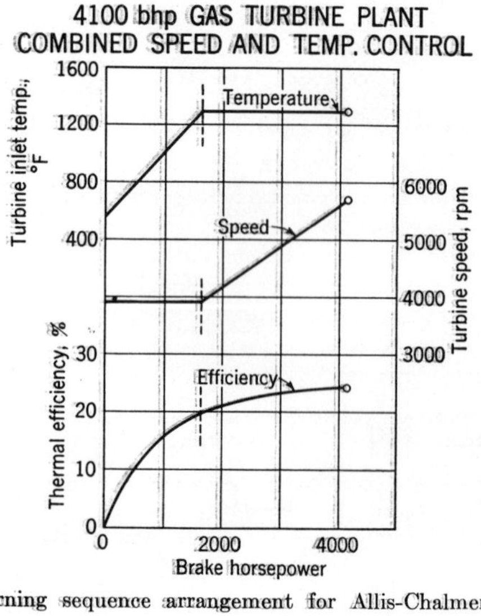

FIG. 8-10. Governing sequence arrangement for Allis-Chalmers locomotive gas turbine.

The turbine element, which can be seen along with its casing in Fig. 8-8, is a six-stage reaction unit designed to deliver 12,243 hp to its coupling when turning at a speed of 5700 rpm at 1300 F inlet temperature and under conditions of a 4.8 pressure ratio in the compressor. The turbine spindle is of fabricated construction consisting of six equal-diameter disk forgings, which are machined and welded together to form the rotor portion of the spindle. Onto this rotor, stub ends are welded to form the bearing and coupling ends. The first four rows of blading are precision-cast, while the last two rows are forged.

Figure 4-17 is a picture of representative blades from this turbine. They are constructed of Allegheny Ludlum S590 material, which consists of 20 per cent each of cobalt, chromium, and nickel and 4 per cent each of tungsten, columbium, and molybdenum, with the remainder ferrous composition. The blades have axially serrated roots to fit in the spindle disks. Tip speeds are respectively 895 fps for the longest and 675 fps for the shortest blades. Blading is designed for radial-flow equilibrium, and each group of blades is warped and tapered. The blades are sufficiently stiff so that bracing wires and shrouds are not required.

A labyrinth-type gland is provided at the high-pressure end of the turbine to keep the hot gases from escaping, while at the same time allowing provision for cooling the turbine spindle. Air from the compressor discharge is admitted to a belt around the turbine gland. This cool air passes through the high-pressure gland and into the first stationary row of blading, while some of it passes along the surface of the turbine spindle to the suction of a small exhauster fan. This arrangement provides a flow of relatively cool air along the spindle, thereby cooling the adjacent bearing, and, at the same time, the air reduces the temperature of the spindle and disk. A Kingsbury thrust bearing is located in the bearing housing at the high-pressure end of the turbine to maintain the turbine spindle in its proper axial position.

The compressor is an axial-flow type with 20 rows each of moving and stationary blades mounted in circumferential grooves. The blades are precision castings. Efficiencies in excess of 85 per cent can be obtained with this compressor. The compressor rotor body with an integral stub at the high-pressure end is made from a single forging. The compressor and the turbine are so arranged that their axial thrusts tend to offset each other.

A fly-ball-type governor is driven from the main reduction gear. This governor, operating in conjunction with the control system, maintains desired speeds by controlling the rate of fuel flow to the combustors on the low-speed range or by controlling generator excitation when operating in the constant-temperature phase. An overspeed governor is provided to stop the flow of fuel to the combustors should this be necessary. A two-shaft generator is used with auxiliaries mounted on top. The high-speed pinion of the turbine turning at 5700 rpm maximum is geared down so that the generator shafts turn at 2000 rpm. Each generator shaft is provided with two armatures so that the turbine drives the equivalent of four generators. Also driven from the main gear is the exciter, a 40-kw d-c generator; and two 175-kw a-c generators. Starting is accomplished by rotating the compressor turbine unit to a starting speed of approximately 1600 rpm by use of two of the generator armatures. A 200-hp diesel-driven generator, located in the auxiliary cab, can supply power for starting, or the locomotive batteries can be used.

8-6. Westinghouse Gas-turbine Unit. A unit for railway and general power use has been designed and constructed by the Westinghouse Electric Corporation. This unit, which is pictured in Fig. 8-11, is designed to develop 2000 net hp (6000 gross hp). It is built in a straight-line arrangement consisting of a double-armature generator, the reduction-gear set, and then the axial-flow compressor, which draws air vertically downward through a bend into a 20-stage compressor. From the compressor the gases flow to 12 unit combustors, which surround the central power shaft of the machine. From the combustors, the hot gases lead directly into the turbine. Leaving the last stage of the turbine, the gases enter into a diffusion section and, with the help of turning vanes, change their direction by 90° so as to exhaust vertically upward. The gases, leaving the

FIG. 8-11. 2000-hp gas-turbine unit for locomotive drive mounted on test stand. (*Westinghouse Electric Corp.*)

turbine at a velocity of some 500 fps, gain a significant pressure increase in the diffuser.

The total weight of the machine and the foundation is 38,000 lb. The unit is 26 ft long, 6 ft high, and 3½ ft wide. It is so narrow that two of them could fit side by side in a locomotive. Such an arrangement would permit the development of 4000 hp in some 30 ft of locomotive length. However, with the addition of working space and accessory equipment the 4000-hp locomotive is 77 ft long.

The compressor of this unit is pictured and described in Chap. 5 (Fig. 5-1). The combustors consist of an inner flame tube with radial air-inlet holes. The unit has been run mainly on No. 4 fuel oil, but satisfactory operation with Bunker C oil and natural gas has also taken place. Each of the 12 combustors is 4½ in. in diameter and 3 ft long. A bellows-type expansion joint takes care of axial expansion. Air-atomizing fuel nozzles are used, and heat releases of approximately 1.2 million Btu/(hr)(ft³)(atm of pressure) have been obtained. The eight-stage reaction turbine is designed for equal enthalpy drops over the stationary and rotating blades at mean diameter. It is pictured in Fig. 4-8 and is designed to operate at inlet temperatures which reach 1350 F at full load and range from 600 to 750 F at no load.

Tests on this system have shown that it was able to develop 2350 net hp with an inlet-air temperature of 31 F. The over-all fuel rate at full load showed a test value of 0.78 lb per net hp per hr. Based on a higher heating value (HHV) of 19,500 Btu per lb, this shows an over-all thermal efficiency of

$$\eta_o = \frac{2545}{(0.78)(19{,}500)} \times 100 = 16.7\%$$

and based on an assumed lower heating value (LHV) of 18,400 Btu per lb,

$$\eta_o = 16.7 \times \frac{\text{HHV}}{\text{LHV}} = 16.7 \times \frac{19{,}500}{18{,}400} = 17.7\%$$

Test performances of the component elements were estimated by pressure, temperature, and flow measurements and by computation. The turbine is believed to have an efficiency ranging from 84 to 86 per cent over the normal operating range, while the compressor efficiency varied from 80 to 86 per cent over its normal operating range. In making performance estimates of the units in the cycle, three methods were employed: (1) direct measurement of temperatures, using shielded-type temperature probes; (2) calculations based on measured air and fuel flow, with a combustion efficiency of 95 per cent assumed and measured combustor inlet temperature; (3) calculations using the measured turbine exhaust pressure and temperature, measured inlet pressure, and then finding turbine work by heat-balance calculation. It was found that a close check could be made by use of methods 2 and 3 but that results based on temperature measurements of the hot gas leaving the combustor were somewhat erratic.

The unit is started by using one generator as a motor. The time required for starting is a function of the power available. With a maximum power input of 35 kw, the unit requires approximately 2½ min. When 80 kw is available, the unit can be started in 1 min. In starting, the acetylene igniters are turned on when the rotor reaches 15 per cent speed, and at 25 per cent speed the fuel is injected. A gas turbine of this type can be operating at capacity within 10 min from cold start if this is necessary.

The effect of inlet temperature on the capacity of any gas turbine is quite striking. This unit is designed for 100 per cent load at 80 and

1350 F turbine inlet temperature. With a 30-F air-inlet temperature, approximately 130 per cent of load can be handled, while with air inlet at 0 F more than 140 per cent of load can be handled. At 100 F the capacity drops to approximately 88 per cent of the rated value.

In many of the tests, the unit was operated without an air filter at the compressor inlet. With dirt-laden industrial air, the compressor blading became excessively dirty after 100 hr of operation. Blade deposits consisted of oily, dirty soot and caused a drop in compressor efficiency of approximately 2 per cent. To clean the blading, the unit was turned over

FIG. 8-12. Westinghouse 2000-hp gas-turbine unit being lowered into locomotive.

slowly with a starting motor and a noncorrosive commercial solvent sprayed into the compressor inlet. After the deposit had been soaked for a few minutes, it was washed off with a steam spray. This cleaning could be accomplished without disassembling any part of the compressor.

The flow capacity of this unit is 25,000 cfm of free air compressed through a pressure ratio of 5:1. Both pressure ratio and flow change when the speed is varied from the normal value of 9200 rpm. Test results have shown that units of this general design can be expected to have reasonable life and economy. It is believed that units of somewhat larger horsepower can also be constructed, and, using a simple cycle with regeneration, fuel rates as low as 0.6 lb per net hp-hr can be expected. Figure 8-12 is a photograph showing how a unit of this type fits into the locomotive.

A 2000-hp unit has been installed in a test station on a natural-gas line at Wilmar, Ark. This unit is operated with natural gas or fuel oil and has shown satisfactory results with either fuel.

8-7. Brown-Boveri Gas-turbine Locomotive.* A Brown-Boveri unit has been constructed for locomotive service in England for the Great Western Line. This gas-turbine locomotive has a continuous rating of 2500 hp. It drives an electric generator through reduction gearing (5800 to 875 rpm). Diesel fuel is used for starting, with change-over to heavy

FIG. 8-13. 40,000-kw gas-turbine power station at Beznau, Switzerland. (*Brown-Boveri Corp.*)

fuel oil after the system is warmed to operating temperature. Gases are supplied to the turbine at 1100 F and leave the compressor at 44 psig. The locomotive is 63 ft long and weighs 115 tons. A maximum tractive effort of 31,500 lb is available from rest up to speeds of 21 mph, with 90 mph being the maximum operating road speed of the locomotive. The system has a power ratio (work ratio) of 0.25, and a regenerator with an effectiveness of 0.4 is employed.

8-8. Brown-Boveri Gas-turbine Power Plant. The largest gas-turbine unit built up to this time, of 27,000 kw capacity, was built by the Brown-Boveri Corporation and installed in the Beznau Station of the Northeast Power Company in Switzerland in February, 1949. This unit and another of 13,000 kw capacity appear in the photograph of Fig. 8-13.

* *Automotive Ind.*, Sept. 15, 1950, pp. 41, 90.

They are used to furnish supplementary power in winter when the output of the low-head hydraulic turbines is reduced because of the low water flow which occurs in freezing weather. The components used in the 27,000-kw plant are essentially duplicate units of the same design as those used in the 13,000-kw plant. In the open cycle of the large plant the air enters two low-pressure compressors, from which after intercooling it passes to an intermediate-pressure compressor. After a second stage of intercooling, it enters a third compression stage. Then it passes to the regenerator and on to the combustor. The gases next enter the power turbine, which drives the third compressor unit and the electric generator. Leaving this turbine, the gases are reheated and pass to the double-flow turbine, which drives the first- and second-stage air compressors. This double turbine is a high-capacity unit, as it should be realized that approximately 65,000 kw is required for gas compression and some two-thirds of the gas-compression power, about 45,000 kw, is absorbed in the first two compression units. The efficiency of the plant ranges between 30 and 34 per cent, based on lower heating value when operated under winter conditions with an ambient temperature of 41 F. These units are designed for the temperature at the inlet to the power turbine of 1110 F and to the two large compressor turbines of 1020 F. The latter turbines are necessarily larger, as they operate at lower gas pressure than those used in the power turbine.

In the left foreground of Fig. 8-13 can be seen the low- and intermediate-pressure compressors of the 13,000-kw unit with their turbine drive at the immediate front. Starting at the extreme far end on the right-hand side from this 13,000-kw unit can be seen the exciter, a starting motor, next in line is the generator and its step-up gearing, and then follows the high-pressure compressor with its final stages covered by insulation and a jacket. The final unit on the shaft is the high-pressure turbine. The two combustion chambers lie behind the compressor and turbine. Air from the compressor enters the first of these, where it is brought to turbine inlet temperature. Leaving the high-pressure turbine, the gases enter the second combustion chamber, where they are reheated to a high temperature and then conducted through a pipe under the floor to the low-pressure gas turbine. The high-pressure turbine shaft runs at constant speed, while the other shaft, which drives only compressors, can run at varying speed.

It is of interest to observe that a unit which develops 13,500 kw at 40 F can be expected to develop only 11,000 kw at 105 F ambient temperature. The thermal efficiency also decreases as ambient temperature increases. The smaller 13,000-kw unit of this station for winter operation is essentially a duplicate of a 10,000-kw unit built by the same company for operation in Peru, where the ambient temperature is appreciably higher.

8-9. Heat-exchange Equipment, General. In gas-turbine systems heat-exchange equipment of various designs is used for many purposes. For example, in the open cycle, regenerators or recuperators are employed to exchange heat between the compressed air, on its way to the combustor, and the expanded gases, which would otherwise be sent to waste at high temperatures. Intercoolers can be used with either open or closed cycles and are employed between stage groups to decrease the temperature of the air or other gas undergoing compression, thereby reducing the over-all shaft work required. The closed cycle requires, in addition, a precooler to cool the gaseous medium on its way from the turbine to the low temperature required for inlet to the compressor. However, the most critical heat-transfer device in the closed cycle is the direct-fired gas heater, where energy is transferred directly from the fuel (combustion gases) into the gaseous working medium of the cycle. Regenerators also are used in closed cycles.

The heat-transfer problem in gas-turbine systems is difficult because it involves in many places transfer of heat from one gaseous medium to a second gaseous medium and coefficients for such heat transfer are extremely low so that for high effectiveness large heat-transfer areas must be employed. In intercoolers and precoolers where water is on one side of the tubes and gas on the other, effective use can often be made of fin-tube surface on the gas side to provide additional surface and thereby offset the effect of the much higher coefficients which exist on the water side of clean tubes.

8-10. Regenerators. A regenerator does not increase the load capacity of a plant. In fact, because of pressure drops on both the gas and the air side of the regenerator the capacity of the plant will even be slightly decreased. However, when conditions within the cycle justify its use, the regenerator can perform an extremely important function in conserving energy which would otherwise be wasted. As there is a direct relationship between cost of fuel and the amount of energy which can be saved in a regenerator, an economic study of the situation can readily determine how much money can be invested in regenerative equipment for a plant of given size. The approach is merely one of figuring how much energy can be saved by the regenerator per kilowatthour produced and on an annual basis finding out how much investment cost in a regenerator this heat saving would justify based on the annual kilowatthour output. As more and more surface is put into a regenerator, thereby increasing its effectiveness, the initial investment cost rapidly increases so that an economic limit is quickly reached. For a rough design basis, in a locomotive, where space is a valuable consideration, an effectiveness of 0.50 would suffice. In land or marine installations, probably an effectiveness of 0.75 might represent a satisfactory design figure. In some cases, values greater or less than either of these might be employed.

The General Electric Company[17] has designed a regenerator to serve a 5,000-kw gas-turbine unit. This is of shell-and-tube construction of carbon steel with the tubes rolled into tube sheets which are attached to the shell at each end. The hot gases are led in through headers into the tubes at one end of the shell and in single pass move through the tubes to the other end, where they are delivered to the waste system. The air to be heated enters the side of the shell at one end and moves to an outlet at the side of the shell on its other end. The air in its passage comes into contact with baffles, which make it take a cross-flow pattern back and forth over the tubes as it moves from the inlet to the outlet end. By admitting the air to be heated at that end which represents the outlet for

TABLE 8-2. PERFORMANCE OF REGENERATOR*

Heat transferred, Btu/hr	11,070,000
Air and gas flow, lb/sec	36
Air pressure entering, psia	128.1
Gas pressure entering, psia	15.69
Air temperature entering, °F	347
Air temperature leaving, °F	690
Gas temperature entering, °F	805
Gas temperature leaving, °F	475
Air-pressure drop, psi	1.7
Gas-pressure drop, psi	0.61
Over-all heat-transfer coefficient, Btu/(hr)(ft²)(F)	7.95
Gas-mass velocity, lb/(ft²)(sec)	4.0
Air mass velocity—cross flow, lb/(ft²)(sec)	6.7
Air mass velocity—parallel flow, lb/(ft²)(sec)	7.2

* G. R. Fusner, Heat-exchange Equipment for a 5000-kw Gas-turbine Generator, *Mech. Eng.*, vol. 72, pp. 316–320, April, 1930.

the gases, a modified counterflow circuit can be attained, although necessarily a cross-flow pattern exists in the system. By removing the heads, it is possible to clean out the exhaust-gas tubes by rodding, steam blasting, or turbining. As the air being warmed on the outer sides of the tubes is essentially clean, no cleaning provision is provided for this side of the regenerator. Specific data on this regenerator are given in Table 8-2. It will be noticed that the pressure drops are not excessive and that a reasonable over-all coefficient of heat transfer is attained.

This particular regenerator contains 13,100 ft² of surface and uses 2500 1-in. OD carbon-steel tubes of 13 gage rolled into 1-in. steel tube sheets. The tube sheets are 20 ft apart and the ⅝-in.-thick shell is 6 ft in diameter. Eight baffles are used inside of the shell to direct the air on its cross-flow passes. The pressure drop on the air side for this particular design is 1.33 per cent of the absolute inlet pressure, and on the gas side it is 3.9 per cent.

It is desirable to use the smallest actual pressure drop on the gas side because the absolute pressure itself is lower, in general but slightly above that of the external atmosphere. It should be realized that pressure drops are expensive in terms of cycle performance, and in any design one must endeavor to keep these to an absolute minimum consistent with over-all economy.

In Fig. 1-8 is shown the regenerator for the Elliott gas-turbine locomotive. It should be observed that in this particular unit the compressed air passes inside of the bent tubes, while the hot gases pass in cross flow across the tubes. The accessibility and open top of the unit make it possible to clean the tubes on the combustion-gas side without undue difficulty in the event they become dirty.

TABLE 8-3. PERFORMANCE OF TWO INTERCOOLER SECTIONS OPERATED IN SERIES*

Heat transferred, Btu/hr	6,921,450
Air flow, lb/sec	36
Air-inlet temperature, °F	316
Air-outlet temperature, °F	95
Air-pressure drop, psi	0.35
Water flow, gpm	400
Water-inlet temperature, °F	70
Water-outlet temperature, °F	104.7
Water-pressure drop, psi	1.7
Over-all heat-transfer coefficient, Btu/(hr)(ft²)(F)	15.0
Water mass velocity, lb/(ft²)(sec)	165.5
Air mass velocity, lb/(ft²)(sec)	5.7

* G. R. Fusner, Heat-exchange Equipment for a 5000-kw Gas-turbine Generator, *Mech. Eng.*, vol. 72, pp. 316–320, April, 1950.

In the regenerator shown at the upper left in the Allis-Chalmers unit of Fig. 8-2, air from the compressor flows inside tubes from left to right, while gases flow outside the tubes in the shell from right to left and then turn to enter the exhaust pipe to waste. This unit has 8500 ft^2 of exchange surface and a design effectiveness in the neighborhood of 60 per cent.

The Allis-Chalmers locomotive regenerator pictured in the foreground of Fig. 8-7 is designed primarily to fit into the space available. High-pressure air is carried through the regenerator in straight tubes, while hot gases pass in cross flow across the tubes with exhaust through the cab roof to atmosphere. The tubes are rolled into tube sheets at each end of the regenerator, and expansion between the shell and the tubes is taken care of by a steel expansion joint at one end of the regenerator shell. Support plates between the tube sheets minimize vibration and hold the tubes in position.

8-11. Intercoolers. Table 8-3 gives expected performance data on two finned-tube intercoolers for the 5000-kw General Electric unit.[17]

Each intercooler contains 150 tubes 6 ft 6 in. long between tube sheets with 2630 ft^2 of surface on the air side. Fins ⅜ in. high spaced nine per inch are spiral-wound and soldered on the copper tubes of the intercooler. Each unit is single-pass, crossflow on the air side and double-pass on the water side.

In addition to the possible heat-transfer devices discussed, plants must usually employ lubricant coolers and sometimes generator air coolers.

Figure 8-14 is a diagrammatic outline of the basic cycle of a General Electric plant for power production using an open cycle and extensive heat-exchange equipment. It will be noticed that intercooling is employed between the low- and high-pressure compressor units.

The low-pressure compressor consists of 9 stages having a maximum blade-tip diameter of 31 in. and turning at 7200 rpm, while the high-pressure unit has 11 stages with a blade-tip diameter of 22 in. and turns at

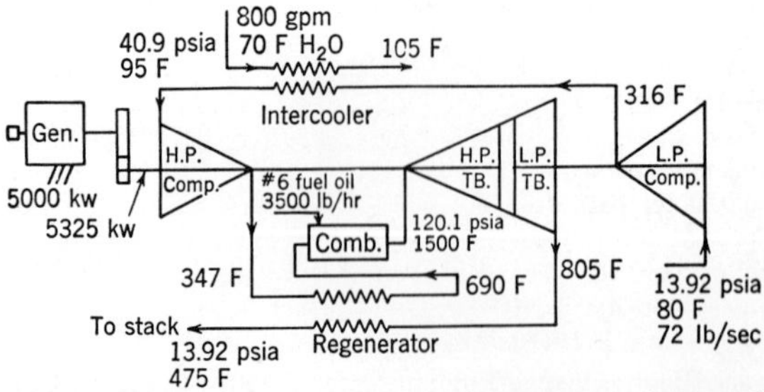

FIG. 8-14. Cycle diagram of General Electric power-plant design with pertinent pressures, temperatures, flows, and output for design rating at 1500 ft altitude.

8695 rpm. The pressure ratios of the low- and high-pressure compressors are, respectively, 3 and 3.1. The data of Fig. 8-14 are presented for an altitude of 1500 ft, for which the corresponding inlet pressure is 13.92 psia.

For this plant designed to develop 5000 kw terminal output, it is expected that a thermal efficiency of 28 per cent based on lower heating value of the fuel can be realized and that the whole plant can be contained in a space approximately 50 by 28 by 13 ft high. The weight should not exceed 60 lb per kw.

8-12. Waste-heat Boilers. Because of the high temperatures at which gases are delivered from a gas-turbine system, there are many conditions under which means of conserving energy in the exhaust should be considered. The waste-heat boiler offers an opportunity of producing steam for process use, and under certain circumstances for power use, often without excessive increase in investment cost for the plant. For a plant generating 3000 kw with a flow of perhaps 80 lb per sec of exhaust gas (air) at 800 F, it can readily be realized that a large amount of heat is available when the gas cools to some 480 F to generate steam at 200 psia (that is, 23,200,000 Btu per hr). An even larger amount of energy would be available if the gas cools to 380 F for generation of 50 psia steam (that is, 30,300,000 Btu per hr). In some cases standard-type boilers already installed might be used in connection with a gas-turbine installation by proper baffle arrangement to improve heat transfer. Newer designs of waste-heat boilers will give better performance in connection with a gas-turbine plant. Enough additional heat-transfer surface should be provided in a waste-heat boiler to bring the temperature of the waste gases close to the temperature of the medium being heated. Design temperature differences of less than 100° are not usually considered.

As is true with all waste-heat boilers, it often happens that the power load and the steam load do not coincide, in which case it may be desirable to have auxiliary burners available so that the required steam flow can be maintained, in the event that the power load on the turbine is small. Particularly in outlying stations, the joint gas-turbine–steam-raising plant holds promise of success. Even in main stations, it has been found that a gas-turbine plant can work effectively into the heat balance of a steam plant in such a way as to improve capacity without much additional cost. For example, the exhaust gases from the gas turbine can warm the feedwater being sent to the boiler, thereby decreasing the amount of the extracted steam required from the main units. It is even possible that moderate-pressure steam generated in a waste-heat boiler could be sent into the lower stages of main turbine units or into separate turbines in such a way as to contribute to the total power output of a station.

8-13. Combination Cycles. Numerous arrangements in which the gas turbine is employed in connection with other power units can be envisaged. The desirability of a composite system is ultimately one of economics set by fuel and investment costs. In a previous section, mention was made of the use of a waste-heat boiler to utilize some of the thermal energy in the exhaust gases. In general, the use of a waste-heat boiler would exclude the use of a regenerator, but in Fig. 8-15 at the left an arrangement is shown whereby the exhaust-gas stream is split and the gases can be directed into the waste-heat boiler or into the regenerator in such manner as best to serve the heat-balance scheme of the plant. For example, if a large steam load is required, it might be desirable to divert all the exhaust gases to the waste-heat boiler and none to the regenerator. On the other hand, if the power requirements are high and steam requirements low, it might be desirable to divert the greater portion of the gases into the regenerator in preference to the waste-heat boiler. Under conditions where the greater flow of gas is through the waste-heat boiler, more fuel is required, as the regenerative saving is definitely reduced.

Figure 8-15 at the right shows yet another possible arrangement. In this system it will be noted that high-temperature gases are used directly in a boiler system before passing to the gas turbine. Under this arrangement less excess air is used in the combustion process so that, for a given burning rate of fuel, the combustion-gas temperature is higher than could safely be employed by the gas turbine. The temperature of the gases is reduced by heat absorption in the boiler passages instead of using a dilution process with excess air to bring about this result. Also indicated in the figure is the possibility of the use of the generated steam as part of the working fluid in the turbine. Various other arrangements have been proposed, with the proposers often applying for patent rights. Many arrangements have merit, and under certain circumstances a heat balance will justify the employment of an unusual arrangement. However, fuel

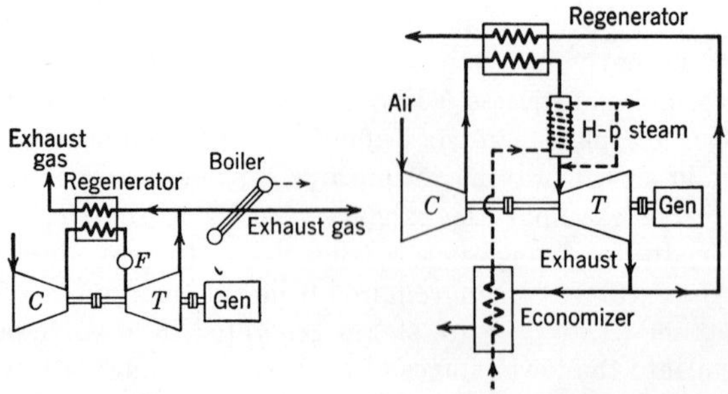

FIG. 8-15. Combination systems employing the open gas-turbine cycle.

saving is often insufficient to compensate for the greater investment costs which accrue because of the additional heat-transfer surface and the numerous control devices required in these systems.

One such arrangement which has some merit envisages the raising of steam in a modified waste-heat boiler and expanding this steam along with the hot gases themselves in some of the turbine stages, thereby contributing additional power to the turbine shaft. Under these conditions it is, of course, desirable to arrange the system so that the temperature of the steam is not appreciably different from the temperature of the gases into which the steam mixes. The preferable arrangement might be to have a separate steam turbine operating and developing power.

8-14. Marine Applications. Ever since the gas turbine received serious consideration as a major power unit, strong interest has been evidenced as to how it might serve for marine propulsion. In Art. 8-2, mention was made of a unit under test for the United States Navy. In Great Britain an extensive program of research is under way leading to the development of marine gas turbines. A marine gas turbine should have a long life, perhaps reaching to 100,000 hr of operation at full power. For applications in fighting craft, a shorter life is acceptable, and a greater stress loading with higher operating temperatures would be envisaged in design.

It is not difficult to design gas-turbine units having higher efficiencies than might be expected in marine steam plants. Although open-cycle gas-turbine units are bulky, it should be possible to arrange the components to fit into the shipboard layout. One favorable factor is the relatively light weight of each of the components involved. The quick response of gas-turbine machinery between starting and full load is in strong contrast to the longer time required by a steam plant. With large-capacity steam units, starting may require hours instead of the minutes needed for the gas turbine. Diesel engines are effective as marine propulsion units. Their high economy is somewhat offset by their weight, and it appears difficult to adapt them when the sizes exceed 10,000 hp. It is reasonable to expect that a gas-turbine unit in the range of 5000 to 10,000 hp would weigh less than a corresponding diesel plant and be comparable with or lower than a marine steam plant in weight. It is possible to design and build gas-turbine plants to realize 30 per cent efficiency based on delivered shaft horsepower. This efficiency can be exceeded by diesel engines but is not attainable with steam plants unless very high steam pressure and high superheat are employed. For large, high-speed ships it is necessary at present to rely on the steam turbine for propulsion to the exclusion of other types of drive. This is particularly true for the range of 30,000 to 200,000 hp.

Many small freighters require in the neighborhood of 6000 to 8000 shaft hp. These powers can be attained readily with diesel engines, with steam turbines, or with open- or closed-cycle gas turbines. Until more data become available, it is impossible to predict which gas-turbine designs will be developed for marine service. It would appear undesirable to operate marine units at excessively high inlet temperatures. Possibly 1350 to 1600 F might represent a top range for present design practice. In regard to type of drive, the turbine could operate through reduction gearing direct to propeller shafting or drive a generator for an electrically driven propeller system. Apparently, the most difficult design problems associated with the marine units are the production of long-life turbines and combustion chambers and the manufacture of regenerative heat-exchanger surface of sufficiently small physical size to fit into the structure of the ship.

The Pratt and Whitney Marine Gas-Turbine Engine, which resembles engines built for aircraft propulsion in many of its design features, is now available for marine use. The unit is capable of producing a maximum continuous horsepower of 25,000 with the normal horsepower at 21,500. It rotates at 3600 rpm and employs 9 stages in the low compressor followed by 6 stages in the high compressor to create a design pressure ratio of approximately 13 to 1. The design airflow through the unit is some 260 lb per sec at 80 F inlet temperature to the compressor. The compressor elements rotate at 8900 rpm for the high-pressure unit and 6600 rpm for the low-pressure unit. The maximum turbine inlet temperatures are 1750 F for short-term intermittent service; 1600 F for peak sustained operation, and 1500 F for base-load continuous operation. The fuel consumption at optimum operating condition is 0.47 lb per hp-hr with liquid fuel of 18,500 Btu per lb lower heating value.

FIG. 8-16. Photograph of 45-hp gas-turbine pump unit. (*Solar Aircraft Co.*)

8-15. Solar Gas-turbine-driven Units. The Solar Aircraft Company, at the request of the United States Navy, developed a gas-turbine-driven pump unit. The complete unit, which is illustrated in Fig. 8-16, has a dry weight (without fuel tank) of 165 lb and occupies a space of essentially 2 by 2 by 2 ft, exclusive of the fuel tank. The turbine with its direct-connected compressor runs at normal full speed of 40,300 rpm and drives a pump through reduction gearing at 4500 rpm. The net rated horsepower output is 45 hp, and the pump itself handles 500 gpm of water with 16 ft suction lift at 100 psig delivery pressure. Diesel fuel or gasoline is burned, and under full load at 80 F ambient temperature 104 lb of fuel per hr is consumed.

Figure 8-17 shows a section through the turbine and compressor. The compressor is a single-stage centrifugal unit with a maximum rotor diameter of 6.97 in. It is machined from a 25 ST aluminum-alloy forging. Blade leading edges are shaped to serve as inducers. A stainless-steel diffuser from the compressor is bolted to the scroll.

A butterfly valve in the scroll outlet can be used to ensure a low air velocity in the combustor to aid ignition during starting. The combustion chamber is built as a right-angled elbow, formed from stainless steel. An 8.3 gal per hr 80° spray angle nozzle admits fuel to the air stream, in the dome section which precedes the flared skirt outlet of the chamber. An aircraft spark plug is used to ignite the mixture during starting. Combustion is largely completed inside the dome area and hot products from this zone mix in with the main air stream below the dome. The combustion chamber, which was developed after extensive experimentation, has been found to give satisfactory operation over a wide range of operating conditions.

A study of Fig. 8-17 will show that the turbine is of unique design, having a one-stage radial-inflow rotor. The rotor, which has an outer diameter of 7.42 in., is encircled by the inlet turbine scroll and nozzle. Hot gases from the combustor pass into the turbine nozzle scroll and are

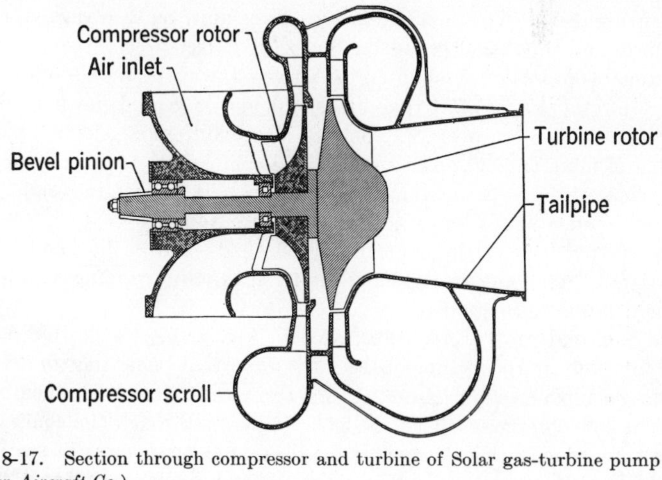

FIG. 8-17. Section through compressor and turbine of Solar gas-turbine pump unit. (*Solar Aircraft Co.*)

directed inward at an appropriate angle by the turbine nozzles. In the turbine blades they turn and leave in an axial direction, passing into a short tail pipe. The turbine rotor is made of a Hastelloy B forging.

It is interesting to note that this unit can be started by hand with one man cranking, although it is desirable to use two men for this purpose. Hand cranking has been found feasible because the unit is small and operates with a relatively low pressure ratio (2.4). To start the unit, the butterfly valve is closed, and cranking is started until the pump speed reaches 500 rpm. Then the throttle is opened, and as the magneto is delivering ignition current to the spark plug, combustion commences and the butterfly valve is opened wide. Cranking is continued, with throttle adjustments being made manually to keep the turbine exhaust temperature within a specified starting range until 1200 rpm pump speed is reached. From this point on the turbine is self-sustaining, and with continued throttle adjustments it will accelerate up to governed speed without further cranking. A maximum design limit of 1300 F was used in connection with operation of the turbine.

Auxiliaries for the unit are driven from the reduction gear at approximately 3300 rpm. These include the fuel pump and a fly-weight-type speed governor mounted on the gear housing. The governor shaft is driven by the pump shaft. Fuel flow to the burner is regulated by the governor so as to hold the turbine speed essentially constant. In addition to the main governor, the unit is provided with a fly-weight-type overspeed trip.

The Solar Division of International Harvester, formerly the Solar Aircraft Company, builds the Solar Titan Engine, shown in diagrammatic form in Fig. 8-18. This is an auxiliary power unit, built in rugged, compact form and capable of producing 80 to 105 horsepower. A single-stage centrifugal compressor is mounted back to back with its radial-inflow turbine. It rotates at 26,700 rpm top speed and is geared down to drive auxiliary systems for purposes such as air conditioning, electric power, starting of main engines on helicopters and commercial airplanes and also to provide auxiliary power for other purposes. It is light in weight and versatile in starting, at temperatures anywhere from −65 F to 130 F. It can be started by electric or pneumatic methods or even by hand. The turbine inlet temperature is 1450 F under maximum running conditions and light liquid fuels are used.

The Solar-Saturn Turbine, illustrated in Fig. 8-19, is a flexible power plant for many purposes, one of the most important being that of driving natural-gas pipeline compressors. In the figure the air can be seen entering the axial-flow compressor from the left. It passes through the combustor to the three-stage turbine, which is rated at 1100 horsepower at sea level with 82 F ambient air. It can produce higher horsepowers when the air

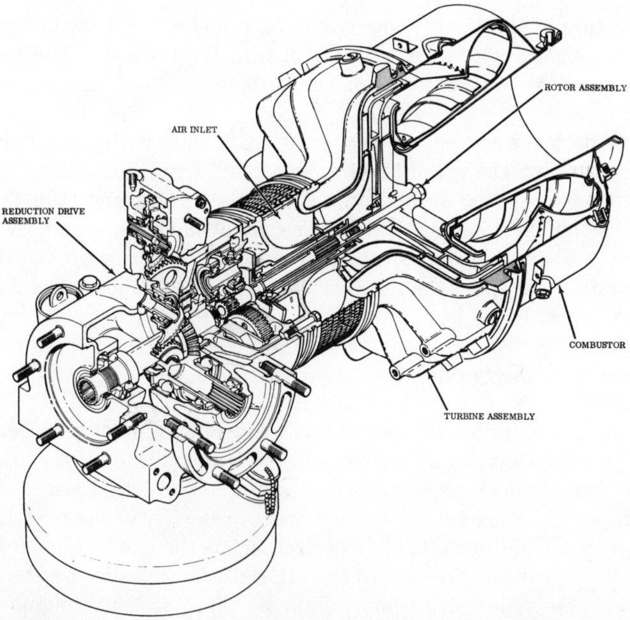

FIG. 8-18. Cutaway view of Solar Titan gas turbine engine. (*Solar Division of International Harvester Co.*)

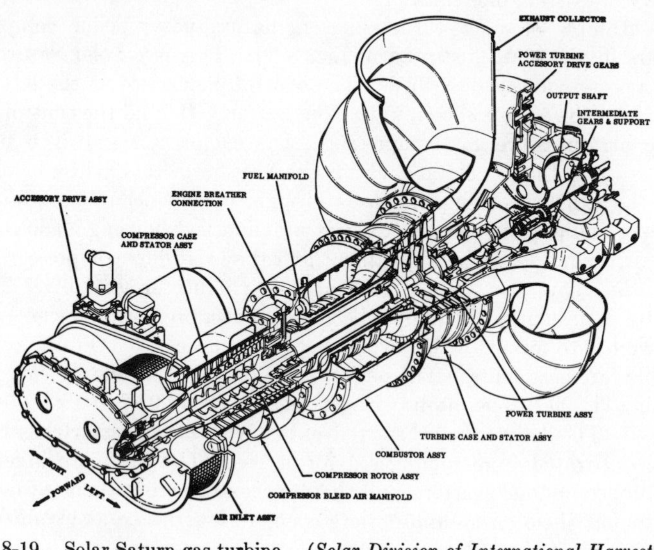

FIG. 8-19. Solar Saturn gas turbine. (*Solar Division of International Harvester Co.*)

temperature is appreciably lower than 82 F and when it is allowed to run at higher than design inlet temperatures to the turbine. The maximum continuous inlet temperature for the turbine is 1450 F and it rotates at 22,300 rpm.

These turbines can be built as single-shaft units with a single common shaft connecting the turbine, compressor and driven member. The two-shaft units are similar in appearance except the first and second turbine stages are on the common compressor shaft while the third stage is independent and connects only to the driven unit. Thus the driven shaft can run at a different speed than the turbine-compressor, and its speed is set in relation to the driven load, and also of course, to the energy provided to it from the gas generator of the main unit. Figure 8-19 shows a two-shaft unit; one can observe how the third turbine stage is separate from the other shaft.

8-16. Gas Turbines in Gas-transmission-line Pumping Service. Indications point to a continuing growth in the United States of the market for natural gas through more extensive use of pipe lines. Some estimates indicate that the market for natural gas will double in the next 10 years. This opens an additional field of utilization, as the gas turbine appears to be ideally suited for pipe-line work. It can economically use the gaseous fuel available, and with small units of high capacity pumping-plant investment costs can be kept to a minimum. Gas-turbine-powered booster pumping stations can be used on new gas lines or to increase the capacity of existing pipe lines.

An example of a gas-turbine-driven, natural-gas pipeline compressor appears in the unit illustrated in Fig. 8-20. This is a Solar Saturn unit of a type previously described. The control panels are at the left front for each of the 5 units shown in the illustration. Behind the control panel is the high-pressure gas compressor. The compressor is driven by the third stage of the gas turbine, all of which is shown covered by insulation with the exhaust passing upward through an expansion joint into the silencing equipment. At the right is a cylindrical housing enclosing the inlet-air, sound-deadening equipment and the axial-flow compressor. Between the compressor and the turbine is the combustor assembly. This is fired by natural gas, which must be supplied at a pressure sufficiently high to permit it to mix with the air leaving the compressor. Oil firing is also possible for these units. The performance of these turbines is very good, reaching 11,500 Btu per hp-hr with gas on L.H.V. and 0.62 lb per hp-hr with fuel oil. The units can be started either with electric starters or with starters operated from compressed air or gas. The units are assembled and shipped on their platform; all that is necessary to put them into service is to mount them on an appropriate concrete foundation and connect the necessary piping and electric controls.

FIG. 8-20. Solar Saturn gas-turbine package units serving in a natural-gas compressor station.

The General Electric Company is also adapting its gas-turbine designs to operate in this field with a 5000-hp turbine driving a centrifugal natural-gas compressor. Figure 8-21 illustrates schematically the layout of a proposed cycle in this connection. The General Electric unit, which is illustrated in Fig. 8-22, employs a 14-stage axial-flow compressor, operating through a pressure compression ratio of 5.5, delivering air to the combustors and to a two-stage high-pressure turbine. The gases from this turbine, at lower temperature and intermediate pressure, flow to the

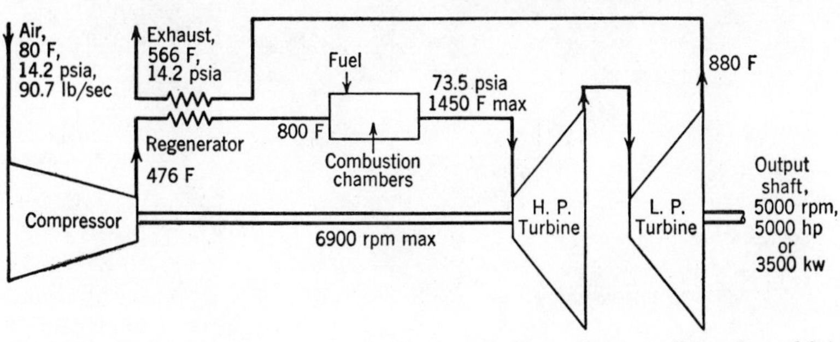

FIG. 8-21. Schematic diagram of 5000-hp regenerative-cycle gas turbine planned by General Electric Company.

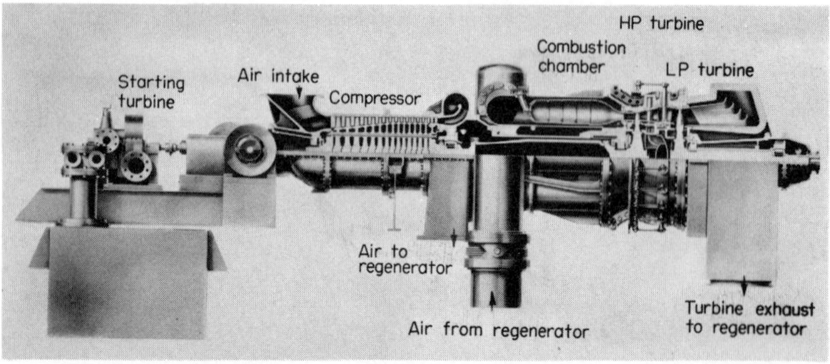

FIG. 8-22. Two-shaft 5000-hp gas turbine with independent load shaft for mechanical drive. (*General Electric Co.*)

single-stage low-pressure turbine which is used to drive the gas-line centrifugal booster compressor. A regenerator is employed, which can be seen in the pictorial layout of the plant in Fig. 8-23. It will be observed that the regenerator, which is cylindrical in form, is arranged for installation outside of the building. The exhaust gases enter at a low point in the

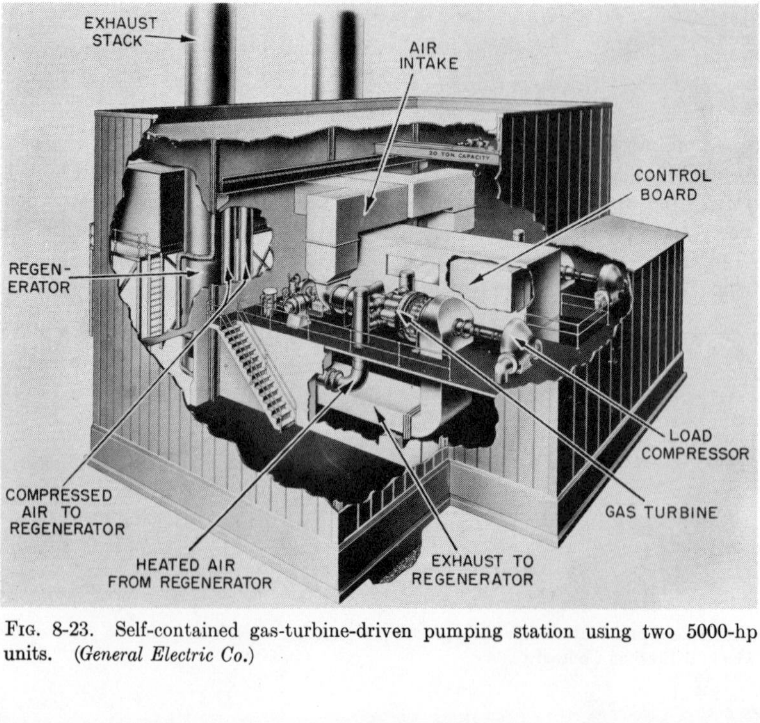

FIG. 8-23. Self-contained gas-turbine-driven pumping station using two 5000-hp units. (*General Electric Co.*)

regenerator and move counterflow to the compressed air, which enters at the side toward the top of the regenerator and leaves through the sides of the regenerator at a low point. Operating data are indicated in Fig. 8-21, but it should be mentioned that the regenerator can raise the temperature of the air to 800 F before entering the six combustion chambers.

The total investment cost for net gas-turbine horsepower is lower than for horsepower supplied by reciprocating engines. Because the plants are cheaper and smaller in over-all size, it appears probable that gas-turbine plants could be better spaced than at the 80- to 100-mile distances now employed for gas lines. For a representative 20,000-hp plant, fewer units would be needed in each station than are required with reciprocating engines, and less labor and attendance would also be required. Another matter of note is that in winter months, when maximum gas capacity is needed, gas-turbine power output also increases from 20 to 25 per cent because of the lower ambient-inlet-air temperatures to the compressor.

8-17. Boeing Gas-turbine Power Plant. The Boeing Aircraft Company has developed a small gas-turbine power plant for use where critical space and weight limitations exist. Figures 8-24 and 8-25 show the schematic arrangement of the turbine components and a photograph of the turbine. The basic power-producing elements (model 500) consist

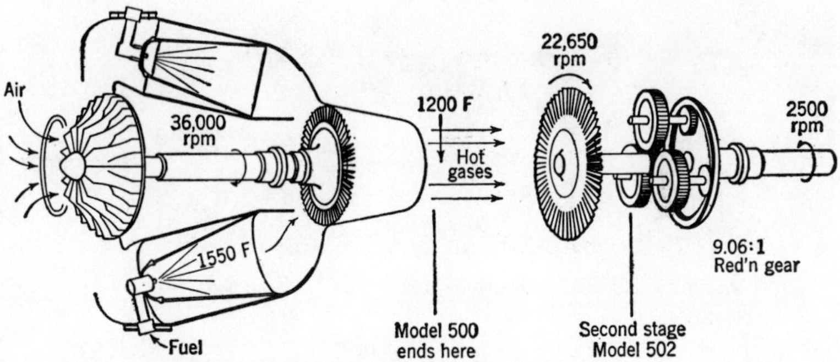

FIG. 8-24. Schematic drawing showing the power-producing and power-output sections of the Boeing gas-turbine engine.

of a centrifugal compressor and a single-stage turbine, which rotate at 36,000 rpm. For a jet unit no excess shaft power over that required for the compressor is produced, and the useful power is a high-velocity jet of hot gas. For producing shaft power, the gas is fed into a second turbine stage. This consists of a single-stage turbine rotating at 22,650 rpm and reduction gearing. Final output shaft speed is 2500 rpm. Pressure ratio across the compressor is approximately 3, and approximately 175 shaft hp is produced. The over-all length of the unit is about 40 in., and the total weight is 240 lb including accessories.

This power plant has been given extensive tests in a large truck, where it replaced a diesel engine which weighed approximately 2000 lb. Although the fuel rate of the turbine is much higher than that of the diesel, there is a possibility that the saving of space and weight which allows the hauling of more cargo may make the turbine a competitive power plant for some hauls. Other applications include the driving of small boats and the generation of electrical power in small auxiliary plants. The primary unit (without the second turbine and gear set) might be used as a jet-propulsion unit for guided missiles.

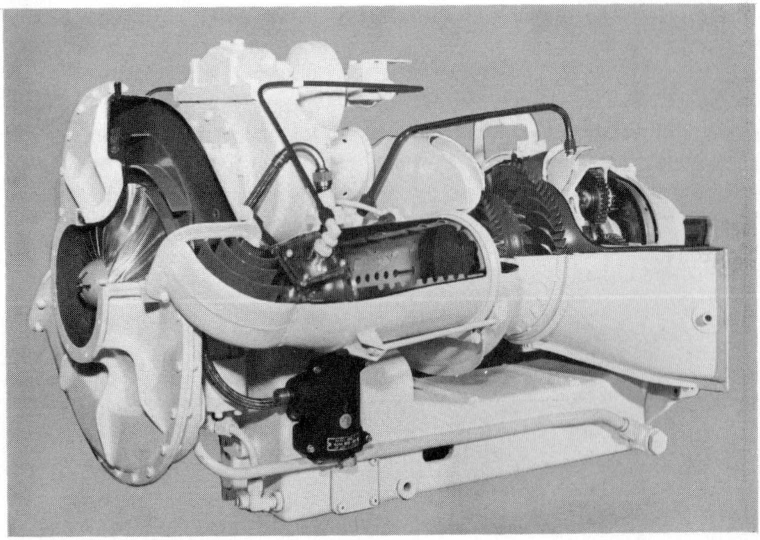

FIG. 8-25. Boeing gas-turbine power unit. (*Boeing Aircraft Co.*)

PROBLEMS

8-1. The 3500-kw Huey Station unit (Art. 8-4) showed the following test data: compressor inlet 14.7 psia at 80 F and 339,000 lb per hr flow; compressor discharge 85.4 psia at 500 F; turbine inlet 78.7 psia at 1290 F, turbine exhaust 14.7 psia at 780 F; fuel used 83,400,000 Btu per hr based on lower heating value, and power delivered 3500 kw. Find (*a*) the thermal efficiency based on lower heating value and (*b*) thermal efficiency based on higher heating value if the ratio LHV/HHV for the fuel is 0.9. (*c*) Find the heat rate on LHV in Btu per kilowatthour. (*d*) Find the air rate in pound per second per net kilowatt. (*e*) Based on compressor inlet and discharge temperatures and flow rate, estimate the compressor horsepower and kilowatts. A mechanical efficiency of 0.97 is assumed. (*f*) Using this and output data, estimate the gross turbine output in horsepower and kilowatts. (*g*) Find the power ratio. *Ans.* (*a*) 14.3; (*b*) 12.9; (*c*) 23,800 Btu per kwhr; (*d*) 0.027 lb per sec-kw; (*e*) 13,980 hp, 10,430 kw; (*f*) 13,930 kw; (*g*) 0.251.

8-2. The regenerator of a gas turbine heats 35 lb per sec of compressed air from 780 to 1100 R. The mean temperature difference Δt acting across the heater surfaces is 200°, and the over-all coefficient of the heat transfer surfaces is $U = 7.0$ Btu/(hr)(ft²)(°F). Find the surface area required in this regenerator to supply the heat requirements. (Note that the simple heat-transfer equation $Q = UA\,\Delta t$ can be applied, where Q is Btu transferred per hour and A is the surface in square feet.) *Ans.* 7120 ft².

8-3. Refer to the data of Fig. 8-21, and assume that the regenerator of this unit operates with a Δt of 80° and the same over-all value of U that applied in the preceding problem. Compute the approximate heat-transfer surface area required for this regenerator. *Ans.* 47,500 ft².

8-4. Making use of the data given in Fig. 8-21, find (*a*) the compressor horsepower absorbed as the 90.7 lb of air per sec increases in temperature from 540 to 936 R. (*b*) For the heat added in the combustor from 1260 to 1910 R, find the fuel-air ratio with No. 6 fuel oil by use of Fig. 9-21. (*c*) Find the exit temperature from the high-pressure turbine. Note that the high-pressure turbine must supply sufficient power to drive the compressor (disregard losses). If the turbine efficiency can be considered 85 per cent, find the exit pressure from the high-pressure turbine. *Ans.* 12,360 hp, $f = 0.0101$, 1558.2 R, 28.0 psia.

8-5. The low-pressure turbine of Fig. 8-21 receives 91.62 lb per sec of hot gases (air) at 1558.2 R and expands the gases to a leaving temperature of 1340 R. (*a*) Compute the horsepower produced. (*b*) Compute the minimum thermal energy required from the fuel. Make use of any data from the preceding problem. (*c*) Allowing for a 5 per cent combustion loss and 5 per cent for thermal losses and the power requirements of plant auxiliaries, find the plant efficiency. This will represent an efficiency based on lower heating value. (*d*) Modify (*c*) to find an efficiency based on higher heating value if for the fuel oil the ratio of LHV/HHV = 0.9. *Ans.* (*a*) 7420 hp, (*b*) 16,070 Btu per sec, (*c*) 29.6%, (*d*) 26.6%.

8-6. Make use of data given in Fig. 8-21, and investigate performance if the regenerator is not used. (*a*) Find the compressor horsepower based on air flow and terminal temperature. (*b*) Find the fuel-air ratio and the energy added to the flow stream in raising its temperature from (476 + 460) R to 1910 R. Use Fig. 9-21. (*c*) Compute the enthalpy and temperature leaving the high-pressure turbine if its total output is absorbed by the compressor. (*d*) On the basis of 85 per cent turbine efficiency, compute the pressure at exit from this turbine.

8-7. The low-pressure turbine of Fig. 8-21 when the regenerator is not used receives 92.1 lb per sec of hot gases at 1560.0 R and expands to a leaving temperature of 1340 R. (*a*) Compute the horsepower produced. (*b*) On the basis of the thermal energy absorbed by the flow stream as found in the preceding problem, compute an ideal thermal efficiency. (*c*) Allowing for a 5 per cent combustion loss and 5 per cent for thermal losses and the power requirements of plant auxiliaries, find the cycle efficiency on a lower-heating-value basis. (*d*) Modify (*c*) to find an efficiency based on higher heating value if for the fuel oil the ratio of LHV/HHV = 0.9.

8-8. *Closed cycle.* The compressor of a closed-cycle system receives air at 100 psia and 80.3 F and compresses it to 400 psia. Leaving the compressor, the air passes through a regenerator with an effectiveness of 60 per cent based on temperatures, through the furnace, and finally enters the compressor turbine at 1800 R and 394 psia. It leaves the turbine at 104 psia, passes through the regenerator and precooler, returning to the compressor at the inlet conditions previously indicated. The compressor and turbine efficiencies can be taken as 84 per cent and 85 per cent, respectively. (*a*) Find the temperature leaving the turbine and regenerator. (*b*) For an air flow of

50 lb per sec, compute the compressor horsepower, the turbine horsepower, and the net horsepower of the system. (c) Find the heat interchange in the regenerator, the energy added from fuel, and the energy rejected at the precooler, and find the over-all internal efficiency of this cycle, assuming that it can be considered adiabatic and disregarding the power required for operation of auxiliaries.

8-9. Work Prob. 8-8, assuming that the inlet temperature to the turbine is changed to 1650 R, with the pressures remaining at the previous values.

8-10. Work Prob. 8-8 if the turbine inlet temperature is changed to 1500 R, with the pressures and other data as before.

REFERENCES

1. Tucker, Rettaliata, and Kottcamp: Experimental 1500 F Unit at Annapolis Yields Valuable High Temperature Data, *Power*, vol. 90, May, 1946.
2. Rettaliata, J. T.: The Gas Turbine, *Allis-Chalmers Elec. Rev.*, September, December, 1941, March, 1942.
3. 3500-HP Gas-turbine Plant, *Mech. Eng.*, vol. 68, pp. 571–574, June, 1946.
4. Howard, A.: Design Features of a 4800-HP Locomotive Gas-turbine Power Plant, *Mech. Eng.*, vol. 70, pp. 301–306, April, 1948.
5. Howard and Buckland: Test of a 4800-HP Gas-turbine Power Plant, *Paper* 48-A-98, *Trans. ASME*, December, 1948.
6. Goldsbury, J.: Gas-turbine Power Plants for Operation with Low-cost Fuel, *Trans. ASME*, vol. 71, pp. 59–64, January, 1949.
7. Rowley and Skrotzki: The Gas Turbine as a Stationary Prime Mover, *Trans. ASME*, vol. 71, pp. 35–41, 1949.
8. Hammond, W. E.: Transmission Systems for Marine Propulsion Gas-turbine Power Plants, *Trans. ASME*, vol. 71, pp. 43–57, 1949.
9. Soderberg, Smith, and Scott: A Marine Gas Turbine Power Plant, *Trans. Soc. Naval Architects and Marine Engrs.*, vol. 53, pp. 949*ff.*, 1945.
10. Tucker, W. B.: Construction of Gas Turbine for Locomotive Power Plant, ASME Gas Turbine Power Division, June 1948.
11. Meyer, A.: Recent Developments in Gas Turbines, *Mech. Eng.*, vol. 69, pp. 273–277, April, 1947.
12. Putz, T. J.: The 2000-hp Gas Turbine on Trial, *Westinghouse Engr.*, vol. 8, pp. 39–41, March, 1948.
13. Commercial American Gas Turbines Begin to Take Shape, *Power*, vol. 93, pp. 114–117, April, 1949.
14. First Worldwide Gas Turbine Survey, *Power*, vol. 93, pp. 95–97, August, 1949.
15. Putz, T. J.: A Compact Locomotive-type Gas Turbine, *Westinghouse Engr.*, vol. 7, pp. 35–39, March, 1947.
16. Putz, T. J.: The Industrial Gas Turbine in Service, *Westinghouse Engr.*, vol. 9, pp. 191–192, November, 1949.
17. Fusner, G. R.: Heat-exchange Equipment for a 5000-kw Gas-turbine Generator, *Mech. Eng.*, vol. 72, pp. 316–320, April, 1950.
18. Schweitzer and Salisbury: Compound Power Plants, *Quart. Trans., SAE*, vol. 3, pp. 657–670, October, 1949.
19. Blake, J. W.: 3500-kw Gas Turbine Raises Station Capability by 6000 Kw, *Power*, vol. 92, September, 1948.
20. First U.S. All-gas-turbine Power Plant—The Edward M. Graham Station, *Power Eng.*, March, 1951.
21. Brown, T. W. F.: Marine Gas-turbine Research in Britain, *Mech. Eng.*, vol. 72, pp. 379–388, May, 1950.

22. Fusner, G. R.: The Gas Turbine with a Waste-heat Boiler, *Mech. Eng.*, vol. 70, pp. 515–518, June, 1948.
23. Peterson and Carlson: Design Features of a 250-KW Gas-turbine Engine, *Mech. Eng.*, vol. 74, pp. 197–201, March, 1952.
24. Goldsbury, J.: Gas-turbine Power Plants for Operation with Low-cost Fuel, *Trans. ASME*, vol. 71, pp. 59–64, 1949.
25. Buckland and Berkey: Design Features of a 5000-HP Gas Turbine, *Paper* 51-A-113, *Trans. ASME*, vol. 74, 1952.
26. Sidler, P. R.: Operating Experiences with Stationary Gas Turbines, *Mech. Eng.*, vol. 74, pp. 381–383, May, 1952.

CHAPTER 9

FUELS, COMBUSTION, AND COMBUSTION CHAMBERS

9-1. The Combustion Process. In heat-engine processes, combustion plays one of the most important roles, as it is through combustion that chemical energy is transformed into a thermal form suitable for the development of work and power. During combustion in the coal-fired steam plant, the burning coal releases energy at high temperature, which is transferred as heat into the water-steam working medium of the system.

In the spark-ignition engine combustion occurs in the engine cylinder when the compressed mixture of gasoline vapor and air, ignited by a spark, undergoes inflammation, with resulting increase in temperature and pressure, and power is produced from the expanding hot gases.

In the diesel engine the combustion process is associated with the burning which takes place when an atomized mist of oil is sprayed into compressed air which is sufficiently hot for ignition and subsequent burning of the oil particles.

In the gas turbine a fuel, usually in liquid form, is injected into a stream of air in a suitable combustion chamber and, after ignition, burns in this stream, usually later mixing with a stream of cooler air which reduces the temperature of the gaseous products to a satisfactory working level.

Each of these combustion processes takes place with extremely high energy releases, as can be seen in Table 9-1. As the number of reacting molecules during combustion increases with pressure, it is customary to report heat releases in terms of atmospheres of pressure whenever combustion processes are carried out at pressures in excess of atmospheric.

Although solid fuels are being considered for gas turbines, essentially all internal-combustion engines and turbines utilize liquid fuels except for gas engines and so-called "dual-fuel" engines. Most of the chemical energy in ordinary fuels exists as combustible carbon and hydrogen, and a significant ratio in connection with liquid and gaseous fuels is the H/C (hydrogen-carbon) ratio. In the usual petroleum fuels, this ratio by weight ranges from about 0.10 to 0.22, with H/C for heavy fuel oil about 0.15 and H/C about 0.18 for gasoline. The carbon in the fuel normally burns to carbon dioxide and remains in gaseous form. Hydrogen, on the other hand, burns to water vapor (steam), and although in combustion processes it usually remains in gaseous or vapor form, this is not necessarily so and, in condensing, an amount of energy equivalent to the latent heat of condensation of the water vapor can be supplied to the system from condensation. Because it is possible for the water vapor in the combustion products to condense or not to condense, there have come into common usage the terms higher heating value and lower heating value, the difference between the two heating values being the latent heat associated with the water vapor in the products resulting from combustion of the hydrogen.

TABLE 9-1. REPRESENTATIVE HEAT RELEASES IN BTU PER HOUR FOR EACH CUBIC FOOT OF COMBUSTION SPACE

Power boiler, pulverized coal	40,000–80,000
Oil-fired	45,000–250,000
Aircraft-engine cylinder	7,000,000–15,000,000 or 600,000–1,200,000 per atm
Diesel-engine cylinder	1,000,000–4,200,000 or 30,000–150,000 per atm
Gas-turbine combustion chamber	500,000–3,000,000 or 125,000–750,000 per atm

In nearly all combustion-engine processes, the gaseous products remain at high temperatures so greatly in excess of the dew-point (condensation) temperature that the higher heating value of the fuel is seldom, if ever, released and it would be more logical to compare internal-combustion-engine performance on lower heating value than on higher heating value. However, much of the past thinking has been so closely associated with higher heating value that this terminology is still frequently employed in reporting internal-combustion-engine performance.

In the following pages certain fundamental combustion calculations will be developed. These will be followed by data on fuels and by detailed combustion problems associated with gas-turbine combustors.

9-2. Combustion Calculations. In addition to carbon and hydrogen, some fuels contain a small amount of sulfur, which contributes slightly to the heating value of the fuel. The air theoretically required to burn a unit weight of a combustible element can be readily calculated by the use of an appropriate chemical equation.

$$C + O_2 \rightarrow CO_2$$
$$12 + 32 = 44$$

Here, 12 lb of carbon require 32 lb of oxygen for complete combustion and yield 44 lb of carbon dioxide, where 12 and 16 are the atomic weights of carbon and oxygen, respectively. Using the more accurate atomic weight of carbon as 12.01, it is evident that 1 lb of C requires

$$\frac{32}{12.01} = 2.66 \text{ lb } O_2$$

As the weight fractions of oxygen and inerts in air are, respectively, 0.2319 and 0.7681, it is obvious that 1 lb of C requires

$$\frac{2.66}{0.2319} = 11.5 \text{ lb air for complete combustion}$$

Similarly,

$$2H_2 + O_2 \rightarrow 2H_2O$$
$$4 + 32 = 36$$

1 lb H_2 requires

$$\frac{32}{4.03} = 7.94 \text{ lb } O_2 \approx 8 \text{ lb } O_2$$

1 lb H_2 requires

$$\frac{7.94}{0.2319} = 34.2 \text{ lb air for complete combustion}$$

By like calculation, 1 lb of sulfur (S) requires 4.3 lb of air for complete combustion to sulfur dioxide (SO_2).

The hydrocarbon fuels contain no oxygen, but coal and the alcohols do. This oxygen in the fuel is already chemically combined with some of the combustible or can readily go into combination. Consequently allowance must be made for this oxygen in computing the weight of air required to burn a unit weight of fuel. The allowance is often made by considering that each pound of this oxygen can combine with ⅛ lb of hydrogen.

Let C, H, O, and S represent the proportional parts by weight of the carbon, hydrogen, oxygen, and sulfur in a fuel. The weight of air A theoretically required to burn 1 lb of the fuel is

$$A = 11.5\text{C} + 34.2\left(\text{H} - \frac{\text{O}}{8}\right) + 4.3\text{S} \tag{9-1}$$

Combustion can hardly ever be carried on with the theoretical amount of air, but a certain amount in excess of that theoretically required must be used to ensure complete combustion in a short time and prevent carbon deposits. One of the ordinary uses for combustion calculations is the finding from an analysis of the waste gases what combustion conditions actually exist. Flue- and exhaust-gas analyses are usually given in volumetric percentages of the so-called "dry" flue gases, CO_2, O_2, CO, and N_2. Generally the most accurate method of making combustion calculations from flue-gas analyses is through a carbon balance.

The volume of any gas multiplied by its density gives the weight of that gas. Since the density of a gas is proportional to its molecular weight, the product of gas volume and molecular weight is also proportional to the weight of a gas. In a gas mixture if the volume of each individual gas is multiplied by its respective molecular weight, the sum of the resulting products is proportional to the weight of the gas mixture. Let CO_2, O_2, CO, N_2 be the percentages or parts by volume of these respective gases in the dry waste gas. The proportional weights of these respective gases and of the mixture are

$$44CO_2 + 32O_2 + 28CO + 28N_2$$

In carbon dioxide (CO_2) atomic weights show that $^{12}/_{44}$ of the total weight is carbon; or the proportional weight of the carbon in the CO_2 is

$$^{12}/_{44} \times 44CO_2, \text{ that is, } 12CO_2$$

Similarly the weight of the carbon in carbon monoxide (CO) is

$$^{12}/_{28} \times 28CO, \text{ that is, } 12CO$$

The weight of the dry waste gases per weight of carbon existing in them must be

$$\frac{\text{Total weight of the gases}}{\text{Weight of carbon in the gases}} = \frac{44CO_2 + 32O_2 + 28CO + 28N_2}{12CO_2 + 12CO}$$

In other words, the pounds of dry waste gases per pound of carbon burned, G_w, is

$$G_w = \frac{44CO_2 + 32O_2 + 28CO + 28N_2}{12CO_2 + 12CO} = \frac{11CO_2 + 8O_2 + 7(CO + N_2)}{3(CO_2 + CO)} \tag{9-2}$$

Equation (9-2) should not be considered a formula. It is simply a statement of the weight of a certain gas mixture divided by the weight of carbon in the mixture. The equation is set up in this fractional form instead of in vertical tabular form so that advantage can be taken of cancellation to reduce arithmetical work.

To find the pounds of dry waste gas per pound of fuel burned, G_w, from Eq. (9-2), must be multiplied by C_g, where C_g is the weight of carbon gasified or burned per pound of fuel. C_g must be determined from test data and the fuel analysis. In the case of liquid hydrocarbon fuels or gaseous fuels, essentially all the carbon in the fuel becomes gasified or is burned unless extremely heavy smoke is present. This is not usually true for solid fuels because some carbon nearly always appears unburned in the refuse.

The weight of air used per pound of fuel and the combustible burned per pound of fuel appear in the waste gases either as *dry flue gas* or as *water vapor*. This mass balance expressed in equational form appears as

Air used + combustible burned = dry flue gas + water vapor

Air used per lb fuel + combustible burned per lb fuel = $G_w \times C_g + 9H$

The combustible burned per pound of fuel consists of all the fuel which entered into combustion. This combustible is the whole original pound of fuel less the ash and moisture in the fuel and also that portion of the combustible matter in solid fuels which remains unburned and appears in the refuse. The term $G_w \times C_g$ has been explained above. For each pound of hydrogen burned 9 lb of water vapor results; thus 9H gives an expression for the weight of water vapor resulting from combustion per pound of fuel (H being the weight fraction of hydrogen in the combustible per pound of fuel). Thus

$$\text{Dry air actually used per lb of fuel} = W_a$$
$$= G_w \times C_g + 9\text{H} - \left(1 - \frac{\%\text{ ash} + \%\text{ moisture}}{100} - \frac{\text{combustible in refuse}}{\text{lb fuel}}\right) \tag{9-3}$$

In the case of a hydrocarbon liquid fuel, this equation usually simplifies to

$$W_a = G_w \times C_g + 9\text{H} - \left(1 - \frac{\%\text{ moisture}}{100}\right) \tag{9-4}$$

The following method of calculating the weight of dry air used per pound of fuel can be used *if the fuel contains a negligible amount of nitrogen.* In the waste gas per pound of fuel the weight of atmospheric nitrogen (containing its associated argon and other inerts) is

$$\left(\frac{\text{lb N}_2\text{ in gas}}{\text{lb C in gas}}\right)\left(\frac{\text{lb C gasified}}{\text{lb fuel}}\right) = \frac{28.16\text{N}_2}{12(\text{CO}_2 + \text{CO})}\,\text{C}_g$$
$$= \frac{7.04\text{N}_2\text{C}_g}{3(\text{CO}_2 + \text{CO})} \quad \text{lb}$$

All this nitrogen must have come from the air (76.8% N_2:23.2% O_2 by weight)* as none was assumed in the fuel.

$$\text{Dry air actually used per lb of fuel} = \frac{7.04\text{N}_2 \times \text{C}_g}{3(\text{CO}_2 + \text{CO})(0.768)} \quad \text{lb} \tag{9-5}$$

Here, as before, N_2, CO_2, and CO are symbols representing relative *volumes* (or moles) of gaseous products in the gas mixture, while C_g is the *weight* of carbon burned per pound of fuel.

The *per cent of excess air* used is evidently

$$\frac{\text{Air actually used} - \text{air theoretically required}}{\text{Air theoretically required}} \times 100 \tag{9-6}$$

The weight of any individual dry waste gas per pound of fuel can be found following the reasoning used to develop Eq. (9-2).

* Note that the ratio of N_2 to O_2 in air by volume is 3.76.

$$\text{Weight of } CO_2 \text{ per lb fuel} = \frac{11CO_2}{3(CO_2 + CO)} C_g \text{ lb} \tag{9-7}$$

$$\text{Weight of } N_2 \text{ per lb fuel} = \frac{7.04N_2}{3(CO_2 + CO)} C_g \text{ lb} \tag{9-8}$$

$$\text{Weight of } O_2 \text{ per lb fuel} = \frac{8O_2}{3(CO_2 + CO)} C_g \text{ lb} \tag{9-9}$$

$$\text{Weight of } CO \text{ per lb fuel} = \frac{7CO}{3(CO_2 + CO)} C_g \text{ lb} \tag{9-10}$$

It should again be emphasized that in all the preceding expressions the gas symbols (CO_2, N_2, O_2, CO) are percentages or proportional parts *by volume* and represent gas-analysis values such as might be obtained by use of an Orsat or by a Bureau of Mines apparatus.

When liquid hydrocarbon fuels burn, it may happen that certain combustible gases stabilize and combustion is not carried to ultimate completion. Hydrogen (H_2) and methane (CH_4) are two gases commonly formed in this way as well as CO, which has already been discussed. In this case, when an equation of the form of (9-2) is written, the additional terms modify the equation to appear as follows:

$$G_w = \frac{11CO_2 + 8O_2 + 7(CO + N_2) + 4CH_4 + 0.5H_2}{3(CO_2 + CO + CH_4)} \tag{9-11}$$

Example 1. The product gases resulting from the combustion of a hydrocarbon fuel (C = 0.851, H = 0.149) showed that the dry flue gas had a composition of 2.8% CO_2, 17.0% O_2, and 80.2% N_2 by volume. Find the weight of product gases per pound of fuel and the weight of air used in burning each pound of fuel.

Solution. Use Eq. (9-2) to find G_w.

$$G_w = \frac{(11)(2.8) + (8)(17.0) + 0 + (7)(80.16)}{(3)(2.8)}$$

$$= 87.1 \text{ lb dry product gases per lb of carbon burned}$$

$$\text{Product gases} = G_w C_g + 9H$$

$$= (87.1)(0.851) + (9)(0.149) = 75.46 \text{ lb product gases per lb of fuel}$$

Use Eq. (9-4) to find W_a.

$$W_a = G_w C_g + 9H - \left(1 - \frac{\% \text{ moisture}}{100}\right) = 75.46 - 1 - 0$$

$$= 74.46 \text{ lb of air per lb of fuel}$$

The answer can also be found by use of Eq. (9-5), as there is no nitrogen in the fuel.

$$W_a = \frac{(7.04)(80.2)(0.851)}{(3)(2.8)(0.768)} = 74.4 \text{ lb air per lb of fuel}$$

Example 2. Develop the combustion relationships for a light hydrocarbon fuel resembling kerosene or No. 1 fuel oil. This fuel has a weight composition, exclusive of impurities, of 85.1% carbon and 14.9% hydrogen with a ratio of H/C = 0.176. The fuel can be represented closely by the chemical formula $C_{18}H_{38}$. Its higher heating value is 19,600 Btu per lb, and its lower heating value 18,250 Btu per lb (133,700 and 124,500 Btu per gal, respectively), and it weighs 6.82 lb per gal.

Solution. The combustion equation for the theoretical air quantity appears, if use is made of the molal method of computation, as

$$C_{18}H_{38} + 27.5O_2 + (3.76)(27.5)N_2 \rightarrow 18CO_2 + 19H_2O + (3.76)(27.5)N_2$$

Let us assume, however, that five times as much air is used as theoretically required, that is, (5) $(27.5)O_2 = 137.5O_2$ moles of O_2 are used per mole of fuel instead of 27.5 moles. Then

$$C_{18}H_{38} + 137.5O_2 + (3.76)(137.5)N_2 \rightarrow 18CO_2 + 19H_2O + 110O_2 + (3.76)(137.5)N_2$$

The constituents consist of

$$1 \text{ mole of fuel} + 137.5 \text{ moles of } O_2 + 517 \text{ moles } N_2$$
$$\rightarrow 18 \text{ moles } CO_2 + 19 \text{ moles } H_2O + 110 \text{ moles } O_2 + 517 \text{ moles } N_2$$

In terms of weights the equation appears as

$$254 \text{ lb } C_{18}H_{38} + 4400 \text{ lb } O_2 + 14{,}559 \text{ lb } N_2$$
$$\rightarrow 792 \text{ lb } CO_2 + 342 \text{ lb } H_2O + 3520 \text{ lb } O_2 + 14{,}559 \text{ lb } N_2 \rightarrow 19{,}213 \text{ lb total}$$

Note that *atmospheric nitrogen* contains argon along with other rare gases, and its resultant equivalent molecular weight is 28.16. This value was used in finding the weight of the nitrogen in the above equation.

The air-fuel ratio is $(4400 + 14{,}559)/254 = 74.6$, or 74.6 lb to 1, or, in moles, 137.5 moles O_2 and 517 moles N_2 to 1 mole of fuel.

The use of moles, as in the foregoing example, is often a desirable method to use in dealing with the products. The molal concentration X_i is directly proportional to partial pressures and to volumetric percentages, if the products are gaseous. If, for example, the gaseous products exist under a total pressure of 60 psi, we have the values shown in the following tabulation.

Moles, n_i	$X_i = \frac{n_i}{n_T}$	$p_i = p_T X_i = 60X_i$
$18CO_2$	0.0271	1.63
$19H_2O$	0.0286	1.72
$110O_2$	0.1657	9.94
$517N_2$	0.7786	46.71
664 total $= n_T$	1.0000	60.00 psi total

Sometimes it is convenient to combine the diatomic gases, O_2 and N_2, and treat them as a single gas resembling air. For this case there are, for n_D, 627 moles of diatomic products per mole of fuel. At times it might be convenient to arrange the O_2 and N_2 in the proper ratio to constitute air and carry the remaining N_2 as nitrogen along with the CO_2 and H_2O in the products.

In the preceding tabulation the X_i column gives the volumetric percentages of the combustion products when these are all in the gaseous state. However, when combustion gases are analyzed to determine their composition, they are cooled to atmospheric temperature and in this process the water vapor (H_2O) largely condenses. Thus it is common to report analyses on a dry basis. To convert the X_i column to a dry basis, divide each item by $1 - X_{iH_2O}$, giving

$$\begin{aligned} 0.0271/(1 - 0.0286) &= 0.0279 \text{ for the } CO_2 \\ 0.1657/(1 - 0.0286) &= 0.1716 \text{ for the } O_2 \\ 0.7786/(1 - 0.0286) &= \underline{0.8015} \text{ for the } N_2 \\ \text{Total} &= 1.0000 \end{aligned}$$

Example 3. The combustion process of Example 2 using excess air was worked out accenting the use of moles. It seems advisable to analyze the same process in an alternative manner.

Solution. The representative fuel oil previously considered consisted of 0.851 part of carbon to 0.149 part of hydrogen. Use of Eq. (9-1) shows that the air required for theoretical combustion is

$$A = (11.5)(0.851) + (34.2)(0.149) = 14.88 \text{ lb air per lb of fuel}$$

One pound of carbon burned yields 3.66 lb of CO_2, and 1 lb of hydrogen yields 8.94 lb of water vapor (usually rounded off to 9).

The weights of products resulting from combustion of 1 lb of this fuel if five times the theoretical air is used are

$$\begin{aligned} CO_2,\ 3.66 \times 0.851 &= 3.11 \\ H_2O,\ 8.94 \times 0.149 &= 1.33 \\ N_2,\ 14.88 \times 0.7681 &= 11.43 \\ \text{Air},\ 14.88 \times 4 &= \underline{59.52} \\ \text{Total} &= 75.39 \end{aligned}$$

Thus 75.39 lb of products per lb of fuel appears, and 74.39 lb of air per lb of fuel is used. These results differ slightly from those found in Example 2 (75.6 and 74.6) because the fractional percentages of C and H in the fuel were not carried to a fourth decimal place.

9-3. Petroleum Compounds. With a few minor exceptions, practically all the liquid fuels used at the present time come from the petroleum which is found in underground pools at many points on the globe. Although there is a great similarity between the crude petroleums found in various sections of the world, each type of crude petroleum has distinguishing characteristics. All of them are basically hydrocarbons; *i.e.*, the molecules consist of carbon and hydrogen, but some crudes have other elements in noticeable amounts. Russian naphthene-base crudes, for example, frequently contain oxygen in significant amounts. Nitrogen may also appear in naphthene crudes. Sulfur is present in crudes from many localities and represents an objectionable element which must be largely removed during the refining process. It is fortunate that in the paraffin-base oils of the eastern United States very small amounts of sulfur occur.

Petroleum Classification. It is customary to classify petroleum crudes into basic families or series, the usual classifications being (1) the paraffin

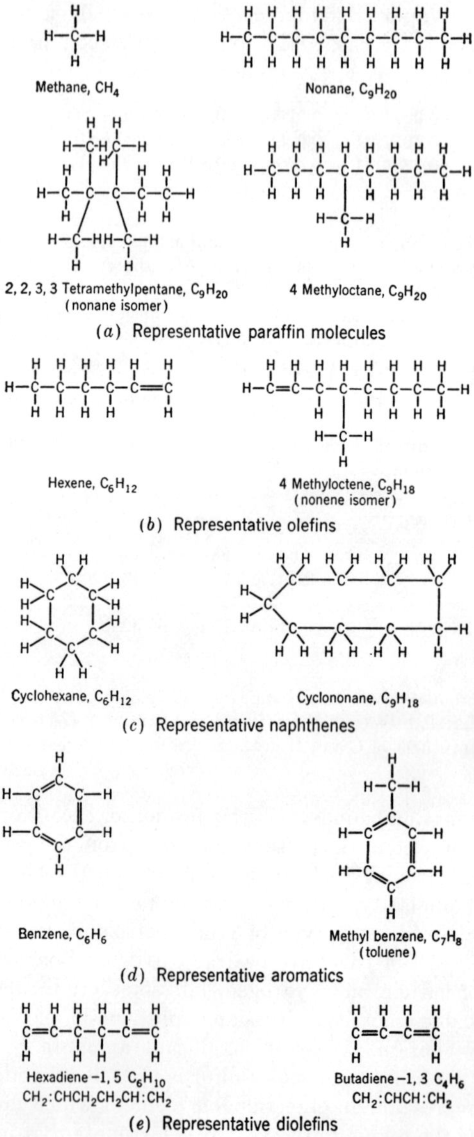

FIG. 9-1. Structural formulas of hydrocarbons.

series, (2) the olefin series, (3) the naphthene series, and (4) the aromatic series. Sometimes (5) the diolefin series is added as a separate series, but frequently this is grouped with the olefin series.

1. *Paraffin series.* This group is characterized by the type formula C_nH_{2n+2}, and the compounds of the series are very stable. Compounds of this series have names which usually end with the suffix "-ane." The first of the series is methane, with the formula CH_4. Nonane, as another example, has the formula C_9H_{20}. A structural diagram of nonane appears in Fig. 9-1, and it can be seen that the structure is that of a relatively long chain. The series ranges from a gaseous methane to the solid paraffin, which has a formula of $C_{20}H_{42}$. In the higher compounds, branches as well as straight chains appear.

In the paraffin series there are many isomers, an isomer being a compound having the same total composition of atoms per molecule but differing both in structural arrangement and in physical properties from other compounds with the same atomic aggregation. One such isomer is shown in Fig. 9-1 for nonane, showing one side branch and known under the specific name of 4-methyloctane. Thirty-five such isomers are known for nonane. With hexadecane (16 carbon atoms) there are theoretically 10,359 isomers. The name for the previously mentioned nonane isomer can be explained in this way: on the fourth carbon atom of the chain there occurs a methyl group, and as this has taken one of the carbon atoms out of the chain, the chain now has but eight atoms and takes the name octane, which is associated with eight.

A more complicated isomer of nonane called 2,2,3,3-tetramethylpentane is also shown in Fig. 9-1. Here it will be noticed that there are pairs of methyl branches on the second carbon atom, giving rise to the "2,2-methyl" part of the name, while the methyl branches on the third carbon atom give rise to the "3,3-methyl" part of the name. The "-pentane" suffix indicates the five basic carbon atoms in the chain, while the "tetra" associated with "four" indicates the total of four methyl groups attached to the basic chain.

Because the Greek and Latin roots and the earlier chemistry prefixes used in naming compounds are not too familiar, it might be well to recall in Table 9-3 the more common suffixes and prefixes used in naming the paraffins and other hydrocarbon groups. Table 9-2 shows the names and characteristics of a large number of hydrocarbon compounds.

2. The *olefin*, or *ethylene*, *series*, with a type formula C_nH_{2n}, consists of unsaturated hydrocarbons. With unsaturated hydrocarbons one or more carbon atoms in the molecule are capable of uniting with other atoms or radicals such as additional hydrogen, chlorine, acid radicals, etc. As the usual valence of carbon is 4, the structural formula of a typical olefin, normal hexene (C_6H_{12}) in Fig. 9-1, shows double bonds connecting two of

TABLE 9-2. HYDROCARBON PROPERTIES*

Hydrocarbon	Formula	Mol. wt	Boiling temp at 14.7 psi	Sp gr liquid at 60 F/60 F	Critical		Sp ht, Btu/(lb)(°F)		$c_p/c_v = k$ for vapor at 60 F	Vapor press., psia at 70 F	Latent heat of vaporization, Btu per lb at 14.7 psia	Higher heat value, Btu per lb at 77 F
					Temp, °F	Press., psia	Liquid at boiling point or at 60 F	Vapor at 60 F				
Paraffins:												
Methane	CH_4	16.04	−258.7	.....	−115.8	673	0.818†	0.526	1.31	....	219.2	23,861‡
Ethane	C_2H_6	30.07	−127.5	0.377§	90.1	708	0.569†	0.409	1.19	559.6	210.2	22,304‡
Propane	C_3H_8	44.09	− 43.7	0.508§	206.2	617	0.69†	0.388	1.13	124.8	183.1	21,646‡
2-Methylpropane (isobutane)	C_4H_{10}	58.12	+ 10.9	0.563§	274.6	528	0.534†	0.387	1.10	45.1	157.5	21,242‡
n-Butane	C_4H_{10}	58.12	31.1	0.584§	305.6	551	0.538†	0.397	1.09	31.2	165.6	21,293‡
2,2-Dimethylpropane (neopentane)	C_5H_{12}	72.15	49.1	0.595§	363	485	0.543†	0.391	1.08	21.9	135.6	20,824
2-Methylbutane (isopentane)	C_5H_{12}	72.15	82.1	0.625	370.0	484	0.562	0.388	1.08	11.6	145.7	20,877
n-Pentane	C_5H_{12}	72.15	96.9	0.631	387.0	485	0.557	0.397	1.07	8.6	153.6	20,914
2,2-Dimethylbutane (neohexane)	C_6H_{14}	86.17	121.5	0.654	422	441	0.511	0.398	1.06	5.3	132.7	20,698
2-Methylpentane	C_6H_{14}	86.17	140.5	0.658	442	...	0.552	0.398	1.06	3.5	138.3	20,743
n-Hexane	C_6H_{14}	86.17	155.7	0.664	454.5	440	0.536	0.398	1.06	2.5	144.8	20,771
2,2-Dimethylpentane	C_7H_{16}	100.2	174.6	0.678	477.9	417	0.518	0.399	1.05	1.7	125.3	20,607
2,2,3-Trimethylbutane (triptane)	C_7H_{16}	100.20	177.57	0.694	496.9	436	0.497	0.399	1.05	1.67	124.6	20,614
3,3-Dimethylpentane	C_7H_{16}	100.20	186.93	0.698	479	...	0.505	0.399	1.05	1.34	126.9	20,625

2-Methylhexane	C_7H_{16}	100.20	194.09	0.683	496.2	400	0.523	0.399	1.05	1.06	132.0	20,645
n-Heptane	C_7H_{16}	100.20	209.17	0.688	512.6	397	0.525	0.399	1.05	0.73	137.5	20,668
2,2,4-Trimethylpentane (isooctane)	C_8H_{18}	114.22	210.63	0.696	520.0	373	0.489	0.400	1.05	0.79	116.7	20,556
2,5-Dimethylhexane (diisobutyl)	C_8H_{18}	114.22	228.39	0.698	530.2	361	.....	0.400	1.05	0.48	123.5	20,551
2,3-Dimethylhexane	C_8H_{18}	114.22	240.10	0.717	....	...	.....	0.400	....	....	126.3	20,581
2-Methylheptane	C_8H_{18}	114.22	243.77	0.702	....	...	.....	0.400	....	....	126.5	20,572
4-Methylheptane	C_8H_{18}	114.22	243.88	0.709	....	...	.....	0.400	....	....	127.6	20,584
3-Ethylhexane	C_8H_{18}	114.22	245.37	0.718	....	...	.....	0.400	....	....	129.0	20,589
n-Octane	C_8H_{18}	114.22	258.20	0.707	565.2	362	0.526	0.400	1.05	0.22	131.7	20,591
2,2,3,4-Tetramethylpentane	C_9H_{20}	128.25	271.42	0.743	....	...	.....	.....	....	....	114.9	
2-Methyloctane	C_9H_{20}	128.25	289.87	0.718	....	...	.....	.....	....	....	122.9	
n-Nonane	C_9H_{20}	128.25	303.44	0.722	613	335	0.523	0.400	1.04	0.08	126.7	20,531
n-Decane	$C_{10}H_{22}$	142.28	345.31	0.734	655.3	312	0.523	0.400	1.04	0.025	122.4	20,483
n-Undecane	$C_{11}H_{24}$	156.30	384.4	0.745	697	292	0.524	0.400	1.03	0.008	118.6	20,443
n-Dodecane	$C_{12}H_{26}$	170.33	421.27	0.753	735.1	272	0.521	0.400	1.03	0.003	115.3	20,410
Olefins:												
Ethene (ethylene)	C_2H_4	28.05	−154.7	.....	49.8	742	.....	0.361	1.24	....	207.6	21,626‡
Propene (propylene)	C_3H_6	42.08	− 53.9	0.521§	196.5	667	0.519	0.360	1.15	151.6	188.2	21,029‡
Butene-1	C_4H_8	56.1	20.7	0.601§	297	588	0.507	0.371	1.11	38.2	167.9	20,830‡
Pentene-1	C_5H_{10}	70.1	85.9	0.646	394	594	.....	0.380	1.08	10.6	.....	20,560
Hexene-1	C_6H_{12}	84.16	146.4	0.678	470	...	0.499	0.383	1.07	3.4	148.6	20,450
Heptene-1	C_7H_{14}	98.18	199.9	0.701	....	...	0.518	0.385	1.06	0.97	148.9	20,380
Octene-1	C_8H_{16}	112.21	250.3	0.720	581	...	0.486	0.386	1.05	0.24	129.1	20,350
Diolefins:												
Butadiene-1,3	C_4H_6	54.09	24.1	0.627§	305.6	627	0.510	0.349	1.12	36.6	178.7	20,217‡
Pentadiene-1,2	C_5H_8	68.11	112.8	0.697								
Naphthenes:												
Cyclopentane	C_5H_{10}	70.13	120.7	0.750	....	...	0.429	.....	.....	5.25	167.5	20,112

TABLE 9-2. HYDROCARBON PROPERTIES*—(*Continued*)

Hydrocarbon	Formula	Mol. wt	Boiling temp at 14.7 psi	Sp gr liquid at 60 F/60 F	Critical		Sp ht, Btu/(lb)(°F)		$c_p/c_v = k$ for vapor at 60 F	Vapor press., psia at 70 F	Latent heat of vaporization, Btu per lb at 14.7 psia	Higher heat value, Btu per lb at 77 F
					Temp, °F	Press., psia	Liquid at boiling point or at 60 F	Vapor at 60 F				
Cyclohexane	C_6H_{12}	84.16	177.3	0.783	538.1	594	0.440	0.421	1.06	1.58	153.9	20,015
Methylcyclohexane	C_7H_{14}	98.18	213.7	0.774	575	...	0.443	0.445	1.05	0.74	138.3	19,978
Ethylcyclohexane	C_8H_{16}	112.21	269.2	0.792	......	...	.....	.....	....	0.20	132.9	19,954
Aromatics:												
Benzene	C_6H_6	78.11	176.2	0.885	551.3	704	0.411	0.342	1.08	1.53	169.3	17,986
Methylbenzene (toluene)	C_7H_8	92.13	231.1	0.872	609.1	611	0.404	0.367	1.06	0.45	156.2	18,245
1,4-Dimethylbenzene (*p*-xylene)	C_8H_{10}	106.16	281.0	0.866	652	514	0.407	.....	.....	0.13	146.1	18,438
n-Propylbenzene	C_9H_{12}	120.19	318.6	0.867	690	475	0.445	0.269	1.07	0.05	136.8	18,667

* These characteristic properties are largely taken from "Hydrocarbons," 3d ed., Phillips Petroleum Company, Bartlesville, Okla., 1949, by permission.

† Measured at boiling-point temperature (14.7 psia).

‡ Higher heating value from gaseous state.

§ Specific gravity at saturation temperature.

the carbon atoms, and combination with other elements could occur there. The olefins have names which end with the suffix "-ene." These compounds are more active than the paraffins, and as they readily react with the acids, they can be removed by acid treatment from crude oils.

3. The *naphthene series* has the same type formula C_nH_{2n} as the olefins. The naphthenes are ring compounds and so for the same generic formula are saturated because the ring is closed. This is illustrated by cyclohexane, which is pictured along with cyclononane in Fig. 9-1. It will be noticed that the double bonds indicating an unsaturated condition are not present here. The prefix "cyclo-" indicates the ring structure. This results in two less hydrogen atoms for the saturated compound than are needed in the corresponding paraffin molecule with the same number of carbon atoms. As is true of the paraffins, many variations in the structure of the naphthenes with the same number of carbon and hydrogen atoms per molecule can occur. Each such isomer will have different

TABLE 9-3. NUMERICAL NOMENCLATURE FOR CHEMICAL COMPOUNDS

1. un- (also meth-)	9. non-
2. bi- or di- (also eth-)	10. dec-
3. tri- (also prop-)	11. undeca-
4. tetra- or quat- (also but-)	12. dodeca-
5. pent-	13. trideca-
6. hex-	14. tetradeca-
7. hep-	15. pentadeca-
8. oct-	16. hexadeca-

characteristics and properties from the other isomers. Naphthene molecules are very common in the diesel-fuel fractions of crude oils, particularly in the Gulf Coast and Southern California crudes, which are highly naphthenic.

4. The *aromatic series*, sometimes called the *benzene series*, has the type formula C_nH_{2n-6}. The aromatics are not saturated and are chemically extremely active, prone to oxidation and the formation of organic acids. In general, the low-boiling aromatics, such as benzene, do not occur in quantity in the crudes, but the high-boiling compounds occur in great quantity, particularly in the crudes of the East Indies. Some of the California crudes also contain aromatics along with other petroleum bases. Figure 9-1 shows representative structural formulas of two aromatics. It will be noticed the methyl benzene consists of a benzene ring to which a paraffinic molecule has been added.

5. The *diolefin series* with type formula C_nH_{2n-2} is similar to the olefins except that two double bonds are present in each molecule, instead of one. The diolefins are chemically extremely active and tend to form gumlike materials of high molecular weight. They are often created during the cracking process of treating crude petroleum. Because of their activity,

they also can be easily removed by acid treatment. The diolefins have names which end in the suffix "-diene."

Beneath the names of the two diolefins hexadiene-1,5 and butadiene-1,3 illustrated in Fig. 9-1 there is shown another way of indicating the chemical structure of a molecule. If this method were applied to the 2,2,3,3-tetramethylpentane of Fig. 9-1, its formula would appear as $CH_3C(CH_3)_2C(CH_3)_2CH_2CH_3$.

9-4. Crude Oil (Petroleum). The crude oils from the eastern United States contain high percentages of paraffinic compounds. Crudes from California and the Gulf Coast region are largely of naphthenic base. The mid-continent crudes are mixed-base crudes, containing significant amounts of both paraffins and naphthenes. Aromatics occur to limited extents in nearly all crudes. However, it is customary to think of what might truly be called aromatic-base crudes as coming largely from fields of the East Indies and from Black Sea fields. It should be mentioned that, in a crude itself, the percentage variation between aromatics, naphthenes, and paraffins varies greatly among the boiling ranges under consideration. For example, in a certain California crude, the fraction boiling between 203 and 252 F showed 6% aromatics, 48% naphthenes, and 46% paraffins, whereas a heavier fraction of the same crude boiling between 302 and 392 F showed 17% aromatics, 61% naphthenes, and 22% paraffins. Similar characteristic variations would exist in other boiling temperature ranges.

Each crude oil, regardless of its basic type, is a mixture of complex hydrocarbon compounds, groups of which boil within certain ranges of temperatures. In addition to the hydrocarbon molecules there are other compounds in the oil which vary from traces to as much as 2 to 3 per cent of the total weight of fuel. These substances can be water, soluble salts, inorganic compounds, and solids in suspension. Sulfur is one of the most undesirable constituents which can appear in a crude oil. When present, it can exist in a variety of forms, frequently in mercaptans. The mercaptans are hydrocarbons with one or more of the carbon atoms replaced by sulfur, and each of them is characterized by an unpleasant odor. Sulfur can also appear in the form of sulfides, disulfides, or alkyl sulfides or as sulfones or sulfonic acids. The latter three of these contain oxygen in their composition in addition to the sulfur. For some uses sulfur must be largely removed from the product. However, for diesel engines and for gas-turbine operation sulfur up to 0.75 per cent is not particularly objectionable. In the case of spark-ignition engines, the responsiveness of a gasoline to tetraethyl lead for reducing detonation is decreased by the presence of sulfur compounds. It is customary to think of the sulfur content as being directly associated with the corrosion- and deposit-forming characteristics of a fuel, although this is not always true. However, when sulfur burns to sulphur dioxide, this material in the presence of water forms an acid which is extremely corrosive to most metals.

9-5. Crude-oil Distillation. Crude oils are distilled to separate them into different fractions, and for a rough classification the following tabulation applies.

Aviation gasoline	80 to 300 F
Gasoline	90 to 410 F
Naphtha	122 to 410 F
Kerosene	350 to 550 F
Domestic fuel oil	370 to 600 F
Gas oil	370 to 750 F
Lubricant oil and residuals	600 F up

This fractional distillation does not give the yield of light fractions and gasolines which the refineries require; so it is customary to crack, or break down, the heavier molecules to give a greater yield of the light fractions. The residual parts remaining from straight distillation or from cracking represent the source of the heavier fuel oils, numbered usually 4, 5, and 6. The lighter fuel oils are made from the gas oil distillates, which are further treated and processed before being classified as a finished fuel. These lighter diesel fuels are usually classified as No. 1 or 2 fuel oil.

9-6. Fuel Uses and Tests. Spark-ignition engines employ gasolines, the light fuel oils are used in most diesel engines and in many gas-turbine installations, while the heavy or residual-base fuel oils are used in a few diesel engines which employ preheating of the fuel and in stationary-type or locomotive-type gas turbines. Residual fuel oils are also used as boiler fuel for power and industrial purposes.

The physical tests that are carried out on hydrocarbon fuels usually cover measurement of the specific gravity or degrees API, viscosity, color, flash point, pour point, sulfur, carbon residue, ash, distillation range, cetane number for diesel fuels, and octane number for spark-ignition fuels.

9-7. Gravity, Degrees API. Gravity is a measure of the density or weight of a unit volume of fuel. It is most frequently expressed as specific gravity or relative weight of a unit volume of fuel compared with an equal volume of water measured at the same temperature. Usually this reference is made to the fuel and water both at 60 F and is written "sp gr 60 F/60 F." Instead of calibrating hydrometers, for testing liquid products, in terms of specific gravity, it is common to calibrate these units in terms of an arbitrarily designated unit, degrees API, where API stands for American Petroleum Institute (Table 9-4). Degrees API is related to specific gravity by the following formula for liquids which are lighter than water. This is true for most petroleum products.

$$\text{Sp gr} = \frac{141.5}{131.5 + {}^\circ\text{API}} \tag{9-12}$$

or

$$°\text{API} = \frac{141.5}{\text{sp gr } 60 \text{ F}/60 \text{ F}} - 131.5 \tag{9-13}$$

Before agreement was reached on this definition of degrees API, use was made of a term called degrees Baumé. The relationship between degrees Baumé and specific gravity follows:

$$\text{Sp gr} = \frac{140}{130 + °\text{Bé}} \tag{9-14}$$

At the present time it is customary to use degrees API to the exclusion of the earlier term.

9-8. Viscosity. Viscosity of a fluid is a measure of the internal friction or resistance to shear of the fluid. The shear stress, under conditions of relative motion, that a fluid can withstand depends on the velocity gradient of shearing. Imagine two parallel plates, one of which is fixed, and let the space between be filled by a fluid. If one of the plates is moved parallel to the other with a velocity u, a small cube of the liquid distorts in an interval of time Δt, as shown in Fig. 9-2. The velocity in the fluid changes from zero at plate A to a value of u at plate B, distance h away. The change in velocity relative to distance z measured perpendicular to the direction of motion, that is, du/dz, represents the rate of shear, and is called the velocity gradient. The assumption, originally made by Newton, that shearing stress τ is proportional to the velocity gradient has been experimentally verified, and thus we can write

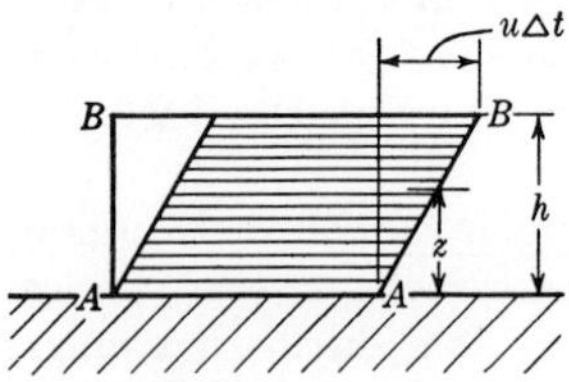

FIG. 9-2. Diagrammatic representation illustrating shear stress proportional to rate of shear.

$$\tau \propto \frac{du}{dz}$$

or

$$\tau = \mu \frac{du}{dz} \tag{9-15}$$

where μ is the proportionality factor. Also shearing stress τ times area A of the plate in question gives the force F. Thus

$$\frac{F}{A} = \tau = \mu \frac{du}{dz} \tag{9-16}$$

$$F = \mu A \frac{du}{dz} \tag{9-17}$$

The factor μ is called the *absolute coefficient of viscosity* of a fluid and has units of

$$\mu \equiv \frac{\text{lb}_f \text{ sec}}{\text{ft}^2}$$

TABLE 9-4. SPECIFIC GRAVITY–DEGREES API WITH CORRESPONDING POUNDS PER GALLON*

°API	Sp gr at 60/60 F	Lb per gal at 60 F	Gal per lb at 60 F	°API	Sp gr at 60/60 F	Lb per gal at 60 F	Gal per lb at 60 F	°API	Sp gr at 60/60 F	Lb per gal at 60 F	Gal per lb at 60 F	°API	Sp gr at 60/60 F	Lb per gal at 60 F	Gal per lb at 60 F
0	1.076	8.962	0.1116	25	0.9042	7.529	0.1328	50	0.7796	6.490	0.1541	75	0.6852	5.703	0.1753
1	1.068	8.895	0.1124	26	0.8984	7.481	0.1337	51	0.7753	6.455	0.1549	76	0.6819	5.676	0.1762
2	1.060	8.828	0.1133	27	0.8927	7.434	0.1345	52	0.7711	6.420	0.1558	77	0.6787	5.649	0.1770
3	1.052	8.762	0.1141	28	0.8871	7.387	0.1354	53	0.7669	6.385	0.1566	78	0.6754	5.622	0.1779
4	1.044	8.698	0.1150	29	0.8816	7.341	0.1362	54	0.7628	6.350	0.1575	79	0.6722	5.595	0.1787
5	1.037	8.634	0.1158	30	0.8762	7.296	0.1371	55	0.7587	6.316	0.1583	80	0.6690	5.568	0.1796
6	1.029	8.571	0.1167	31	0.8708	7.251	0.1379	56	0.7547	6.283	0.1592	81	0.6659	5.542	0.1804
7	1.022	8.509	0.1175	32	0.8654	7.206	0.1388	57	0.7507	6.249	0.1600	82	0.6628	5.516	0.1813
8	1.014	8.448	0.1184	33	0.8602	7.163	0.1396	58	0.7467	6.216	0.1609	83	0.6597	5.491	0.1821
9	1.007	8.388	0.1192	34	0.8550	7.119	0.1405	59	0.7428	6.184	0.1617	84	0.6566	5.465	0.1830
10	1.0000	8.328	0.1201	35	0.8498	7.076	0.1413	60	0.7389	6.151	0.1626	85	0.6536	5.440	0.1838
11	0.9930	8.270	0.1209	36	0.8448	7.034	0.1422	61	0.7351	6.119	0.1634	86	0.6506	5.415	0.1847
12	0.9861	8.212	0.1218	37	0.8398	6.993	0.1430	62	0.7313	6.087	0.1643	87	0.6476	5.390	0.1855
13	0.9792	8.155	0.1226	38	0.8348	6.951	0.1439	63	0.7275	6.056	0.1651	88	0.6446	5.365	0.1864
14	0.9725	8.099	0.1235	39	0.8299	6.910	0.1447	64	0.7238	6.025	0.1660	89	0.6417	5.341	0.1872
15	0.9659	8.044	0.1243	40	0.8251	6.870	0.1456	65	0.7201	5.994	0.1668	90	0.6388	5.316	0.1881
16	0.9593	7.989	0.1252	41	0.8203	6.830	0.1464	66	0.7165	5.964	0.1677	91	0.6360	5.293	0.1889
17	0.9529	7.935	0.1260	42	0.8155	6.790	0.1473	67	0.7128	5.934	0.1685	92	0.6331	5.269	0.1898
18	0.9465	7.882	0.1269	43	0.8109	6.752	0.1481	68	0.7093	5.904	0.1694	93	0.6303	5.246	0.1906
19	0.9402	7.830	0.1277	44	0.8063	6.713	0.1490	69	0.7057	5.874	0.1702	94	0.6275	5.222	0.1915
20	0.9340	7.778	0.1286	45	0.8017	6.675	0.1498	70	0.7022	5.845	0.1711	95	0.6247	5.199	0.1924
21	0.9279	7.727	0.1294	46	0.7972	6.637	0.1507	71	0.6988	5.817	0.1719	96	0.6220	5.176	0.1932
22	0.9218	7.676	0.1303	47	0.7927	6.600	0.1515	72	0.6953	5.788	0.1728	97	0.6193	5.154	0.1940
23	0.9159	7.627	0.1311	48	0.7883	6.563	0.1524	73	0.6919	5.759	0.1736	98	0.6166	5.131	0.1949
24	0.9100	7.578	0.1320	49	0.7839	6.526	0.1532	74	0.6886	5.731	0.1745	99	0.6139	5.109	0.1957
												100	0.6112	5.086	0.1966

* From *Natl. Bur. Standards (U.S.) Circ.* C410.

as can be seen from Eq. (9-17), since

$$\mu = \frac{F\,dz}{A\,du} \equiv \frac{\text{lb}_f \times \text{ft}}{\text{ft}^2 \times \text{ft/sec}} \equiv \frac{\text{lb}_f \times \text{sec}}{\text{ft}^2} \tag{9-18}$$

In the metric system with F, force, expressed in dynes

$$\mu \equiv \frac{\text{dyne} \times \text{sec}}{\text{cm}^2}$$

This latter unit has been named the *poise*.

Using the proper transformation constants and recalling $F = Ma$, it is possible to set up the following tabulation of viscosity equivalents:

$$\begin{aligned} 1 \text{ poise} = 100 \text{ centipoises} &= \frac{1 \text{ dyne} \times \text{sec}}{\text{cm}^2} = \frac{\text{gram mass}}{\text{cm} \times \text{sec}} \\ &= 0.0672\,\frac{\text{poundal} \times \text{sec}}{\text{ft}^2} = 0.0672\,\frac{\text{pound mass}}{\text{ft} \times \text{sec}} \\ &= 0.00209\,\frac{\text{lb}_f \times \text{sec}}{\text{ft}^2} = 0.00209\,\frac{\text{slug}}{\text{ft} \times \text{sec}} \end{aligned} \tag{9-19}$$

Quantitative values of the viscosity of different substances are tabulated in many handbooks and textbooks. It should be realized that the viscosity of both liquids and gases varies greatly with temperature and to a lesser extent with pressure. The viscosity of a liquid decreases with increasing temperature; the viscosity of a vapor or gas is just the opposite and increases as the temperature increases, except at very high pressures, when the gas behaves more like a liquid. In the case of nitrogen this inversion point for viscosity change with temperature occurs at about 4300 psi.

Kinematic viscosity is defined as the ratio of the absolute viscosity to the density of a fluid at a given temperature (and pressure). The Greek letter ν (nu) or sometimes y is used for kinematic viscosity.

$$\nu = \frac{\mu}{\rho} \equiv \frac{\dfrac{\text{lb}_f \times \text{sec}}{\text{ft}^2}}{\dfrac{\text{lb}_m}{\text{ft}^3}} = \frac{\dfrac{1}{32.17}\,\dfrac{\text{lb}_m}{\text{ft} \times \text{sec}}}{\dfrac{\text{lb}_m}{\text{ft}^3}} \tag{9-20}$$

or dimensionally in the English system

$$\nu = \frac{\text{ft}^2}{\text{sec}} \tag{9-21}$$

In the metric system

$$\nu = \frac{\mu}{\rho} = \frac{\text{poises}}{\dfrac{\text{gram}}{\text{cm}^3}} = \frac{\text{gram}}{\text{cm} \times \text{sec}} \times \frac{\text{cm}^3}{\text{gram}} = \frac{\text{cm}^2}{\text{sec}} \tag{9-22}$$

Kinematic-viscosity values in the metric system are called *stokes* or more frequently *centistokes,* where the centistoke = 0.01 stoke. Neither viscosity nor kinematic viscosity is named in the English system except that some use has been made of a unit called the reyn, where

$$1 \text{ reyn in lb}_f\text{-sec per in}^2. = 6.9 \times 10^6 \text{ centipoises} \tag{9-23}$$

Kinematic-viscosity values of oils and other liquids are determined by means of the Saybolt viscosimeter. This consists of a standard tube (ASTM D88-44) as illustrated in Fig. 9-3, which is mounted in a controlled-temperature unit. The liquid sample is placed in the tube and the thermostat set for the desired temperature. When this is reached, the surplus liquid is sucked out and the test performed. The test merely consists in carefully timing the seconds required for 60 milliliters of the test liquid to flow through the standard oil tube. The result is then reported as the Saybolt Universal seconds at the temperature employed. The most usual temperatures employed are 100, 130, and 210 F, but −40 F may also be required for low-temperature test work. The Saybolt Universal outlet tube is 0.1765 cm in diameter and 1.225 cm long. This may be too small for conveniently testing very viscous liquids, and for these a Saybolt Furol tube is used. This is like the former tube but has a diameter of 0.315 cm. Table 9-5 (ASTM D446-39—1949) is a conversion table for Saybolt Universal seconds to kinematic viscosity in centistokes. Table 9-6 (ASTM D666—1944) gives similar data for Saybolt Furol viscosity conversion at the ordinary temperature of 122 F used with this type of tube. Very roughly for the same kinematic viscosity Saybolt Furol seconds are about one-tenth as great as Saybolt Universal seconds.

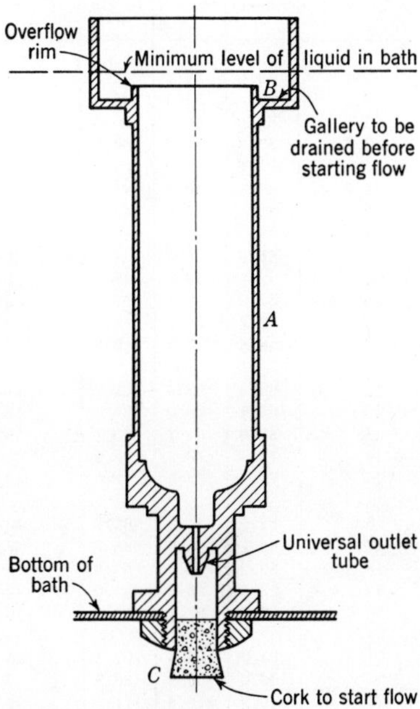

FIG. 9-3. Sectional view of standard oil tube for Saybolt viscosimeter.

Viscosity temperature charts of the type shown in Fig. 9-4 (ASTM D341-43—1949) are useful in furnishing viscosity data at temperatures other than those used in the test of a liquid. To graph a given liquid, it is good to have the available test points as far apart as possible.

TABLE 9-5. VALUES FOR CONVERTING KINEMATIC VISCOSITY TO SAYBOLT UNIVERSAL VISCOSITY

(From ASTM D446-39—1949)

Kinematic viscosity, centistokes	Equivalent Saybolt Universal viscosity, sec			Kinematic viscosity, centistokes	Equivalent Saybolt Universal viscosity, sec		
	At 100 F* (basic values)	At 130 F	At 210 F		At 100 F* (basic values)	At 130 F	At 210 F
2	32.6	32.7	32.8	31	145.3	145.6	146.3
2.5	34.4	34.5	34.6	32	149.7	150.0	150.7
3	36.0	36.1	36.3	33	154.2	154.5	155.3
3.5	37.6	37.7	37.9	34	158.7	159.0	159.8
4	39.1	39.2	39.4	35	163.2	163.5	164.3
4.5	40.7	40.8	41.0				
5	42.3	42.4	42.6				
6	45.5	45.6	45.8	36	167.7	168.0	168.9
7	48.7	48.8	49.0	37	172.2	172.5	173.4
8	52.0	52.1	52.4	38	176.7	177.0	177.9
9	55.4	55.5	55.8	39	181.2	181.5	182.5
10	58.8	58.9	59.2	40	185.7	186.1	187.0
11	62.3	62.4	62.7	41	190.2	190.6	191.5
12	65.9	66.0	66.4	42	194.7	195.1	196.1
13	69.6	69.7	70.1	43	199.2	199.6	200.6
14	73.4	73.5	73.9	44	203.8	204.2	205.2
15	77.2	77.3	77.7	45	208.4	208.8	209.9
16	81.1	81.3	81.7	46	213.0	213.4	214.5
17	85.1	85.3	85.7	47	217.6	218.0	219.1
18	89.2	89.4	89.8	48	222.2	222.6	223.8
19	93.3	93.5	94.0	49	226.8	227.2	228.4
20	97.5	97.7	98.2	50	231.4	231.8	233.0
21	101.7	101.9	102.4	55	254.4	254.9	256.2
22	106.0	106.2	106.7	60	277.4	277.9	279.3
23	110.3	110.5	111.1	65	300.4	301.0	302.5
24	114.6	114.8	115.4	70	323.4	324.0	325.7
25	118.9	119.1	119.7				
26	123.3	123.5	124.2	Over 70	Saybolt seconds = centistokes × 4.620	Saybolt seconds = centistokes × 4.629	Saybolt seconds = centistokes × 4.652
27	127.7	127.9	128.6				
28	132.1	132.4	133.0				
29	136.5	136.8	137.5				
30	140.9	141.2	141.9				

* To obtain the Saybolt Universal viscosity equivalent to a kinematic viscosity determined at t F, multiply the equivalent Saybolt Universal viscosity at 100 F by $1 + (t - 100) \times 0.000064$; for example, 10 centistokes at 210 F are equivalent to 58.8×1.0070, or 59.2 Saybolt Universal seconds at 210 F.

TABLE 9-6. VALUES FOR CONVERTING KINEMATIC VISCOSITY TO SAYBOLT FUROL VISCOSITY AT 122 F
(From ASTM D666-44)

Kinematic viscosity, centistokes	Saybolt Furol viscosity, sec	Kinematic viscosity, centistokes	Saybolt Furol viscosity, sec	Kinematic viscosity, centistokes	Saybolt Furol viscosity, sec
48	25.2	100	48.5	200	94.7
50	26.0	105	50.8	210	99.3
52	26.9	110	53.1	220	104.0
54	27.8	115	55.3	230	108.7
56	28.7	120	57.6	240	113.3
58	29.6	125	59.9	250	118.0
60	30.5	130	62.2	260	122.7
62	31.4	135	64.5	270	127.4
64	32.3	140	66.9	280	132.0
66	33.2	145	69.2	290	136.7
68	34.1	150	71.5	300	141.3
70	35.0	155	73.8	310	146.0
72	35.9	160	76.1	320	150.7
74	36.8	165	78.4	330	155.4
76	37.7	170	80.7	340	160.1
78	38.6	175	83.1	350	164.8
80	39.5	180	85.4	360	169.5
82	40.4	185	87.7	370	174.1
84	41.3	190	90.0	380	178.8
86	42.2	195	92.4	390	183.5
88	43.1				
				400	188.2
90	44.0				
92	44.9			Over 400	Saybolt seconds = centistokes × 0.470
94	45.8				
96	46.7				
98	47.6				

Increasing interest has arisen for the use of calibrated glass tubes (ASTM D445-46T), sometimes known as Ostwald pipettes, for timing at a controlled temperature the flow of a test fluid through a calibrated capillary section of the instrument. The test results in seconds can be transformed into kinematic viscosity. Various calibrating media are available for such instruments. For example, one such medium is water, which at 68 F has a kinematic viscosity in centistokes of 1.007, at 100 F of 0.689, and at 130 F of 0.518.

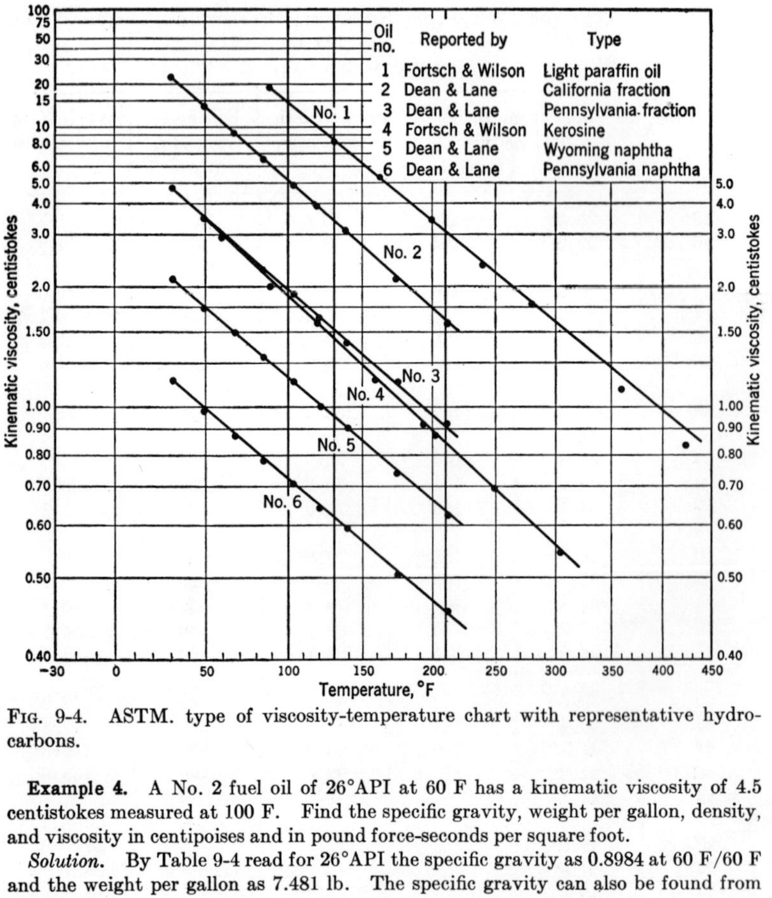

FIG. 9-4. ASTM. type of viscosity-temperature chart with representative hydrocarbons.

Example 4. A No. 2 fuel oil of 26°API at 60 F has a kinematic viscosity of 4.5 centistokes measured at 100 F. Find the specific gravity, weight per gallon, density, and viscosity in centipoises and in pound force-seconds per square foot.

Solution. By Table 9-4 read for 26°API the specific gravity as 0.8984 at 60 F/60 F and the weight per gallon as 7.481 lb. The specific gravity can also be found from Eq. (9-12).

$$\text{Sp gr} = \frac{141.5}{131.5 + 26} = \frac{141.5}{157.5} = 0.8984$$

To compute the weight per gallon use the density of water at 60 F (which as found from water-steam tables is 62.35 lb per ft³) and

$$62.35 \times \frac{231 \text{ in.}^3 \text{ per gal}}{1728 \text{ in.}^3 \text{ per ft}^3} = \frac{62.35}{7.48} = 8.33 \text{ lb per gal for water}$$

Then $8.33 \times 0.8984 = 7.484$ lb per gal for the oil.

The density of the fuel oil is

$$62.35 \times 0.8984 = 56.02 \text{ lb per ft}^3 \text{ at } 60 \text{ F}$$

In the metric system the density of water at 60 F is 0.9990 gram per cm³ so that the fuel oil has a density of

$$0.9990 \times 0.8984 = 0.8975 \text{ gram per cm}^3$$

Light fuels have a coefficient of expansion of closely 0.0004 per degree Fahrenheit. Thus specific volume at t°F can be represented as

$$\text{Sp vol}_{tF} = \text{sp vol}_{60F}\,[1 + 0.0004(t - 60)]$$

$$\text{Sp vol}_{tF} = \frac{1}{0.8975}\,[1 + 0.0004(100 - 60)] = \frac{1.016}{0.8975} \text{ cm}^3 \text{ per gram}$$

$$\text{Density at 100 F} = \frac{0.8975}{1.016} = 0.8834 \text{ gram per cm}^3$$

The kinematic viscosity at 100 F is 4.5 centistokes, or 0.045 stoke, and by Eq. (9-22)

$$\nu = \frac{\mu}{\rho} = \frac{\mu}{0.8834} = 0.045$$

$$\mu = 0.03975 \text{ poise} = 3.975 \text{ centipoises}$$

Equation (9-19) is used to change the units of μ.

$$\mu = 0.03975 \times 0.00209 = 0.0000831 \text{ lb}_f\text{-sec per ft}^2$$

For this fuel, Table 9-5 shows that, for 4.5 centistokes at 100 F, the viscosity measured in Saybolt Universal seconds is 40.7.

9-9. Flash- and Fire-point Tests. For petroleum products, these tests are made either in the so-called Cleveland open cup or the Pensky-Martens closed tester. In the Cleveland open-cup test, the cup which holds the sample fits snugly in a protecting holding plate and is heated from the bottom by a convenient heat source—gas burner, electric heater, or alcohol lamp. The oil sample in the cup is some ⅞ in. deep and comes to within ⅜ in. from the top edge of the cup. The cup itself is approximately 2¹¹⁄₁₆ in. in diameter. A thermometer is suspended in a vertical position in the sample with its bulb approximately ¼ in. from the bottom of the cup and midway between the center and back of the cup. A pilot test flame is available which can be made to pass over the surface of the heated oil in the cup. In operation, the sample is placed in the cup to the filling line and heat applied. The oil is heated at a rate not exceeding 30 F rise per minute until the temperature is approximately 100 F below the probable flash point. Thereafter the rate of heating is diminished and should not exceed 11 F per min. The test flame is applied to the oil at each 5-F rise in temperature by passing it in a straight line across the center of the cup. The flash point is reached when a combustion flash appears at any point on the surface of the oil, and the temperature at that instant must be read. After determining the flash point the heating is continued until that temperature is reached at which the oil continues to burn for at least 5 sec. The temperature which is read at the time of flame application for continuous burning is recorded as the fire point of the material.

The Cleveland Cup Test, ASTM designation D92-46, is used for determining flash and fire points of petroleum products with the exception of fuel oils and those having open-cup flash temperatures below 175 F.

The Pensky-Martens closed cup is used for determining the flash point of fuel oils and more volatile products. The placing of the sample and the method of temperature measurement are similar to the arrangements used with the open cup. A stirrer is provided to move the sample in the cup. The cover of the cup has suitable openings through which the thermometer can be inserted, and a shutter opens to make the flash trials. The cup itself is similar to that of the open-cup test previously described. This type of test carries the ASTM designation of D93-46.

9-10. Reid Vapor Pressure. This test, which is carried out in accordance with ASTM designation D323-49 is a means for determination of the vapor pressure of volatile, nonviscous petroleum products with the exception of liquefied petroleum gases. As the hydrocarbon fuels are mixtures of many chemical compounds, a unique vapor pressure such as exists for a single chemical at a given temperature is not attainable. The vapor-pressure test method provides for partial air saturation of the products for vapor pressures below 26 psig, although no air saturation is provided for products above 26 psig. The Reid vapor-pressure bomb consists of the liquid petroleum (gasoline) chamber into which, under prescribed conditions, a chilled sample of the product is poured so as completely to fill the chamber. An air chamber and gage are then attached, and the unit is placed in the controlled-temperature bath, where the liquid is at least 1 in. above the top of the air chamber and the bath is kept at a constant temperature at 100 F $\pm$ 0.2 F.

Great care must be exercised in making vapor-pressure measurements as it is necessary to prevent the more volatile fractions of the fluid from evaporating when transferring the sample into the bomb. The volume ratio of the air chamber to the liquid chamber is kept within the limits of 3.8 to 4.2. Various corrections must also be applied to account for barometric pressure and for initial temperature of the air chamber, and the bourdon-type gages used on the tester must be calibrated frequently. Reid vapor pressures give some indication of the volatility of a hydrocarbon fuel.

9-11. Cloud and Pour Point. This test concerns the low-temperature characteristics of lubricants or heavy fuel oils. The cloud point is that temperature at which the paraffin wax and other solid substances in a petroleum oil start to crystallize out or separate from the solution when the oil is chilled under test conditions. The pour point is the lowest temperature at which the oil will pour or flow when it is chilled without disturbance. In making such a test, a sample is placed in a suitable cooling bath under prescribed conditions (ASTM designation D97-47) and chilled, with careful measurements of temperature being taken until either the cloud or the pour point is reached. Freezing mixtures of ice and water or ice and various salts or carbon dioxide and acetone are used in the cooling bath.

9-12. Cetane Number of Fuel Oils. The cetane rating of a fuel is one measure of its inherent ignition characteristics. Cetane number is particularly important in the case of fuels used in diesel engines. When a fuel is injected into the cylinder of an engine, a delay period occurs before the fuel ignites and burns. The time lag varies with the fuel and also with the shape and character of the combustion chamber of the engine. If the ignition delay is long, it is possible for a large quantity of atomized fuel to accumulate within the combustion chamber before inflammation starts. This can result in a large amount of fuel burning at once, causing an extremely rapid pressure rise and sometimes resulting in knock. A fuel with a high cetane number has a small ignition delay, while a low cetane number is associated with excessive ignition delay. Cetane numbers constitute an arbitrary scale with which to compare the performance of a fuel with that of reproducible reference fuels. The reference fuels used are cetane, which is arbitrarily said to have a cetane rating of 100, and a second reference fuel, alpha methylnaphthalene, which has a reference value of zero cetane number. An unknown fuel which performs in a standard test engine in the same manner as would a mixture of 45 parts of cetane and 55 parts of alpha methylnaphthalene is said to have a cetane number of 45.

Cetane is a hydrocarbon paraffin specifically known as *n*-hexadecane with a structural formula $CH_3(CH_2)_{14}CH_3$, molecular weight of 226.4, melting point of 65 F, and boiling point of 549.5 F. Alpha methylnaphthalene structurally consists of the naphthalene double ring (*viz.*, two benzene rings joined at a common side) to which a methyl CH_3 radical is added at the top of the right ring. Its formula is $C_{10}H_7CH_3$. It boils in the range 464 to 469 F and solidifies at 7.6 F.

A good diesel fuel has ignition qualities which are at variance with those desired for gasoline. A gasoline with high antiknock quality for use in a spark-ignited gasoline combustion engine has a high spontaneous-ignition temperature and long ignition lag so that, as the mixed charge of fuel and air burns progressively after ignition, the burning gases in expanding do not compress the remaining unburned charge to such an extent that self-ignition conditions are reached.

In gas-turbine combustion chambers, the cetane number is not too important a characteristic in indicating how a fuel will burn and perform. It is important, however, that the fuel can easily be pumped through the fuel lines and distributed into the combustion zone, after which the combustion-delay period should be as short as possible for rapid combustion

TABLE 9-7. DETAILED REQUIREMENTS FOR FUEL OILS*
(U.S. Department of Commerce Commercial Standard CS12-48, ASTM Designation D396-48T)

Grade†	Flash point, F deg	Pour point, F deg	Water and sediment, %	Carbon residue on 10% residuum	Ash, %	Distillation temps F deg: 10% point	Distillation temps F deg: 90% point	Distillation temps F deg: End point	Viscosity: Saybolt Universal at 100 F, sec		Viscosity: Saybolt Furol at 122 F, sec		Viscosity: Kinematic centistokes at 100 F		Viscosity: Kinematic centistokes at 122 F		Gravity, API deg
	Min	Max	Max	Max	Max	Max	Max	Max	Max	Min	Max	Min	Max	Min	Max	Min	Min
No. 1. A distillate oil intended for vaporizing pot-type burners and other burners requiring this grade§	100 or legal	0	Trace	0.15	....	420	...	625	...	...	...	..	2.2	1.4	...	..	35
No. 2. A distillate oil for general-purpose domestic heating for use in burners not requiring No. 1	100 or legal	20‡	0.10	0.35	....	‖	675	...	40	...	...	..	4.3	....	...	..	26
No. 4. An oil for burner installations not equipped with preheating facilities	130 or legal	20	0.50	....	0.10	...	...	...	125	45	...	..	26.4	5.8			
No. 5. A residual-type oil for burner installations equipped with preheating facilities	130 or legal	...	1.00	....	0.10	...	...	...	...	150	40	..	....	32.1	81		
No. 6. An oil for use in burners equipped with preheaters permitting a high-viscosity fuel	150 or legal	...	2.00Δ	....	....	...	...	...	...	...	300	45	....	....	638	92	

* Recognizing the necessity for low-sulfur fuel oils used in connection with heat-treatment, nonferrous metal, glass and ceramic furnaces, and other special uses, a sulfur requirement may be specified in accordance with the following table:

Grade of fuel oil	*Sulfur, max, %*
No. 1	0.5
No. 2	1.0
Nos. 4, 5, 6	No limit

Other sulfur limits may be specified only by mutual agreement between the buyer and seller.

† It is the intent of these classifications that failure to meet any requirement of a given grade does not automatically place an oil in the next lower grade unless in fact it meets all requirements of the lower grade.

‡ Lower or higher pour points may be specified whenever required by conditions of storage or use. However, these specifications shall not require a pour point lower than 0 F under any conditions.

§ No. 1 oil shall be tested for corrosion for 3 hr at 122 F. The exposed copper strip shall show no gray or black deposit.

‖ The 10 per cent point may be specified at 440 F, maximum, for use in other than atomizing burners.

△ The amount of water by distillation plus the sediment by extraction shall not exceed 2.00 per cent. The amount of sediment by extraction shall not exceed 0.50 per cent. A deduction in quantity shall be made for all water and sediment in excess of 1.0 per cent.

in a small space. There appears to be no difficulty in burning either gasolines or lighter fuel oils. However, special care is needed to obtain flexible performance with heavy fuel oils.

9-13. Fuel Oils. These hydrocarbon fuels are extensively used in gas turbines. They range from light kerosene at about 0.8 specific gravity to the No. 6 fuel oils, which can have specific-gravity values in excess of 1.0. Expressed in degrees API (see Table 9-4) this range is from 45 to less than 10. A commercial standard, reproduced as Table 9-7, gives representative classifications of the usual fuel oils. In this listing it should be noted that No. 1 fuel oil is essentially kerosene. No. 6 fuel oil, which is a residual from the refining process, is often known as Bunker C oil. Its composition and character are quite variable, with water and sediment frequently in evidence. Table 9-8 gives representative data on heavy fuel oils.

TABLE 9-8. RESIDUAL-FUEL-OIL DATA*
(Bunker C or No. 6 Grade)

	A	*B*	*C*	*D*	*E*	*F*	*G*	*H*	*I*
Gravity, ° API	8.0	9.2	11.3	14.3	15	16.5	16.9	17.8	22
Sp gr	1.014	1.006	0.991	0.968	0.966	0.956	0.954	0.948	0.922
Viscosity at 122 F, Saybolt Universal sec	......	......	......	......	1280	......	580	630	
Viscosity at 212 F, Saybolt Universal sec	......	......	......	......	170	......	81	145	
Flash point, °F	......	......	......	......	300	......	250	290	
Hydrogen-carbon ratio, H/C	0.107	0.113	0.116	0.117	0.128	0.107	0.131	0.134	0.127
Wt composition:									
Carbon, %	88.3	88.4	86.5	87.60	86.7	87.50	87.2	86.6	87.40
Hydrogen, %	9.5	9.95	10.04	10.27	11.1	10.17	11.4	11.6	11.10
Sulfur, %	1.2	0.68	1.49	0.70	1.5	1.14	0.4	1.4	0.42
Water, %	0.05	0.04	0.03	0.05	0.02	0.05	0.0	0.1	0.05
Undetermined and ash, %	0.95	0.81	1.67	1.38	0.68	1.14	1.0	0.3	1.03
HHV, Btu per lb	18,084	18,274	18,088	18,454	19,400	18,319	19,110	19,440	18,778
LHV, Btu per lb	17,179	17,324	17,128	17,480	17,750	17,351	17,790	17,820	17,723

* Data from various sources.

The heating values of fuel oils are directly related to their densities, and for this reason various semiempirical formulations for heating value have been proposed. The Sherman and Kropf formulations for higher heating value are usually given in three ranges.

For gasoline:

$$H_H = 18{,}320 + 40\ (\text{API} - 10) \tag{9-24}$$

For kerosene and light fuel oil:

$$H_H = 18{,}440 + 40\ (\text{API} - 10) \tag{9-25}$$

For heavy fuel oils:

$$H_H = 18{,}650 + 40\ (\text{API} - 10) \tag{9-26}$$

where H_H is the higher heating value in Btu per pound and API represents the degrees API. It should be realized that these relations are only approximate, as factors other than gravity affect the heating value, one of the most significant being the type of basic crude from which the refined product is made.

Precise heating-value formulations require knowledge of the basic crudes from which a fuel is refined. The paraffin- and olefin-base crudes have the lowest heating value per gallon of fuel (117,000 to 125,000 Btu per gal higher heating value); the naphthenic-base crudes have a greater heating value, about 127,500 Btu per gal; and the aromatics have the highest heating values, ranging from 128,000 to 131,000 Btu per gal. However, even in each crude-oil family the type of chain branches can affect the resultant heating values. The deviations, however, are not so serious as to prevent the use of general formulations for heating-value determinations and for making estimates on representative fuel types. Figure 9-5 is a plot of the lower heating value of hydrocarbons in Btu per pound against degrees API. It will be noticed that there is a considerable scatter to the test heating values from these commercial fuels. In general, the heating value of these fuels on a weight basis increases with increasing values of degrees API or in terms of decreasing values of specific gravity. On the other hand, the reverse condition holds when the heating value is considered on a volume (gallon) basis, with the greatest heat of combustion in Btu per gallon being associated with the denser residual fuel oils and being least with the gasolines. As liquid fuels are usually sold on a volume basis, this can be significant.

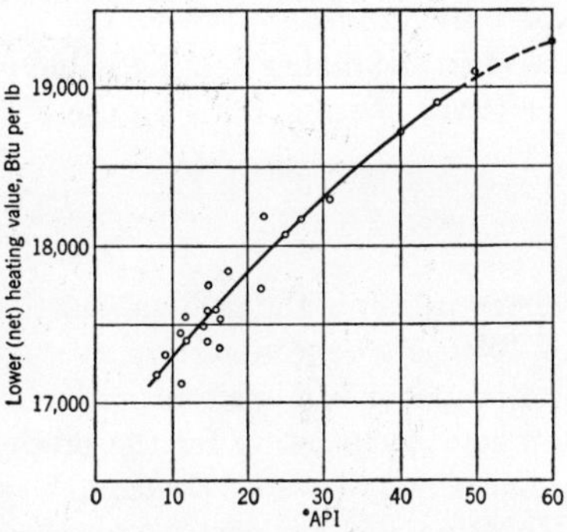

FIG. 9-5. Lower heating value of hydrocarbon fuels as related to degrees API.

In addition to the Sherman-Kropf formulas for hydrocarbon fuels, several other formulations should be mentioned. The National Bureau of Standards has proposed a formula primarily for fuel oils as follows:

$$H_H = 22{,}320 - 3780 \times (\text{sp gr})^2 \tag{9-27}$$

The following formula for lower heating value, proposed by the National Bureau of Standards, applies specifically to kerosenes and gasolines:

$$H_L = 19{,}960 + 1360 \times (\text{sp gr}) - 3780 \times (\text{sp gr})^2 \tag{9-28}$$

Turner and Lord[2] propose a formulation for lower heating value in terms of the hydrogen-carbon ratio (H/C):

$$H_L = 15{,}935 + 15{,}800(\text{H/C}) \tag{9-29}$$

One should not be too concerned at minor differences obtained when the preceding formulations are used to determine heating value. In particular, variable heating values are obtained in connection with the residual fuel oils because of the uncertainty of their composition from extraneous items which enter during the refining process and because of variations in the basic crude.

When the higher heating value of a fuel is known, it is possible to find the lower heating value by use of the following relation:

$$H_L = H_H - \frac{9\text{H}}{100} \times 1030 \tag{9-30}$$

where H/100 is the decimal fraction of hydrogen in the fuel. The value of 1030 is a representative average of the latent heat interchange. In a mixture of combustion products condensation starts at the dew-point temperature existing for the products and continues as the temperature progressively lowers. Inasmuch as there is no single temperature of condensation, it is not possible to select the latent heat at any given temperature and use it to represent the range through which condensation occurs.

There is no universally accepted standard temperature for the heating value of hydrocarbon fuels, although certain of the ASME test codes indicate 68 F as the standard. However, in much of the literature associated with use of hydrocarbon fuels, 60 F has been used, and it is this value which will be employed in the computation work throughout the rest of this chapter.

The boiling-point range of fuel oils is indicated in Table 9-7 at least to the extent of showing a flash-point temperature and the 10 per cent point of the lighter oils. In the case of the heavier oils, the temperature range is so indefinite as to have little significance. With these heavy oils it should be realized that except for a trivial amount of light ends the greater part of normal vaporization would occur at temperatures in excess of

100 F, with some of the heavy ends running above 700 F. Reference to this table also gives viscosity values for certain of the fuel oils. Viscosity plays an important role when the question of pumping the oil or atomizing it through nozzles becomes necessary. The No. 4, 5, and 6 oils may be so heavy that it is necessary to preheat them in order to have them flow through the system and nozzles. In this connection Table 9-9 gives the customary preheating temperatures required with fuel oils in order to make them flow satisfactorily.

TABLE 9-9. PREHEATING TEMPERATURES FOR FUEL OILS

°*API*	*Suggested preheat temperature range, °F*
10–12	275–325
12–14	220–275
14–16	175–250
16–18	150–200
18–20	140–160
20–22	100–140
22–24	70–100
24–26	70– 80

9-14. Combustion-energy Equation. Combustion in the gas turbine can be considered to be a steady-flow process. Each pound of air with enthalpy h_1 and kinetic energy $C_1^2/2g_cJ$ enters the combustor where f lb of fuel with its chemical energy of H_L, Btu per pound lower heating value, combines with some of the air to produce the $1 + f$ lb of products which leave with enthalpy h_{p2} Btu per lb and kinetic energy of $C_2^2/2g_cJ$. In equational form for this adiabatic process,

$$\left(h_1 + \frac{C_1^2}{2g_cJ}\right) + f(h_f + H_L) = (1 + f)\left(h_{p2} + \frac{C_2^2}{2g_cJ}\right) \tag{9-31}$$

where h_f represents the *sensible* enthalpy of the liquid fuel with respect to the heating-value datum temperature, which will be taken here as 600 R. A specific heat of 0.5 can be used without great error for liquid petroleum fuels. The combustion efficiency, which expresses the fraction of the heating value actually turned into useful temperature rise through combustion, in this equation is initially taken to be 100 per cent.

For a better understanding of what is involved in combustion, consider the process indicated in Fig. 9-6. Here, under steady-flow conditions, 1 lb of dry air and f lb of fuel are burnt at constant pressure in such manner that the hot exhaust products, upon cooling to the original temperature without condensation of water vapor in the products, transfer from the system as heat an amount of energy Q. This heat energy can be used to measure the lower heating value, H_L Btu per lb, for the given test conditions. The steady-flow energy equation for the process shows (assuming the kinetic-energy terms as negligible)

$$1(h_1) + f(h_f + H_L) = Q_{out} + (1 + f)(h_{p2}) \tag{9-32}$$
$$Q_{out} = fH_L + \Sigma h_{reactants} - \Sigma h_{products}$$

or to a close approximation

$$fH_L = Q_{out} \tag{9-33}$$

If the air used relative to f is greatly in excess of that required for combustion, it would be possible with most solid and liquid fuels for the combustion products to be above the dew-point (condensation) temperature and for the combustion products to remain completely in a gaseous state.

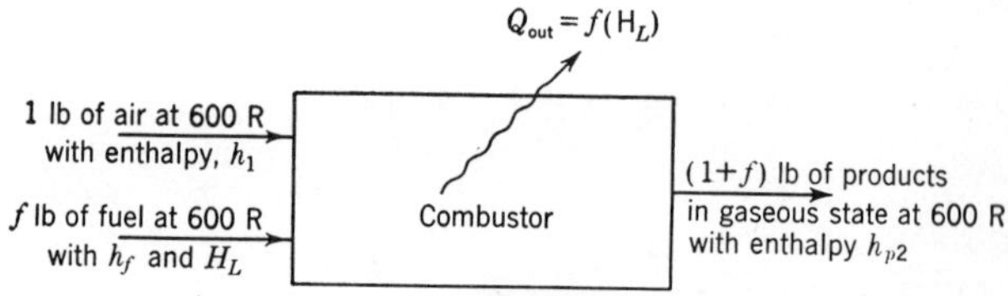

FIG. 9-6. Combustion device operating as a calorimeter.

Now consider a combustion process, under steady-flow conditions, taking place in an adiabatic combustor from which the products are necessarily delivered at an elevated temperature. The production of high-temperature products is the end purpose of combustion systems of gas

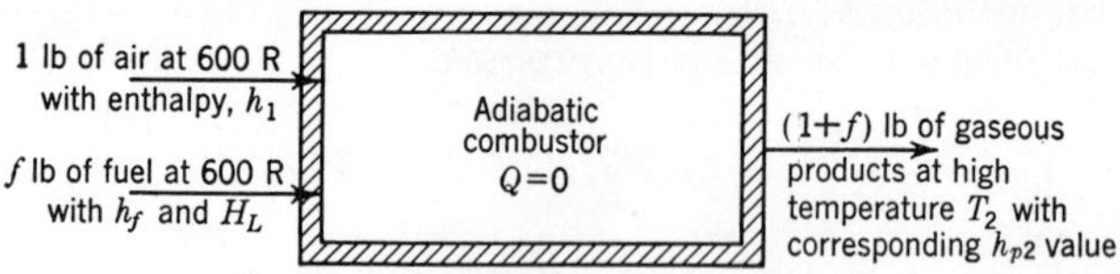

FIG. 9-7. Adiabatic combustor.

turbines. Comparisons of Figs. 9-6 and 9-7 in diagrammatic form illustrate that if the heat, represented by the lower heating value H_L, is not removed during the combustion process then an amount of energy equivalent to this is added to and remains associated with the high-temperature combustion products. This Q or H_L in the products is actually part of the original internal energy of the fuel. Thus the enthalpy of entering fuel can be considered to be composed of two parts:

1. *Sensible enthalpy*, which can be changed merely by heating and cooling the fuel, etc., but without molecular or chemical change. The symbol used will be h_f Btu per lb above a standard datum.

2. *Chemical enthalpy*, or energy which can be liberated in thermal form to raise the product-gas temperature. This is H_L Btu per lb. With this subdivision of the enthalpy of the entering fuel, the steady-flow equation for the combustion chamber may be written as follows (again disregarding the kinetic-energy terms, which most frequently cancel out with negligible error, in the energy equation):

$$1(h_1) + f(h_f + H_L) = (1 + f)(h_{p2}) \tag{9-34}$$

When lower heating values are tabulated for other than a 600 R datum, it has been observed that indicated values increase by about 20 Btu per lb per 100° rise in datum temperature. However, small error arises in calculations when heating-value determinations based on a different datum are used. This is obvious with the realization that the lower heating values of hydrocarbon fuels range around 17,000 to 19,000 Btu per lb.

Now let it be assumed that combustion is not complete. In this case the chemical energy actually released is the lower heating value H_L times a combustion efficiency η_b. For this condition the steady-flow energy equation for Fig. 9-6 would show

$$Q_{\text{out}} = \eta_b f H_L + \Sigma h_{\substack{\text{reactants} \\ \text{at 600 R}}} - \Sigma h_{\substack{\text{products} \\ \text{at 600 R}}} \tag{9-35}$$

Assume, as before, that Q is not released but is retained (or fed back in) to elevate the temperature of the products. Although the *composition* of the products is not the same as it would be for complete combustion, an energy equation of the same form as Eq. (9-34) applies, *i.e.*,

$$1(h_1) + f(h_f + \eta_b H_L) = (1 + f)h_{p2} \tag{9-36}$$

In this equation there is, again, the tacit recognition that the energy actually released was originally part of the chemical energy of the fuel.

In Eq. (9-31) and its more simplified forms (9-34) and (9-36) it should be realized that the combustion products contain CO_2, H_2O, and other gases, besides the large amounts of nitrogen and usually of excess air. Consequently h_{p2} can be found only approximately by reference to air tables. To make more accurate combustion calculations possible English and Wachtl[1] have developed a set of charts which use a factor ψ_h which accounts for the rise in temperature of the gases following combustion as taking place with the true combustion products instead of with air only. Although complete combustion is assumed, it is true that the enthalpy of the products of combustion is little affected by small differences in composition caused by incomplete combustion. Combustion efficiencies are ordinarily in excess of 90 per cent, and values above 95 per cent are common.

The factor ψ_{h2} is defined by the following equation:

$$h_{p2} = h_2 + \frac{f}{1+f}\psi_{h2} \tag{9-37}$$

This equation states that the enthalpy of 1 lb of *products*, h_{p2}, is considered to be the enthalpy of 1 lb of *air* at exhaust conditions, h_2, plus a correction factor, $\frac{f}{1+f}\psi_{h2}$, used because the 1 lb of products is not exclusively air. The units of ψ_{h2} should be noted as being Btu in exhaust products per pound of fuel.

Equation (9-36) can be written as follows:

$$\eta_b f H_L = (1+f)h_{p2} - h_1 - fh_f \tag{9-38}$$

Substitute h_{p2} from Eq. (9-37) in Eq. (9-38), which yields

$$\eta_b f H_L = (1+f)\left(h_2 + \frac{f}{1+f}\psi_{h2}\right) - h_1 - fh_f$$

Upon rearranging,

$$f = \frac{h_2 - h_1}{(\eta_b H_L + h_f) - (h_2 + \psi_{h2})} \tag{9-39}$$

Because absolute internal-energy values are unknown, a datum must be assumed for measurement of enthalpies and ψ_h. The charts prepared by English and Wachtl (Fig. 9-8) assume that at 600 R (= 460 + 140 F) the enthalpy h of air is 48 Btu per lb, and both the ψ_{h2} and h_f of liquid fuel are zero. Consequently to use Table 2-4, which has an h of 33.6 for air at 600 R, it is necessary to add 14.4 to h_2 in the denominator of Eq. (9-39), giving

$$f = \frac{h_2 - h_1}{(\eta_b H_L + h_f) - (h_2 + 14.4) - \psi_{h2}} \tag{9-40}$$

where Eq. 9-40 requires the use of Table 2-4 and Fig. 9-8. Equation (9-40) will be used to the exclusion of (9-39) in this text.

It sometimes happens that a second combustion process is carried out with gaseous products from a previous combustion (reheating). For this case it is possible by the previously shown methods to develop the following relation for the second combustion process, in which $f_2 - f_1$ lb of fuel is added to give a final over-all fuel-air ratio of f_2. Referring to Fig. 9-9,

$$\eta_{b2} H_L (f_2 - f_1) = (1+f_2)h_{p3} - (1+f_1)h_{p2} - (f_2 - f_1)h_{f2} \tag{9-41}$$

The factor ψ_{h3} for this second combustion process is defined as follows,

$$h_{p3} = h_3 + \frac{f_2}{1+f_2}\psi_{h3} \tag{9-42}$$

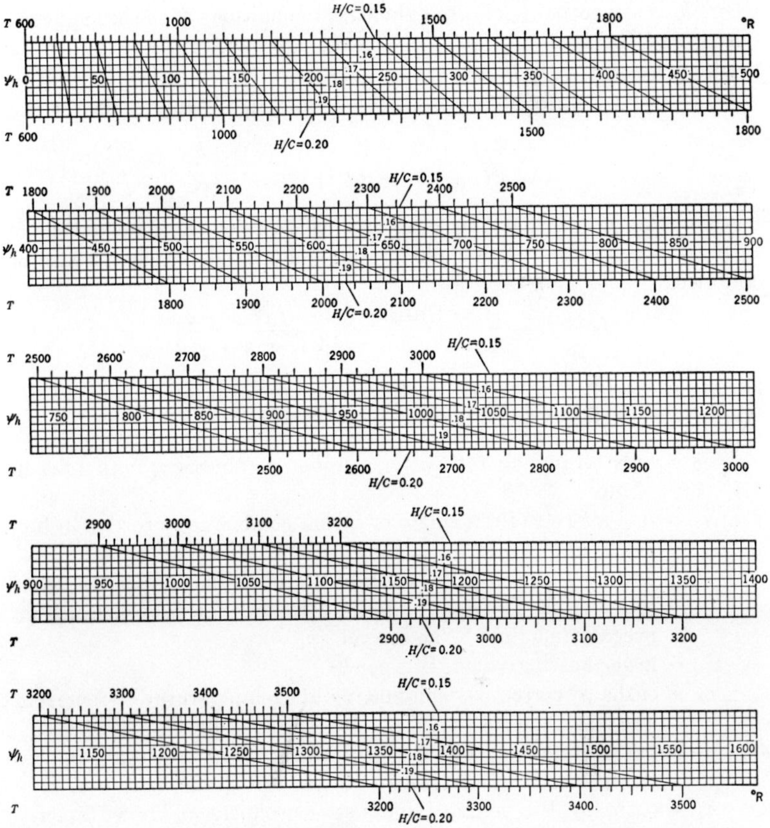

FIG. 9-8. Combustion function ψ_h for various temperatures. (*After English and Wachtl, NACA Tech. Note* 2071, *Ref.* 1.)

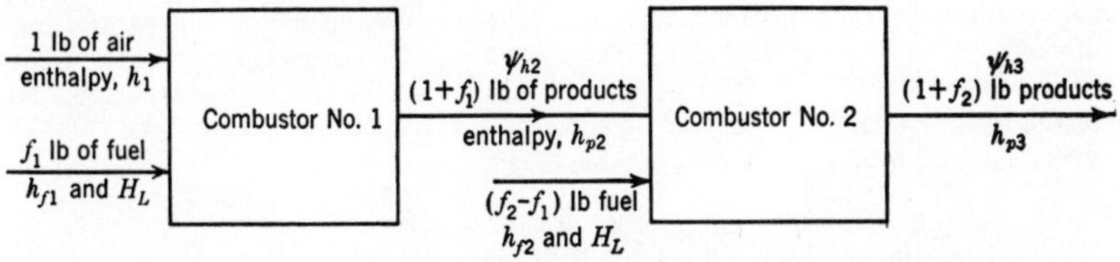

FIG. 9-9. Two-step combustion (reheating). This process utilizes the products from a previous combustion to provide the oxygen required for combustion in the second step.

where h_3 is the enthalpy of *air* at the outlet conditions of the second combustor and the second term involving ψ_{h3} corrects for the fact that the products are not entirely air.

Introducing the parameter ψ_{h3} and solving Eq. (9-41) for f_2 gives

$$f_2 = \frac{\eta_{b2}H_Lf_1 + h_3 - (1 + f_1)h_{p2} + f_1h_{f2}}{\eta_{b2}H_L - (h_3 + 14.4) - \psi_{h3} + h_{f2}} \tag{9-43}$$

It should be noted that the correction of 14.4 for the zero datum of enthalpy for air has been included. The following equation can also be obtained:

$$f_2 - f_1 = \frac{(1 + f_1)(h_3 - h_2) + f_1(\psi_{h3} - \psi_{h2})}{\eta_{b2}H_L - (h_3 + 14.4) - \psi_{h3} + h_{f2}} \tag{9-44}$$

Equations (9-43) and (9-44) are to be used in connection with values taken from Table 2-4 and Fig. 9-8.

f_2 = over-all fuel-air ratio after a second combustion, lb fuel per lb of air

f, or f_1 = fuel-air ratio for a single or initial combustion process, lb fuel per lb of air

h_3, h_2 = enthalpy values, Btu per lb of the pure air component, to be taken from Table 2-4. State 2 follows the first combustion process and state 3 the second

H_L = lower heating value, Btu per lb

h_f = enthalpy correction for liquid fuel at a temperature different from the 600 R datum $= c_f(T_{\text{fuel}}\ °\text{R} - 600) = 0.5(T_{\text{fuel}}\ °\text{R} - 600)$

ψ_{h3}, ψ_{h2} = parameters from Fig. 9-8, to account for enthalpy of combustion products in contrast to air [see Eqs. (9-37) and (9-42)]

h_p = enthalpy, Btu per lb, of combustion products not known directly except by use of Eqs. (9-37) and (9-42) or by extensive computations from the original fuel-air mixture

Example 5. Air leaves a compressor at 440.3 F and enters a gas-turbine combustor, where it enters into combustion with a light hydrocarbon fuel having a lower heating value of 18,937 Btu per lb and a H/C ratio of 0.19. The gases finally leave the combustor exit at 1440.3 F. If the efficiency of burning is 0.95 and the fuel is at 220.3 F, find the fuel-air ratio.

Solution. Equation (9-40) is applicable.

h_2 = 367.56 Btu per lb at 1900 R (Table 2-4)

h_1 = 106.46 Btu per lb at 900 R (Table 2-4)

ψ_{h2} = 530 Btu per lb at 1900 R (Fig. 9-8)

$$f = \frac{367.56 - 106.46}{(0.95)(18{,}937) + 0.5(680.0 - 600) - (367.56 + 14.4) - 530}$$

$$f = \frac{261.10}{17{,}019} = 0.0154$$

Example 6. Under certain test conditions, measurements on the combustor of Example 5 showed that a fuel-air ratio of 0.018 held, with inlet conditions the same as before and η_b assumed as 0.95. Find the probable outlet temperature if the combustor is assumed to be adiabatic.

Solution. It is most convenient to assume a final temperature and use a cut-and-try method to reach the correct solution. Start with Eq. (9-38), and solve for h_{p2}, which for this case gives

$$(0.95)(0.018)(18{,}937) = (1 + 0.018)(h_{p2}) - 106.46 - (0.018)(40)$$
$$h_{p2} = 423.0$$

The value of h_{p2} can now be used in connection with Eq. (9-37), giving

$$423.0 = \frac{0.018}{1 + 0.018}\psi_{h2} + h_2$$

To aid in arriving at a first approximation for T_2, find in the air table the temperature corresponding to the enthalpy h_{p2}. As we are concerned with products of combustion with higher specific-heat values than those for air, it is obvious that the real T_2 is less than that given by the air tables and so a somewhat lower trial temperature should be selected. For an h of 423, Table 2-4 shows $T_2 = 2100$ R, reduce this, say to 2050 R for the first trial temperature. Then for 2050 R, as $\psi_{h2} = 610$ and $h_2 = 409.13$, we get

$$423 \neq (0.01768)(610) + 409.13 = 419.9$$

Try 2062 R, and

$$423 \equiv (0.01768)(615) + 412.48 = 423.3$$

Thus 2062 R is closely the required temperature.

Example 7. A motor gasoline to be burned as a jet-engine fuel has a lower heating value $H_L = 18{,}800$ Btu per lb and a H/C ratio of 0.168. Consider combustion efficiency as 0.95, and find the fuel rates, f, for a 500° temperature rise if the initial air temperatures are (*a*) 140.3 F, (*b*) 540.3 F, and (*c*) 1040.3 F. Consider the fuel initially at 100.3 F.

Solution. By Eq. (9-40), for part (*a*) read from Table 2-4, $h_1 = 33.61$ for $T_1 = 140.3 + 459.7 = 600$ R and $h_2 = 156.23$ for $T_2 = 1100$ R. From Fig. 9-8 read $\psi_{h2} = 152$. Substitute in Eq. (9-40)

$$f = \frac{156.23 - 33.61}{(0.95)(18{,}800) + 0.5(560 - 600) - 156.23 - 14.4 - 152}$$
$$f = 0.0699 \text{ lb per lb of air}$$

(*b*) For $T_1 = 540.3 + 459.7 = 1000$ R, $h_1 = 131.20$, $T_2 = 1500$ R, $h_2 = 259.50$. $\psi_{h2} = 308$

$$f = \frac{259.50 - 131.20}{(0.95)(18{,}800) + 0.5(560 - 600) - 259.50 - 14.4 - 308}$$
$$f = 0.0743 \text{ lb per lb of air}$$

(*c*) For $T_1 = 1040.3 + 459.7 = 1500$ R, $h_1 = 259.50$, $T_2 = 2000$ R, $h_2 = 395.22$. $\psi_{h2} = 536$

$$f = \frac{395.22 - 259.50}{(0.95)(18{,}800) + 0.5(560 - 600) - 395.22 - 14.4 - 536}$$
$$f = 0.0803 \text{ lb per lb of air}$$

Example 8. If in part (*c*) of Example 7 the heating from 1040.3 to 1540.3 F is accomplished with residual fuel oil with a lower heating value $H_L = 17{,}720$ Btu per lb and H/C = 0.13, find the fuel-air ratio f. Consider 95 per cent combustion efficiency and the fuel to be at 560 R.

Solution. Read $h_2 = 395.22$ for 2000 R and $h_1 = 259.50$ for $T_1 = 1500$ R. Figure 9-8 can be used to find ψ_{h2}, but in using it a correction must be made for the H/C ratio. Note that at 2000 R the ψ_{h2} values vary by $(538 - 492) = 46$ for a range in H/C from 0.15 to 0.17. For a range from 0.13 to 0.15 essentially the same change in magnitude would occur so that for 2000 R read $\psi_{h2} = 492$ at 0.15 and subtract 46, namely, $492 - 46$, to obtain the 2000 R value at a H/C of 0.13, giving $\psi_{h2} = 446$. Then by Eq. (9-40) substitution yields

$$f = \frac{395.22 - 259.50}{(0.95)(17{,}720) + 0.5(560 - 600) - 395.22 - 14.4 - 446}$$

$$f = 0.0850 \text{ lb per lb of air}$$

9-15. Combustion Chambers—General Considerations. The problems which must be solved in connection with combustion chambers of gas-turbine units are numerous and severe, as the temperatures attained reach and even surpass safe life limits for metals. Stability of combustion with associated flame maintenance presents an additional consideration. The liberation of heat in the combustion zone of a combustion chamber reaches extremely high values, ranging from 500,000 to 3,000,000 Btu/(hr)(atm)(ft^3) of combustion space. Finally, the mixing of the main stream and the secondary air must be brought about with minimum loss in pressure and the avoidance of premature flame quenching.

Present-day metals have an extremely short life if they remain at temperatures much above 1600 F. Consequently, it is necessary to operate with high air-fuel ratios, ranging from about 50:1 to as high as 150:1, to reduce the combustion temperatures. For most hydrocarbon fuels, as the air ratio for combustion ranges from 14:1 to 16:1, it is necessary to isolate the main combustion stream of primary air. Later in the chamber secondary air is provided, not only to assure complete combustion, but primarily to reduce temperature.

In the primary combustion zone of a typical burner, wide variations occur in the amount of fuel which is supplied. If the load increases, requiring more fuel, unless proper adjustment is made the combustion takes place on the rich side, possibly so rich in fuel that a condition known as *rich blowout* can occur. On the other hand, at low loads on the turbine plant with reduced fuel, the mixture may become so lean (*i.e.*, scant in fuel) that combustion cannot be maintained and a condition known as *lean blowout* can occur. Each combustor must be able to operate over a range of fuel-air conditions and throughout the range maintain good efficiency of combustion.

Figure 9-10 represents a typical design of a combustion chamber in which the main combustion zone is close to the fuel nozzle—and into this zone a limited amount of primary air is supplied. The combustion reaction continues for some distance farther down in the chamber, with the secondary air, which enters through the sides of the chamber, contributing to the combustion process and also assisting in cooling the main stream.

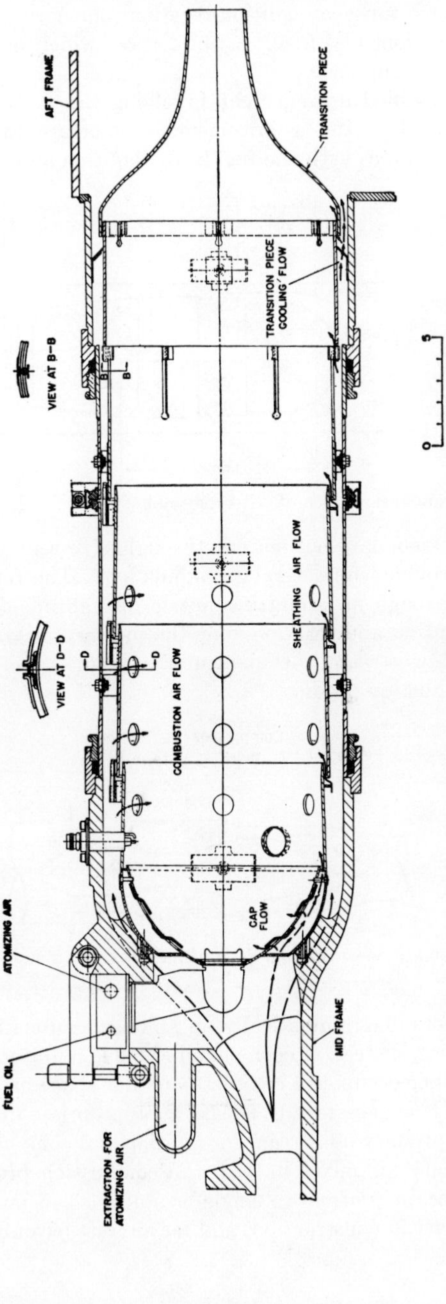

FIG. 9-10. Gas-turbine combustion chamber in longitudinal section. (*General Electric Co.*)

There are thus three zones, an ignition-reaction zone, a secondary zone of continuing combustion, and finally a third zone, which might be called the quenching or mixing zone.

A somewhat modified arrangement involving the same principles is illustrated in Fig. 9-11. In this arrangement, as before, the primary air for combustion combines with the fuel in the ignition and reaction zone,

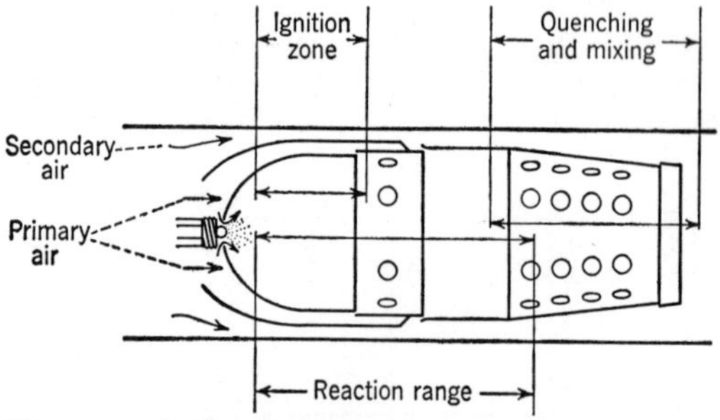

Fig. 9-11. Diagrammatic sketch of cell-type combustor showing various zones.

following which secondary air enters through side-wall cooling slots. Holes are also provided for lateral air admission. The fuel is injected under pressure through an atomizing nozzle just ahead of the ignition zone. In the ignition and reaction zone the primary air and fuel alone react. Farther downstream, turbulent mixing occurs as air enters laterally into the combustion products.

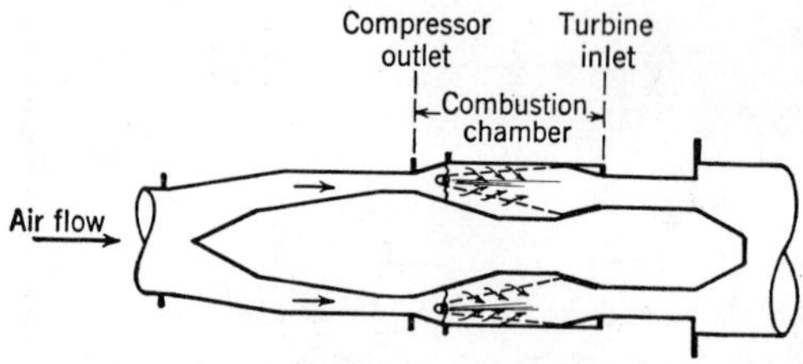

Fig. 9-12. Diagrammatic view through gas turbine showing an annular combustion chamber.

Figure 9-12 shows diagrammatically an annular combustion chamber. In this arrangement, instead of each chamber being a unitary device, the combustion chamber occupies a complete annular space around the turbine. As before, it is necessary to isolate an appropriate quantity of air to constitute the primary air stream for ignition and combustion, following which secondary air mixes in with the combustion products. The annular combustor, in contrast to single-or multiple-cell burners, is generally more compact in construction, and for aircraft jet units this makes possible a smaller over-all diameter for the unit. Although multiple ignition points are usually provided for an annular combustor, it is possible to operate with one ignition spark plug and the flame progresses throughout the total annular space.

The liner surface area for a given cross section for flow is usually less in an annular chamber than with a group of multiple chambers having the same flow cross section. On the other hand, uniform mixing and cooling may be more difficult to accomplish, and it may also be difficult to vary the primary and secondary air as precisely as is possible with unitary chambers. In Art. 9-17 a detailed description of the annular combustion chamber as used on a jet-turbine unit is discussed.

No matter which type of combustion chamber is provided, certain generalizations can be made. When primary air velocity rises above certain critical values, there is an increasing tendency for blowouts to occur. The stability of the flame is affected by the conditions of the walls; with a relatively cold wall, the flame is extinguished if it approaches too close to the wall. Carbon formations can be objectionable, as under certain conditions carbon can deposit in the chamber, while at other times it can enter into active combustion with the air stream. This secondary combustion can result in high wall temperatures and interfere with air and fuel distribution. Carbon deposition can accumulate to such an extent as even to interfere with the flame. If surface temperatures surrounding the combustion chamber are high, there is minimum tendency for carbon to deposit. Porous surfaces through which air enters tend to reduce carbon deposits. It is also possible to have the air stream arranged in such manner as to sweep over the surface, dislodging loose particles of carbon into the active combustion stream, where they can burn. In addition to the carbon problem, before combustion is completed the liquid fuel may strike against the walls, producing a wet condition, sometimes with resulting tar formation. However, under most conditions with a hot wall, the fuel usually vaporizes before coming into contact with the wall so that it continues in the air stream without objectionable wall effect.

9-16. Combustors of Westinghouse Land Turbine. On the unit pictured in Fig. 8-11 there are 12 cell-type combustors. Figure 9-13 shows a detail of the combustor in a longitudinal cross section. Each combustor is 4½ in. in diameter and 3 ft long. The combustor casing is made of carbon-steel pipe with a bellows-type expansion joint welded to one end with tapered flanges for anchoring the combustor in position. The flame tube is of chromium-nickel-alloy sheet rolled into a series of circular sections spot-welded to each other. The fuel nozzles operate on Bunker C (No. 6) fuel oil, which in this particular design can be ignited without the necessity of starting with a lighter grade of fuel oil. The oil is injected through the air-atomizing spray nozzles, which are located at the end of the flame tube. After the oil has once been ignited, combustion continues in the primary combustion zone, to which only a portion of the air is admitted. The rest of the combustor is surrounded by the secondary air, which mixes in with the primary combustion products farther down in the combustion tube. To produce this mixing, the secondary air enters the center section through holes in the flame tube.

Ignition of the fuel is carried out by use of an acetylene flame which is directed from the flame tube shown above and to the right of the fuel-inlet tube. The acetylene gas and air mixture is supplied at a pressure 5 to 7 psi in excess of the compressor discharge pressure. The acetylene-air mixture which flames into the combustion space is itself ignited by a spark plug. A gas flame for ignition was used to eliminate the spark-plug point fouling which has occurred in some cases when the ignition points

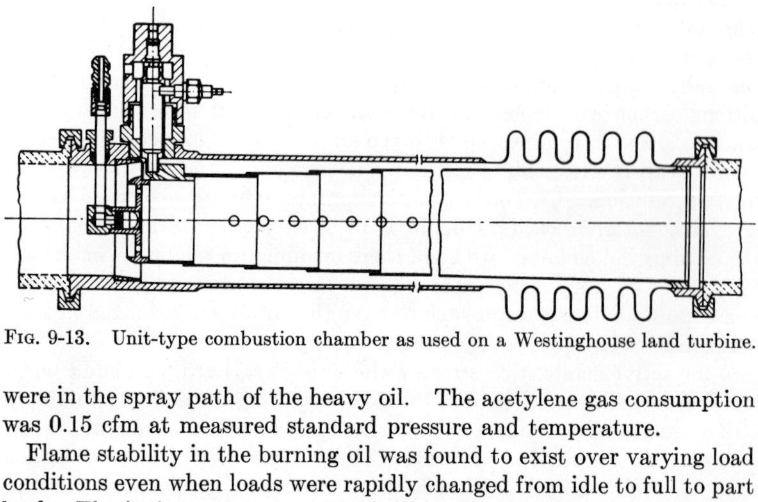

FIG. 9-13. Unit-type combustion chamber as used on a Westinghouse land turbine.

were in the spray path of the heavy oil. The acetylene gas consumption was 0.15 cfm at measured standard pressure and temperature.

Flame stability in the burning oil was found to exist over varying load conditions even when loads were rapidly changed from idle to full to part load. The igniter was not required during the load changes once combustion was established. At full load a typical combustor handled an air flow of 2.6 lb per sec and a fuel flow of 0.034 lb per sec. The temperature increased from 450 to 1350 F, and the pressure in the combustor was 58.8 psig. Between 24 and 72 cfm of atomizing air (measured at standard atmospheric conditions) was used in connection with the fuel delivery. Over-all air-fuel ratios for the combustor ranged from 70 to 171, and combustion efficiencies in excess of 96 per cent were obtained in some tests.

9-17. Annular Combustion Chamber for a Jet Unit. The combustion chamber which is used on the J-34 Westinghouse turbojet unit is illustrated in a cutaway section in Fig. 9-14. This turbojet uses an 11-stage axial-flow compressor. In this figure it will be noted that compressed air from the compressor enters a diffusion passage and passes from this into the combustor. The central part of the combustor is not employed as an air passage. In fact the connecting shaft between the turbine and compressor is located in this space.

The combustor itself is made of fabricated stainless steel. In it the air flow is divided into annular paths. Next to the external housing is an outer liner. Air passes radially inward through the outer liner and later mixes with the combustion gases. A similar inner liner also exists to keep the central core of the unit at reduced temperature.

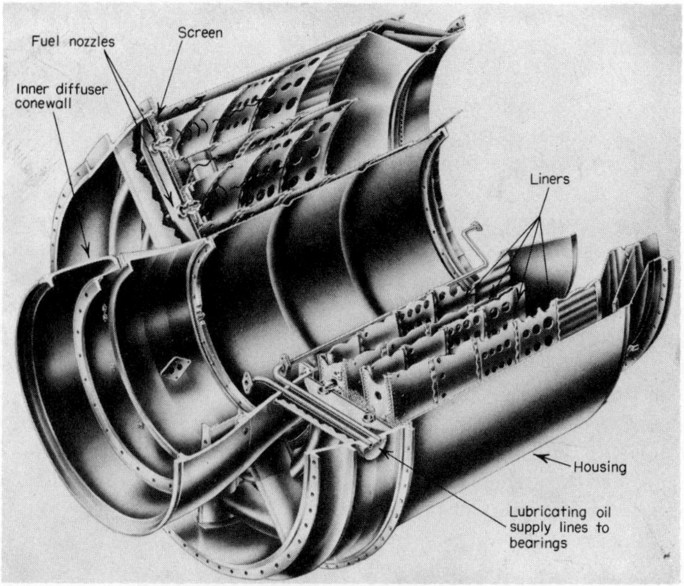

FIG. 9-14. Annular combustion chamber of Westinghouse J-34 jet engine.

The fuel nozzles spray into two annular combustion sections. Here the fuel, atomized under pressure, burns in the limited primary air which enters these two annular sections near the compressor end of the unit. Secondary air enters downstream through the radial openings in the liners. A careful selection of hole sizes in the combustion liners has been found necessary to achieve combustion stability under the wide range of altitude and flight-speed conditions which necessarily arise in connection with operation of a jet engine.

Ignition is accomplished by igniter spark plugs, which are served from duplicate ignition coils. The fuel system employs a pump and governor combination and is supplied with an additional emergency fuel pump.

The fuel is supplied to the spray nozzles from two fuel-line manifolds located at the forward end of the combustion chamber. An automatically operated dump valve drains the system of fuel on shutdown. Normally gasoline is used as fuel.

9-18. Combustion-chamber, 4800-horsepower General Electric Locomotive Turbine. Figure 9-10 is a cross section of one of the six combustion chambers of the locomotive gas turbine pictured in Fig. 1-9. This chamber is designed for operation on Bunker C oil, after starting from cold on a lighter oil. Two types of oil-atomizing fuel nozzles showing the air and fuel paths are illustrated in Figs. 9-15 and 9-16. The oil-pressure drop through the nozzles is in the neighborhood of 400 psi. The fuel oil is supplied by an 18-cylinder variable-displacement pump so that each of the six fuel nozzles is fed separately by three of the pump cylinders. Atomizing air mixes with the fuel and directs it into the combustion chamber. This air is supplied by a centrifugal compressor, which further compresses the air delivered from the compressor before employing it as atomizing air. The ratio of atomizing air to fuel ranges from 0.51 down to 0.11 lb of fuel per lb of air, and the air-nozzle pressure is about 1.6 times the combustion-chamber pressure. The Bunker C oil is preheated in cooling the compressed air supplied to the air-atomizing compressor. Combustion air enters the cap of the unit and also through 24 1-in. holes in the first three liner sections. Following this, the sheathing air flow enters to decrease the temperature of the combustion gases and assist in cooling the liner of the combustion chamber.

Some test operating data on this unit indicate an 800° temperature rise, producing an outlet-gas temperature of 1300 F, a chamber pressure of 87 psia, air flow through each chamber of 15.6 lb per sec, a pressure drop of 4 to 6 per cent of total inlet pressure, a combustion efficiency ranging from 95 to 98 per cent, and heat releases of 5,900,000 Btu/(hr)(ft^3). The liner thickness is 0.125 in., and its weight is 65 lb.

In the whole series of tests, the temperature of the liner ranged from 1200 to 1500 F. Tests on this chamber were made with kerosene as well as with Bunker C fuel, and it was observed that the fuel fired has little or no effect on the temperature pattern in the combustion chamber. Circumferential nonuniformities in fuel distribution did have a marked effect on chamber temperatures. It was found that Bunker C could be burned without soot and coke formation, but proper atomization was found to be very important. Deposits occurred mainly under idle-speed and no-load conditions in the chamber. Two things account for this, *viz.*, the atomizing air is lower in quantity, and the combustion air is colder. The smoke formation, although noticeable, was not considered to be objectionable. It was somewhat less dense in using kerosene than in using the heavier fuel. Spark plugs were used to ignite the diesel fuel which was used dur-

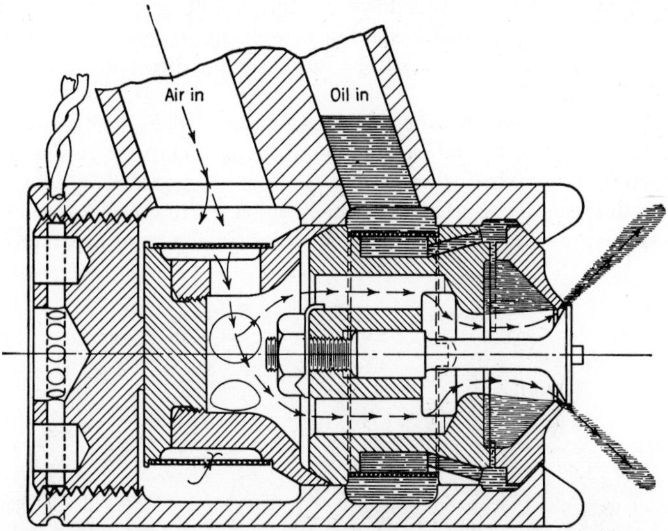

FIG. 9-15. Section through pintle vortex-type air-atomizing fuel nozzle for a General Electric combustor.

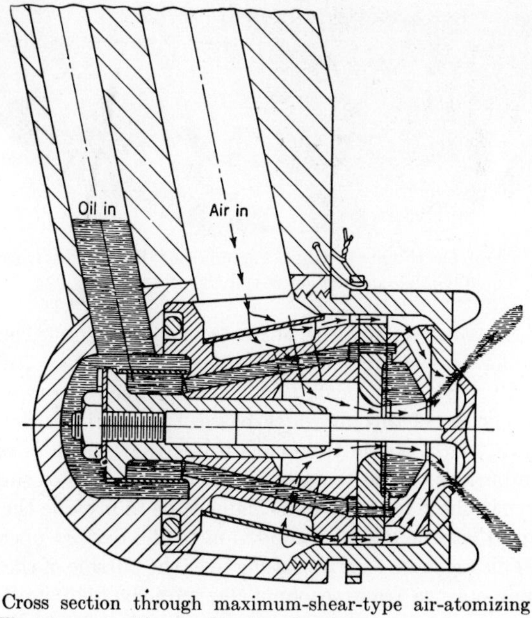

FIG. 9-16. Cross section through maximum-shear-type air-atomizing fuel nozzle for General Electric gas-turbine combustion chamber.

ing starting. When shutting down the unit, a shift-over to diesel fuel was made so that the oil passages could be completely purged of Bunker C oil. As far as ignition was concerned, the Bunker C and diesel fuel were equally easy to ignite.

9-19. Combustion Chamber of Ghost Engine. The Ghost engine (Art. 7-7), manufactured by the De Havilland Engine Company, in Fig. 9-17 is pictured in a rear view which shows the turbine preceded by 5 of its 10 combustion chambers. Combustion-chamber parts are pictured in Fig. 9-18. It can be seen that the chamber consists of a front and rear cover

FIG. 9-17. View of De Havilland Ghost engine from the rear, showing the turbine and 5 of the 10 double-inlet combustion chambers.

section and an inner flame tube of heat-resistant steel. The cast inlet head (at the left) is provided with a duplex entry, and air feeds over the fuel-spray valve, which is located centrally between the two entry passages. The portion of this air which passes through the scoop section in the dome head of the flame tube constitutes that which is required for primary combustion. The remainder and greater part of the air enters through holes in the side of the inner flame tube to provide the secondary air and dilute the products of combustion to the desired operating temperature. This secondary air in passing over the outside of the flame tube keeps its temperature from reaching dangerously high limits. It also protects the outer shell of the combustion chamber from the extremely high temperatures. The flame tube can expand both radially and axially, as it is supported in the chamber on resting piers or plugs. Two fuel pumps supply fuel to the burners in the combustion chambers. Ignition is accomplished by two spark-plug-type igniters, which ignite the fuel entering two of the combustion chambers. The flame from these ignition points spreads from combustion chamber to combustion chamber through interconnecting pipes.

In starting this engine an electric starter is used. After approximately 8 sec of engine turning, the igniters ignite the fuel being delivered from the fuel pumps. Continued application of starter power assists in speeding

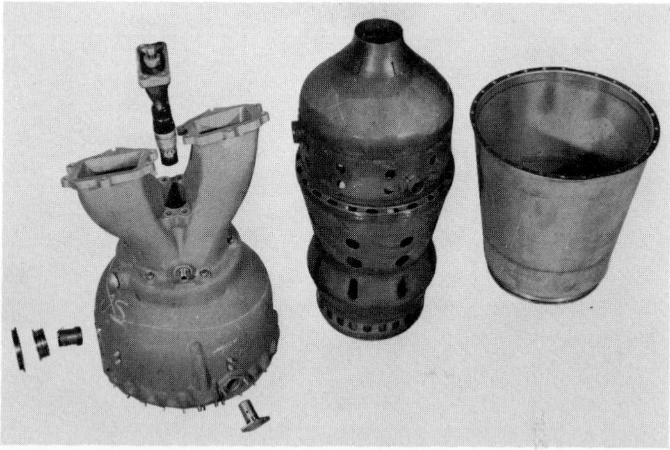

FIG. 9-18. Details of De Havilland Ghost combustion-chamber assembly, showing front outer casing with burner and flame-tube plug, flame tube in center, and rear outer casing at right.

up the engine in the shortest possible time, with the engine itself accomplishing the latter part of the acceleration to idling speed. This whole starting process can be accomplished within 30 sec after pressing the starting button.

It will be noticed that these combustion chambers are essentially straight-flow-through-type units so that a minimum of pressure loss is required in their operation. Turbulence is accomplished by giving the secondary air a slight swirl as it enters the inner flame tube. The rear end of each combustion chamber feeds into the turbine entry duct. A separate sealing ring allows for expansion at this point.

9-20. Fuel-Air Ratio Charts. In making combustion computations, it is frequently desirable to know the fuel-air ratio associated with a given temperature rise. Figures 9-19 to 9-21 are plotted from basic calculations to give desired information of this type. In using them, subtract the temperature of the air leaving the compressor from the desired temperature of the gas products leaving the combustor, to find the temperature rise. Starting with the temperature-rise value on the left axis, run horizontally over until intersection is made with one of the base temperature lines on the chart. From the intersection with the base temperature line, drop vertically down to the bottom of the chart, and read the fuel-air ratio required to produce the desired temperature rise. For example, consider gasoline as a fuel. Using Fig. 9-19, if 800 R air leaving the compressor is to be raised to a final temperature of 1800 R leaving the combustor, it will be found that for this 1000° rise the fuel-air ratio is 0.0150.

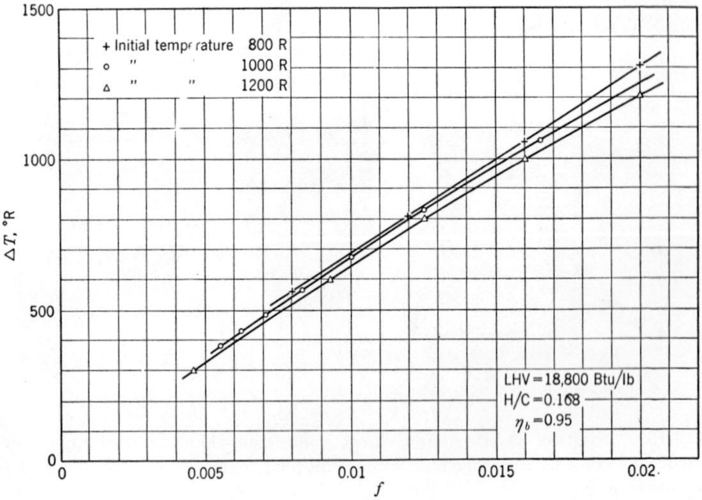

FIG. 9-19. Fuel-air ratio, temperature-rise graphs for a representative gasoline.

It will be observed that if the base temperature is higher than 800 R, a greater fuel-air ratio would exist. This is true because the specific heat of the gaseous products increases as the temperature level is increased and thus relatively more fuel is required.

The information for these graphs was obtained by use of the computational methods developed in Art. 9-14, and the graphs are based on a combustion efficiency of 95 per cent, with representative lower heating values being used for the gasoline, light fuel oil, and residual fuel oil of the respective charts of Figs. 9-19 to 9-21.

Figure 9-22 has been prepared to show graphically how the temperature of exit combustion gases can be reduced by increasing the air-fuel ratio, *i.e.*, by lowering the fuel-air ratio f. This graph is plotted for a single

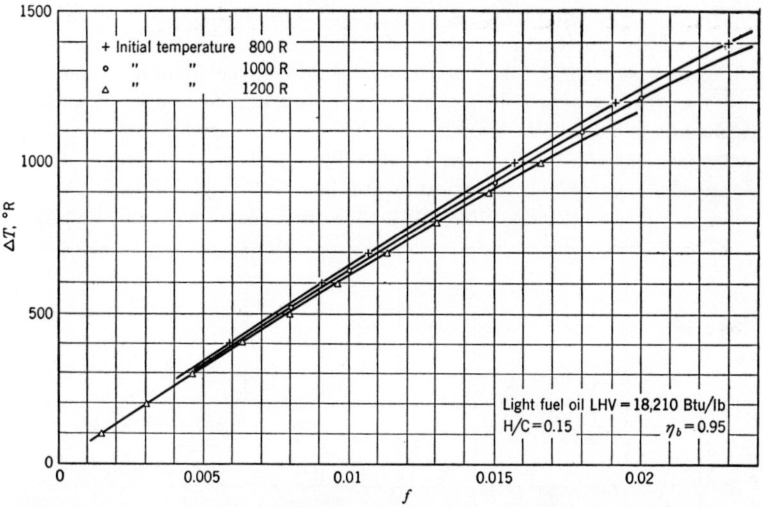

FIG. 9-20. Fuel-air ratio, temperature-rise graphs for light fuel oil (diesel fuel).

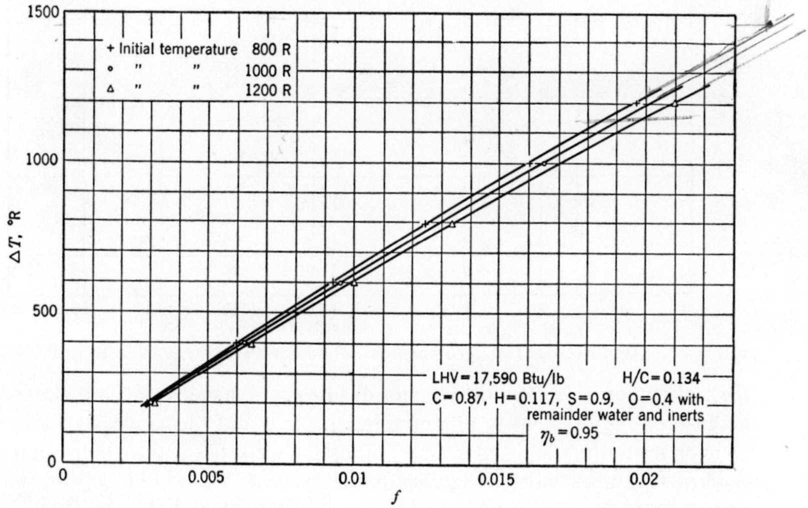

FIG. 9-21. Fuel-air ratio, temperature-rise graphs for residual fuel oil (Bunker C).

datum temperature at inlet to the combustor (namely, 1000 R) but is representative of the shape such graphs would take. For example, for an exit temperature of 1860 R, an air-fuel ratio of 80 is required, whereas to produce the lower temperature of 1470 R relatively much more air is required at an air-fuel ratio of 160:1.

The charts covered by Figs. 9-19 to 9-22 represent the most usual fuels employed by gas-turbine units. In the case of jet engines, gasoline formerly was used almost exclusively because it represented a good fuel and was readily available. However, the jet engine is not sensitive to variations in good volatile fuels, and so there is a growing tendency to set fuel specifications toward less volatile fuels and with no regard to the octane number, which is so important in aviation gasolines. The less volatile the fuel, the less possibility there is of vapor loss and boiling at

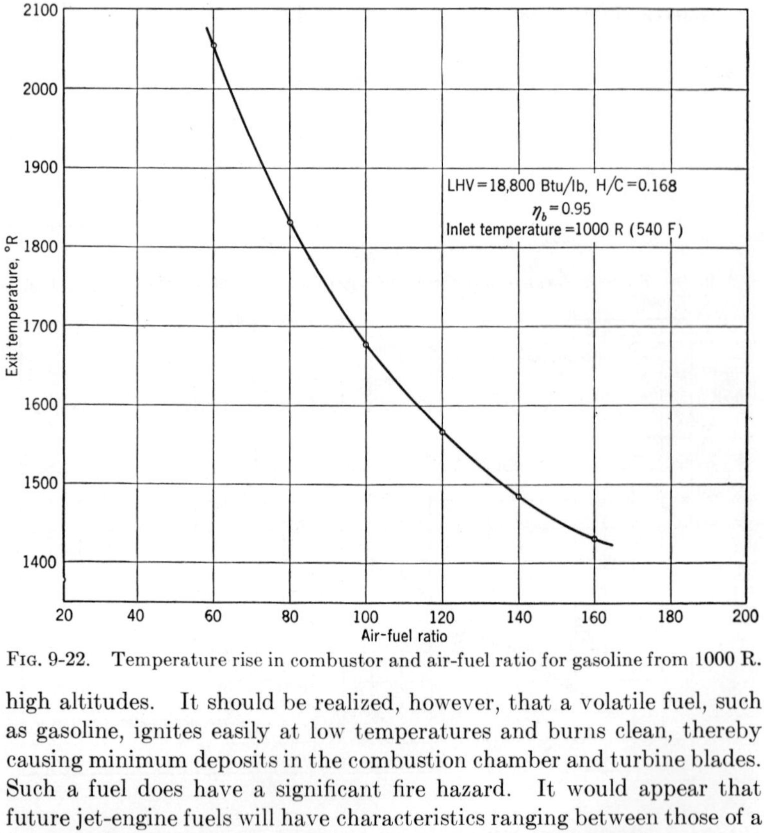

FIG. 9-22. Temperature rise in combustor and air-fuel ratio for gasoline from 1000 R.

high altitudes. It should be realized, however, that a volatile fuel, such as gasoline, ignites easily at low temperatures and burns clean, thereby causing minimum deposits in the combustion chamber and turbine blades. Such a fuel does have a significant fire hazard. It would appear that future jet-engine fuels will have characteristics ranging between those of a low-volatile gasoline and a high-volatile kerosene. This would mean that the end point in a distillation test for the fuel would rise from the approximate 400-F value for a volatile gasoline to some 600 F for a representative jet fuel.

In the case of large engines for stationary and railway service, it appears that these will use a kerosene or light fuel oil for starting, followed by change-over to a heavy or residual fuel oil. The lighter fuel gives cleaner combustion, but the heavier, cheaper fuel offers no serious difficulty to effective gas-turbine performance. Natural gas may also become an important fuel in certain types of installation.

9-21. Solid Fuels and Gas Turbines. Solid fuels in one form or another represent the largest reserves of stored energy available to the world. Consequently, it is not surprising that much thought has been given to the possibility of using solid fuels in connection with gas turbines. Research is under way in this country by the Locomotive Development Committee of Bituminous Coal Research, Inc. This research has been primarily concerned with the problem of coal pulverization, combustion, and ash separation. The latter aspect of the problem has in many ways caused the greatest difficulty, as ash carried into the turbine could cause clogging of passages and erosion. Otherwise it is possible that conventionally designed gas-turbine equipment can be employed with but minor modification and the addition of coal-processing equipment. With a closed-cycle unit, coal-fired equipment need only consider the problem of heat transfer through the metal surface and recovery of residual energy from the stack gases. Results have even been reported on a 500-hp closed-cycle gas-turbine engine fired by peat. This unit is in test operation at John Brown and Company in Glasgow, Scotland. Vast deposits of lignite also offer an additional energy source which might be adapted to gas-turbine operation.

PROBLEMS

9-1. A fuel having weight fractions of 0.874 parts of carbon, 0.111 parts of hydrogen, and 0.015 parts of inert material is burned with no excess air. (*a*) Find the weights of products of combustion and nitrogen. (*b*) Find the weight composition of the wet products of combustion. (*c*) Find the composition by weight of the dry products. (*d*) Find the composition on a volumetric basis of the dry products.

9-2. Work Prob. 9-1 when 500 per cent excess air is employed.

9-3. For the hydrocarbon, normal dodecane ($C_{12}H_{26}$), burned with no excess air, write the combustion equation. (*a*) State the resulting moles of products for each constituent per mole of fuel. (*b*) Find the volumetric composition of the dry products of combustion. (*c*) What is the air-fuel ratio?

9-4. Work Prob. 9-3 if six times the theoretical air is used in combustion. Find also the air-fuel ratio (by weight).

9-5. The dry products of combustion from a natural gas analyze 13.4% of CO_2, 4.6% O_2, 82.0% N_2. The fuel by weight contains 63.2% C, 22.1% H, 11.7% N_2, 3.0% O_2. (*a*) Find the pounds of dry flue gas per pound of fuel; (*b*) the pounds of air per pound of fuel; (*c*) the fuel-air ratio.

9-6. Write the combustion equation for octane (C_8H_{18}) when using 100 per cent excess air. (*a*) Find the volumetric analysis of the wet products of combustion. (*b*) Find the volumetric analysis of the dry products of combustion. (*c*) Find the air-fuel ratio (by weight).

9-7. A light fuel oil for use in a jet engine has a lower heating value of 18,210 and a H/C ratio of 0.15. The combustion efficiency can be considered as $\eta_b = 0.95$. Plot a curve using fuel-air ratio as the abscissa and as ordinate temperature rise in the combustor when inlet conditions to the combustor are (*a*) 700 R; (*b*) 900 R. Velocity variations on entering and leaving the combustor need not be considered. Consider fuel to be at 600 R.

9-8. A motor gasoline is used as a fuel in a jet engine. The lower heating value of the gasoline is 18,800, and the H/C ratio is 0.168. The combustion efficiency can be considered as $\eta_b = 0.95$. Plot a curve using fuel-air ratio as the abscissa and as ordinate temperature rise in the combustor when inlet temperature to the combustor is (*a*) 800 R; (*b*) 1200 R. Velocity variations on entering and leaving the combustor need not be considered. Consider fuel to be at 600 R.

9-9. A mid-continent residual fuel oil of No. 6 grade has an analysis of C = 0.87, H = 0.117, S = 0.09, O = 0.04, with the remainder water and inerts. This fuel has a higher heating value of 18,612 Btu per lb and a lower heating value of 17,590 Btu per lb. Its degrees API are 11.4, and the H/C ratio is 0.134. This fuel is used in a stationary-type gas-turbine combustor with a combustion efficiency of 0.95. Plot a curve using fuel-air ratio as the abscissa and as ordinate temperature rise in the combustor when inlet conditions to the combustor are (*a*) 900 R; (*b*) 1100 R. Velocity variations on entering and leaving the combustor need not be considered, nor should the sulfur and other extraneous items in the fuel oil enter in the calculation. Consider fuel to be at 600 R.

9-10. Work Prob. 9-9 for inlet air to the combustor at 700 R.

REFERENCES

1. English and Wachtl: Charts of Thermodynamic Properties of Air and Combustion Products, *NACA Tech. Note* 2071, April, 1950.
2. Turner and Lord: Thermodynamic Charts for the Computation of Combustion and Mixture Temperatures at Constant Pressure, *NACA Tech. Note* 1086, 1946.
3. Hershey: Development and Testing of a Gas Turbine Combustor, *Trans. ASME*, vol. 69, no. 8, pp. 859–867, November, 1947.
4. Buckland and Berkey: Combustion System for Burning Bunker C Oil in a Gas Turbine, *ASME Paper* 48-A-109, 1949.
5. Putz: 2000 Horsepower Gas Turbine Generator Set, *ASME Paper* 46-A-67, 1947.
6. Childs, McCafferty, and Surine: Effect of Combustor-inlet Conditions on Performance of an Annular Turbojet Combustor, *NACA Rept.* 881, 1947.
7. Pinkel and Shames: Analysis of Jet-propulsion-engine Combustion-chamber Pressure Losses, *NACA Rept.* 880, 1947.
8. Pinkel and Karp: A Thermodynamic Study of the Turbojet Engine, *NACA Rept.* 891, 1947.
9. ASTM Standards, each with a specific designation, American Society for Testing Materials, 1916 Race St., Philadelphia, Pa.
10. Way, Stewart: Turbojet Combustion Chamber Problems, *Aero Digest*, vol. 60, pp. 52–56, 87–88, February, 1950.
11. Howes and Rampton: Properties of Hydrocarbon Gas Turbine Fuels, *J. Inst. Petroleum (London)*, vol. 35, pp. 419–435, June, 1949.
12. Symposium on Combustion Reactions in Relation to Gas Turbine Practice, *Jr. Inst. Petroleum (London)*, vol. 37, pp. 487–581, September, 1951.
13. "Hydrocarbons," 3d ed., Phillips Petroleum Company, Bartlesville, Okla., 1949.
14. Lewis, Hottel, and Nerad: "3d Symposium on Combustion and Flame and Explosion Phenomena," The Williams & Wilkins Company, Baltimore, 1949.
15. Nelson, W. L.: "Petroleum Refinery Engineering," 3d ed., McGraw-Hill Book Company, Inc., New York, 1949.
16. Smith and Stinson: "Fuels and Combustion," McGraw-Hill Book Company, Inc., New York, 1952.

CHAPTER 10

STRESSES IN TURBINE-PLANT ELEMENTS

10-1. Mechanical Problems. In translating the thermodynamic principles of gas-turbine analysis into operating machines, many practical difficulties arise. There are problems of cost, materials, weight, space, tolerances, vibration, noise, thermal and mechanical stresses, and safety. In the service which is required of the machine, further difficulties may be encountered in connection with gyroscopic effects, speed requirements of the load, selection of couplings and gears, and additional possibilities of vibration. Many of these practical problems are interrelated, and the designer cannot always isolate one problem at a time. For example, noise and vibration can arise simultaneously.

Some of the basic mechanical problems of gas turbines will now be discussed in a brief and general way. The attempt here is not necessarily to *solve* all the problems which arise but to *recognize* and *present* them. Such a brief survey is valuable in focusing attention on the types of problems encountered and serves as a base from which more detailed studies may be made. It will be found, as might be expected, that many of the problems of gas turbines are similar to those encountered with steam turbines.

The present chapter will point out the main forces and stresses which must be considered in a gas-turbine unit which is operating at a steady speed and load and with no vibration of any parts. Then the next chapter will consider the vibration problems which arise to be superimposed on the steady-load picture.

10-2. Stress Calculations. The purpose of stress calculations is to predict operating stresses, with the ultimate aim the prevention of failure. Some knowledge of the behavior of the materials of construction is necessary in order that a calculation method may be selected. For example, is the material going to behave elastically in service so that the stress equations used may assume a straight-line relationship between stress and strain? Or is plastic behavior expected, with necessary modifications in calculation procedure?

Once stress calculations have been performed, the stresses found must be compared with stresses which the material can withstand. Now another problem appears. What should be the basis for comparison? Should the calculated stresses be compared with yield point, ultimate strength, creep-rupture strength, fatigue limit, or stress producing a specified amount of creep in a specified time at a specified temperature? It seems obvious that the type of load (steady or fluctuating) and a criterion of failure must be decided upon.

For each element of a gas-turbine unit, four factors are of importance; life expected of the part, temperature, elongation, and stress. For example, a high-performance military jet engine is not expected to last as long as a stationary unit used for power generation, and so higher temperature and stresses are used in the jet-turbine unit.

Two criteria of failure may be recognized, (1) complete rupture and (2) a specified limiting amount of deformation. Complete rupture is failure, of course, but how much deformation constitutes functional failure? This question has two aspects. Excessive deformation can cause physical contact between moving parts (as turbine blades rubbing against the casing); or the deformation can produce warping of blade shapes, nozzle vanes, or other parts, with a resulting loss in performance of the gas-turbine unit. Severe loss of performance is certainly *functional* failure, although no mechanical or kinematic failure is involved.

It appears that the yield strength of many gas-turbine materials should be high, in order that excessive deformation may be prevented, and that for hot parts creep resistance should also be high, so that prolonged operation under load at elevated temperatures is possible. If loads fluctuate, then fatigue strength at operating temperature becomes important.

The purpose of the foregoing discussion is to provide a brief background which can serve as a basis for selection of a reasonable calculation method. Stress calculations are often made on the basis of the theory of elasticity, in which it is assumed that stress and strain are proportional. If the operating stress does not exceed the elastic limit (very close to the yield strength), such calculations serve very well. It should be realized that stresses computed by this theory cannot always predict failure. If calculated stresses are higher than the elastic limit of the material, then the elastic equations do not apply, and it should be realized that yielding will occur and stresses will be lower than those calculated by assuming continued elastic action. Local yielding may occur in many places during operation without causing failure, and sometimes this local yielding is actually desirable in order to smooth out stress peaks and obtain a more uniform distribution of stress.

The long-time permanent deformation of creep at high temperature under load can occur with low stress, and elastic theory may be used to estimate the working stress. *Failure*, however, depends upon time, temperature, and allowable deformation as well as upon the stress or load.

In specifying materials for gas-turbine parts, many properties must be considered. The subject of materials is treated in more detail in Chap.

12. Certainly two items of great importance are the yield strength and creep strength, and operating stresses usually are specified at low enough values so that stress equations can be used which assume elastic behavior. At the same time, it is realized that in localized areas some stresses may exceed yield strength, and so enough ductility is desired to allow deformation to take place without local cracking. If cracking develops, it spreads rapidly and usually leads to rupture.

Elastic-stress equations are sufficient for proportioning parts for uniformity of stress distribution, resulting in shapes which carry the necessary loads with a minimum amount of material. One outstanding example is the turbine disk, which usually approximates a constant-stress design, as will be discussed later in the chapter.

In the following articles, which deal with the calculation of stresses in the critical parts of gas turbines, the stress equations which are used assume elastic action—stress and strain are proportional. Suitable modification of the calculated stresses and suitable factors of safety (sometimes called factors of ignorance) are always necessary in the light of operating experience.

10-3. The Steady Loads in Gas Turbines. It will be assumed that a perfectly balanced turbine and compressor unit is driving a shaft-connected load under steady-state conditions. There is no vibration of parts resulting from variations of pressure or from any other cause, and rpm is constant.

There are forces applied to all parts, of course, but not all will be mentioned. For example, the load shaft must carry the load, and the casing, mounting arrangements, bolting, and any heat-exchange equipment must be sufficiently strong and rigid. Proper bearings and lubrication must be supplied. These are problems not entirely peculiar to the gas turbine. The most critical parts are the rapidly moving parts within the turbine and compressor, and the discussion will deal primarily with these parts—the rotors and blades.

10-4. Steady Loads on the Blades. There are the possibilities of gas, inertia, and thermal forces and stresses. The gas forces, in general, have tangential (F_T) and axial (F_A) components which act on the blades of the rotor. These components result from gas-momentum changes and from pressure differences across the blades. The velocity diagrams and thermodynamic relationships previously developed indicate how these gas forces can be determined. If the gas forces are distributed uniformly across a blade, then the resultant acts through the centroid of the area. Such a distribution is not common, however.

The blades are subject to large centrifugal forces and stresses at the high speeds of rotation employed. The centrifugal forces are usually much greater than the gas forces and are often a limiting factor in design.

The expression for total centrifugal force on a blade is

$$f_c = \frac{w}{g_c} r\omega^2 \tag{10-1}$$

where f_c = centrifugal force, lb_f
w = weight (mass) of blade, lb
g_c = 32.17 lb per slug (see footnote in Art. 2-1)
r = radius from center of rotation to center of gravity of blade, ft
ω = speed of rotation, rad per sec

Obviously, the force is minimized if the blade mass is as small as possible and if the center of gravity is as close as possible to the center of rotation. Blades are often tapered to achieve these aims.

Tapered blades are desirable from another viewpoint, too. It should be recognized that the stress caused by centrifugal force diminishes from a maximum at the root to a minimum of zero at the tip in a blade of uniform cross section. For example, a cross section halfway between tip and root would be subject only to the centrifugal force of the blade mass lying *outside* the cross section. It follows that the middle cross section need not be as large as the root cross section for the same stress. It also follows, then, that blades should be tapered in order to utilize the least material, and that most effectively. Constant-stress design is possible and seemingly desirable. Equations based on constant stress show that the blade shape should be a curved taper, but this shape is not always practical with respect to *flexural* strength. A straight-line approximation is sometimes used, giving a linear taper in cross-sectional area from root to tip. Figure 10-1 illustrates a tapered blade.

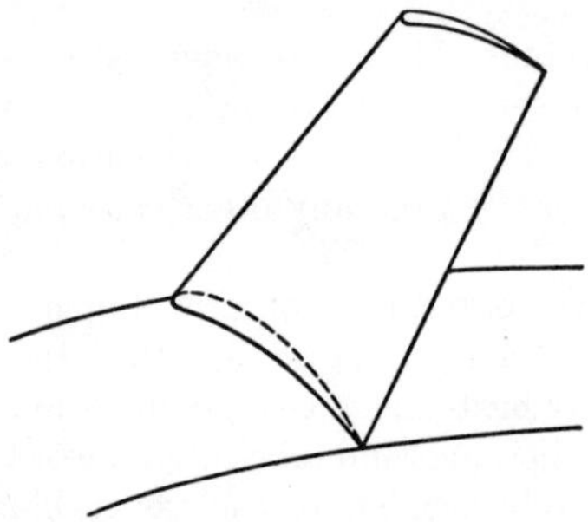

FIG. 10-1. Blade tapered from root to tip in order to reduce centrifugal stress.

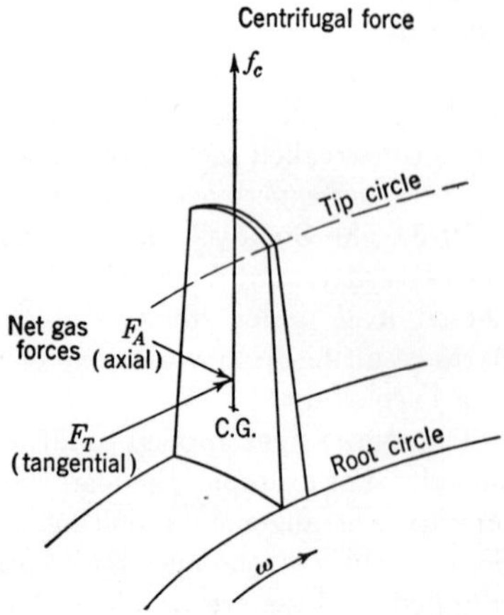

FIG. 10-2. Steady loads imposed on a turbine or compressor blade during operation.

Figure 10-2 shows a blade with very little twist from root to tip, with its typical steady loading. Probably the stress caused by steady axial loading is not usually very large, although it could become important for some sections in a blade with a large twist from root to tip. The summation of all axial forces produces a net axial *thrust*, of course, which must be balanced by suitable thrust bearings or by air-pressure balancing diaphragms or pistons.

Let us focus attention on f_c and F_T. Figure 10-3 shows the simplified picture. The critical section for stress is at the root, which must take the

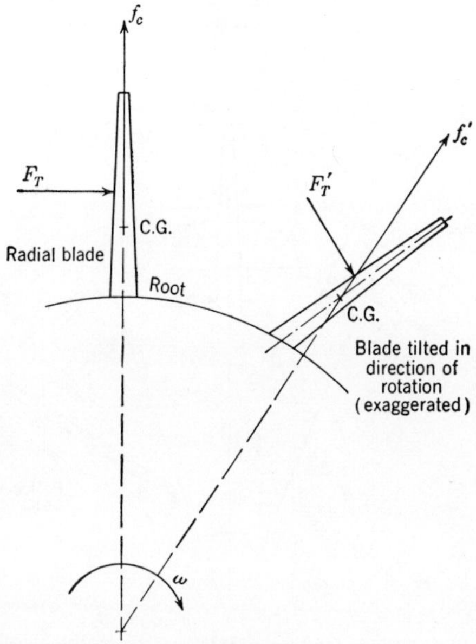

FIG. 10-3. Tilting the blade gives a centrifugal moment about the root which counteracts the moment of the tangential gas force.

full centrifugal force of the whole blade mass and also the bending stress from the gas force. Note that the bending moment of the gas force can be exactly canceled by the moment of the centrifugal force if the blade is tilted in the direction of rotation. Such a design results in exact cancellation for only one set of gas-flow conditions occurring simultaneously with any one certain rpm. The method may be used to design for a particular high operating load and speed. Of course for other operational conditions the bending stress is also then reduced but not entirely canceled.

Approximate stresses can be found from the usual stress formulas for bending moment and for direct load, and the equations need not be repeated here. A simple example will be used to present and review the fundamentals.

Example 1. A radial tapered blade of small twist with given dimensions, gas torque load, and rotor speed is pictured in Fig. 10-4. Only the loading in a plane perpendicular to the axis is considered to be significant. Assume the root section to be approximately rectangular, with a tangential thickness of 0.3 in. and axial length of 1 in. as shown. Let the blade weight be 0.18 lb, and assume the resultant tangential gas force caused by gas-momentum change to be 40 lb acting through the blade centroid as shown. Find the maximum stress in the critical root section.

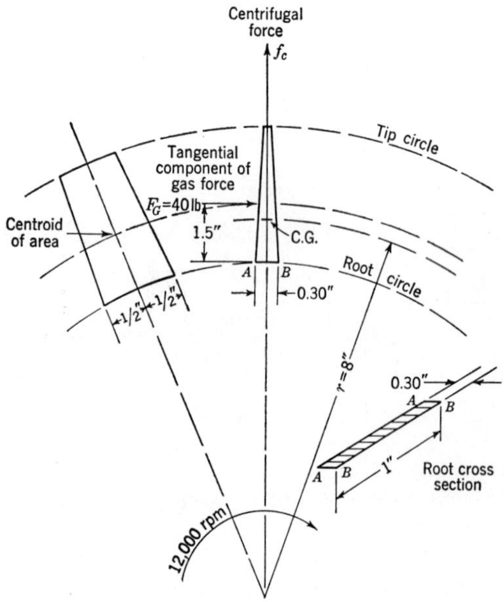

FIG. 10-4. Blade loading and dimensions for Example 1.

Solution. Find the centrifugal force first by use of Eq. (10-1).

$$
\begin{aligned}
f_c &= \frac{w}{g_c} r\omega^2 \\
&= \frac{0.18}{32.17}\frac{8}{12}\left(\frac{12{,}000}{60}2\pi\right)^2 \\
&= \frac{0.36}{96.5}(400\pi)^2 \\
&= 7860 \text{ lb}
\end{aligned}
$$

Find the direct tensile stress caused by centrifugal force.

$$
s_c = \frac{f_c}{A} = \frac{7860}{0.3} = 26{,}200 \text{ psi}
$$

For gas bending stress, use

$$s_g = \frac{MC}{I}$$

$$M = (40)(1.5) = 60 \text{ in.-lb}$$

$$C = 0.15 \text{ in.}$$

$$I = \tfrac{1}{12}bh^3 = \tfrac{1}{12}(1)(0.3)^3$$

$$= \tfrac{1}{12}(0.027) = 0.00225 \text{ in.}^4$$

Therefore,

$$s_g = \frac{(60)(0.15)}{0.00225} = 4000 \text{ psi}$$

The two stresses add along the edge AA, giving a total maximum tensile stress of

$$s = 30{,}200 \text{ psi}$$

The average shear stress across the root section will next be calculated.

$$s_s = \frac{40}{0.3} = 133.3 \text{ psi}$$

It is evident that the shear stress is usually small enough so that it can be neglected. It is also evident that the largest stress usually results from the centrifugal force on the blade. If the blade is tilted in the direction of rotation so that the centrifugal moment developed balances the gas moment (Fig. 10-3), then the only remaining stress of any consequence is the direct centrifugal stress. Design values for this stress vary in gas turbines from about 6000 psi up to a maximum around 30,000 psi, dependent upon the temperature, material used, and the life expectancy of the machine. The lower temperature compressor blades may be stressed to above 40,000 psi if made of steel or to 20,000 or a little higher if made of aluminum.

10-5. Blades with Sharp Curvature and Twist. Blades with appreciable curvature and appreciable twist from root to tip pose additional calculation problems. In calculating the gas bending moment at the root, it is convenient to divide the blade into narrow strips, each at substantially constant radius, and to determine the gas force on each strip from the gas flow through the flow area bounded on one edge by that strip together with the velocity and thermodynamic relationships which originally established the blade shape. The system of forces might look something like that shown in Fig. 10-5*a*. There are, in each strip, tangential and axial components of appreciable magnitude, plus a small radial component which can be neglected. The gas forces come about from two sources: gas-momentum changes; and when reaction is present, pressure difference across the blade acting on the projected blade area.

The resultant gas force on each strip is broken into the respective tangential and axial components, and moments about centroidal axes in the root section or about centroidal axes in the sections at the strip dividing lines can be calculated. There will be two moments, one about the AA axis and one about the TT axis of Fig. 10-5, when moments at the root are considered. A summation of the moments about root axis AA of tangential force components yields the moment for the gas forces on the whole blade about that axis, and similarly a summation yields the moment for the whole blade about the root axis TT.

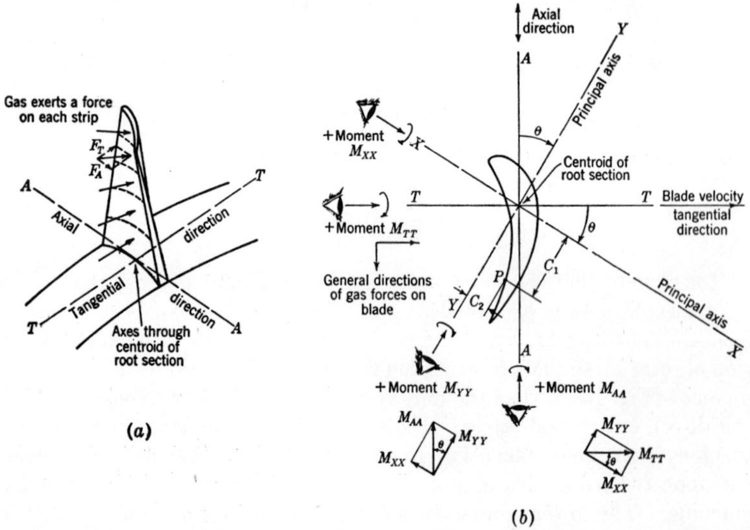

FIG. 10-5. (*a*) A twisted blade is analyzed as a series of strips with a gas force and a centrifugal force exerted on each strip. Centrifugal forces are not shown, but they act radially outward through the centers of gravity of the strips. (*b*) Moments must be resolved about the principal axes for calculation of bending stresses.

With the moments in the root section now known, the next problem is the calculation of stresses. The principles of mechanics require that the bending moments be obtained about the two principal axes in the root section. These axes are mutually perpendicular axes through the centroid about which the second moments of area are, respectively, a maximum and a minimum. For symmetrical areas, axes of symmetry are always principal axes. Figure 10-5*b* illustrates an example of an unsymmetrical area—the usual turbine or compressor blade root section. The principal axes are turned an angle θ from the tangential and axial directions about which moments have been determined.

Let the known moments about the tangential and axial axes be M_{TT} and M_{AA}, respectively. If M_{TT} is resolved into components about axes xx and

yy (Fig. 10-5b), there results

$$M_{xx} = M_{TT} \cos \theta$$
$$M_{yy} = M_{TT} \sin \theta$$

Similarly, for the resolution of M_{AA},

$$M_{xx} = -M_{AA} \sin \theta$$
$$M_{yy} = M_{AA} \cos \theta$$

Therefore, the total moments about the principal axes are

$$M_{xx} = M_{TT} \cos \theta - M_{AA} \sin \theta \tag{10-2}$$
$$M_{yy} = M_{TT} \sin \theta + M_{AA} \cos \theta \tag{10-3}$$

The algebraic signs should be correlated with the directions shown in Fig. 10-5b.

Now that the moments about the principal axes are determined, normal stresses may be approximated from $s = MC/I$ for each point in the cross section. For example, at point P, the stress caused by the moment about axis xx is

$$s_1 = \frac{M_{xx}C_1}{I_{xx}} \tag{10-4}$$

where s = normal bending stress, psi
M_{xx} = bending moment about the principal axis xx, in.-lb$_f$
C_1 = distance from P to the xx axis, in.
I_{xx} = moment of inertia (second moment) of the entire cross-sectional root area about axis XX, in.^{4}

Similarly, at point P the stress caused by the moment about axis yy is

$$s_2 = \frac{M_{yy}C_2}{I_{yy}} \tag{10-5}$$

The total bending stress at point P is then the algebraic sum of the two, provided this sum does not exceed the elastic limit.

$$s_P = s_1 + s_2 \tag{10-6}$$

It is usually relatively easy to visualize individually whether s_1 and s_2 are compressive or tensile and thus to check whether the total bending stress s_P is compressive or tensile.

The item of primary concern is, of course, the location and amount of the *maximum* stress. The extreme positions on the profile should be checked. The maximum stress does not always occur there, however, and the stresses at several other points may have to be determined. Variations in the profile shape produce variations in the location of the point of maximum stress.

Example 2. Refer to Fig. 10-5*b*. Assume the angle θ indicating the location of the principal axes to be 30°. The following additional information is known.

Gas force moment about axis AA = 10 lb-in. (in + direction as shown)

Gas force moment about axis TT = 4 lb-in. (in + direction as shown)

$$I_{xx} = 0.0025 \text{ in.}^4$$
$$I_{yy} = 0.0006 \text{ in.}^4$$
$$C_1 = 0.5 \text{ in.}$$
$$C_2 = 0.1 \text{ in.}$$

At point P, find the stress caused by bending.

Solution. First, resolve the moments into moments about the principal axes. Equations (10-2) and (10-3) apply.

$$\begin{aligned} M_{xx} &= M_{TT} \cos \theta - M_{AA} \sin \theta \\ &= 4 \cos 30° - 10 \sin 30° \\ &= 4(0.866) - 10(0.500) \\ &= 3.464 - 5.000 \\ &= -1.536 \text{ lb-in.} \\ M_{yy} &= M_{TT} \sin \theta + M_{AA} \cos \theta \\ &= 4(0.500) + 10(0.866) \\ &= 2.00 + 8.66 \\ &= 10.66 \text{ lb-in.} \end{aligned}$$

The signs should be interpreted by inspection of Fig. 10-5*b*. It should be noted that compression is produced at P by M_{yy}, and tension by M_{xx}.

Equations (10-4) and (10-5) may be used for determining the bending stresses.

$$\begin{aligned} s_1 &= \frac{M_{xx}C_1}{I_{xx}} = \frac{(-1.536)(0.5)}{0.0025} \\ &= -307 \text{ psi} \\ &= 307 \text{ psi (tension)} \\ s_2 &= \frac{M_{yy}C_2}{I_{yy}} = \frac{(10.66)(0.1)}{0.0006} \\ &= 1776. \text{ psi (compression)} \end{aligned}$$

Therefore the total stress at point P caused by bending produced by the gas forces is

$$\begin{aligned} s_P &= -307 + 1776 \\ &= 1469 \text{ psi (compression)} \end{aligned}$$

There may also be stresses at P caused by direct centrifugal loading and possibly also bending stresses produced by moments of centrifugal force. Consider the centrifugal force acting on the mass in each of the blade strips of Fig. 10-5*a*. The centrifugal force on each strip will act radially outward through the center of gravity of the strip. The line of action of the centrifugal force for any one strip, when extended, will not necessarily pass through the centroid of the root section, and thus bending moments about the AA and TT centroidal axes are present as well as the direct centrifugal tensile force on the root. Thus, centrifugal forces as well as gas forces can cause bending moments. The centrifugal bending moments can be found by summation of moments produced at the root section by centrifugal forces in each strip, a procedure similar to that for the gas-moment determination. When the complete centrifugal bending moments about the two axes are determined, stress calculations are identical with those for the gas moments.

A logical thought is to design the blades so that the moments at the root section caused by centrifugal forces cancel those caused by the gas forces. By careful design in properly positioning the successive centers of gravity of the blade strips forward in the direction of rotation and off to the side in an axial direction, the gas moments about the AA and TT axes (Fig. 10-5) may be canceled. Exact cancellation can occur only for one set of gas-flow conditions at any given rpm, obviously, but the moments are *reduced* at all loads and speeds by proper design. For exact cancellation, only the direct centrifugal stress will be carried by the root section, and the stress calculation is relatively easy. In the early design stage, this simplification of calculation is very helpful. It should be noted that, during operation, the gas forces tend to deflect the blades forward, thus automatically tending to shift the centers of gravity in the proper direction for producing a centrifugal force moment which tends to balance the gas moment.

In a highly twisted blade, it is also possible that the gas forces acting on the successive small strips (into which the blade is considered to be divided) can produce *twisting* (torsion) of the blade. This torsional action is usually small by comparison with the other loads and stresses, when all loads are steady.

It should not be implied that in *all* gas-turbine units all the foregoing principles of blade modification are applied. The basic thermodynamic or flow shape and size should be maintained, of course. However, the amount of care used in shaping and positioning the blading for stress control depends upon size, speed, cost of manufacture, and whether an optimum high-performance design is necessary. Usually a compromise is adopted. It should be recognized, too, that very precise calculations are not warranted in the light of the inexactness of our knowledge concerning actual operational conditions, properties of the materials at operating conditions, and the reproducibility and uniformity of those properties. The possibilities for stress control have been surveyed, however, in a general way.

10-6. Blade Shrouds. Shrouds should be mentioned. A single shroud may be attached to the tip of each blade to minimize end leakage and turbulence near the blade tip. Other possibilities are the use of a shroud band, which forms a continuous ring fastened to the blade tips, or a separate band for a group of blades, say six or eight. In any instance, additional mass and centrifugal force are added, and shrouds are undesirable from the stress viewpoint. For this reason, they have not been widely used in high-speed gas turbines with long blades. For short blades, however, gas leakage at the blade tips is of more concern, and the tip radius is smaller. Shrouding is used to advantage on short blades. The problem of predicting stresses becomes more difficult when shrouds are present, for the addition of a shroud band leaves a stronger but statically indeterminate structure to be solved. Shrouding also changes the elasticity and mass of blades or blade groups, thus possibly enabling the avoidance of resonance and severe vibration, which could result from periodic variations in loading on the blades. This, of course, is not a problem for our present hypothetical situation, in which all loading is exactly steady.

10-7. Thermal Stresses in Blades. If the temperature of a blade is always uniform throughout and the blade is free to expand or contract as its temperature is changed, there will be no thermal stresses in the blade material. Thermal stresses arise because of temperature gradients in the material and are functions of those gradients. It seems obvious that even with steady flow of gas past the blade there will still be temperature gradients in the blade because not all the blade surroundings are at the same temperature. Rapid temperature changes resulting from rapid load fluctuation are particularly deleterious to blade life, causing a sort of thermal fatigue. The prediction of temperature gradients and accompanying stresses and distortions in blading is very difficult, in general, and much operating experience is necessary. If some sort of blade-cooling method is used, the problem is even more complicated. However, thermal stresses in blades probably have not been as large as those in the turbine disks and consequently have been somewhat less troublesome, at least during steady-load operation. Moderate overspeeding can cause rapid failure, however, if normal speed operation is near the strength limit of the material. At higher rpm, temperatures and centrifugal forces increase, while the strength of the blade material may decrease. Overspeed governors are used to prevent accidental overspeeding, which might quickly ruin a turbine.

10-8. Steady Centrifugal Stresses in Solid-drum Rotors. Next let us consider a solid-drum rotor to which the blades might be attached. Figure 10-6 illustrates a long, rigid solid-drum rotor rotating at high speed so that the weight of the rotor is small by comparison with centrifugal forces developed. Unfortunately for simplicity in explanation, there is no direct and easy way of developing the stress equations for such a long, thick rotor. Recourse to the mathematical theory of elasticity is necessary. The reader who is not familiar with the procedure will have to accept the following stress equations as resulting from the assumptions of equilibrium, continuity of the homogeneous isotropic material being stressed, and strain proportional to stress. The equations are published in treatises on the theory of elasticity.

The following formulas (10-7) and (10-8) give the centrifugal stresses for the condition in which adjacent planes of the rotor are parallel. Thus strains are considered to be the same in the axial direction at any radius. This condition is approximately realized at interior planes of a thick rotor —planes not too close to the ends. The case of plane stress, where *stress* rather than *strain* is considered to be constant along the thickness, is given in Art. 10-9.

$$s_t = \frac{3 - 2\mu}{8(1 - \mu)} \rho\omega^2 \left(b^2 - \frac{1 + 2\mu}{3 - 2\mu} r^2\right) \frac{1}{144} \tag{10-7}$$

$$s_r = \frac{3 - 2\mu}{8(1 - \mu)} \rho\omega^2 \left((b^2 - r^2)\right) \frac{1}{144} \tag{10-8}$$

where ω = speed of rotation, rad per sec

$\rho = \dfrac{\text{lb of material per ft}^3}{g_c}$ = slugs per ft³

μ = Poisson's ratio, dimensionless

b = outer radius of drum, ft

r = any radius at which stress s is being computed, ft

s_t = tangential stress (positive sign indicates tension), psi

s_r = radial stress (positive sign indicates tension), psi

The equations are presented to indicate the manner in which the stresses are affected by the variables involved, so that the discussion may proceed from some reasonable basis.

Such a solid rotor might be found in a stationary axial-flow compressor, with the blades attached to the periphery, but this construction is not usual. For light weight, the disk construction to be discussed later is more common, although hollow drums have also been used.

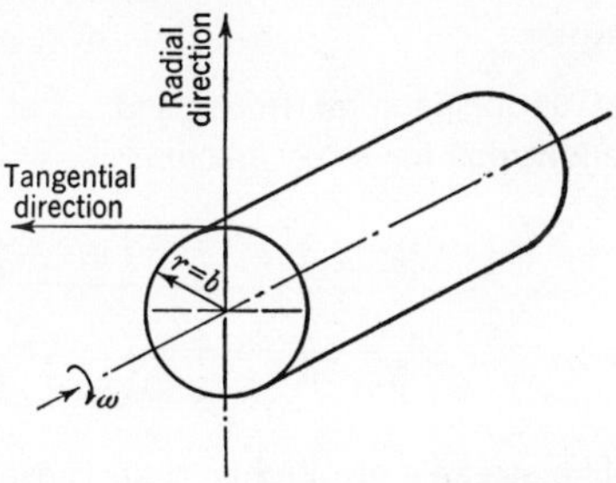

FIG. 10-6. A revolving solid-drum rotor.

It appears that both tangential and radial stresses for the solid drum are maxima at the axis of the drum, where $r = 0$, and that they are *equal* along the axis. In any cross section at radii other than zero, the tangential stress s_t is always larger than the radial stress s_r. Obviously the stress is kept low by using low speeds, material of low density, and as small a radius as possible.

The centrifugal stresses and thermal stresses (which are discussed in a later section) are normally the major items of importance, except possibly in the local regions on the periphery where blades are attached.

10-9. Steady Centrifugal Stresses in Disks. Now assume the drum to be much shortened so that it becomes a disk such as might be found as the rotor of one of the turbine stages, carrying just one row of blades. The assumption is that the stress is constant across the disk thickness at any radius, which is reasonable for a relatively thin disk. The equations for the thin disk of constant thickness become

$$s_t = \rho\omega^2 \left(\frac{3 + \mu}{8} b^2 - \frac{1 + 3\mu}{8} r^2 \right) \frac{1}{144} \tag{10-9}$$

$$s_r = \rho\omega^2 \left[\frac{3 + \mu}{8} (b^2 - r^2) \right] \frac{1}{144} \tag{10-10}$$

It is evident that s_t and s_r both have maximum values and are equal at the center where $r = 0$. For other radii, the tangential stress s_t is always larger than the radial stress s_r. Therefore the tangential or hoop stress is the more important one to consider. If solid disk construction is used, the design may call for a stub shaft integral with the disk or for provision for suitable bolting connections to the disk. Boltholes, of course, cause concentrations of stress.

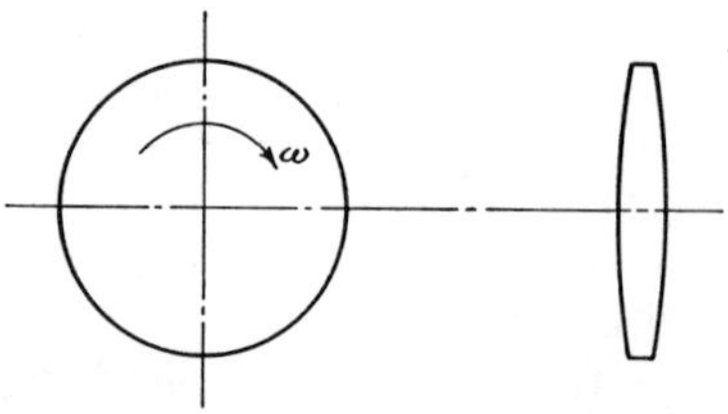

FIG. 10-7. Approximate shape of a revolving disk for constant centrifugal stress.

Now suppose the disk to have a circular hole in the center for mounting on a shaft. This type of construction is sometimes used. Let the radius of this hole be a ft. The equations for stress become

$$s_t = \rho\omega^2 \frac{3 + \mu}{8} \left(b^2 + a^2 + \frac{a^2b^2}{r^2} - \frac{1 + 3\mu}{3 + \mu} r^2 \right) \frac{1}{144} \tag{10-11}$$

$$s_r = \rho\omega^2 \frac{3 + \mu}{8} \left(b^2 + a^2 - \frac{a^2b^2}{r^2} - r^2 \right) \frac{1}{144} \tag{10-12}$$

The stresses are higher than those existing when no hole is present.

Again, at any radius r, tangential stress is higher than radial stress. Maximum tangential stress occurs at the inner boundary of the hole, where $r = a$. This maximum stress is

$$s_{t\,\max} = \rho\omega^2 \frac{3 + \mu}{4} \left(b^2 + \frac{1 - \mu}{3 + \mu} a^2 \right) \frac{1}{144} \tag{10-13}$$

The discussion and equations presented for disks indicate that the stresses vary with the radius and that tangential stresses are larger than radial stresses at any given radius. This suggests that the disk should not be constructed of uniform thickness but should be tapered to produce essentially constant stress at each radius for economy in use of material and also to minimize weight, thereby further reducing centrifugal stress. Figure 10-7 shows the approximate outline of a properly designed disk. The equation describing the profile for a disk of uniform strength is developed in Art. 10-10.

Bursting speeds for disks of ductile material can be approximated from calculations in which stress is considered to be constant across a diameter. Figure 10-8 shows the situation, with the resultant centrifugal force on each half acting through the center of gravity of each half. A balance of forces on one half in the vertical direction gives

$$f_c = As(144)$$

where A = cross-sectional area along a diameter, ft²
s = constant stress on this area, psi

But the total vertical centrifugal load f_c is

$$f_c = \frac{W}{g_c} r_G \omega^2$$

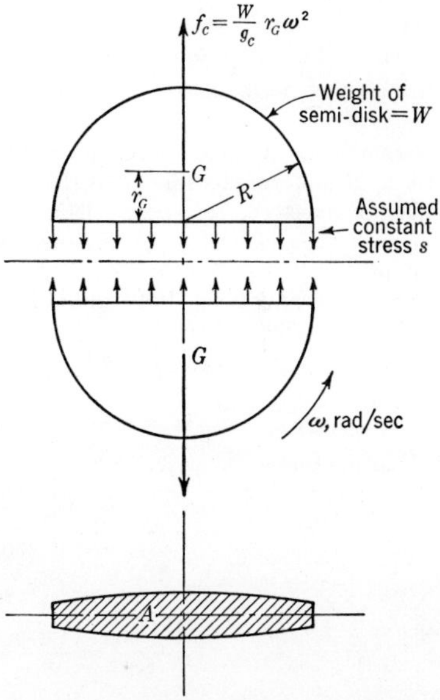

FIG. 10-8. Bursting speeds calculated using the assumption of constant stress along a diameter correlate well with actual experimental bursting speeds.

where W = weight (mass) of semidisk, lb
r_G = radius to the center of gravity of the semidisk, ft
ω = rotational speed, rad per sec. Equating the two expressions for f_c,

$$As(144) = \frac{W}{g_c} r_G \omega^2$$

$$s = \frac{W r_G \omega^2}{A g_c (144)} \qquad (10\text{-}14)$$

To use this equation for determining bursting speed, solve for ω.

$$\omega = \sqrt{\frac{A g_c s(144)}{W r_G}} \quad \text{rad per sec} \qquad (10\text{-}15)$$

The contouring of the disk determines A and r_G, and also W for a given density of material. The stress s is the ultimate stress at bursting.

The foregoing equations indicate by simple calculation approximate safe design limits. For any of the rotor types discussed, very high speeds will produce high *local* stresses. Local plastic flow will occur for sufficiently ductile material, with stress relief and a tendency toward stress uniformity. Eventually the speed gets so high that the *average stress* [Eq. (10-14)] exceeds the ultimate strength of the material and the wheel bursts. This is a very real danger and puts a limit on the rotor speeds which may be used.

Example 3. Assume a solid disk of 14 in. diameter revolving at 15,000 rpm. The disk is of uniform thickness and thin enough so that the stress may be considered constant across the thickness. Disk material has a density of 0.30 lb per in.^{3} and Poisson's ratio of 0.29. Calculate the radial and tangential centrifugal stresses at radii of 0, 1, 2, 3, 4, 5, 6, and 7 in.

Solution. Equations (10-9) and (10-10) may be used.

$$s_t = \rho\omega^2\left(\frac{3+\mu}{8}b^2 - \frac{1+3\mu}{8}r^2\right)\frac{1}{144}$$

$$s_r = \rho\omega^2\left[\frac{3+\mu}{8}(b^2 - r^2)\right]\frac{1}{144}$$

Evaluating terms,

$$\rho\omega^2 = \frac{(0.3)(1728)}{32.17}\left[\frac{15{,}000}{60}(2\pi)\right]^2$$

$$= 39{,}800{,}000$$

$$\frac{3+\mu}{8} = \frac{3.29}{8} = 0.411$$

$$\frac{1+3\mu}{8} = \frac{1.87}{8} = 0.234$$

The equations for stresses become with radius in inches and stresses in psi,

$$s_t = 1918(20.14 - 0.234r^2)$$

$$s_r = 1918(0.411)(49 - r^2) = 788(49 - r^2)$$

The calculations are summarized in the table below.

Radius, in.	s_t, psi	s_r, psi
0	38,600	38,600
1	38,200	37,800
2	36,800	35,400
3	34,600	31,500
4	31,450	26,000
5	27,400	18,900
6	22,450	10,240
7	16,620	0

Example 4. For the solid disk of Example 3, calculate the approximate bursting rpm. Assume the rupture strength of the disk material for a selected life and temperature to be 50,000 psi.

Solution. Equation (10-15) may be used.

$$\omega = \sqrt{\frac{Ag_c s(144)}{W r_G}} \quad \text{rad per sec}$$

$$A = (\text{diameter})(\text{thickness}) = Dt$$

$$\frac{W}{g_c} = \frac{1}{2}\left(\frac{\pi D^2}{4} t\rho\right), \qquad \text{where } \rho = \frac{\text{lb per ft}^3}{g_c} = \frac{\text{slugs}}{\text{ft}^3}, \text{ as before}$$

$$r_G = \frac{4r}{3\pi} \text{ for semicircle (see any handbook)}$$

$$= \frac{2D}{3\pi}$$

Therefore,

$$\omega = \sqrt{\frac{Dts(144)(8)(3\pi)}{\pi D^2 t\rho(2D)}} = \sqrt{\frac{12s(144)}{\rho D^2}}$$

$$= \sqrt{\frac{12s(144)}{\rho(4R^2)}} = \sqrt{\frac{3s(144)}{\rho R^2}}$$

$$= \sqrt{\frac{(3)(50{,}000)(144)}{\dfrac{(0.3)(1728)}{32.17}\,\dfrac{49}{144}}} = 1984 \text{ rad per sec}$$

$$= 18{,}970 \text{ rpm}$$

The addition of blade-attachment rings and seats and the blades themselves all add weight at the peripheries of disks or drums. This increases the centrifugal loading, which increases the stresses in the rotors. Rotor dimensions must be altered so that this additional load can be carried safely. After such alterations have been made at the design stage by use of good judgment and based upon experience, what is left is a disk of arbitrary profile. It is often desirable to give a final check to the proposed design for stresses before it is built.

The centrifugal-stress analysis of a disk of arbitrary profile is a more difficult matter than that for a disk of uniform thickness or a disk of uniform strength. One method is to assume the disk divided into rings of constant thickness, each at a constant mean radius. Equations of equilibrium for each ring may be written, involving unknown stresses at boundaries between adjacent rings. By using known conditions, for example that radial stress is zero or some known value at the outermost disk radius, eventual determination of the stresses at the ring boundaries is possible. Details will not be presented here, but references are given at the end of the chapter. If the stresses found are reasonably low, the design is probably safe. If some stresses are too high, the stress calculations help to guide the designer in altering the profile to lower the excessive stresses and possibly to obtain a more uniform distribution of stress.

10-10. The Disk of Uniform Strength. Disk construction of rotors is widely used, and the disk of uniform strength is always a desirable design for economy of material and minimum stress. Actual rotors usually approximate such a design. Because of its importance, the equation expressing the proper form of tapering will be derived. The disk is considered to be thin enough so that stress is constant across the thickness. Shear forces will be neglected.

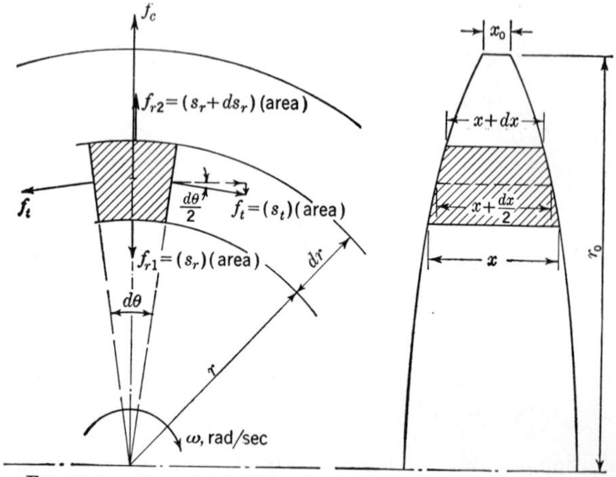

FIG. 10-9. Forces on element of disk of varying thickness. Disk assumed thin enough so that stress is constant across thickness at any radius.

In Fig. 10-9 is shown a rotating disk, tapered in thickness. A small element of the disk must be in equilibrium under the forces pictured. Therefore, for equilibrium in the radial direction,

$$f_c + f_{r2} = f_{r1} + 2f_t \sin \frac{d\theta}{2}$$

or, for the small angle $d\theta/2$,

$$f_c + f_{r2} = f_{r1} + 2f_t \frac{d\theta}{2} \tag{10-16}$$

Each term will now be evaluated.

$$f_c = \text{centrifugal force} = \frac{(\text{volume})(\text{density})}{g_c}\left(r + \frac{dr}{2}\right)\omega^2$$

$$= \left(r + \frac{dr}{2}\right) d\theta\, dr \left(x + \frac{dx}{2}\right)\left(r + \frac{dr}{2}\right)\rho\omega^2$$

$$= \left(r + \frac{dr}{2}\right)^2 \left(x + \frac{dx}{2}\right) d\theta\, dr\, \rho\omega^2$$

$$f_{r2} = (r + dr)\, d\theta\, (x + dx)(s_r + ds_r)$$

$$f_{r1} = r\, d\theta\, x s_r$$

$$f_t = dr\left(x + \frac{dx}{2}\right)s_t$$

Substituting into the equation for equilibrium in the radial direction, there results,

$$\left(r + \frac{dr}{2}\right)^2 \left(x + \frac{dx}{2}\right) d\theta\, dr\, \rho\omega^2 + (r + dr)\, d\theta\, (x + dx)(s_r + ds_r)$$
$$= r\, d\theta\, x s_r + dr \left(x + \frac{dx}{2}\right) s_t\, d\theta$$

The $d\theta$'s cancel, and each term can be expanded. Then if the terms containing second-order differential products are thrown out, since those products are very small, the expanded and simplified equation becomes

$$xr^2\rho\omega^2 + s_r x - s_t x + s_r r \frac{dx}{dr} + rx \frac{ds_r}{dr} = 0 \qquad (10\text{-}17)$$

Now, because uniform stress is desired throughout, s_r will be set equal to s_t and both called s. The equation is then

$$xr^2\rho\omega^2 + sr \frac{dx}{dr} + xr \frac{ds}{dr} = 0$$

Dividing by r,

$$xr\rho\omega^2 + s \frac{dx}{dr} + x \frac{ds}{dr} = 0$$

$$xr\rho\omega^2 + \frac{d(xs)}{dr} = 0$$

$$x\rho\omega^2 r\, dr + d(xs) = 0$$

Integrating, for a constant stress s,

$$\rho\omega^2 \frac{r^2}{2} + s \ln x + \text{constant} = 0 \qquad (10\text{-}18)$$

The only variables established by the disk form are r and x. The constant can be evaluated if a value of thickness x is known or assigned at a certain value of radius r. Assume that the outside thickness is x_o at the outside radius r_o. The constant then becomes $-\rho\omega^2 r_o{}^2/2 - s \ln x_o$, and Eq. (10-18) becomes

$$\frac{\rho\omega^2}{2} (r^2 - r^2{}_o) + s(\ln x - \ln x_o) = 0$$

This can be written as

$$x = x_o e^{(\rho\omega^2/2s)\,(r_0{}^2 - r^2)} \qquad (10\text{-}19)$$

Any consistent units can be used in Eq. (10-19). However, if distances are measured in feet, density ρ is measured in slugs per cubic foot, and it is desired to have stress expressed in pounds per square inch, then Eq. (10-19) becomes

$$x = x_o e^{\frac{\rho\omega^2}{2s(144)}(r_o^2 - r^2)} \tag{10-20}$$

where x = disk thickness at any radius r, ft
x_o = outermost disk thickness, ft
ρ = density, slugs per ft^3 = lb per ft^3/32.17
ω = rotational speed, rad per sec = $2\pi(\text{rpm})/60$
s = constant stress at any radius, psi
r_o = outermost radius, ft
r = radius to any point, ft

Equation (10-20) gives the thickness x in terms of outer-rim thickness x_o and radius r when ρ, ω, r_o, and stress s are known or assigned. The density ρ is usually known, and the proper tangential blade velocity (and therefore disk outer-rim velocity) is determined in conjunction with the gas-jet velocity out of the nozzles. The disk outer-rim velocity is $r_o\omega$ and is somewhat less than the blade velocity because it is closer to the center of rotation. Evidently r_o and ω can both be adjusted, with experience as a guide, to give what appear to be reasonable size and reasonable rpm. Next a reasonable design stress s is selected. The disk-profile equation still cannot be used because the outer-rim thickness x_o is unknown. How can a proper thickness be estimated?

If the disk is to become the rotor of a centrifugal compressor, there is nothing attached to the outer rim and so it may be made quite thin. Manufacturing ease, presence of vane ends, the rigidity necessary, and past experience all must be considered.

If blades are to be attached to the outer rim, then there usually must be a widened area into which the blade roots may be inserted, plus the blade masses, plus the metal involved in generous fillets used to minimize stress concentrations. All these masses will be outside of and attached to the outer disk radius r_o. The disk outer thickness x_o must be large enough to carry the load imposed by the additional masses. Previous experience is very helpful in estimating this thickness. The rim load produces load in the disk, too, which adds to disk stresses which would be present in the rotating disk alone without the rim load.

To illustrate the profile of a disk of uniform strength, an example will be presented.

Example 5. Assume an outer thickness x_o of 0.5 in. at a disk radius of 6 in. Disk material weighs 0.3 lb per in.^{3}, and speed of rotation is 10,000 rpm. Let the working stress be 15,000 psi. Find the disk thickness at 1-in. intervals of radius.

Solution. Equation (10-20) is the basic equation.

$$x = x_o e^{\frac{\rho\omega^2}{2s(144)}(r_o^2 - r^2)} \quad \text{ft}$$

But

$$x_o = \frac{0.5}{12}\text{ ft}$$

$$\frac{\rho\omega^2}{2s(144)} = \frac{(0.3)(1728)}{32.17}\left(\frac{10{,}000}{60}2\pi\right)^2\frac{1}{(2)(15{,}000)(144)}$$

$$= 4.09$$

Therefore,

$$x = \frac{0.5}{12}e^{4.09\left(\frac{36}{144}-r^2\right)}$$

If it is desired to express x and r in inches, then the modified equation becomes

$$\frac{x}{12} = \frac{0.5}{12}e^{4.09\left(\frac{36}{144}-\frac{r^2}{144}\right)}$$

or

$$x = 0.5e^{0.0284(36-r^2)}$$

Tabulated results appear below.

Radius, in.	*Thickness, in.*
6	0.5
5	0.68
4	0.88
3	1.08
2	1.24
1	1.35
0	1.39

The plot of the profile appears as the solid line in Fig. 10-10.

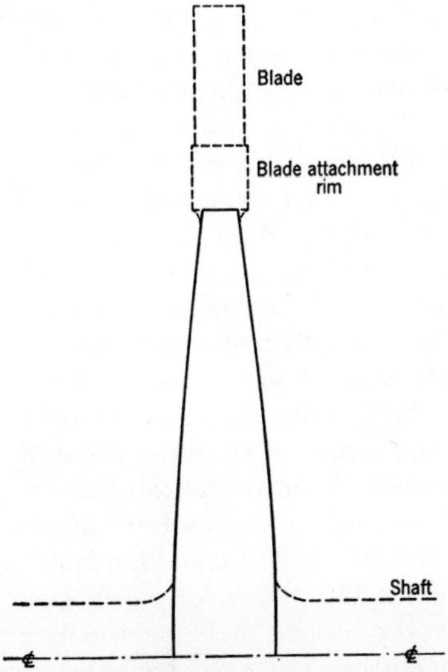

FIG. 10-10. Profile of disk of uniform stress of Example 5 (to scale—solid line). Dotted line illustrates necessary modification for use as turbine rotor.

As built, compromise must be adopted. Straight-line taper is easier to manufacture and may approximate the correct shape. Provision must be made for a shaft or shaft attachment with associated fillets and for vanes or blades which might have to be fastened to the disk or built as part of it. The dotted line in Fig. 10-10 shows a possible design after modification for these items which must be added to the basic disk. Obviously the simple stress calculation is now not correct, and good judgment must be applied in altering the profile. A final stress check would then be made on a disk of varying and arbitrary profile. This matter was discussed briefly in the previous Art. 10-9. As in most machinery, the ultimate check of a design comes with the operation and test of the equipment after it is built.

10-11. Steady Stresses in Centrifugal-compressor Rotors. In the centrifugal type of compressor, centrifugal forces and stresses are the important limiting factors. The rotor may be considered to be a disk such as has already been discussed, plus attached vanes for guiding the air flow. The most widely used vanes have been radial, with perhaps a curved section at the inlet eye close to the center to minimize inlet turbulence. In calculating centrifugal stresses, the approximation is sometimes made that the vane material is distributed over the basic disk to which the vanes are attached. The equivalent rotor is then just a disk, usually of varying thickness to approximate constant-stress design. Stress calculations are made on the equivalent disk, the procedure being similar to that already mentioned for disks.

The attached vanes are subjected to forces from the gas flow, but these forces are not large by comparison with the centrifugal forces, and the vanes can be made quite thin. Vane trouble (and many of the troubles with other parts of gas turbines, too) has arisen mostly from vibration and not from steady loads imposed. Much experimental work has been necessary to arrive at satisfactory shapes for the vanes to direct the gas flow properly and to prevent mechanical failure. Figure 6-5 shows one successful design of centrifugal-compressor rotor.

An impeller disk as built obviously necessarily deviates considerably from a disk of constant strength. Fillets, vanes, and the necessity for attachment at the hub all alter the profile. The rotor preliminary dimensions and form are often dictated by previous experience when that is available and by sound fundamental considerations such as those already discussed. The stress calculation then may be made afterward as a check on the design. What is usually involved, then, is a calculation of centrifugal stress in a disk of arbitrary profile, a matter already discussed.

In the method in which it is assumed that the vane material is distributed over the disk area, the vane material adds to the mass and centrifugal force of the basic disk. However, it is assumed that the added disk cross section carries none of the load. Thus, the vane material adds to the centrifugal load, but not to the strength.

If the vanes are not assumed to be distributed, calculations may still be made but the process is more difficult and lengthy. Assumptions of some sort are still necessary, so that the stresses calculated for the more complex shape are still only approximations. A designer must always face the problem of when to stop calculating and begin building and testing.

10-12. Thermal Stresses in Turbine Rotors. The hot turbine rotor experiences temperature gradients and corresponding stresses and distortions which are very difficult to predict accurately. At the time of preliminary design, it is necessary to assume some manner of variation of temperature in the turbine disk under proposed operating conditions. Then an estimate of the thermal stresses can be made. Precise calculations are very difficult, but estimates can be made which are better than wild guesses. One very simple case will be presented here.

Assume a thin solid disk of uniform thickness, with temperature constant across the thickness at any radius. Temperature of the disk varies along the radius only. Within the elastic range of the material, the radial and tangential (hoop) thermal stresses are given by the following equations, taken from the theory of elasticity.[1]

$$s_r = \alpha E \left(\frac{1}{b^2} \int_o^b Tr\,dr - \frac{1}{r^2} \int_o^r Tr\,dr \right) \tag{10-21}$$

$$s_t = \alpha E \left(-T + \frac{1}{b^2} \int_o^b Tr\,dr + \frac{1}{r^2} \int_o^r Tr\,dr \right) \tag{10-22}$$

where s_r = radial stress, psi

s_t = tangential stress, psi

α = coefficient of linear thermal expansion of disk material, (assumed constant) $\dfrac{\text{in.}}{\text{in.}^\circ\text{F}}$

E = modulus of elasticity of disk material (assumed constant), psi

T = temperature, at the radius r, above some initial uniform temperature at which there are no thermal stresses, F

b = outermost radius of disk, in.

r = radius at which stresses are being computed, in.

An example will illustrate the calculations.

Example 6. Assume the same solid rotor disk of 14 in. diameter that was used in Example 3, but now stationary and with temperatures of 960 F at the outer rim and 300 F at the center. (These are reasonable values for an actual gas-turbine disk under some conditions.) Assume the disk to be of uniform thickness and thin enough so that temperature and stresses vary only with radius, and not along the thickness. The linear coefficient of expansion of the disk material is 8×10^{-6}, and the average modulus of elasticity for the temperature range is 29×10^6. Assume the temperature to vary with the fourth power of the radius. Find the thermal stresses at radii of 0, 1, 2, 3, 4, 5, 6, and 7 in.

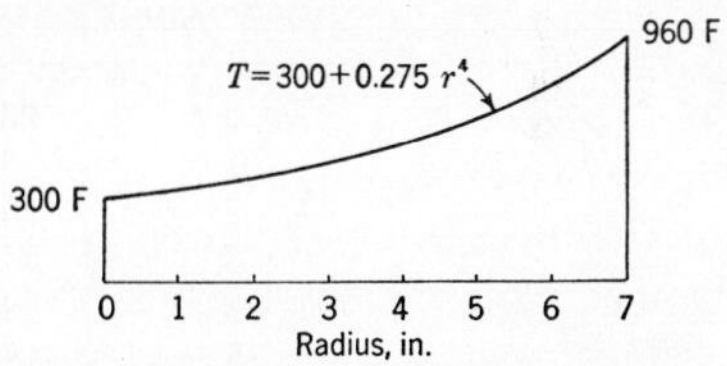

FIG. 10-11. Variation of temperature with radius (not to scale) for disk of Example 6.

Solution. Equations (10-21) and (10-22) are applicable. It is necessary to express temperature T in terms of radius r in order to integrate. Figure 10-11 illustrates the temperature distribution. Because there are no thermal stresses in the disk when the temperature is uniform throughout [see Eqs. (10-21) and (10-22) for T = constant], then T may be measured above any arbitrary temperature level. Let this level be 0 F. The temperature distribution can be expressed as

$$T = 300 + kr^4$$

[1] Timoshenko, Ref. 1.

To evaluate k, use the known temperature of 960 F at the radius of 7 in. Then

$$960 = 300 + k(49)(49)$$
$$k = \frac{660}{2400} = 0.275$$

Therefore,

$$T = 300 + 0.275r^4$$

Now the terms in Eqs. (10-21) and (10-22) will be evaluated.

$$\begin{aligned}
\frac{1}{b^2}\int_o^b Tr\,dr &= \frac{1}{49}\int_0^7 (300r + 0.275r^5)\,dr \\
&= \frac{1}{49}\left[\frac{300}{2}(49) + \frac{0.275}{6}(49)(49)(49)\right] \\
&= 260 \\
\frac{1}{r^2}\int_o^r Tr\,dr &= \frac{1}{r^2}\int_o^r (300r + 0.275r^5)\,dr \\
&= \frac{1}{r^2}\left(\frac{300r^2}{2} + \frac{0.275r^6}{6}\right) \\
&= 150 + 0.0459r^4 \\
\alpha E &= (8 \times 10^{-6})(29 \times 10^6) = 232
\end{aligned}$$

Equations for stresses are

$$\begin{aligned}
s_r &= 232[260 - (150 + 0.0459r^4)] \\
&= 232[110 - 0.0459r^4] \\
s_t &= 232[-(300 + 0.275r^4) + 260 + (150 + 0.0459r^4)] \\
&= 232[110 - 0.299r^4]
\end{aligned}$$

The calculated stresses are summarized below.

Radius, in.	s_r, psi	s_t, psi
0	25,500	25,500
1	25,490	25,450
2	25,400	24,700
3	24,700	21,200
4	22,800	11,910
5	18,860	−7,660
6	11,730	−43,400
7	0	−102,000

Positive signs mean tension, and negative signs indicate compression. The calculated stresses should be compared with those of Example 3.

The thermal stresses look high—very high—and some discussion is necessary.

During operation of the rotor, the centrifugal stresses are always tensile stresses. Therefore the tangential stresses at the outer rim caused by centrifugal action oppose those caused by temperature gradients, and the net operating stress at the rim is reduced. At the disk center, of course, thermal and centrifugal stresses add. The equations used assume that the material remains within the elastic limit. With stresses of the magnitude indicated, plastic flow could result in some regions, and stresses could be relieved somewhat, not reaching the calculated values. If plastic flow occurs, then, upon cooling of the disk, stresses reverse so that there remain in the cooled disk compressive tangential stresses near the hub and large tensile tangential stresses near the rim. Radial stresses near the hub might also be reversed, leaving residual compression. It is apparent that this plastic flow giving stress relief and residual reversed stress is a desirable thing. When the unit is again operated, the residual stresses are first relieved and stresses pass through zero before they begin to increase. At full operating conditions, hot running stresses (thermal plus centrifugal) are then considerably lower than those indicated by superposed values calculated by elastic theory. Of course, it is necessary to take plastic flow into account in specifying running clearances. Also, at turbine-overhaul periods, clearances can be checked and corrected as gradual long-time yielding (creep) of disks and blades takes place. It appears that less operating trouble with disks has been experienced than originally was anticipated.

It should be remembered that an actual rotor is not the simple disk of uniform thickness assumed for the example problem and that the temperature distribution and consequent thermal stresses may be altered considerably from one design to another. Various disk-cooling arrangements are in use, and temperature distribution depends upon the cooling system. The disk of the example certainly is not an optimum design.

The admittedly oversimplified problem presented does serve to indicate that thermal stresses cannot be disregarded in designing gas-turbine disks. Calculations for actual disks of arbitrary and varying profile can be made. References are given at the end of the chapter. The calculations usually involve considering the disk to be composed of a series of rings, each of constant thickness. The procedure is quite lengthy and tedious, but reasonable approximations of the thermal stresses involved are possible. Figure 10-12 shows the stress distribution in the disk of one turbine after allowing for plastic yielding.

10-13. Fastenings for Blades. The fastenings are very critical parts of the gas-turbine unit, and many schemes have been proposed. The attachment must be designed to carry the centrifugal load of the blade (often 2 or 3 tons or more) and keep the blade properly aligned. The temperature in the turbine unit is high in the region of some of the blade roots, and there is a good possibility of high local stresses and subsequent plastic flow which might misalign the blade

Welding has been proposed and tried, but of course this makes it difficult to remove old blades and insert new ones. Nevertheless, the method has been used satisfactorily. Proposals have even been made to machine the blades as an integral part of the rotor, but this presents obvious difficulties.

Experience with steam turbines has been helpful, and present successful fastenings have been evolved utilizing this experience. Figure 10-13 presents some forms of gas-turbine root fastenings. In the fir-tree design, the blade root fits rather loosely in its socket, with edge lugs peaned over

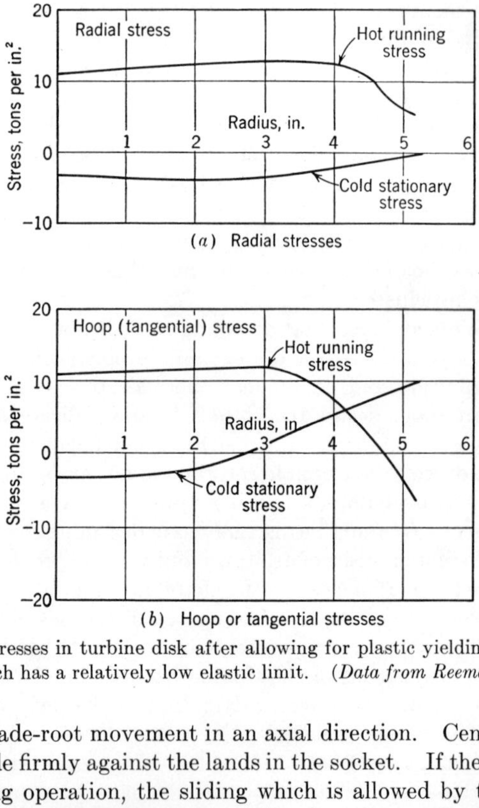

FIG. 10-12. Stresses in turbine disk after allowing for plastic yielding. Material is Stayblade, which has a relatively low elastic limit. (*Data from Reeman, Ref.* 7.)

to prevent blade-root movement in an axial direction. Centrifugal force seats the blade firmly against the lands in the socket. If the blade should vibrate during operation, the sliding which is allowed by the relatively loose fit is claimed to give appreciable damping action to limit the vibration.

The space occupied by the fastenings limits the number of blades that can be placed on any given periphery.

10-14. Gyroscopic Loading. Before the discussion of steady loading in gas turbines is ended, the load imposed by gyroscopic action should be mentioned. The rotor of a gas-turbine unit is in effect a large top, and tilting of the axis gives rise to a gyroscopic couple. In stationary units, of course, no gyroscopic couple can exist. In transportation units, and particularly in jet airplanes, the gyroscopic effect can become appreciable, and the engine bearings and mountings must be designed to handle this load.

It is convenient to use a right-hand rule in representing angular momentum as a vector. Thus, as in Fig. 10-14, the fingers wrap the axis in the direction of rotation, and the thumb points in the direction of the vector.

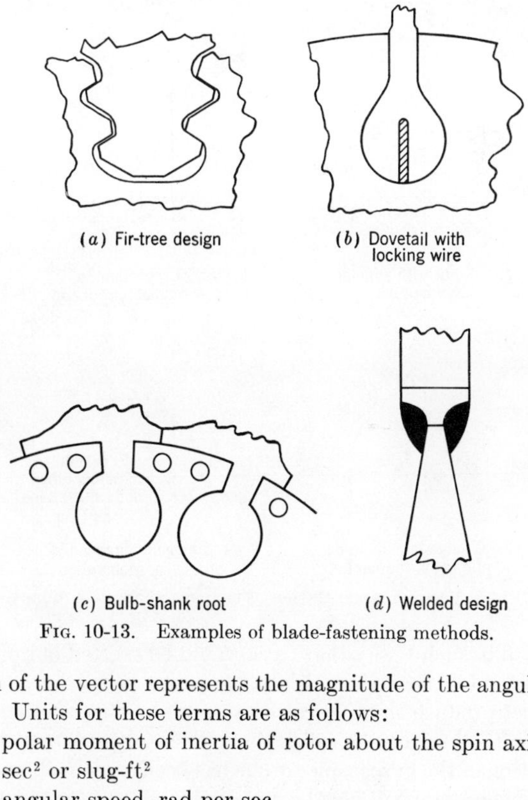

FIG. 10-13. Examples of blade-fastening methods.

The length of the vector represents the magnitude of the angular momentum $J_{zz}\omega$. Units for these terms are as follows:

J_{zz} = polar moment of inertia of rotor about the spin axis zz, ft-lb$_f$-sec^2 or slug-ft^2

ω = angular speed, rad per sec

The angular momentum can change in two ways, in magnitude and in direction. For a magnitude increase the vector is lengthened, and the change in momentum is represented as an extension of the original vector (Fig. 10-14). If the thumb of the right hand is pointed in the direction of $\Delta(J_{zz}\omega)$, the fingers wrap the axis in the direction in which a couple must have been applied to produce the given change of momentum.

For a change in the direction of angular momentum, tilting of the axis is necessary. In Fig. 10-14*d*, the change of angular momentum $\Delta(J_{zz}\omega)$ is shown for an axis tilt of $\Delta\theta$ rad. If the thumb of the right hand is pointed along the direction of the change of angular momentum $\Delta(J_{zz}\omega)$, the fingers wrap the change vector in the direction in which a couple must be applied to the rotor as the given direction change of the axis takes place.

As an illustration, consider that the tilt of (*d*) in Fig. 10-14 is applied to the rotor pictured in (*a*). Accompanying the tilt must have been a

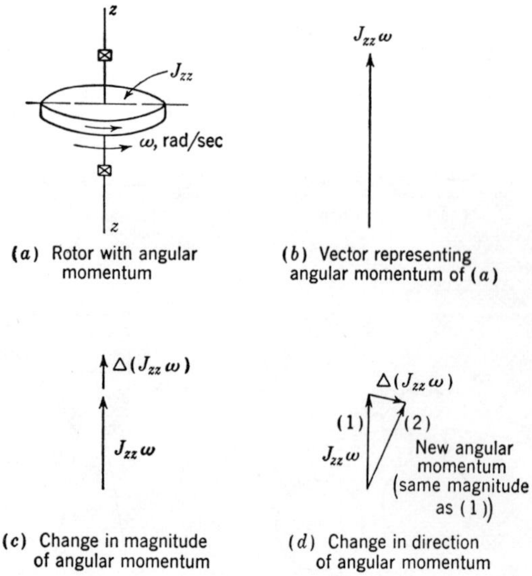

FIG. 10-14. Vector representation of changes in angular momentum.

couple applied in such a way that force would be exerted out of the paper on the upper end of the axis and into the paper on the lower end. The bearings must produce these forces on the axis. The reaction, the force of the axis on the bearings, is called the gyroscopic torque or couple.

The direction of the gyroscopic torque has been explained, and an equation for the magnitude will now be developed. From Fig. 10-14*d*,

$$\Delta(J_{zz}\omega) \cong J_{zz}\omega\,\Delta\theta$$

and this change takes place in time Δt. In the limit,

$$\frac{d(J_{zz}\omega)}{dt} = J_{zz}\omega\,\frac{d\theta}{dt}$$

But according to the principles of mechanics, a couple must be applied to produce a change in angular momentum, and the couple must be equal to the time rate of change of momentum, all expressed in consistent units.

For the present instance, therefore,

$$C = J_{zz}\omega\Omega \tag{10-23}$$

where C = gyroscopic couple, ft-lb$_f$

J_{zz} = polar mass moment of inertia about axis of spin, ft-lb$_f$-sec^2 or slug-ft^2

ω = angular speed, rad per sec

$\Omega = \dfrac{d\theta}{dt}$ = rate of tilt of axis, rad per sec

The magnitude and direction of gyroscopic torque have been explained, and an example will illustrate the use of the principles.

Example 7. A gas-turbine unit in a jet airplane revolves at 10,000 rpm, counterclockwise when viewed from behind the airplane. The airplane executes a climbing arc at a rate of 45° in 5 sec. The rotors are approximately equivalent to a solid steel drum 12 in. in diameter and 4 ft long. Find the gyroscopic couple during the climb and the effect on the operation of the airplane.

Solution. The magnitude of the gyroscopic couple is

$$C = J\omega\Omega$$

$$J = \frac{1}{2}\frac{W}{32.17}r^2 \text{ for a solid disk (see any handbook)}$$

$$\begin{aligned} W &= \text{weight of rotor, lb} \\ &= (\pi r^2 h)(\text{density}) \\ &= \pi(\tfrac{1}{4})(4)(490) = 1540 \text{ lb} \end{aligned}$$

$$J = \frac{1}{2}\frac{1540}{32.17}\frac{1}{4} = 5.98 \text{ lb}_f\text{-ft-sec}^2 \text{ or slug-ft}^2$$

$$\omega = \frac{10,000}{60}(2\pi) = 1045 \text{ rad per sec}$$

$$\Omega = (\tfrac{45}{360})(\tfrac{1}{5})(2\pi) = 0.157 \text{ rad per sec}$$

$$\begin{aligned} C &= (5.98)(1045)(0.157) \\ &= 982 \text{ lb}_f\text{-ft} \end{aligned}$$

The bearings and engine mounts must withstand this gyroscopic torque. The effect of the gyroscopic torque is to tend to turn the airplane to the pilot's left.

Published specifications concerning the gyroscopic effect for one actual jet engine are given below.

J for compressor and turbine system = 103 lb-ft^2
= 3.205 slug-ft^2

Gyroscopic couple at 12,300 rpm and 3½ rad per sec = 173,000 lb$_f$-in.
= 14,410 lb$_f$-ft

PROBLEMS

10-1. An axial-flow compressor operates at a speed of 8000 rpm. The outer diameter of one of the disks to which blades are attached is 16 in. A blade has a constant cross-sectional area of 0.4 in.^{2} along its length of 4 in. The blade is oriented radially so that the centers of gravity of all cross sections lie on the same radial line. The blade material has a density of 0.286 lb per in.^{3} Calculate (*a*) the total blade centrifugal force which the root must withstand; (*b*) the average direct centrifugal stress on the root area, on the mid-blade cross-sectional area, and on the tip area; (*c*) the forces and average stresses of (*a*) and (*b*) for a blade of aluminum alloy weighing 0.1 lb per in.^{3} *Ans.* (*a*) 8310 lb; (*b*) 20,780 psi, 11,400 psi, 0 psi; (*c*) 2905 lb, 7260 psi, 3985 psi, 0 psi.

10-2. Same conditions as in Prob. 10-1, except that the blade is tapered linearly from a root area of 0.4 in.^{2} to a tip area of 0.2 in.^{2} Find the total centrifugal force on the blade and the average centrifugal stress on the root area. Comment on the effect of tapering the blade. *Ans.* (*a*) f = 6050 lb; (*b*) stress at root area = 15,130 psi.

10-3. The part of a turbine blade that lies outside the root attachment is 3 in. long and weighs 0.3 lb. The center of gravity is 1.2 in. from the root, and the disk diameter is 19.6 in. Turbine speed is 10,000 rpm. Assume that gas forces produce a moment at the root about the axial axis of 40 in.-lb$_f$. Further assume that the gas forces do not *bend* the blade, *i.e.*, the blade stays straight. Find the tilt of the blade (in degrees from the original radial line through the center of gravity) for the centrifugal moment about the axial axis to cancel the gas moment about that axis. *Ans.* Tilt in direction of rotation = 0.203°.

10-4. Using free-vortex flow conditions, develop equations giving the gas force moments at the root of an axial-flow compressor blade about tangential and axial axes. Carefully define terms and symbols used.

10-5. For an axial-flow compressor using the constant-reaction principle and symmetrical blading, develop equations giving the gas force moment at the blade root about tangential and axial axes. Carefully define terms and symbols used. For n blades in a row, what is the axial *thrust* per row?

10-6. A solid cylinder is being considered as a rotor for an axial-flow compressor to operate at 7000 rpm. It is to be 24 in. long and 12 in. in diameter. The density of the proposed material is 0.29 lb per in.^{3} For 2-in. intervals of radius, calculate and plot the radial and tangential centrifugal stresses at the mid-plane. Assume elastic behavior of the material and a Poisson's ratio of 0.30. Where are the stresses greatest? (Note that this is a plane-strain problem, in which adjacent planes are essentially parallel.)

Ans.

Radius, in.	s_t, psi	s_r, psi
0	6225	6225
2	5770	5540
4	4380	3460
6	2070	0

10-7. Assume the cylinder of Prob. 10-6 to be reduced in length to 1.5 in., retaining the same diameter of 12 in. The rotor is now thin enough so that *stresses* will be assumed to be constant across the thickness, rather than *strain*. For the same rpm, density of material, and Poisson's ratio as used in Prob. 10-6, calculate and plot the radial and tangential centrifugal stresses at 2-in. intervals of radius. Compare the stresses obtained with those from Prob. 10-6. In which case is the maximum stress higher?

Ans.

Radius, in.	s_t, psi	s_r, psi
0	6000	6000
2	5620	5340
4	4460	3330
6	2550	0

10-8. For the disk of Prob. 10-7, assume that the material is sufficiently ductile so that the stress is approximately uniform over a diametral cross section at bursting. Estimate the bursting rpm for a rupture stress of 60,000 psi. Repeat for diameters of 24 and 36 in. *Ans.* 10,080 rpm for disk of 12 in. diameter.

10-9. Consider a single-sided centrifugal compressor with vanes which are curved near the inlet eye to guide the incoming air. It is desirable that no centrifugal bending stress exist in the vanes during operation. Discuss how this might be accomplished. Sketch the disk and one vane.

10-10. A rotor disk for a turbine is tapered linearly from a hub thickness of 2.5 in. to a rim thickness of 0.75 in. Rotor diameter is 18 in. Assume reasonably ductile material of construction having a density of 0.29 lb per in.^{3} and a rupture strength of 50,000 psi, and calculate the approximate bursting rpm.

10-11. A turbine rotor disk is to be designed for constant centrifugal stress of 12,000 psi at 8000 rpm. The diameter of the disk is to be 16 in., and the outer-rim load establishes a disk thickness of 0.80 in. at the outer radius. The disk alloy weighs 0.3 lb per in.^{3} (*a*) Determine the disk thickness at 1-in. intervals of radius, and draw a diametral cross section to scale. (*b*) Discuss briefly a method or methods by which the bursting rpm of this disk could be approximated, assuming reasonably ductile material.

Ans.

Radius, in.	*Disk thickness, in.*
0	3.44
2	3.14
4	2.385
6	1.512
8	0.80

10-12. A disk for one stage of an axial-flow compressor is to be designed for constant centrifugal stress at an rpm of 9000. The disk diameter is 16 in., and the outer-rim thickness is 0.60 in. The density of the disk material is 0.3 lb per in.^{3} (*a*) Plot the disk profile (thickness) for each of two stresses, 10,000 psi and 20,000 psi. (*b*) For 10,000 psi stress and for rpm of 12,000 calculate the disk profile, and compare it with that determined for 9000 rpm. Does this appear to be a practical shape?

Ans. (*a*) 9,000 rpm.

10,000 psi stress		20,000 psi stress	
Radius, in.	Disk thickness, in.	Radius, in.	Disk thickness, in.
0	5.47	0	1.81
2	4.75	2	1.69
4	3.14	4	1.37
6	1.58	6	0.975
8	0.6	8	0.6

10-13. Repeat Example 6 of the chapter, but assume the temperature to vary as the square of the radius instead of as the fourth power. Compare the thermal stresses obtained with those of Example 6, and comment on the effect of the temperature gradients. In which temperature distribution is there a higher maximum gradient?

Ans.

Radius, in.	s_r, psi	s_t, psi
0	38,400	38,400
2	35,200	29,100
4	25,900	940
6	10,160	−45,900
7	0	−76,500

10-14. Consider a thin disk of constant thickness and 20 in. in diameter. Thermal coefficient of expansion is 8×10^{-6}, and modulus of elasticity is 29×10^{6}. Center temperature is 300 F, and outer-rim temperature is 960 F. Assume linear variation of temperature with radius, and calculate the radial and tangential thermal stresses at radii of 0, 2, 4, 6, 8, and 10 in.

Ans.

Radius, in.	s_r, psi	s_t, psi
0	51,000	51,000
2	40,900	30,650
4	30,650	10,200
6	20,400	−10,220
8	10,200	−30,650
10	0	−51,000

10-15. A 16-in.-diameter thin disk of constant thickness, so that stress is assumed to be constant across the thickness at each radius, is subjected to a rim temperature of 1000 F. The temperature varies with the square of the radius, dropping to a value of 500 F at the center. The coefficient of expansion of the disk material is 9.6×10^{-6} in./(in.) (°F), and the modulus of elasticity is 28×10^{6} psi. (*a*) Calculate and plot the radial and tangential *thermal* stresses for 2-in. intervals of radius. (*b*) For a rotor speed of 6000 rpm, with disk material of density of 0.29 lb per in.^{3} and Poisson's ratio of 0.29, calculate and plot the *centrifugal* stresses for 2-in. intervals of radius. (*c*) Plot the *combined* stresses (thermal + centrifugal). Carefully identify tension and compression. (*d*) What assumption is made when the sum of the thermal and centrifugal stresses is considered to be the actual stress?

Ans. (*c*)

Radius, in.	s_t, psi	s_r, psi
0	41,420	41,420
2	34,842	38,870
4	15,110	31,110
6	−17,780	18,140
8	−64,020	0

(+ signifies tension)

10-16. Assume that the rim load attached to the outer periphery of a disk is distributed substantially uniformly around the periphery. Let the rim mass be m lb per in. of rim at the radius r of the center of the rim. If it is assumed that the rim centrifugal force is eventually carried uniformly over a disk diametral cross section of area A in.^{2}, derive an expression for the stress on the cross section caused by the rim load. Define terms and symbols used.

10-17. The mass moment of inertia of a jet-engine rotor about its rotational axis is 150 lb-ft^2 (or 4.66 slug-ft^2). An airplane in which this engine is installed pulls out of a vertical dive to a horizontal line of flight at a uniform rate during a time of 5 sec. The rotor turns at 9000 rpm, in a clockwise direction as viewed from behind the plane. Determine (*a*) the magnitude of the gyroscopic torque, (*b*) the direction of the torque applied to the rotor shaft by the bearings, caused by the gyroscopic effect (use a sketch), and (*c*) the effect on the operation of the airplane. *Ans.* (*a*) 1,380 ft-lb$_f$. (*c*) Couple tends to turn airplane to *right.*

10-18. A mounting arrangement for a jet engine is being checked. It is determined that, after other forces have been considered, a gyroscopic torque of 20,000 lb$_f$-ft can be withstood successfully by the mounting. The jet-engine rotor turns at 10,000 rpm and has a weight of 600 lb with a radius of gyration of 6 in. What rate of tilt of the rotor axis in flight is allowable? *Ans.* 4.1 rad per sec.

10-19. A jet-engine rotor turns counterclockwise when viewed from the front of the airplane in which it is mounted. In what direction does the gyroscopic action tend to move the airplane for each of the following maneuvers: (*a*) a turn to the left with the rotor axis always in a horizontal plane, (*b*) a similar turn to the right, (*c*) a pure roll of the airplane about a straight line of flight, (*d*) a dive during which the rotor tilts from horizontal to vertical, (*e*) a climb in which the rotor tilts from horizontal to vertical, (*f*) a simultaneous climb and turn to the right. *Ans.* Couple tends to make the airplane (*a*) climb; (*d*) turn left.

REFERENCES

1. Timoshenko, S.: "Theory of Elasticity," McGraw-Hill Book Company, Inc., New York, 1934.
2. Bressman, J. R.: Radial Stress in Linearly Tapered Turbine Blades, *Product Eng.*, vol. 20, no. 2, p. 155, February, 1949.

3. Timoshenko, S.: "Strength of Materials," Part II, D. Van Nostrand Company, Inc., New York, 1941.
4. Guins, V. G., and G. H. Heiser: Disk Stresses, *Machine Design*, February, 1948, p. 144.
5. Leopold, W.: Centrifugal and Thermal Stresses in Rotating Disks, ASME *J. Applied Mech.*, December, 1948, p. 322; Rotating Disks at High Temperature, *Machine Design*, December, 1948, p. 141.
6. Beck, Frank: Centrifugal Stresses in Disks, *Machine Design*, May, 1949, p. 137.
7. "Development of British Gas Turbine Jet Unit," Institution of Mechanical Engineers, London, reprinted for American Society of Mechanical Engineers, New York, 1947 (8 authors).
8. Colwell, A. T., and R. E. Cummings: Turbine Engine Blading: Manufacturing Technique and Fastening Methods, *Trans. SAE*, vol. 2, no. 3, July, 1948.
9. Emmert, H. D.: Current Design Practices for Gas Turbine Power Elements, *Trans. ASME*, vol. 72, no. 2, February, 1950.
10. Fonda, L. B.: High Temperature Disk-forging Developments for Aircraft Gas Turbines, *Trans. ASME*, vol. 70, p. 1, 1948.
11. Robinson, E. L.: Safety Margins and Stress Levels in High-temperature Equipment, *Trans. ASME*, vol. 73, no. 1, p. 89, January, 1951.
12. Woodhouse, Harold: Impeller Blade Design, *Machine Design*, February, 1950, p. 111; Turbine Blade Fastenings, *Machine Design*, February, 1951, p. 148.
13. Manson, S. S.: Stress Investigations in Gas Turbine Disks and Blades, *Trans. SAE*, vol. 3, no. 2, p. 229, April, 1949; The Determination of Elastic Stresses in Gas Turbine Disks, *NACA Tech. Note* 1279, 1947.
14. Smith, Ronald B.: Problems in the Mechanical Design of Gas Turbines, *Trans. ASME*, vol. 69, p. A-99, 1947.
15. Holms, A. G., and R. D. Faldetta: Effects of Temperature Distribution and Elastic Properties of Materials on Gas-turbine-disk Stresses, *NACA Tech. Note* 1334, 1947.
16. Stodola, A.: "Steam and Gas Turbines," McGraw-Hill Book Company, Inc., New York, 1927.
17. Church, E. F.: "Steam Turbines," 3d ed., McGraw-Hill Book Company, Inc., New York, 1950.

CHAPTER 11

VIBRATION AND BALANCING

11-1. General Observations Concerning Vibration. In a general sense, vibration can occur in any system in which there is mass (inertia) and elasticity. The mass and elasticity (and their distribution within the system) determine the natural or resonant frequencies of the system. The presence of certain types of damping also affects these frequencies, but usually to a lesser degree unless the damping is considerable. A reasonably small amount of damping affects the amplitude of a vibration much more than it affects the natural frequency.

If a periodic disturbance is applied to a system, the disturbing frequency obviously should not be near any of the natural frequencies which it might excite. In other words, resonance must be avoided if large amplitudes of vibration are to be avoided. Consider a small force which, if applied statically, produces a negligibly small deflection and accompanying stress. This same force, when applied periodically at or near a natural frequency, will add energy to the system with each application and build up the magnitude of the deflection. Thus the action may be thought of as the application of a dynamic-magnification factor to the static deflection. One of two things can happen. The amplitude can continue to increase (1) until failure occurs or (2) until the energy supplied per cycle equals the energy dissipated per cycle by damping in the system. Even this equilibrium amplitude may be large enough so that the system may ultimately fail prematurely (from fatigue).

Figure 11-1 shows the steady-state dynamic-magnification factor for deflection plotted as a function of the frequency ratio, which is the ratio of the applied disturbing frequency to the natural (resonant) frequency of the system. This natural frequency is the free-vibration frequency of the system *without* consideration of any system damping. The various lines in the figure apply for various amounts of damping, expressed as fractions of the "critical" damping of the system. If a body is displaced from equilibrium and then released to vibrate freely, the critical damping is the minimum amount of damping which will prevent the body from traveling past the equilibrium point—in other words, prevent free oscillation. The assumptions for Fig. 11-1 are steady-state vibration, linear spring constants, damping force proportional to velocity (viscous damping), and a one-degree-of-freedom system with a force applied sinusoidally with time. These assumptions limit the precise use of Fig. 11-1, but the principle, importance, and nature of the variables involved in the dynamic-magnification factor are clearly evident.

Example 1. A turbine or compressor blade may be thought of as a cantilever beam. If such a beam is displaced from equilibrium and released, it will vibrate freely with a certain natural frequency. With a *steady* gas pressure applied to the blade, no vibration results. The straight horizontal line in the figure accompanying this example depicts the steady total gas force. But in real turbines and compressors this "average" force is applied in buffets of gas force applied by gas flow through successive nozzle vanes during rotation of the turbine or compressor rotor. The wavy line

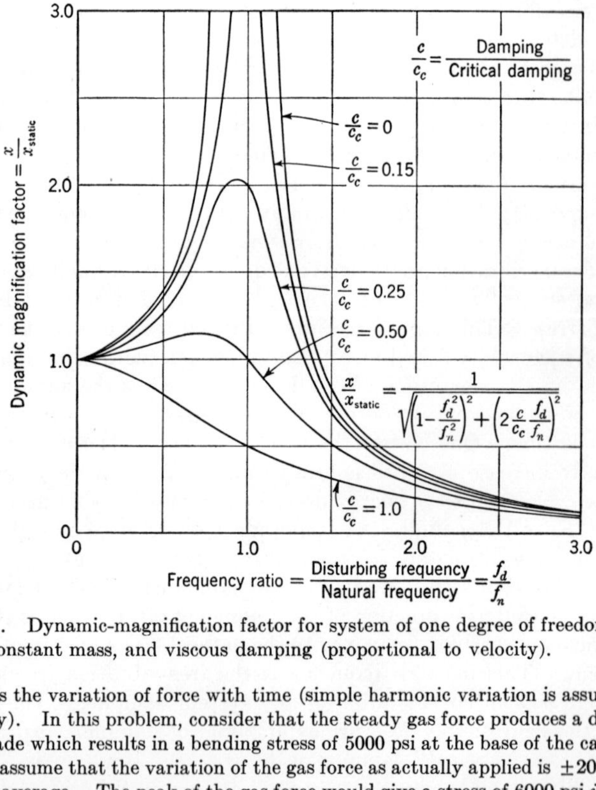

FIG. 11-1. Dynamic-magnification factor for system of one degree of freedom, linear spring, constant mass, and viscous damping (proportional to velocity).

illustrates the variation of force with time (simple harmonic variation is assumed, for simplicity). In this problem, consider that the steady gas force produces a deflection of the blade which results in a bending stress of 5000 psi at the base of the cantilever. Further, assume that the variation of the gas force as actually applied is ± 20 per cent from the average. The peak of the gas force would give a stress of 6000 psi *if applied as a steady load* (elastic behavior of the beam material assumed).

But the gas force is not applied as a steady load. The deflection (and stress) resulting from the force *variation* may be greater or less than the deflection (and stress) produced by the assumption of static-force application, depending upon nearness to resonance and the amount of damping present. In other words, the *dynamic-magnification factor* must be applied to the assumed *static* deflection (and stress) caused by the load.

Now let it be assumed that the gas-force variations on the blade occur at a frequency equal to 0.75 of the natural blade frequency, and assume blade damping to be 1 per cent of critical damping. What then will be the stress induced by dynamic action (vibration), and what will be the *total* stress during operation? Assume stress proportional to deflection, and use Fig. 11-1.

Solution. From Fig. 11-1, the dynamic-magnification factor is approximately 2.2. Therefore, the blade deflection is 2.2 times the static deflection. If stress is propor-

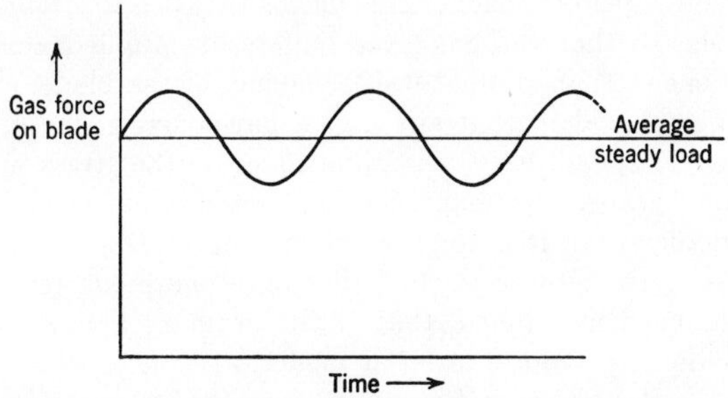

FIGURE FOR EXAMPLE 1. Gas force on blade.

tional to deflection, then the original 1000-psi variation from the average actually becomes 2200 psi. The total maximum blade stress then becomes

$$5000 + 2200 = 7200 \text{ psi}$$

Without consideration of the dynamic magnifier, the stress would have been

$$5000 + 1000 = 6000 \text{ psi}$$

Therefore, the added stress attributable to vibration is

$$7200 - 6000 = 1200 \text{ psi}$$

This is a stress increase of $^{2200}/_{5000} = 44$ per cent of the average steady-load stress, caused by variation of gas force on the blade. Another viewpoint might be that there is a stress increase of $^{1200}/_{6000} = 20$ per cent of the maximum stress which would be present without dynamic magnification.

For the magnitude of stresses involved in the foregoing example, the stress increase might not appear to be serious. However, if a turbine blade is designed for a certain life dependent upon stress, temperature, and deformation (and associated failure from creep or fatigue), then an increase of 25 per cent (or even less) in the bending stress can result in a greatly reduced life of the blade before failure. This is a serious problem and illustrates why even small amounts of vibration can have unfortunate consequences when a blade is designed for high stresses based on steady gas and centrifugal loads only. Figure 11-1 does not extend beyond a magnification factor of 3, but in practice much higher values obviously could be attained under unfavorable operating conditions.

Additional information can be gained from a further inspection of Fig.

11-1. At higher frequency ratios the magnification factor decreases markedly, declining to a value less than 0.2 at the ratio 3.0. This means, for the values used in the example, that the bending stress would be approximately 5000 + (0.2)(1000) = 5200 psi instead of the value of 7200 psi obtained at the frequency ratio of 0.75. Obviously with very high frequency ratios, the stress approaches the 5000-psi stress produced by the average "steady" load. This means that it is desirable to design, when possible, so that the gas-force buffets are applied at very much higher frequencies than the natural frequency of the blade. When this is done, it also is evident that any higher harmonics of the fundamental gas-force frequency will have negligible effect on the stress produced by the vibration because the magnification factor is even smaller for the higher harmonics than it is for the fundamental.

The reader must here be warned that only one mode (or manner) of vibration has been mentioned, that of the ordinary bending, or "flap," type. A blade may vibrate in other modes, several of which are shown in Figs. 11-5 and 11-6, and the natural frequencies are different for all these modes. For each mode, the same general considerations apply as have already been discussed for the fundamental, or flap, mode. If a higher harmonic of the gas-force variations can be neglected for the fundamental mode, this is no guarantee that it will not cause trouble, for it might be near resonance for a higher mode. The avoidance of resonance or near resonance with all modes obviously is not a simple problem. The subject is further discussed later in the chapter.

Emmert[4] indicates that a "ripple factor" of about 0.1 (or 10 per cent) should be applied to the average bending stress in blades. This factor in effect says that the gas force on the blade usually varies about ±10 per cent from the average. (Note that this corresponds to the ±20 per cent value used in Example 1). If the bending stress is computed from the average steady load and then additional stress is obtained from the 10 per cent variation and a suitable stress-concentration factor (usually not over 1.5), the sum of the stresses is the working bending stress, *provided* that there is no dynamic magnification of the variable (±10 per cent) part of the stress.

The foregoing discussion accents the basic factors in any vibration problem. Undesirably large amplitudes of vibration can be minimized in two ways, (1) avoidance of resonance, and (2) the use of proper damping.

Resonance may be avoided by the following general methods:

1. Eliminate the periodic forcing or disturbing actions which are the sources of vibration. For gas turbines, this involves balancing, elimination of gas-pressure fluctuations against the parts, and a perfectly steady nonvarying load. Perfection here is, in general, impossible.

2. Change the frequencies of the driving actions. Speed changes and changes in numbers of nozzles and blades are examples.

3. Change the resonant frequencies of the system. This involves changes in mass and elasticity of the system and in the distribution of these things within the system.

A certain amount of damping is present in any system whether or not special damping devices are introduced. Friction between moving parts and friction within any part as it deflects (hysteresis) are examples. In the gas turbine and compressor, there is a certain amount of aerodynamic damping action, too, caused by movement of the blades in a gas field. Usually it is desirable to use materials with high internal friction so that damping action is inherent, but this may not always be compatible with other desired characteristics of the materials. The amount of damping action attainable is limited, also, so that in some cases additional damping action is necessary and must be supplied by auxiliary devices.

The fundamental material which has been discussed in a general way is treated in considerable detail in many textbooks written on the subject of vibration. The application of this material to gas turbines is the objective here, and it will be discussed in the following sections.

Example 2. Assume that the *average* gas force on a turbine blade produces a bending stress of 3000 psi at the root. Assume a ripple factor of 0.10 and a stress-concentration factor of 1.5. Calculate the maximum bending stress, assuming negligible effects from dynamic magnification (vibration).

Solution.

$$\begin{aligned}\text{Maximum bending stress} &= 3000 + (0.10)(1.5)(3000)\\ &= 3000(1 + 0.15)\\ &= 3450 \text{ psi}\end{aligned}$$

11-2. Balancing of Flexible and Rigid Rotors. The objective of balancing is the elimination of a source of vibration of the whole machine. A flexible rotor is one which undergoes appreciable deflection between bearing supports because of centrifugal forces. This deflection moves the center of gravity of the deflected portion off the bearing center line (and, with imperfect balancing, the center of gravity is usually off the bearing center line even without rotation). At any given rotational speed a centrifugal force acts through the center of gravity, which tends to produce more deflection, which then produces more force. The action continues until the elasticity, or spring action, of the bent rotor counteracts the centrifugal force. Then a steady whirling of the rotor remains at any constant rpm. Usually there is some damping action (air resistance, internal hysteresis), which helps to limit the magnitude of the vibration.

Any appreciable amount of whirling produces misalignment, stresses, bearing wear, and vibration, which usually cannot be tolerated. It is unfortunate that a flexible rotor can be counterbalanced to eliminate vibration at only *one* speed. For a constant-speed machine this may not seem to be serious, but a machine must always be brought up to speed in starting it and down again in stopping it. Severe vibration can occur in these periods if they last for appreciable lengths of time.

A *rigid* rotor may be balanced at *any* speed, and it remains in balance at *all* speeds because there is no deflection of the rotor by centrifugal

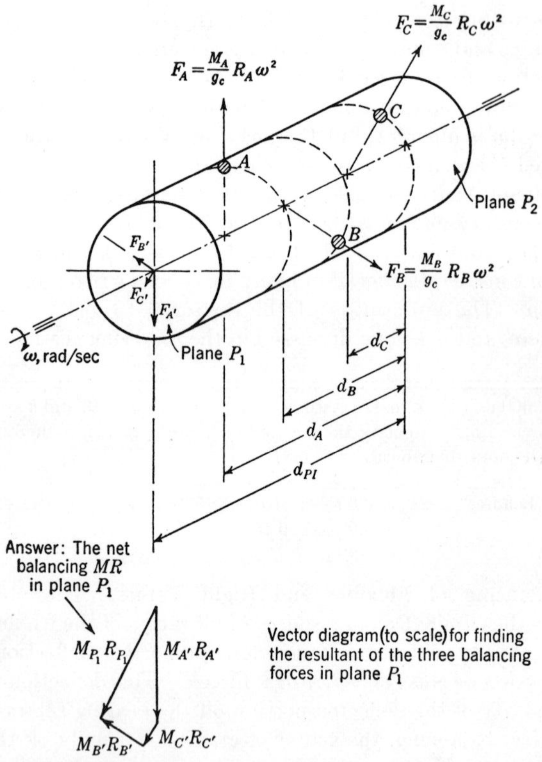

FIG. 11-2. Dynamic balance of a rigid rotor of known unbalances.

forces. Many gas-turbine rotors are made stiff enough to minimize whirling and may be considered to be rigid rotors. In the next article the balancing of such rotors will be discussed.

11-3. Dynamic and Static Balance of Rigid Rotors of Known Unbalance. Figure 11-2 shows a rotor with three *known* unbalances giving centrifugal forces F_A, F_B, and F_C. In general, there could be any number of unbalanced masses, but three will serve to illustrate the principles involved. This rotor (and *any* rigid rotor) can be put in complete balance by adding two masses, one in each of two arbitrarily selected reference planes.

Suppose we select the reference planes at the ends of the rotor as convenient locations at which to add balancing weights and designate those planes as P_1 and P_2 (Fig. 11-2). Now consider adding a force in P_1 to balance the moment of F_A about P_2. The needed force $F_{A'}$ in P_1 can be obtained from the moment equation

$$F_{A'}d_{P1} = F_A d_A$$

Similarly, $F_{B'}$ and $F_{C'}$ in plane P_1 will balance moments of forces F_B and F_C about plane P_2. But, $F_{A'}$, $F_{B'}$, and $F_{C'}$ can be combined to give one resultant force F_{P1} in plane P_1, and this single force can be obtained from the centrifugal effect of a single mass which we shall call M_{P1}.

$$F_{P1} = \frac{M_{P1}}{g_c} R_{P1}\omega^2$$

where F_{P1} = force, lb_f
M_{P1} = mass, lb_m
g_c = 32.17 lb_m per slug
R_{P1} = radius to center of gravity of M_{P1}, ft
ω = speed of rotation, rad per sec

Either the balancing mass or the radius is selected, and the other is calculated. Moments about plane P_2 are now in balance.

The procedure is then repeated, taking moments about plane P_1 and locating a balancing force and weight in plane P_2.

The addition of the two balancing weights now leaves the rotor in perfect dynamic, or running, balance. No moment remains about either of the two reference planes. Therefore, according to the principles of mechanics, moments are balanced about all other planes, and *forces* also are balanced. Therefore all vibrational disturbing effects caused by unbalance have been eliminated.

Instead of the second moment balance about plane P_1, it would have been perfectly permissible to find a force which would make all the centrifugal *forces* add to zero. This force would be added in plane P_2 in order that the previous moment balance about P_2 remain unchanged. The final result, the balancing mass in P_2, would be exactly the same as that obtained by taking moments about P_1.

Unbalances are usually specified in inch-ounces. In the expression for centrifugal force, $F = (MR/g_c)\omega^2$, the value of MR indicates the amount of unbalance, and it makes no difference how the value is broken down into values for M and for R.

Example 3. For Fig. 11-2, consider the following conditions of unbalance to exist:
(a) Unbalances: At A, 8 in.-oz
At B, 6 in.-oz
At C, 10 in.-oz
(b) Distances: $d_{P1} = 24$ in.
$d_A = 20$ in.
$d_B = 12$ in.
$d_C = 4$ in.
(c) Angles or positions of unbalances are shown in the figure.
Find the amount and location of required balancing weights.
Solution. First, take moments about plane P_2.
For A:

$$F_{A'}d_{P1} = F_A d_A$$

$$\left(\frac{M_{A'}}{g_c} R_{A'}\omega^2\right)(d_{P1}) = \left(\frac{M_A}{g_c} R_A\omega^2\right)(d_A)$$

$$M_{A'}R_{A'} = M_A R_A \left(\frac{d_A}{d_{P1}}\right)$$

$$= 8(^{20}\!/_{24}) = 6.667 \text{ in.-oz}$$

Note that $F_{A'}$ $\left(\text{or } \frac{M_{A'}}{g_c} R_{A'}\omega^2, \text{ and therefore } M_{A'}R_{A'}\right)$ must be opposite F_A (and $M_A R_A$), as shown in the end view (plane P_1).
For B:

$$M_{B'}R_{B'} = M_B R_B \left(\frac{d_B}{d_{P1}}\right)$$

$$= 6(^{12}\!/_{24}) = 3 \text{ in.-oz, directed as shown in the end view } (F_{B'})$$

For C:

$$M_{C'}R_{C'} = 10(^{4}\!/_{24}) = 1.667 \text{ in.-oz directed as shown } (F_{C'})$$

The resultant of these three balancing MR's may be obtained graphically or analytically. The graphical method has been used in the figure, and the final single vector for MR and its direction are shown in the vector diagram of Fig. 11-2.

$$M_{P1}R_{P1} = 7.3 \text{ in.-oz}$$

The division between M_{P1} and R_{P1} is arbitrary. Practical considerations such as space and weight must be considered.

Moments are now in balance about plane P_2. To complete the balancing, moments must be canceled about plane P_1. The procedure is the same as that explained and will not be described again.

It is not always necessary or desirable to *add* weights to balance. Metal may be *removed* opposite the spot where the balancing weight would have been required in order to get the same effect.

If a rotor is so thin that all centrifugal forces lie substantially in one plane, then only centrifugal *forces* need to be balanced. When this is done, the center of gravity of the rotor necessarily lies on the bearing center line and this is a circumstance which can be checked by a static test. One such simple test consists in using level hardened rails to support the rotor on its shaft and allowing the rotor to roll until the heavy side is down. A balancing weight is added opposite the heavy side.

Eventually the rotor is balanced, so that it stays wherever it is put. The center of gravity is on the shaft center line, and the summation of MR's for all particles of the rotor is zero. The rotor is said to be in *static balance*.

However, if this same rotor has appreciable thickness, it may *not* be in *dynamic* balance. Although the centrifugal forces cancel each other, the moment arms of these forces about some arbitrary reference plane may be such that shaking *moments* occur during operation. The rotors for gas turbines are usually thick rotors and require the complete dynamic-balancing procedure. It should be noted that if a rotor is in dynamic balance, then it is automatically in static balance also.

FIG. 11-3. An axial-flow compressor rotor in the cradle of a dynamic balancing machine. (*Gisholt Machine Co.*)

11-4. Machines for Determining the Unbalance. It has been established that any rigid rotor may be put in dynamic *and* static balance by the addition of one weight in each of two arbitrarily selected planes through the rotor. Therefore any number of unbalances which exist can be replaced by two equivalent unbalances, one in each of the arbitrary planes. The problem of balancing a rotor is therefore concerned with:

1. Selecting the two planes in which it is convenient to make the balancing correction.
2. Determining the amount and location of each of the two equivalent unbalances.

3. Applying corrections, either removing metal to remove the unbalance or adding metal opposite the unbalance weight to counterbalance it.

There are many machines available for determining the amount and location of unbalance in specified planes. These machines rotate the part to be balanced in special elastically supported frameworks and pick up the vibration of the frameworks caused by centrifugal forces in the unbalanced part. By proper calibration of the setup, the unbalance may be determined. There are automatic machines which indicate directly both the amount and the angular location of equivalent unbalance.

FIG. 11-4. Complete rotor assembly of centrifugal compressor, shaft, and gas turbine in place for the balancing operation. (*Gisholt Machine Co.*)

Correction is applied by removing metal or by adding counterweights. For gas-turbine rotors rotary files or hand grinders are often used to remove metal. Corrections are applied by this method in order to minimize stress-concentration effects, which would contribute to the existence of excessively high local stresses and possible failure during high-speed operation.

Figure 11-3 shows a detailed view of an axial-compressor rotor for a gas turbine in position for the balancing operation in a balancing machine. In Fig. 11-4 a complete rotor assembly is being balanced in the De Havilland (British) jet-engine plant. The assembly consists of a single-stage turbine, a single-sided centrifugal compressor, and a rigid tubular connecting shaft. Individual units are sometimes balanced separately and then the completely assembled rotor balanced also if necessary.

11-5. Vibration Control by Suitable Mountings. The gas-turbine unit and connected load form a mass of material which must be supported on some kind of base. Usually a mounting pad of some kind is used—rubber, steel spring, cork, felt, or other material—so that there is effectively a mass supported on springs. Such a system has many natural frequencies of vibration; for example, there are vertical, lateral, rocking, and pitching frequencies, and they may all be different. Resonance with any of them can produce severe vibration, even if the magnitude of the disturbing force is small.

Presumably the rotor is balanced, but obviously *perfect* balance cannot be achieved, and a small residual unbalance is always present. When a turbine unit has a connected load, that load itself is a possible source of vibrational disturbances at frequencies not necessarily the same as those from the small residual of the gas-turbine rotor. Evidently chances are good for resonance at some speeds of operation. Mountings must be selected so that resonance is avoided during normal operation. Studies of vibration theory and practice show that to reduce most effectively the magnitude of vibration of an elastically supported mass, and to isolate that vibration from the connected base, a "soft" spring support should be used. That is, the natural frequency of the mass and spring arrangement should be much lower (say one-half or one-third or less) than the frequency of the disturbing forces (Fig. 11-1). In most gas turbines, speeds (and therefore disturbing-force frequencies) are quite high. However, a load may operate at a much lower speed. Also, in gaining or reducing speed, critical-frequency ranges may be passed through. This must be done quickly to prevent the attainment of large amplitudes. The presence of damping in the elastic supporting material itself or the introduction of additional damping devices is helpful in limiting vibration during resonant periods. In general, mountings should be sufficiently soft so that, during normal operation, effective isolation and vibration reduction are obtained. The lowest disturbing frequency and the highest natural frequency are important factors (see Fig. 11-1). Sometimes it is not possible to have all natural frequencies below the forcing frequencies, as is desirable for maximum isolation. Resonance or near resonance must still be avoided during operation, and the unit then should run normally *between* resonant speeds.

The mountings obviously must be sufficiently strong to support the weight of the machine and be placed to give it proper stability. The placement affects the natural frequencies which result.

11-6. Vibration of Turbine and Compressor Blades. Three things are of primary importance:

1. The natural frequencies of the blades
2. The frequencies and magnitudes of the forcing or exciting pulsations
3. The amount of damping present

The basic consideration at the design stage is the avoidance of resonance, and this involves predictions of natural frequencies of blades and of periodic disturbing frequencies.

The mathematical prediction of all the natural frequencies of a given blade is a difficult (if not impossible) matter, and experimental techniques

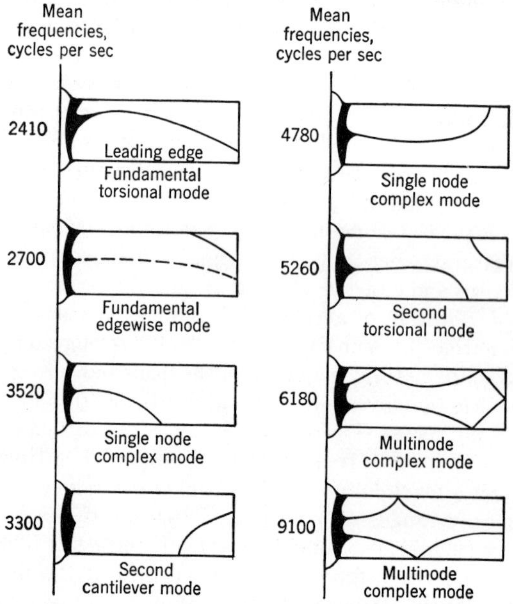

FIG. 11-5. Natural frequencies and modes of vibration of cool turbine blades for W2/800 turbine (British). Room temperature. Fundamental flap mode is not shown. (*From Voysey, Ref.* 1.)

have been widely used. One such technique is to apply exciting vibrations to a blade at controlled frequencies and to observe the patterns formed by sand sprinkled on the vibrating blade. Sand remains at rest at nodal positions (positions of no motion), and so the manner of vibration can be visualized. By changing the controlled vibrating frequency, maximum vibration and therefore resonant frequencies can be determined. The method therefore gives the critical frequencies and the manner of vibration at each of these frequencies. Figure 11-5 shows the principal modes of vibration determined at *room temperature* for the turbine blades of one British jet engine. It should be mentioned here that blades from the same row of some turbines and axial compressors have exhibited natural-frequency differences of as much as 20 per cent. With such a spread in any one row, and with different blade shapes in succeeding rows, the probability is slim that resonance or near resonance can everywhere be avoided during operation. The presence of shrouds adds to the difficulty of natural-frequency predictions.

The reader should note that, if the fundamental flap frequency as a simple cantilever is near the frequency of the first torsional mode, combined bending and twisting produce *flutter* as a possibility. It should also be noted that many of the natural frequencies lie within the frequency range of normal hearing. A high temperature has the effect of lowering the natural frequencies, while the centrifugal forces tend to stiffen the blades, giving higher frequencies.

What are the exciting actions for the blades? In the turbine there are probably three important sources of variation. One is the combustion chambers. If there are 10 chambers evenly spaced around the turbine periphery, then for every revolution of the turbine wheel each blade will receive 10 cycles of applied forces caused by variation in the gas field between combustion chambers. This is termed a tenth-order excitation. Harmonics of this order could be dangerous, too, particularly the next higher one, the twentieth. A combustion chamber which is a complete and continuous ring around the engine supposedly should create no periodic disturbances.

Next there are the stationary turbine nozzles. If a turbine has 48 nozzles, then each blade is subjected to 48 cycles of pressure and force variation in one revolution, a forty-eighth-order excitation.

Finally, there is the pressure variation produced from the compressor. For example, a centrifugal-compressor impeller having 29 vanes produces at each outlet diffuser 29 cycles of gas-pressure fluctuation per revolution, which may persist through the combustion chambers and turbine nozzles and reach the turbine rotating blades.

Figure 11-6 illustrates the correlation between the order of excitation and turbine speed, with the principal nodal patterns and natural frequencies of the *hot* turbine blades (700 C or 1292 F) for the same engine and blades as pictured in Fig. 11-5. The natural frequencies for the hot and cold blades should be compared. The possibilities of resonance at different turbine speeds are easily visualized from diagrams such as Fig. 11-6. From inspection of the nature of fatigue failures which already have occurred, the nodal patterns are helpful in establishing the modes of vibration and the critical frequencies responsible for the failures. Design changes may then be suggested which will avoid resonance in the future. For example, if the failures are caused by excitation from the nozzles, then a different number of nozzles could be used in a future design. An alternative would be to alter the blades to change the natural frequency which caused the resonance.

Excitation for the axial-flow compressor blades can arise because of the presence of a support spider across the air intake. If a spider has four arms, then each blade in the first row of the compressor will be buffeted four times in each revolution—a fourth-order excitation. The disturbance will have less magnitude as it encounters blade rows downstream from the first, but if resonance occurs, serious vibration (and noise as well) can still result. Another source of periodic disturbance is the stationary guide vanes, corresponding to the stationary nozzles of the turbine.

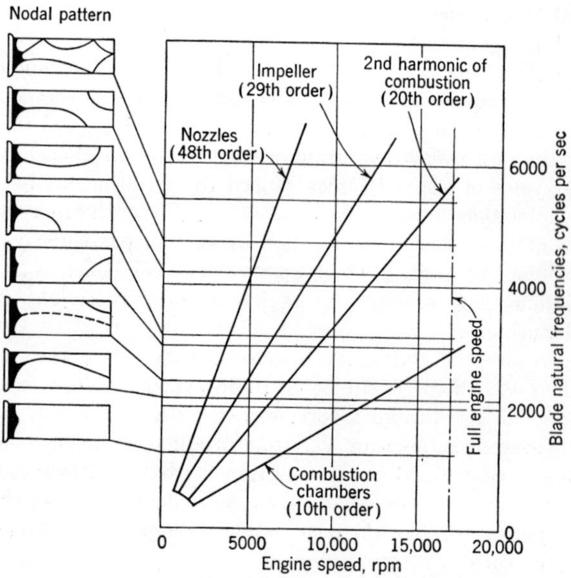

FIG. 11-6. Natural and exciting frequencies for hot turbine blades for W2/800 turbine (British). Temperature 1292 F. (*From Voysey, Ref.* 1.)

The centrifugal forces produced during rotation tend to stiffen blades and thus raise their natural frequencies. Frequencies determined from sand patterns are thus too low for operating conditions, but this fact does not destroy the usefulness of information such as that presented in Fig. 11-6 if the designer keeps the stiffening effect in mind. For example (Fig. 11-6), at full engine speed the designer might anticipate tenth-order trouble with blade modes second and third from the bottom, although the sand nodal patterns show the blade frequencies to be too low for resonance. The frequency variation caused by centrifugal action can be calculated,[1] and the stiffening effect usually raises natural frequencies not more than 10 to 20 per cent. Percentagewise, the lower frequency modes are affected much more than the higher ones.

Damping of blade vibration comes mainly from three actions. They are internal friction (hysteresis), aerodynamic or fluid damping from the gas surrounding the blade, and damping at the blade-root fastening. There is evidence that in some instances well over half the total blade

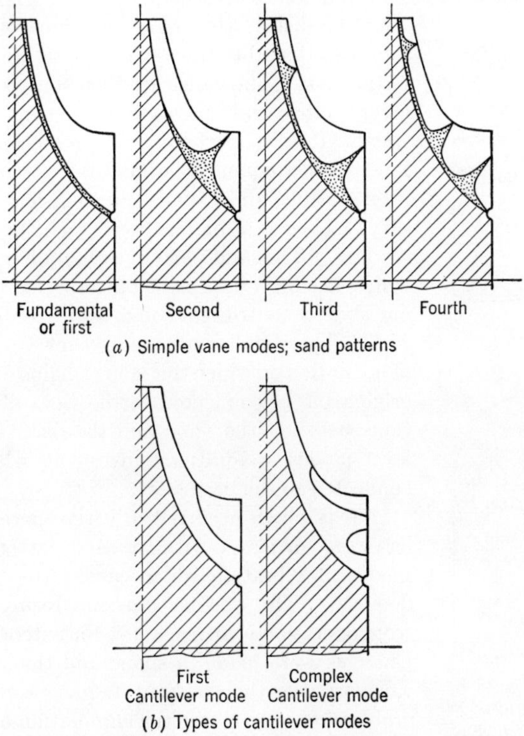

FIG. 11-7. Vibration modes of centrifugal impeller vanes. (*After Voysey, Ref.* 1.)

damping is derived from the aerodynamic effects. Even a small amount of damping is very helpful in limiting the magnitude of vibration, and much thought has been given to the subject. The rather loose sliding fit of the fir-tree type of blade-root fastening is claimed to produce appreciable damping action. For the cool compressor blades, experimental coatings of varnish have resulted in doubling the apparent internal-damping capacity. Experimental work has been done with plastic compressor

[1] Timoshenko, Ref. 13, and Emmert, Ref. 4.

blades with high internal-damping properties. Requirements other than damping properties obviously often rule out the use of certain types of materials, however. Avoidance of resonance (although not always completely possible) is usually a better *aim* than running near resonance with dependence on damping for limiting vibration and preventing fatigue failure.

11-7. Vibration in the Centrifugal-compressor Rotor. The rotor may be considered to be a disk with vanes attached. First, it will be assumed that the disk is very rigid by comparison with the vanes so that the vanes may vibrate independently of the disk. The vanes are then some sort of cantilever plates and may vibrate in many complex ways. Precise mathematical prediction of all the modes and frequencies of natural vibration has not been feasible, and experimentation has been necessary. The situation has been much the same as for blading. The sand-pattern technique is one method which has been used, and patterns for the vanes of one style of centrifugal compressor are shown in Fig. 11-7. The simple vane modes show nodes (regions of no motion on which the sand remains undisturbed) originating or connected to the vane fillet at junctions between the vane and the disk. The cantilever modes in addition show nodes which are not connected to the base fillet.

Cracks which occurred in early operating impellers were correlated with the sand patterns, and the modes of vibration which caused the cracks were deduced. The cracks were the result of fatigue from operation near resonance for extended periods. Changes were made in shape and thickness of the vanes to alter their natural frequencies. The basic problem, of course, was the elimination of resonance during normal operation. Not all modes of vibration have caused trouble because resonance has not occurred with all modes.

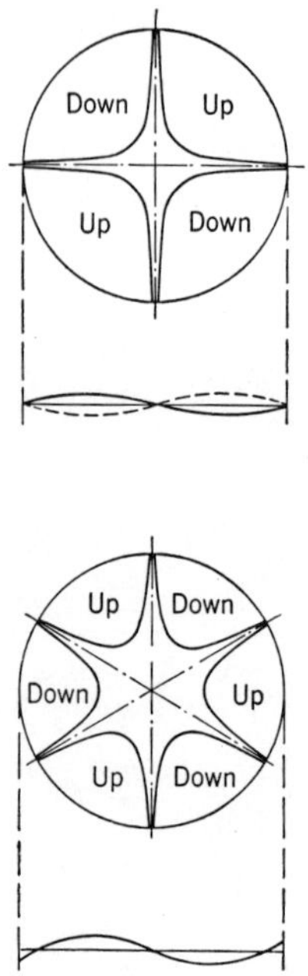

FIG. 11-8. Some modes of vibration of a disk with no vanes attached.

Next the disk flexibility will be considered. The disk by itself (without vanes) has certain natural modes of vibration, some of which are shown in Fig. 11-8. Nodal lines are diameters, and there is always an even total number of "up" and "down" regions (for modes higher than the first). Therefore it is desirable that an odd number of vanes be used to avoid possible coincidence of vane roots and nodal lines. Build-up of this mode of vibration is thus inhibited. Another type of natural frequency of a disk is a ring mode, in which nodal lines are concentric circles. These modes usually have caused no trouble.

Finally, there is the usual instance in which the flexible vanes are attached to a flexible disk. Vibration of one then can affect the other. Consider a double-sided centrifugal impeller, with the attached vanes vibrating. Figure 11-9 pictures an edge view of the rotor and shows the ways in which the vanes on the two sides of the disk can vibrate, either in phase or out of phase. The out-of-phase style causes the disk to flex, and continued excitation of the motion conceivably could cause serious vibration and stressing. However, with an odd number of vanes usually

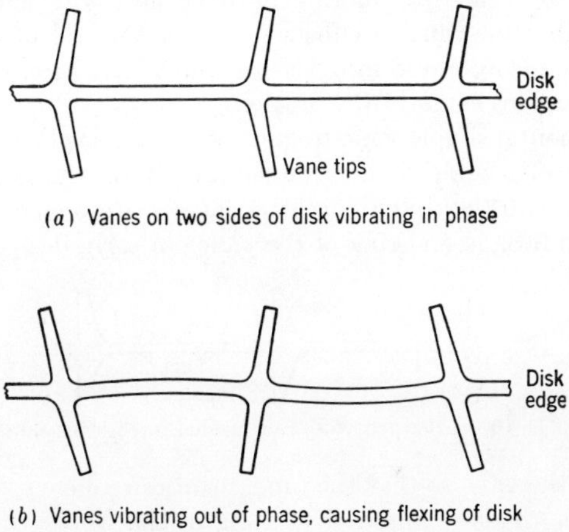

FIG. 11-9. Edge view of double-sided centrifugal impeller, showing effects of vane-tip vibration on disk edge.

no trouble is encountered. The same general considerations apply to the single-sided impeller.

It appears that many of the vibration-fatigue failures which have occurred have been attributed to the first two vane modes mentioned, the *simple vane* modes and the *cantilever* modes. The problem has been the avoidance of resonance with these modes. Practically, only the lowest two or three frequencies of each of these two styles have been of importance, and the fundamental simple vane mode has given most of the trouble. In some designs, normal operation occurs with excitation frequency between the fundamental, or lowest, natural frequency and the next higher one.

The most important periodic disturbances which excite vibration in the centrifugal-compressor rotor are the varying pressure fields encountered by the vane tips as they pass the diffusers. If there are 10 diffusers, then each vane tip suffers 10 cycles of pressure variation for each rotor revolution. The periodic forces at the tip may be transmitted to the main part of the vane (closer to the inlet) in one of two ways, either down the vane itself by bending of the vane, or to the disk first and down the disk to the other parts of the vane. The second method requires the vane tip to be enlarged and stiffened. For the first method, each vane vibrates individually. For the second method, the lower part of each vane receives forces from the disk, which itself is receiving impulses from each vane tip. With proper consideration for numbers of vanes and diffusers, disk stiffness, and vane-tip rigidity, the summation of impulses received at a vane lower region from buffets at the vane tips can be made zero. The "decoupling" of vane-*tip* buffeting into the disk with subsequent cancellation of vibration-causing effects for the lower and main part of the vanes has been incorporated into designs which have been very successful in resisting troubles caused by vibration.

The fundamental simple vane frequency (which has been probably the biggest source of trouble) is determined largely by the large vane body near the inlet, with the tip having little effect. Resonance with this frequency caused fatigue cracking of the vanes in early designs. One cure

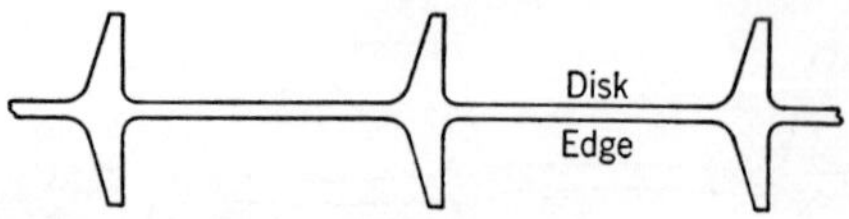

FIG. 11-10. Enlarged vane tips needed for "decoupling."

was to stiffen the vane so that the fundamental frequency was above the running range. This required thickening of the whole vane, and aerodynamic performance suffered. Another and better cure was to use the thin vanes, which gave higher performance (but also a critical frequency in the running range), and "decouple," as has been discussed to prevent excitation of the lower vane regions. The latter method has been very successful. Some thickening of the vane tips has been necessary along the lines shown in Fig. 11-10.

The estimation of the fundamental frequency of the simple vane mode of vibration is an item of considerable importance at the design stage. A vane is a complicated shape, and for usual design work some sort of simplifying approximation is very helpful. Voysey[1] has published a method which is simple and allows calculation of the fundamental within 5 per cent, within his experience. He searched for a "typical" section in the vane such that a simple frequency calculation using that section as a clipped cantilever beam would correlate well with the actual observed critical frequency of the vane. The section he found is shown as OA in Fig. 11-11, a section at 35° with the horizontal and through the inlet-eye vane tip at the widest part. The section is considered to be a linearly tapered cantilever beam (see Fig. 11-11) of constant thickness. The natural frequency of such a beam is

$$f = \frac{A}{2\pi} \frac{b}{L^2} \sqrt{\frac{Eg_c}{3\rho}} \tag{11-1}$$

where f = natural frequency, cps
A = frequency coefficient (dimensionless); it depends upon the taper ratio R and is to be taken from Fig. 11-12
b = half the beam width at the wall (Fig. 11-11), ft
L = beam length (Fig. 11-11), ft
E = modulus of elasticity in tension, psf = psi × 144
g_c = 32.17 lb_m per slug
ρ = density of beam material, lb per ft^3

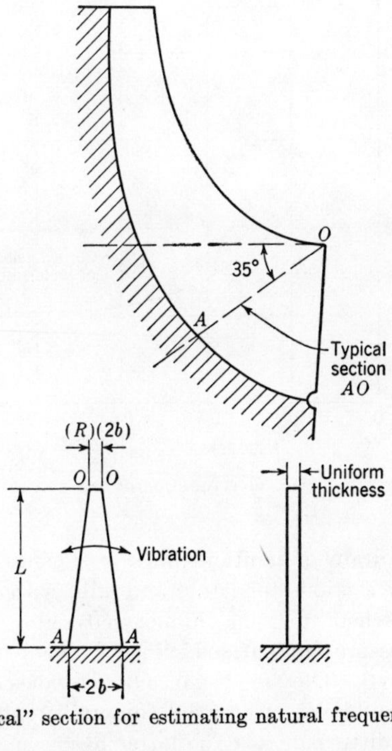

FIG. 11-11. "Typical" section for estimating natural frequency of impeller vane.

The ratio of actual measured frequency to calculated frequency using the typical section and the above formula showed very good correlation for a large number of rotors tested. The ratio varied from about 1.7 to less than 1.9. By using an average value (say 1.8), the designer can estimate the natural frequency to within about 5 per cent—much better than guesswork.

$$f_{\substack{\text{actual}\\\text{measured}}} = 1.8 f_{\substack{\text{calc. from}\\\text{typical section}}} \tag{11-2}$$

Later designs based upon this procedure have proved its worth.

The next higher mode (the second) is usually in the range of 1.5 to 3 times the fundamental, according to Voysey.

11-8. Vibration of the Combustion Chambers. The combustion chambers have usually been made of thin sheet metal, and periodic pressure variations inside have sometimes built up large vibrations of the shell. Some have buckled inward when the magnitude became sufficiently large. Other troubles have been fatigue cracking near heavy sec-

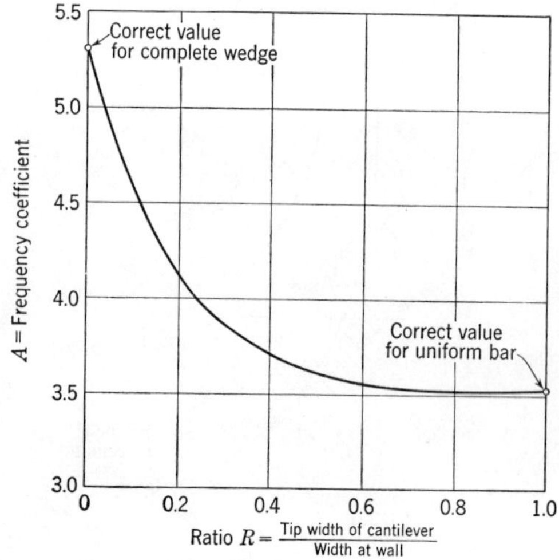

FIG. 11-12. Frequency coefficient as a function of taper ratio R at the typical vane section. (*After Voysey, Ref.* 1.)

tions which carry drain or ignition plugs or flanges for attachment of ducts. The heavy sections tend to stand still, with subsequent severe bending in the attached vibrating thinner material.

The periodic pressure fluctuations evidently occur from three sources, according to Voysey.[1] They are the number of vanes of the impeller (or, correspondingly, the blades of an axial compressor), break-away flutter, and a combustion disturbance. The latter has seemed to be related to fluctuations received from the fuel pump.

When trouble has occurred, it has usually not been too difficult to correct. Local cracks which have developed because of resonance fatigue in certain regions have been corrected by changing the mass and elasticity of those regions. Such redesign changes the natural critical frequencies. Much yet remains to be known about the nature of the excitation and the calculation of vibrating frequencies before trouble can be eliminated at the design stage. It has been relatively easy and economical to correct the chambers vibrationally only when trouble has occurred, and acceptable designs have been attained mainly with experience in eliminating the weak spots one by one.

11-9. Shaft Critical Speeds. A shaft-and-disk arrangement exhibits a certain fundamental lateral natural frequency of vibration, plus higher frequencies of different modes. If two disks on the ends of a shaft are very light and small, their effect on the shaft frequency is small. In the limit, if the disks are infinitely large, then the shaft is effectively built-in at the ends. This stiffens the shaft and raises the lateral natural frequency (Fig. 11-13).

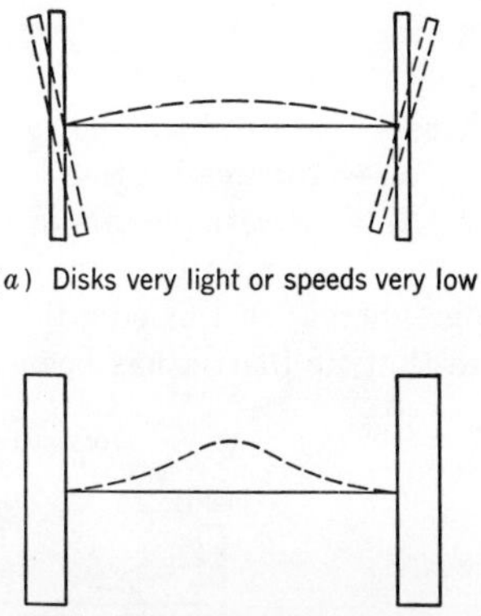

(*a*) Disks very light or speeds very low

(*b*) Disks very heavy or light disks revolving at very high speeds. The shaft ends approach a built-in condition, resulting in a stiffer spring action and higher natural frequency.

FIG. 11-13. Vibration of shaft and disks.

When the rotor composed of shaft and disks revolves, resonance occurs when the speed of revolution equals the lateral natural frequency of vibration. The shaft bows out and is said to "whip." At or near resonance and with little damping, the amplitude of whip can destroy the shaft and bearings.

In gas turbines, the shaft connecting the turbine and compressor is usually a large, very rigid and rugged hollow tube that has very little flexibility (Fig. 11-4). Great rigidity is desired and necessary in order that the critical frequencies be kept above the running speed. In addition, the masses of the turbine and compressor disks tend to keep each disk running in one plane (gyroscopic action), particularly at high speeds. This causes the shaft to approach a built-in condition and consequently become stiffer. The gyroscopic effect is appreciable, and the shaft itself is a gyro, too. Each of its transverse slices tends to maintain its plane of rotation. The net effect of high operating speeds, therefore, is to raise critical frequencies appreciably.

Because of proper design of shafts, they have caused relatively less trouble in gas turbines than many other parts. References are given at the end of the chapter which present methods of approximating critical whipping speed. It has been necessary, of course, to avoid operation at these resonant speeds. Normal operating speed for some jet-engine rotors lies above the first critical, so that a resonant speed must be passed through in bringing the rotor up to normal rpm. If this is not done quickly, serious damage can occur from the build-up of vibration (whip).

11-10. Torsional Vibration. Consider a system of two disks connected by a shaft. The jet engine of Fig. 11-4 is such a system. The torsional stiffness and the mass moments of inertia of the disks determine the natural frequency of torsional oscillation. The equation for natural frequency is

$$f = \frac{1}{2\pi}\sqrt{\frac{K_t(J_1 - J_2)}{J_1 J_2}} \qquad \text{(11-3)}$$

where f = natural frequency, cps
K_t = torsional spring constant, in.-lb per rad
J_1, J_2 = polar moments of inertia of the vibrating disks about the axis of vibration, in.-lb$_f$-sec^2

It is inherent in this equation that the shaft has no inertia of its own or else that its inertia has been replaced by effective inertias added to the

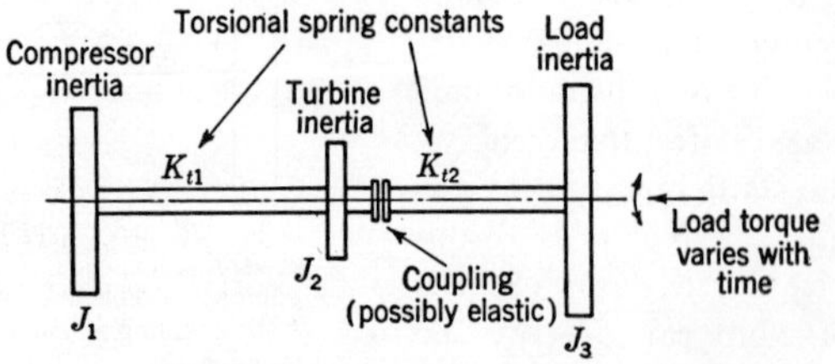

FIG. 11-14. Gas-turbine unit connected to a driven load.

disks. For given disks, high torsional stiffness gives a high natural frequency.

If torque variations are applied periodically to either disk at a frequency near the natural frequency, large and perhaps destructive torsional vibration can result if damping is small. In the usual gas turbine the torque applied to the disks is very smooth because of the number and overlapping effects of nozzles and vanes, and torsional vibration is not a serious problem within the engine itself.

Suppose, however, that the turbine engine is to drive a shaft-connected load such as a generator, possibly through a gear set and flexible couplings. A possible setup is shown in Fig. 11-14, without gears. The load itself may be the source of periodic torsional variations which can excite the natural frequencies of the system.

The system of three disks and two shafts has two critical frequencies, and the estimation of these two is reasonably simple, although a bit more difficult than the estimation for the simpler system within the turbine compressor unit. For higher numbers of disks and shafts an analytical solution is much more difficult, and methods are used which involve successive approximations. The Holzer tabulation method is one useful method of analysis. Many books on vibration theory cover this method and others, and references are given at the end of the chapter.

The foregoing discussion accents the fact that the *application* of a turbine engine is a determining factor with regard to torsional vibration. As always, resonance must be avoided, and this involves knowledge of the natural frequencies of the system and the frequencies of the disturbing impulses.

For the instances involving gearing or branched systems, the references include information on prediction of critical frequencies.

One of the serious effects of torsional vibration is the reduced life of gear sets and couplings which are present. If the coupling is elastic torsionally, then a change of coupling offers a convenient method of altering torsional rigidity to avoid resonance. A very "soft," flexible coupling helps to isolate load variations from reaching the turbine unit or gear set when it is present. Misalignment is not so serious a problem when elastic couplings are used as when the connections are rigid. However, if a flexible coupling is to be used at high speed and is required to flex unduly because of bad alignment, its life will undoubtedly be shortened.

11-11. Noise. Sound is produced when the ear receives pulsations of air pressure which lie in the audible range of frequencies and are of sufficient magnitude. In other words, sound is dependent upon vibration. The frequency range for hearing varies among individuals, but a range between 30 cps and 16,000 cps probably includes most people. Many do not have so wide a spread.

In gas turbines, many vibrations can occur in the audible spectrum. There are possibilities of vibration of flowing gases and of vibration of mechanical parts which eventually produce air-borne sound waves. A glance at Figs. 11-5 and 11-6 shows that blade natural frequencies are often squarely in the audible range, and sufficient vibration will produce noise. The vibration of centrifugal-compressor vanes can also become audible. Some early designs were tested by running at a speed which produced a loud howl, caused by operating at resonance with one of the natural vibration frequencies of the vanes. This was a severe test meant to cause failure in a short time so that different designs could be compared quickly. During the test, speed could be regulated to produce the loudest howl at all times, and this served as a very good measurement of resonant frequency. Final operation in a completed engine should not be near any such "howl" speeds, of course.

Blades and compressor vanes produce rapid pressure fluctuations in the fluid stream as they pass stationary nozzles, guides, or diffusers. These fluctuations may be heard as noise.

Vibration of various parts of auxiliaries has produced noise, too. One notable example is the vibration of tubes or plates in heat exchangers excited by aerodynamic effects. The problem is always one of first locating the vibration which is the source of the noise and then reducing the amount of that noise by making appropriate changes in the vibrating system.

The plague of screaming noise that occurs with the rush of gases into and out of gas-turbine engines remains a problem. For some applications (military jet airplanes, for example) the noise may not be particularly important. For others (stationary installations, locomotives, ships) excessive noise is definitely undesirable. Silencers (mufflers) vary in their effectiveness, and they always introduce pressure losses which may be very detrimental to the performance of an engine.

Noise is usually considered as a secondary problem at the present stage of development of the gas turbine. Its elimination is desirable; but except in those instances in which it has become completely unbearable (and there have been a few such instances) or in which it has indicated excessive mechanical vibration of parts, noise has not received as much attention as some of the other problems. Work now being done indicates that in most instances satisfactory methods of quieting can ultimately be developed.

It is well known that ultrasonic waves (sound waves above the audible frequencies) can be injurious to animal tissue. Experiments have been performed in which small animals have been killed by ultrasonic waves of sufficient intensity, and in relatively short times—in seconds or minutes. Man, too, is undoubtedly affected adversely by ultrasonic irradiation, and there is evidence that the high-frequency waves emitted during jet-engine operation have produced symptomatic reactions in some individuals, notably dizziness and loss of equilibrium. The effects of jet-engine ultrasonic noise on human beings are being further investigated at the present time.

PROBLEMS

11-1. A single-stage gas turbine has 40 stationary nozzles, 61 blades on the wheel, and 6 combustion chambers. Construct a diagram showing the disturbing frequencies in cycles per second applied to the blades as a function of speed for a speed range of 0 to 15,000 rpm. Include first and second harmonics of the combustion chamber and stationary nozzle frequencies.

11-2. Air enters an axial-flow compressor, passing first through 40 stationary blades and then to the first moving row. Find the fundamental harmonic exciting frequencies in cycles per second applied to the first row of moving blades for rotor speeds of 5000, 10,000, 15,000, and 20,000 rpm. *Ans.* 3333 cps; 6667; 10,000; 13,333.

11-3. A centrifugal compressor with 17 vanes discharges into 12 diffusing sections. At a rotor speed of 10,000 rpm, determine the fundamental disturbing frequency applied to each vane. Also find the pulsations per second into each diffuser section. *Ans.* 2000 cps, 2840 cps.

11-4. Estimate the critical vane frequency for a centrifugal compressor from the following information: At the "typical" vane section (Art. 11-7 and Fig. 11-11), the base width, $2b$, is 0.80 in., and the tip width is 0.15. The length L (for the typical section of Fig. 11-11) is 3 in. The material used for the vane has a density of 0.10 lb per in.^{3} and a modulus of elasticity of 10×10^6 psi. For 12 diffusing sections, what rpm would give resonance with the vane frequency? *Ans.* 6000 cps, 30,000 rpm.

11-5. A single-stage gas turbine turns at 6000 rpm, and produces 3000 shaft hp. The turbine wheel has 75 blades, and gas admission to the blades takes place completely around the periphery (full admission) from 60 nozzles. Find (*a*) the horsepower per blade; (*b*) the average force per blade at a blade radius of 12 in.; (*c*) the ratio of disturbing to natural frequency for a blade natural frequency of 2000 cps as a cantilever; (*d*) the dynamic magnification factor for blade deflection, resulting from the fact that the gas force is being applied in pulses rather than as steady force, (use Fig. 11-1); (*e*) the rpm for resonance with the given blade natural frequency. Assume small damping. *Ans.* (*a*) 40; (*b*) 35; (*c*) 3; (*d*) approx. 0.13; (*e*) 2000.

11-6. In the fir-tree type of blade root, the introduction of a solid lubricant between the blade root and the mating socket in the wheel rim has resulted in reduced vibration of the blade. Explain.

11-7. By use of experimentation in which blades were vibrated at controlled known frequencies, some of the critical frequencies for a certain type of compressor blade were found to occur at 900, 2300, and 3500 cps. In the axial-flow compressor, this type of blade is to be used in the first stage. Across the air inlet immediately before this stage is a support in the form of a cross which divides the inlet cross-section into four approximately equal areas. (*a*) At what compressor rpms could this support become the cause of serious blade vibration? (*b*) If one of the four areas were completely plugged by ice, what new speeds might become troublesome? *Ans.* (*a*) 13,500 rpm; 34,500; 52,500. (*b*) 54,000 rpm; 138,000; 210,000.

11-8. The *average* gas force applied to a blade is 45 lb, but it is applied in *puffs* varying ±15 per cent from the average value, at a frequency of 2500 cps. (*a*) If the simple cantilever frequency of the blade is 1000 cps, find the dynamic-magnification factor for the blade deflection. Assume damping equal to 0.05 of critical. (*b*) Does a moderate amount of damping have an appreciable effect on the magnification factor for the frequencies involved? (*c*) Find the peak value of the equivalent static gas force which would produce the same maximum blade deflection as occurs for the given conditions. (Assume elastic behavior of blade material.) *Ans.* (*a*) 0.19, (*c*) 46.28 lb.

11-9. Repeat Prob. 11-8*a* and *c*, but for a blade natural frequency of 2700 cps and negligible damping. Use the equation given with Fig. 11-1. *Ans.* (*a*) 7.1, (*c*) 93 lb.

11-10. In a rigid rotor 3 ft long, unbalanced masses are present as follows: (1) in a transverse plane through the left end of the rotor, 4 in.-oz at 30° clockwise from vertical (viewed from left end of rotor); (2) in a plane 1 ft from the left end, 5 in.-oz at 180° from vertical; (3) in a plane 2 ft from the left end, 2 in.-oz at 270° clockwise from vertical (viewed from left end of rotor). (*a*) Find the two equivalent unbalances and their locations in planes at the ends of the rotor, where correction is to be applied. (These would be the unbalances indicated by a balancing machine.) (*b*) Find and show on a sketch the amount and location of proper balancing weights, if weights are to be added at radii of 6 in. *Ans.* (*b*) In *left* end, add 0.223 oz at 264½° clockwise from vertical.

11-11. A turbine blade is subjected to a large alternating stress which fatigue tests have shown will produce failure in 2×10^8 cycles for the stress involved. If the blade receives 40 stress variations per wheel revolution, how many revolutions will occur before the blade fails from fatigue? This represents how many operating hours at a rotor speed of 6000 rpm? *Ans.* 5×10^6 revolutions; 13.9 hr.

11-12. A gas-turbine power plant drives a connected load through a gear set. Normal operating turbine rpm is 6000. The compressor inlet has a single support bar across the face. The centrifugal impeller has 15 vanes discharging into 10 diffusers. The gas leaving the diffusers enters 10 combustion chambers uniformly spaced around the turbine periphery. The hot gas goes through 68 stationary nozzles, then impinges on the 77 moving blades of the single-stage turbine, and finally is exhausted. The load runs at one-sixth of turbine speed and operates with torque variations in the shaft. First and fourth harmonics are particularly noticeable, the first with frequency equal to the rpm of the load shaft.

List the exciting frequencies which are applied to the various parts, and briefly explain the nature and cause of each excitation.

REFERENCES

1. "Development of the British Gas Turbine Jet Engine," Institution of Mechanical Engineers, London, reprinted for American Society of Mechanical Engineers, New York, 1947 (8 authors).
2. Den Hartog, J. P.: "Mechanical Vibrations," 3d ed., McGraw-Hill Book Company, Inc., New York, 1947.
3. Manson, S. S., A. J. Meyer, H. F. Calvert, and M. P. Hanson: Factors Affecting Vibration of Axial-flow Compressor Blades, *Proc. Soc. Exptl. Stress Anal.*, vol. 7, no. 2, 1950.
4. Emmert, H. D.: Current Design Practices for Gas Turbine Power Elements, *Trans. ASME*, vol. 72, no. 2, p. 189, February, 1950.
5. Wilkes, G. B., Jr.: Changes in Internal Damping of Gas-turbine Materials Due to Continuous Vibration, *Trans. ASME*, vol. 71, p. 631, 1949.
6. Nolan, R. W.: Vibration of Marine-turbine Blading, *Trans. ASME*, vol. 72, no. 4, May, 1950.
7. Hasenblaus: Pressure Distribution Measurements on Turbine Rotor Blade Passing behind Turbine Nozzle Lattice, *NACA Tech. Mem.* 1173, July, 1947.
8. Manson, S. S.: Stress Investigations in Gas Turbine Disks and Blades, *Trans. SAE*, vol. 3, no. 2, April, 1949.
9. Ghose, S. C.: A Comparative Study of the Noise from Turbojet and Reciprocating Aircraft Engines in Flight, *J. Roy. Aeronaut. Soc.*, November, 1950.
10. Moody, A. M. G.: The Axial Vibration of Turbine Disks, *J. Applied Mechanics*, March, 1945.
11. Stodola, A.: Steam and Gas Turbines, McGraw-Hill Book Company, Inc., New York, 1927.
12. Dorey, R. N.: Dart Turboprop Design Accents Long-life Features, *J. SAE*, November, 1950.
13. Timoshenko, S.: "Vibration Problems in Engineering," D. Van Nostrand Company, Inc., New York, 1937.

CHAPTER 12

MATERIALS AND METALLURGY

12-1. Factors Influencing Selection of Materials. The severity of operating conditions in gas turbines runs the gamut from regions of low temperature and moderate stress to regions of very high temperature and high stress. In some applications the engine inlet region confines a subzero hurricane, while only a short distance away (in and after the combustion chambers) a raging inferno must be enclosed and regulated. In addition, close dimensional control must be maintained, and this requires consideration of deformation and rigidity. The service life of the engine, its size and weight, and the cost of the materials and the methods which must be used to process them are related problems.

Figure 12-1, taken from Evans,[4] gives a graphic means of presenting some of the basic factors which the designer must consider simultaneously as he approaches the problem of mechanical design and selection of materials for any part of the gas-turbine engine (or for any machine, for that matter).

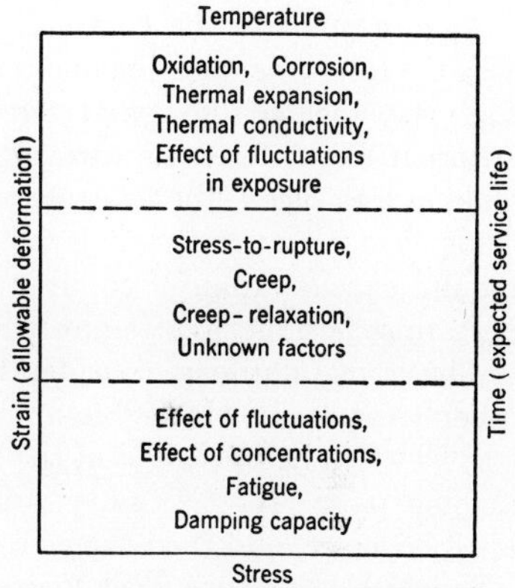

FIG. 12-1. Metallurgical considerations (inside square) to satisfy the designer's demands (outside of square). (*From Evans, Ref.* 4.)

The designer obviously is not always free to follow the dictates of the thermodynamic and aerodynamic desirabilities. If thin sections are desirable in order to minimize resistance to air flow and stresses and vibration require thick sections, then a compromise must be adopted. If a thermodynamic analysis indicates higher performance with higher temperatures, then the available materials will necessarily put a limit on performance. The designer must work with the materials which are available to him.

One promising avenue of approach to the improvement of gas-turbine thermal efficiency and performance is the development of suitable materials, particularly those which will perform satisfactorily at high temperatures. Much effort is being expended in this direction.

What should be the criterion for failure—fracture or a specified deformation? There is no one answer to this question for all situations. For a turbosupercharger, efficiency is not so important as for a larger power unit, and clearances can be large. Complete rupture may be the criterion for failure here, while in a unit of higher efficiency with small clearances, there is a limited amount of stretch allowable before parts begin to rub. Another possibility is that of deformations which do not cause rubbing or contact but which do warp carefully contoured profiles (blades, nozzles), leading to reduced efficiency. It seems obvious that the proper criterion for failure is that set of circumstances which seriously impairs or ends the useful function of the machine. The method of forming and finishing of the material must also be considered. For example, surface imperfections or casting defects may be starting points for early failure. Finally, cost enters the picture. It is always desirable to use the lowest cost, most readily available materials which will perform satisfactorily.

At any point in the continuing history of development, the designer asks the following question about a material: For a given temperature and a given expected life, what stress can be carried within the assigned criterion for failure (rupture or specified allowable deformation)? In making a decision, the designer must know the type of loading. Static loads lead to considerations of static strength, and fluctuating loads require consideration of endurance, or fatigue, strength.

In this chapter, no attempt will be made to present a full discussion of the basic metallurgical considerations and methods of heat-treatment or other treatments which *determine* the properties of materials. Rather, a survey will be presented of the required properties of materials for gas-turbine parts, the necessary property information which the designer must consider, and the properties and compositions of some of the materials which have been used successfully. The focus falls quite naturally upon the materials for parts which must operate at elevated temperatures.

12-2. Requirements for the High-temperature Materials. Because the hot materials are the critical ones in gas turbines, the discussion is directed mostly toward them. Turbine blades and disks, turbine nozzle vanes, diaphragms, combustion chambers, and certain ducts and bolts are exposed to very high temperatures, accompanied by high stresses in some instances. Following is a lengthy list of properties about which information is needed by the designer for use in actual design and for comparison of materials so that a proper choice may be made:

1. Creep and creep-rupture strength; creep-relaxation characteristics
2. Yield strength
3. Fatigue, or endurance, strength
4. Ductility
5. Modulus of elasticity and Poisson's ratio

6. Thermal coefficient of expansion
7. Thermal conductivity
8. Density
9. Hardness, erosion resistance
10. Impact resistance
11. Oxidation and corrosion resistance
12. Effects of temperature cycles; resistance to thermal shock
13. Effects of notches
14. Damping capacity
15. Availability and cost
16. Methods, effects, and costs of processing
17. Consistency in reproduction of properties

Information is desired concerning many of the above properties over a wide range of temperature, for parts must be able to withstand reasonably rough handling during assembly and inspection as well as to operate satisfactorily at high temperatures. Data are required, too, upon changes in properties that take place after prolonged operation at high stress and high temperature.

The foregoing list is probably not complete, but it certainly accents the fact that gas-turbine designers are putting a burden on the metallurgists. Seldom has there been such a concerted demand for materials to withstand such severe duty, and much remains to be known about the behavior of the promising materials during processing and during actual operation at the conditions encountered.

It should not be implied that all the parts subjected to high temperatures should possess desirable attributes of the listed properties to the same degree. The combustion chambers, for example, carry little stress by comparison with the turbine blades. The same is true of the nozzle vanes and ductwork. The really critical parts are the blades and rotors. The tabulation on page 444 gives a quick view of the temperature and some of the important material properties which must be considered for specific parts of a typical jet engine.

Failure of high-temperature cast materials has occurred along the grain boundaries, indicating that these are locations of weakness. Intergranular corrosion has often been a factor. Precipitation of weak constituents at grain boundaries within the material must be minimized during processing and operation. Very large grain size leaves long boundaries as planes of weakness. Very small grain size allows more creep because of the viscous action at the large number of boundary locations, but this action may increase resistance to fatigue. Fatigue failures often occur within the grains and steady-load failures along the grain boundaries. At least this statement appears to be true for some materials.

In forged materials, grain size and direction are controlled during the

Part	Temperature, °F	Material properties
Turbine wheel	400 to 500 at center 1200+ at rim	Ductility Yield strength Creep
Turbine buckets	1200 to 1500	Fatigue strength Stress rupture Ductility Corrosion Creep Thermal shock
Diaphragm	900 to 1600	Corrosion Thermal shock Strength
Combustion liners	900 to 1800	Corrosion Thermal shock Ductility Thermal properties (conductivity, expansion) Strength

forging operation as one step in producing the desired properties. In either forged or cast materials, the importance of *uniformity* cannot be overemphasized, and this is a major problem, particularly in large parts.

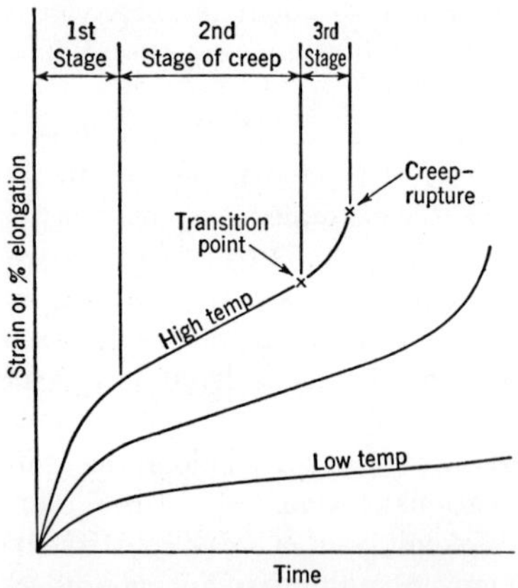

FIG. 12-2. Effect of temperature upon creep of metals. Stress is the same for all curves. Temperature is constant along each curve.

The pressures of wars and the international political situation have spurred developments in the field of high-temperature metallurgy which otherwise would not have had equivalent attention for many years. It seems, according to the old adage, that necessity is again the mother of invention.

12-3. Creep and Creep Rupture. The slow permanent deformation (creep) which occurs after prolonged operation under stress has been one of the most important considerations in assigning a design stress. The nature of creep will be reviewed briefly. Figure 12-2 illustrates some typical creep curves, with the three stages of creep indicated. The initial rapid straining occurs with localized high stresses and plastic flow, with a subsequent more uniform stress distribution. The material becomes work-hardened. Deformation becomes more difficult, and the curve flattens into the second stage of creep. In the third stage, deformation becomes rapid again, and rupture eventually takes place. The beginning of the third stage (called the transition point) may often be considered as a prelude to early failure.

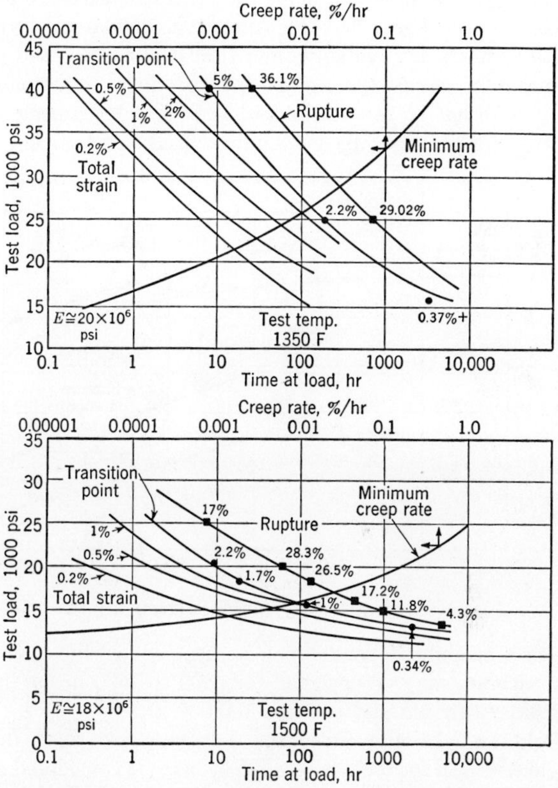

FIG. 12-3. Design curves for Refractalloy 70. (*After Scott and Gordon, Ref.* 3.)

Figure 12-2 is for one stress only. If the effects of different stresses are determined, a set of useful design curves for a material can be plotted. Figure 12-3 shows such a set for two temperatures for one high-temperature alloy. Such curves might be considered as typical design curves for presenting creep and creep-rupture properties. Several curves at different temperatures should be available.

The square relationship of Fig. 12-1 should be kept in mind in examining Fig. 12-3. It is seen that the designer has information on the four items of primary concern, stress, deformation, temperature, and life. The test is a long-time static test, however, and gives no guide as to the endurance, or fatigue, strength. Additional information is necessary for the conditions in which the criterion of failure is the endurance strength.

How important is the creep strength at rupture? Emmert[7] states that the rupture stress for a specified service life is one of the two basic measures of strength in use, the other being the yield strength for 0.2% offset.* It should be mentioned here that this statement concerning creep assumes that vibration effects are negligible and that it is *rupture* and not just a certain amount of creep *deformation* that produces functional failure. Presumably deformation can be checked and corrected during inspection and overhaul—for example, by machining off the blade tips.

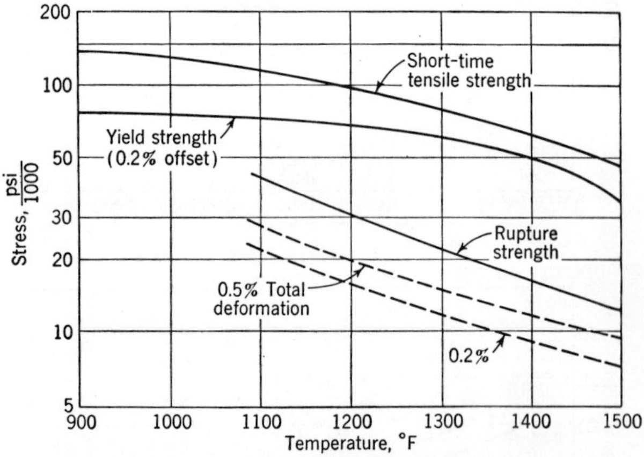

FIG. 12-4. Materials-data curve; wrought S-590; 10,000 hr. (*After Emmert, Ref.* 7.)

The life of a machine depends upon its use. Short-life machines such as aircraft turbines may be designed for 1000 to 1500 hr, while long-life power machinery should last perhaps 100,000 hr.

12-4. Yield and Fatigue Strengths. As was indicated in the last article, yield strength for 0.2 per cent offset may be considered as the second of the two basic criteria for strength. Figure 12-4 shows the drop in yield strength at higher temperatures for one material. Why should yield strength be high? It prevents permanent deformation of parts caused by short-time overstressing (not by creep). Another consideration is that if stresses are lower than the yield strength, the stress calculations based on elastic behavior may be applied with some confidence. In

* The yield strength at 0.2 per cent offset is determined from the stress-strain curve as follows: From a point on the strain axis of 0.2 per cent strain, a line is drawn parallel to the initial straight-line portion of the stress-strain curve of the material. This parallel line intersects the stress-strain curve at a certain value of stress. This stress is the yield strength at 0.2 per cent offset, a stress value somewhat higher than the elastic limit.

any instance, of course, such stress calculations are only approximations, and stress-concentration factors and factors of safety based on experience must be applied.

Yield strength should be high, and yet it should be realized that a certain amount of ductility is also required in order that localized plastic flow can take place in regions of high stress concentration. Brittle materials break because of inability to flow and relieve stress. There seems to be a variance of opinion as to just exactly how much ductility is required for the different uses involved in gas turbines.

In selecting a design stress, an operating temperature and service life are selected. Then the yield strength and rupture strength are determined for the material decided upon (see Fig. 12-4). Proper stress-concentration factors and factors of safety (from experience) are applied, and a design stress is chosen, with consideration of both yield strength and creep-rupture strength as possible limiting conditions.

The fatigue, or endurance, limit seems to have been left out of the picture, at least momentarily. For the *blades* and vanes of turbines and compressors, fatigue strength must sometimes be considered as a limiting stress because of the possibility of vibration of the blades caused by gas buffeting.

Failures have occurred from vibration, and the best cure for this kind of trouble is to redesign to avoid resonance. However, it is not always possible to run sufficiently far from all resonant speeds to eliminate all vibration completely; so the designer must always consider the fatigue strength as a possible limit for blades. In general, though, fewer data seem to be available on fatigue strength than on other properties for many materials, although this situation is being corrected. Often gas bending stresses are kept low and blade frequencies high to avoid trouble from fatigue and to compensate for lack of knowledge of fatigue strength. Table 12-1 gives an indication of fatigue resistance of Inconel X, and it will be noted that the general trend is a decrease in fatigue resistance at

TABLE 12-1. FATIGUE RESISTANCE OF INCONEL X*
(Typical composition: 73% Ni, 15% Cr, 7% Fe, small amounts of C, Mn, Si, Cb, Ti, Al)

Temp, °F	Stress for failure, psi		
	10^6 cycles	10^7 cycles	10^8 cycles
1200	67,800	60,100	53,800
1350	51,500	49,500	48,500
1500	53,000	39,500	35,000

* Data from Crawford, Nickel-chromium Alloys for Gas-turbine Service, *Trans. ASME*, August, 1947, p. 609.

higher temperatures. No true endurance limit exists for the high-temperature alloys in common usage. That is, there is no particular value of stress which can be carried indefinitely without failure, as is the case for some other materials. Some representative values (from Clark[13]) of stresses producing failure in fatigue at specified temperatures are as follows:

Cast S-816	:1500 F, 33,000 psi at 10^8 cycles of reversed bending at 120 cps
Cast Vitallium (not heat-treated)	:1200 F, 44,000 psi at 10^8 cycles
	1500 F, 33,000 psi at 10^8 cycles
19-9 DL	:1200 F, 43,000 psi at 10^8 cycles
Hastelloy B	:1200 F, 66,000 psi at 10^8 cycles
	64,000 psi at 2.5×10^8 cycles
	1900 F, 34,000 psi at 10^8 cycles

It should be remarked here that different treatments of these same materials could produce marked differences in fatigue properties.

It is an interesting fact that many materials appear to suffer a loss in effective ductility during long-time tests at high temperatures, resulting in lessened deformation at creep rupture than is obtained in shorter tests at higher stress (see Fig. 12-3). Probably this loss of ductility contributes to fatigue failure.

The high-temperature strength of materials can often be improved by increasing the hardness, but this also may result in lowered ductility. Because both strength and ductility are necessary, compromises must be adopted. Much improvement in gas-turbine materials could be realized if ductility could be maintained as materials are made harder and stronger.

12-5. Other Properties. Creep strength, yield strength, fatigue strength, and ductility have been discussed briefly as of primary importance. Certainly these are not the only items which must be considered, but they serve at present as indications of load-carrying capacity. The properties listed in Art. 12-2 must *all* be considered in selecting materials. The stress-calculation formulas *require* knowledge of many of the properties. Other properties allow comparison as to resistance to abrasion by solid particles which enter with the inlet air (sand, for example); resistance to loss of metal and surface deterioration by corrosion and oxidation; relative costs and availability; methods and costs of processing; and relative effects of thermal shock resulting from rapid temperature changes. All these things are in the picture at the time of selection of materials. Once a material has been selected, then the load-carrying criteria of strength are applied and running stresses are calculated.

12-6. Processing. The *composition* of materials is of importance, of course, but of no less importance is the manner of *processing* them.

Various alloying elements are used to impart desired properties—for example, chromium for corrosion resistance and titanium or molybdenum as hardening agents. Once the composition is fixed, there are still the possibilities of various kinds of heat-treatment and work hardening, and the final properties are greatly affected by these physical treatments. Grain size and direction, for example, are two important factors which need to be controlled. Finally, there is the serious matter of "scatter" of properties, or the lack of uniformity and reproducibility of desirable attributes. The designers always want to know the *minimum* stress at which failure can occur, not some average obtained from tests on many specimens which showed a wide variation from the average. It is desirable to have the minimum and the average close together.

This matter of uniformity is of great importance, and the processing method is a significant factor. Rolled and wrought materials exhibit the most consistent reproduction of properties, with minimum flaws and rejections. The materials requiring precision casting (some turbine-blade materials) seem to have given the most trouble with a high percentage of rejects, sometimes 30 per cent or more. Very critical inspection is necessary, and the final cost may be high.

It would appear that rolling and forging are advantageous methods, and this is true. For high-temperature operation in a turbine, however, a material is required which is strong, hard, and resistant to deformation; and these properties are the very reverse of those required for easy forging and rolling (and also machining). Further, it must be recognized that the temperature difference between that for processing and that for operation in the turbine is being continually narrowed. In a rather fantastic limiting situation, it would appear that we would need *simultaneously*, at *one* high temperature, a soft, plastic material for ease of processing and a hard, strong material for satisfactory running performance.

The foregoing statement gives some idea why cast materials have entered the picture and probably will gain in acceptance as temperatures go still higher. The material can be strong and hard as processed (cast), but now it must be recognized that it will be very difficult to machine or grind. What is required is *precision* casting as close as possible to the final dimensions, with an absolute minimum of finish grinding required. The precision investment method has been widely and successfully used for producing turbine blades. The general technique is not new, having been used for many years by dentists in producing gold fillings for teeth. The use of the technique for different materials has been a very real problem, however, because of differences in temperatures, expansion, and shrinkage.

A process which has become very important in inducing strength in parts which can be forged (such as turbine disks) has been the so-called hot-cold working process. Rolling or forging is carried on at high temperatures, but not high enough to induce recrystallization of the metal.

12-7. Some Typical Materials. It would be well at this point, after having briefly surveyed the material properties of interest, to examine some high-temperature materials that have been used successfully. Tables 12-2 to 12-4 present pertinent information about some typical American alloys. These alloys are essentially austenitic in structure. Table 12-2 shows the alloy bases to be iron, nickel, nickel-cobalt, and cobalt. Some alloys are available in wrought form, and some in precision-cast form. Hardening of the base metal is obtained by addition of hardening agents, by suitable heat-treatment, and by working. The specific treatment depends upon the metal involved. The criterion for temperature rating is arbitrary and is based upon 100 hr rupture life at a load of 25,000 psi. Design stresses would be made very much less than this in units intended for long life, and metal temperatures would also probably be lower than those listed.

In Table 12-3, the compositions of the alloys are listed. It will be noted that practically all the materials contain an appreciable amount of chromium, regardless of the base metal. The chief reason for the chromium is the resistance to oxidation and corrosion which is gained by its presence. The other materials are added to control other properties, such as hardness and strength.

Table 12-4 indicates the heat-treatments which are used for the wrought alloys and also the resulting grain sizes and hardnesses. In general, fine grain size is desired for good ductility and high fatigue resistance, but usually creep strength is reduced as fatigue resistance and ductility are increased. A compromise is necessarily adopted.

Tables 12-5 and 12-6 give a more complete listing of some of the important design properties for a variety of high-temperature materials including the properties of one alloy of the Nimonic series, which is widely used by the British. These tables represent a rather detailed study made under government sponsorship. Much is to be gained from a close inspection of the information presented. It will be noted that data are given on tensile strength, on creep-rupture stress for specified life periods up to 1000 hr, on yield strength for 0.2 per cent offset, on elongation at rupture, which is some indication of ductility, and on the treatments used in producing the properties recorded. The tables give a reasonably good picture of the status of high-temperature metallurgy shortly after the Second World War and a hint of the great number of possibilities of attaining desirable properties by changes in composition and treatment. Of course, there have been some modifications and improvements since the publication of the tabulated information, and continuing development is to be

TABLE 12-2. IDENTIFICATION AND CLASSIFICATION OF HEAT-RESISTANT ALLOYS FINDING APPLICATION UNDER HIGH STRESSES AT TEMPERATURES OVER 1000 F*

Alloy name	AMS specification	Development organization	Alloy base	Hardening agents		Temperature rating, °F†
				Chemical	Physical‡	
Available in Wrought Form						
Uniloy 19-9DL	5722A	Universal-Cyclops	Fe	C	HCW	1300
EME	5730	Midvale	Fe	C, N	CW	1250
16-25-6	5727	Timken	Fe	Mo, N	HCW	1350
Discaloy	5733	Westinghouse Electric	Fe	Ti	PH	1300
Inconel X	5668B	International Nickel	Ni	Ti	PH	1400
Hastelloy B	5752	Electro Metallurgical	Ni	Mo	ST	1350
K-42-B	......	Westinghouse Electric	Ni + Co	Ti	PH	1400
Refractalloy 26	5760	Westinghouse Electric	Ni + Co	Ti	PH	1500
Multimet (N155)	5768	Electro Metallurgical	Ni + Co	Mo, N	PH	1400
S590	5770	Allegheny Ludlum	Ni + Co	Mo, C	PH	1440
Refractalloy 70	......	Westinghouse Electric	Ni + Co	Mo	PH	1450
S816	5765A	Allegheny Ludlum	Ni + Co	Mo, C	PH	1550
Available in Precision Cast Form						
Haynes Stellite 27	5378	General Electric and Haynes Stellite	Ni + Co	C, Mo	.....	1450
Haynes Stellite 30	5380	General Electric	Ni + Co	C, W	.....	1525
Haynes Stellite 31	5382A	Haynes Stellite	Ni + Co	C, Mo	.....	1550
Refractalloy 80	......	Westinghouse Electric	Ni + Co	Mo	PH	1450
Haynes Stellite 21 (Vitallium)	5385	Haynes Stellite	Co	C, Mo	.....	1400
Haynes Stellite 23	5375	Haynes Stellite	Co	C, W	.....	1475

* From Scott, Gas Turbine Alloys, 10 Years Later, *Metal Progress*, October, 1950, p. 503, by permission from *Metal Progress*.

† Based on a 100-hr rupture life at 25,000 psi load.

‡ Code: CW = cold-worked
HCW = hot-cold-worked
PH = precipitation-hardened
ST = solution-treated

TABLE 12-3. NOMINAL COMPOSITION OF HEAT-RESISTANT ALLOYS*

Alloy name	Group I (basic elements)						Group II (intermediate)			Group III (hardeners)			
	Ni	Co	Cr	Fe	Mn	Si	Mo	W	Cb	C	N	Ti	Other
Uniloy 19-9DL	9	..	19	66	1.1	0.5	1.4	1.4	0.4	0.31	....	0.2	
EME	12	..	19	63	0.5	0.5	....	3.2	1.0	0.15	0.15		
16-25-6	25	..	16	51	1.0	0.5	6.2	...	...	0.08	0.15		
Discaloy	26	..	13.5	54	0.8	0.8	3.2	...	...	0.03	....	1.6	0.1Al
Inconel X	73	..	15	7	0.7	0.4	....	...	1.0	0.04	....	2.5	0.7Al
Hastelloy B	66	..	....	5	0.6	0.2	28	...	...	0.08	....	...	0.3V
K-42-B	42	22	18	14	0.7	0.7	....	...	...	0.03	....	2.1	0.2Al
Refractalloy 26	38	20	18	16	0.8	1.0	3.2	...	...	0.03	....	3.0	0.2Al
Multimet (N155)	20	20	21	30	1.5	0.6	3.0	2.5	1.0	0.12	0.15		
S590	20	20	20	27	1.5	0.6	4.0	4.0	4.0	0.43			
Refractalloy 70	21	30	20	14	2.0	0.3	8.0	4.2	...	0.04			
S816	20	41	20	4	1.5	0.6	4.0	4.0	4.0	0.37			
Haynes Stellite 27	36	30	26	1	0.5	0.4	6	...	...	0.42			
Haynes Stellite 30	15	51	26	1	0.5	0.4	6	...	...	0.42			
Haynes Stellite 31	10	54	25	1	0.6	0.7	....	7.5	...	0.50			
Refractalloy 80	20	30	20	14	0.6	0.7	10	5	...	0.10			
Haynes Stellite 21	2	64	26	1	0.3	0.6	5.5	...	...	0.20			
Haynes Stellite 23	2	65	25	1	0.3	0.6	....	6	...	0.38			

* From Scott, Gas Turbine Alloys, 10 Years Later, *Metal Progress*, October, 1950, p. 503, by permission from *Metal Progress*.

expected. Many of the materials listed, however, are still in widespread use.

Tables 12-7 and 12-8 list densities and coefficients of expansion of many of the much-used critical gas-turbine materials. Knowledge of these items is required in the calculation of inertia and thermal stresses. It should be noted that the high-temperature alloys are all heavier than iron and steel.

TABLE 12-4. HEAT-TREATMENT OF WROUGHT ALLOYS (BARS AND FORGINGS)*

Alloy name	Stress relief		Solution		Aging		ASTM grain size	Brinell hardness
	Temp	Time	Temp, °F	Time, hr	Temp, °F	Time, hr		
Uniloy 19-9DL....	1200	4 + hr	None	...	None	.....	...	228–277
EME†...........	1200	4 + hr	None	...	None	.....	...	269–321
16-25-6§..........	<1210	4 hr per in.	None	...	None	.....	...	235–293
Discaloy..........	None	....	1800	1	1350	20		
					1200	20	4+	248–302
Inconel X.........	None	....	2100	2–4	1550	24		
					1300	20	...	262–341
Hastelloy B.......	None	....	1900	24	None	.....	4–8	210–285
K-42-B†..........	None	....	1750	1	1300	24	5+	269–341
K-42-B†..........	None	....	1950	1	1350	20	3+	221–302
Refractalloy 26	None	....	1800	1	1350	44	6+	302–375
Refractalloy 26....	None	....	1950	1	1500	10		
					1350	20	3+	262–341
Refractalloy 26....	None	....	2100	1	1500	20		
					1350	20	1+	241–311
Multimet (N155)..	None	....	2175	1	1500	4	...	193–241
S590.............	None	....	2200	1	1400	10	...	248–331
Refractalloy 70....	None	....	2250	4	1600	16	5+	293–352
S816.............	None	....	2150	1	1400	12–16	1+	248–331

* From Scott, Gas Turbine Alloys, 10 Years Later, *Metal Progress*, October, 1950, p. 503, by permission from *Metal Progress*.

† EME and K-42-B bars.

§ 16-25-6 forgings.

12-8. Manufacture of Turbine and Compressor Blades. There are many ways in which turbine and compressor blades and vanes have been produced. Colwell and Cummings[10] list the following seven:

1. Forging
2. Lost wax (precision casting)
3. Machining
4. Fabrication (sheet-stock forming)
5. Powder metallurgy
6. Rolling
7. Mercast (precision casting with frozen-mercury patterns)

The method of processing cannot, of course, be divorced from the material

TABLE 12-5. TENSILE PROPERTIES OF HIGH-TEMPERATURE ALLOYS FROM 70 TO 2000 F*

Trade name	Temp °F	Tensile stress, psi	Yield stress, psi	% elongation	% reduction of area	Treatment†
Forged Alloys						
			0.2% Yield			
LC N-155	Room	146,750	121,000	25.5	47	CW, 1200 F
	1200	116,000	....	13	23	CW, 1200 F
			0.1% Yield			
	1350	80,940	65,650	24	37	Hot-worked
	1500	49,400	29,000	34.0	35.0	2300 F, ¾ hr, WQ 1500 F, 4 hr, FC
HC N-155	Room	138,000	70,600	16.0	12.6	2200 F, 1 hr, WQ
	1500	58,000	29,000	34.2	28.8	1500 F, 4 hr, FC
			0.2% Yield			
S-590	Room	160,500	89,500	10	10.5	2270 F, 1 hr, WQ 1400 F, 16 hr
	1200	81,600	49,000	27	31	2300 F, 1 hr, WQ 1400 F, 16 hr, AC
	1350	66,875	58,500	27	33	2270 F, 1 hr, WQ 1400 F, 16 hr
			0.1% Yield			
K-42-B	Room	162,500	97,200	31.0	39.4	1750 F, 2 hr, WQ 1200 F, 72 hr, FC
	1200	128,000	91,250	10.0	13.3	
	1350	101,000	79,700	6.5	10.2	
	1500	71,000	41,250	2.5	4.2	
	1700	23,900	....	40.5	56.0	
Hastelloy B	Room	139,200	57,000	49.5	51.1	1950 F, 2 hr, AC
	1500	77,500	42,500	27.0	21.2	1950 F, 2 hr, AC
	1600	52,500	39,700	45.5	51.8	1950 F, 2 hr, AC
	1800	31,500	16,100	70.0	53.0	1950 F, 2 hr, AC
	1900	18,500	9,800	54.5	46.9	1950 F, 2 hr, AC
	2000	13,100	7,600			1950 F, 2 hr, AC
			0.2% Yield			
S-816 forged	Room	175,000	73,000	39	45	As rolled
	1200	120,200	....	17	22	2300 F, 1 hr, WQ 1400 F, AC
	1350	98,900	....	15	22	2300 F, 1 hr, WQ 1400 F, AC
	1500	78,260	....	12	21	2300 F, 1 hr, WQ 1400 F, AC

TABLE 12-5. TENSILE PROPERTIES OF HIGH-TEMPERATURE ALLOYS FROM 70 TO 2000 F*—(*Continued*)

Trade name	Temp °F	Tensile stress, psi	Yield stress, psi	% elongation	% reduction of area	Treatment†
	1600	59,770	....	18	20	2300 F, 1 hr, WQ 1400 F, AC
	1700	46,180	....	14	17	2300 F, 1 hr, WQ 1400 F, AC
Nimonic 80.......	Room	153,000	83,800	36.5	33.6	1950 F, 2 hr, WQ 1300 F, 16 hr
	1200	97,500	77,000	14.5	20.8	1950 F, 2 hr, WQ 1300 F, 16 hr
Inconel X........	Room	192,000	136,000	26	45	Hot-rolled, aged
	1200	118,000	....	12	14	Hot-rolled solution-treated, aged
	1500	70,000	....	3	10	Hot-rolled, solution-treated, aged
			0.1% Yield			
Gamma Columbium	Room	107,500	46,000	26.0	30.1	2250 F, ¾ hr, OQ 1500 F, 50 hr
	1500	43,500	23,000	38.0	42.2	2250 F, ¾ hr, OQ 1500 F, 50 hr
19-9DL..........	Room	140,750	115,000	29	48.1	2100 F, 1 hr, AC
	1200	91,000	81,000	16	38.8	Finish roll 1200 F Stress relieve 1200 F, AC
Timken 16-25-6...	Room	162,250	143,500	15.5	33.8	2100 F, 1 hr, AC
	1200	107,500	94,000	13.0	27.9	Finish roll 1200 F, AC Stress relieve 1200 F, AC
	1350	60,100	27,000	40.0	35.5	Preheat 1500–1550 F 2150 F, WQ
	1500	48,300	31,400	27.5	26.2	2160 F, 2 hr, WQ
	1800	21,000	12,000	22.0	19.5	2160 F, 2 hr, WQ
			Cast Alloys			
			0.2% Yield			
Vitallium.........	Room	110,000	65,000	10	12	As cast
	Room	120,000	....	5.1	6.6	1350 F, 48 hr
	1200	71,000	35,600	18.4	41.5	As cast
		93,200	71,500	1.8	5.6	1350 F, 50 hr

TABLE 12-5. TENSILE PROPERTIES OF HIGH-TEMPERATURE ALLOYS FROM 70 TO 2000 F*—(*Continued*)

Trade name	Temp °F	Tensile stress, psi	Yield stress, psi	% elongation	% reduction of area	Treatment†
	1350	78,000	59,300	3.5	8.2	1350 F, 50 hr
		82,700	63,700‡	4.0	9.8	1350 F, 50 hr
	1500	57,300	47,350‡	8.3	19.7	1350 F, 50 hr
		61,400	50,700	5.3	4.5	1350 F, 50 hr
	1600	49,200	35,000	9.8	15.8	1350 F, 50 hr
		48,600	36,800	9.5	19.9	1350 F, 50 hr
	1700	42,470	....	27	52.4	As cast
	1800	33,265	....	35	52.4	As cast
		32,910	....	49	63.1	1700 F, 16 hr, AC
6059............	Room	82,550	46,900§	7	10.3	As cast
		76,800	41,100§	3	3.4	1700 F, 16 hr, AC
	1200	46,300	34,400	3.3	7.7	As cast
		51,500	35,600	4.2	8.2	As cast
	1500	51,400	38,700	12.0	21.8	1350 F, 50 hr
		51,000	37,700	8.3	7.0	1350 F, 50 hr
	1700	42,950	....	23	26.5	As cast
		45,430	....	16	34	1700 F, 16 hr, AC
	1800	33,400	....	24	50.3	As cast
		33,690	....	26	41.7	1700 F, 16 hr, AC
422-19...........	Room	98,100	55,100§	5	11.9	As cast
		94,500	56,200‡	1	0.65	1700 F, 16 hr, AC
	1200	58,300	36,900	7.2	15.7	As cast
		62,200	38,400	7.0	10.3	As cast
	1350	74,800	59,000	2.0	2.1	1350 F, 50 hr
		80,700	63,800	1.6	3.3	1350 F, 50 hr
	1600	49,200	35,000	9.8	15.8	1350 F, 50 hr
	1700	45,180	....	17	26	As cast
		47,135	....	18	33.3	1700 F, 16 hr, AC
	1800	36,290	....	24	33.7	As cast
		37,800	....	21	38.7	1700 F, 16 hr, AC
61...............	Room	103,400	58,350§	7	11.2	As cast
		108,500	53,900§	6	14.2	1700 F, 16 hr, AC
	1200	79,500	39,000	14.0	18.4	As cast
		103,400	74,000	2.1	5.8	1350 F, 50 hr
	1350	73,600	61,000	1.2	5.0	1350 F, 50 hr
		85,600	65,200	2.8	7.9	1350 F, 50 hr
	1600	46,800	34,200	9.6	14.8	1350 F, 50 hr
		44,800	32,000	10.0	18.8	1350 F, 50 hr
	1800	33,115	....	32	40.6	As cast
		33,050	....	27	39.5	1700 F, 16 hr, AC

TABLE 12-5. TENSILE PROPERTIES OF HIGH-TEMPERATURE ALLOYS FROM 70 TO 2000 F*—(*Continued*)

Trade name	Temp °F	Tensile stress, psi	Yield stress, psi	% elongation	% reduction of area	Treatment†
X-60	1500	60,000	53,500	1.0	None	1500 F, 50 hr
	1600	54,000	39,200	1.5	2.1	1600 F, 50 hr
		50,000	36,000	1.4	1.1	1500 F, 50 hr
X-40	Room	101,000	74,100	11	14	As cast
	1200	76,800	37,400	18	28.1	As cast
		77,300	37,800	20	28.6	As cast
	1350	82,200	53,400	7.0	8.1	1350 F, 50 hr
		69,200	53,800	2.3	7.0	1350 F, 50 hr
	1500	57,500	42,000	14.0	15.4	1350 F, 50 hr
		61,400	47,500‡	5.0	12.1	1350 F, 50 hr
	1600	48,500	35,200	14.3	18.1	1350 F, 50 hr
S-816	Room	112,000	59,000	5	13	As cast

* Abstracted from Grant, Frederickson, and Taylor, A Summary of Heat Resistant Alloys from 1200 F to 1800 F, *Iron Age,* Mar. 18, Apr. 8, 15, 1948, by permission from *The Iron Age.*

† Code: AC = air-cooled
OQ = oil-quenched
WQ = water-quenched
CW = cold-worked
FC = furnace-cooled

‡ Estimated.

§ 0.02 per cent yield.

under consideration, and not all the above processes can be used with all materials. Neither are all the above processes of equal importance in present manufacturing practice.

The *forging process* was one of the first to be used and has been very successful in producing uniform properties in finished blades and vanes for both turbines and compressors. Equipment has been developed which allows forging to very close tolerances, although some finish machining is often required. Die life depends upon the material being forged, of course, and dies for compressor blades last much longer than those for turbine blades because of the difference in the necessary materials. Some of the successful forgeable materials are indicated in the tables presented earlier in this chapter.

The *lost-wax process* is an investment-casting method, and allows the use of materials which cannot easily be forged or machined. Many of the promising high-temperature materials fall into this category, notably

TABLE 12-6. STRESS TO RUPTURE AND ELONGATION VALUES OF HIGH-TEMPERATURE ALLOYS*

For 10, 100, and 1000 Hr for Temperatures from 1200 to 1800 F

Trade name	Temp, °F	10 hr		100 hr		1000 hr		Treatment†
		Psi	%	Psi	%	Psi	%	
Forged alloys								
HC N-155	1200	79,000	....	60,000	....	45,500	....	Forged 2075–1600 F 2100 F, 1 hr, AC Finish-rolled, 1700 F Stress-relieved, 1200 F, AC
	1350	51,000	18	36,000	10	26,000	8	2320 F, WQ Aged 50 hr, 1500 F
	1500	32,000	12	24,500	10	18,600	4	2320 F, 1 hr, WQ 1500 F, 4 hr, AC
	1600	21,300	17	15,500	7	11,400	3	As hot-worked + 1 hr, 2300 F, WQ 1500 F, 4 hr, AC
LC N-155	1200	75,000	15	59,000	15	46,000	13	Hot-worked
	1350	47,000	....	36,000	15	27,000	11	Hot-worked
	1500	25,800	26	20,000	28	15,800	....	2320 F, 1 hr, WQ 1500 F, 4 hr, AC
	1700	12,500	50	7,600	19	4,700	7	2280 F, ½ hr, AC
	1800	8,700	20–26	4,900	11	2,750	5.5	2280 F, ½ hr, AC
S-590	1200	69,000	33	52,000	35	40,000	37	2325 F, 1 hr, WQ 1400 F, 16 hr, AC
	1350	42,000	28	31,500	24	23,500	19	2300 F, 1 hr, WQ 1400 F, 10 hr, AC
	1500	25,500	16–20	20,000	16–20	15,500	6–12	2250–2325 F, 1–2 hr, WQ 1500 F, 16–50 hr, AC
	1600	19,000	40–48	14,000	28	10,500	20–21	2300 F, 1 hr, WQ 1600 F, 16 hr, AC
	1800	9,000	56	5,500	28	3,400	....	2270 F, 1 hr, WQ 1800 F, 16 hr
19-9DL	1200	70,000	7.0	60,000	4.3	50,000	2.5	Forged 2055–1300 F 2100 F, 1 hr, AC
	1350	60,000	4.0	35,000	4.0	20,500	4.0	Rolled 1200 F Stress relieved 1200 F, AC
	1500	17,500	5.0	13,300	4.0	10,000	3.0	2250 F, ½ hr, OQ 1500 F, 50 hr
Timken 16-25-6	1200	78,000	12	57,000	8.5	42,000	4.3	Forged 2065–1650 F 2100 F, 1 hr, AC
	1350	42,000	8.5	28,000	4	19,000	1.8	Rolled 1200 F Stress relieved 1200 F, AC
	1500	19,500	60	13,300	35	9,000	20	Preheat 1550–2150 F, WQ
K-42-B	1200	95,000	1	62,000	1	39,500	0.5	1950 F, WQ 1350 F, 20 hr
	1350	49,500	1	36,500	1	26,500	1	1950 F, WQ 1350 F, 20 hr
	1500	33,800	2.5	22,500	2.5	15,000	2.5	2100 F, WQ or OQ 1500 F, 20–50 hr

TABLE 12-6. STRESS TO RUPTURE AND ELONGATION VALUES OF HIGH-TEMPERATURE ALLOYS*—(*Continued*)

Trade name	Temp, °F	10 hr		100 hr		1000 hr		Treatment†
		Psi	%	Psi	%	Psi	%	
Hastelloy B	1200	100,000	2.1	58,000	2.2	33,200	8.5	2000 F, 1 hr, WQ 1200 F, 4 hr, FC
	1350	48,000	....	34,000	22.5	25,000	....	1950 F, 2 hr, AC
	1500	25,600	28	17,000	59	11,300	....	1900 F, 24 hr, AC
	1600	16,700	....	11,700	....	8,300	....	Not known
S-816 forged	1200	83,000	14	64,000	8	49,200	4	2250–2300 F, 1 hr, WQ 1400 F, 6–16 hr, AC
	1350	55,000	18	40,000	8	29,500	6	2300 F, WQ; 16 hr, 16 hr, 1400 F
	1500	35,100	10	27,000	10	21,100	10	2300–2350 F, 1 hr, WQ 1500–F, 16–50 hr
	1600	21,700	16	14,800	15	10,000	12	2300 F, 1 hr, WQ 1600 F, 50 hr
	1700	14,900	15	9,800	6	6,400	....	2300 F, 1 hr, WQ 1700 F, 16 hr
	1800	8,800	18	5,300	10	3,100	4	2300 F, 1 hr, WQ 1800 F, 16 hr
Refractalloy 26	1200	....	....	74,000	....	....	....	Not known
	1350	....	....	48,000	....	....	....	Not known
	1500	47,000	20	29,000	18	18,000	16	2100 F, 1 hr, OQ 1500 F, 20 hr, AC 1350 F, 20 hr, AC
Nimonic 80	1200	78,000	3	60,000	2.5	45,000	1.8	1950 F, WQ 1290 F, 16 hr
	1350	56,500	1	36,000	1	23,300	1	1950 F, 4 hr, WQ 1400 F, 50 hr
	1500	23,000	13	16,300	9.5	11,800	0.5	1950 F, 4 hr, WQ 1500 F, 50 hr
Inconel X	1200	64,000	1	59,000	1	55,000	1.5	2100 F, 24 hr, AC 1300 F, 6–20 hr, AC
	1350	58,000	40–60	48,000	15–20	40,000	2	2100 F, 24 hr, AC 1300 F, 6–20 hr, AC
	1500	45,000	25 ± 5	29,000	11	18,500	2.6	2100 F, 4–24 hr, AC 1300 F, 6–20 hr, AC
Gamma Columbium	1200	56,000	....	40,000	28	28,800	20	Preheat ½ hr, 1550 F 2250 F, 45 min, OQ Aged 50 hr, 1500 F
	1500	23,500	26	16,000	23	11,200	9	Preheat ½ hr, 1550 F 2250 F, 45 min, OQ Aged 50 hr, 1500 F
	1600	14,000	33	10,500	24	8,000	5	Preheat ½ hr, 1550 F 2250 F, 45 min, OQ Aged 50 hr, 1600 F
Refractalloy B	1500	19,500	29	13,000	16.8	8,700	4.5	2280 F, 30 min, OQ 50 hr, 1500 F
Refractalloy	1500	24,500	20	19,000	18	15,000	11	2350 F, 4 hr, OQ 1500 F, 24 hr
	1600	16,000	38	12,500	14	9,600	6.4	2350 F, 4 hr, OQ 1500 F, 24 hr

TABLE 12-6. STRESS TO RUPTURE AND ELONGATION VALUES OF HIGH-TEMPERATURE ALLOYS*—(*Continued*)

Trade name	Temp, °F	10 hr		100 hr		1000 hr		Treatment†
		Psi	%	Psi	%	Psi	%	
Cast Alloys								
Vitallium	1200	68,000	....	54,000	....	43,000	....	As cast
	1350	48,000	....	36,500	....	28,000	....	Aged 50 hr, 1500 F
	1500	31,000	15	22,000	10	16,000	5	Aged 50 hr, 1350–1500 F
	1600	23,800	34	18,400	18	14,500	11	Aged 50 hr, 1500 F
	1700	16,400	38	13,100	10	10,500	5	As cast
	1800	12,200	42	8,800	13	6,400	3	As cast
6059	1200	66,000	....	55,000	....	46,000	....	Not known
	1500	30,500	23	24,000	22	18,800	15	Aged 50 hr, 1500 F
	1700	16,200	29	12,000	16	8,600	10	As cast
	1800	12,000	27	9,200	14	7,000	8	As cast
422-19	1350	60,800	30	47,000	23	36,300	11.5	Aged 50 hr, 1500 F
	1600	25,000	22–26	19,000	6	14,700	0.5–1	Aged 50 hr, 1600 F
	1800	14,400	27	10,100	18	7,100	2	As cast
61	1200	71,000	....	58,000	....	47,000	....	Not known
	1350	46,000	6	35,200	5	27,000	5	Aged 50 hr, 1500 F
	1600	25,100	10–15	17,100	5.5	11,800	2.5	Aged 50 hr, 1600 F
	1800	13,500	20	8,700	12	5,600	5	As cast
X-40	1200	66,000	....	55,000	....	46,000	....	Not known
	1350	59,000	20	44,800	31	34,000	37	Aged 50 hr, 1350 F
	1600	25,000	36	21,300	19	18,500	7.5	Aged 50 hr, 1600 F
	1800	13,300	46	11,300	24	9,800	8	As cast
S-816 cast	1200	72,000	....	58,000	....	46,000	....	As cast
	1350	54,000	....	40,000	16	29,000	....	As cast
	1500	35,500	15	28,500	3	22,000	1	2300 F, 1 hr, WQ Aged 16–50 hr, 1350, 1500 F
	1600	24,500	19	18,000	17	13,000	11	As cast
	1700	18,000	7	14,800	6	12,000	2	As cast
	1800	14,000	16	10,500	10	8,000	6	As cast
Co-Cr base (9W)	1500	42,600	....	32,000	....	24,200	....	Aged 50 hr, 1350 F
Co-Cr-Ni base	1350	49,000	28	42,000	21	35,000	12	Aged 50 hr, 1350 F

* Abstracted from Grant, Frederickson, and Taylor, A Summary of Heat Resistant Alloys from 1200 F to 1800 F, *Iron Age*, Mar. 18, Apr. 8, 15, 1948, by permission from *The Iron Age*.

† Code: AC = air-cooled
OQ = oil-quenched
WQ = water-quenched

some of the heavy materials for turbine blades and nozzle vanes. The tables in the early part of the chapter list many of the alloys which are usually cast. The method has been used very successfully, but there have been problems in obtaining the close tolerances and uniformity required. Production rejects have been high. The blades are often used in the as-cast condition, with an absolute minimum of finish machining or grinding. The as-cast blades can also be heat-treated to give them longer service life. One effect is the elimination of the islands of carbide precipitation which occur in the as-cast condition. In general, the perform-

TABLE 12-7. DENSITY VALUES OF HEAT-RESISTANT ALLOYS*

Alloy	Density,† G per cm³	Alloy	Density,† G per cm³
Forged Alloys			
19-9DL	7.933	Nimonic 80	8.192
Refractalloy B	8.022	Inconel X	8.3
Timken 16-25-6	8.059	S-495	8.260
Gamma Columbium	8.064	HC N-155	8.269
K-42-B	8.152	S-590	8.313
LC N-155	8.199	Refractalloy	8.529
		S-816	8.587
		Hastelloy B	9.24
Cast Alloys			
Vitallium	8.298	61	8.542
422-19	8.314	X-40	8.608
6059	8.381	X-50	8.855
X-41	8.472		

Iron has a density of 7.87.

* From Grant, Frederickson, and Taylor, A Summary of Heat Resistant Alloys from 1200 F to 1800 F, *Iron Age*, Mar. 18, Apr. 8, 15, 1948, by permission from *The Iron Age*.

† For density in units of pounds per cubic foot, multiply the listed values by 62.4. For units of pound per cubic inch, multiply by 0.0361.

ance of cast blades is closely related to the microstructure of the cast material, as might be expected.

In the lost-wax process, a wax or suitable plastic pattern is invested in a mold of some refractory material such as silica used with suitable bonding materials. The mold is heated at a low temperature to drive off moisture and melt out the wax, and then the mold is fired to give it strength. A conventional casting process follows. Centrifugal casting may be used to force the molten metal to fill the mold. The problems of attaining close dimensional control of the final casting are obvious. Changes in temperature and in moduli of expansion of the various materials involved must be closely correlated. Finished parts can be held to close tolerances by good practice—of the order of magnitude of 0.003 in. per in.

Machining and grinding have been used to produce blades and vanes for both turbines and compressors for materials which have allowed these methods of working. The use of bar stock allows use of a uniform material, and finished parts exhibit good consistency and uniformity. However, the cost of a large number of machine tools for large-volume produc-

TABLE 12-8. COEFFICIENTS OF EXPANSION FROM 70 F TO TEMPERATURES LISTED*
(Values given should be multiplied by 10^{-6}. Units are inch per inch per degree Fahrenheit)

Alloy	800 F	1000 F	1200 F	1500 F	1600 F
S-495	9.0	9.11	9.29	9.46	
Timken 16-25-6	9.29	9.36	9.52	9.69	
LC N-155	8.89	9.10	9.40	9.77	9.90
HC N-155	....	8.7	....	.....	9.9
S-497	8.08	8.26	8.50	8.80	
19-9DL	9.59	9.78	9.97	10.01	
S-590	8.43	8.54	8.61	8.97	9.20
S-816	....	8.0	....	.....	9.4
K-42-B	....	8.4	....	.....	10.0
Hastelloy B	6.39	6.52	6.90	.....	7.49
Nimonic 80	7.42	7.47	7.56	.....	7.99
Inconel X	....	8.2	8.4	9.0	9.2
Vitallium	7.96	8.18	8.38	8.68	8.72
61	7.96	8.18	8.48	9.24	
6059	7.74	7.98	8.14	8.6	8.76
422-19	7.86	7.91	8.07	8.42	8.54
X-40	7.88	7.98	8.18	8.43	8.79

* From Grant, Frederickson, and Taylor, A Summary of Heat Resistant Alloys from 1200 F to 1800 F, *Iron Age*, Mar. 18, Apr. 8, 15, 1948 by permission from *The Iron Age*.

tion tends to make this method of processing more expensive than some others, and some of the best high-temperature materials simply are not suitable for processing by machining.

Fabrication of sheet stock into hollow vanes and blades is a process offering several advantages, both from the standpoint of the turbine engine and from the standpoint of cost. Hollow vanes and blades can be made readily and cheaply, and these blades are much lighter in weight than their solid counterparts. Centrifugal loads may thus be reduced, and there is ready provision for flow of cooling air streams. At present, most of the blades operating at the highest temperatures in turbines are made in solid form from high-temperature materials.

Increased use of hollow blades can probably be expected in the future. The cooling provisions obtained allow the use of less critical materials because the metal temperatures can be lowered significantly. In addition, if further development produces high-temperature materials which can be processed satisfactorily in sheet form, the use of much higher gas temperatures may become feasible.

The techniques of *powder metallurgy* have been economically used to produce stator axial-compressor vanes in quantity. In one iron-base material, powdered iron is pressed to the desired shape in a mold, then removed and sintered at about 2000 F, and finally coined to shape. Then it is impregnated with copper at a high temperature, the copper filling the voids in the iron compact and forming an alloy with the iron. The blade is then given a corrosion-resistant case of chromium or nickel-chromium and heat-treated to produce additional strength. The finished blades may be polished or not, depending upon the tolerances required. The powder-metallurgy process can probably be used to advantage to produce parts other than the stator vanes of axial compressors, and development of the process to produce rotor blades is probably feasible.

The *rolling process* is a logical method for production of compressor blades and vanes, because satisfactory strip material has been available. The strip is rolled between rolls and dies of the proper shape to form the desired contour. For a constant blade thickness and no twist, a continuous strip emerges which can be cut to proper lengths. Sections of variable thickness and twist can be rolled, but the process becomes more expensive. In either instance, an attachment for the blade base must be considered. The base may be upset to form some kind of root shape, or some form of welding attachment may be used.

The *Mercast process* is a precision-casting technique of recent origin. It is essentially the same kind of method as the lost-wax precision-investment process, except that frozen mercury is used as the pattern instead of wax. Liquid mercury is poured into a master mold, where it is frozen at temperatures below −40 F. Then it is removed and coated with a cold refractory slurry to a thickness of ⅛ in. or more. The refractory shell is dried at low temperature, then the shell and mercury are brought to room temperature, and the mercury is melted out. The refractory shell is fired to give it strength and then is used as the mold for a usual casting process. Complicated parts can be made by use of the Mercast process, and very close tolerances and excellent surface finish can be obtained. The cost is higher than that of some other methods, however.

The most widely used processes for production of turbine and compressor blades up to the present time have been *forging* (with finish machining if required) and lost-wax *precision casting*. Examples of turbine-blade materials which have been used in production engines are Vitallium, S-816, Hastelloy B, and Haynes Alloy 36, a recent cast alloy having a nominal composition of 18% Cr, 15% W, 10% Ni, 0.4% C, 0.05% B, and the balance cobalt. In Britain, the Nimonic series of alloys is in widespread use.

12-9. The Turbine Disk. This part must carry high stresses and is subjected to temperature variations. The outer rim is the hottest region, as it is closest to the hot combustion gas and hot turbine blades, while the inner region runs at a much lower temperature. Streams of cooling air bled from the compressor are often directed toward the hot faces (sides) of the disk in order to limit the temperature there. It would appear that a certain amount of ductility would be desirable in order to accommodate the different amounts of thermal expansion which must take place. High strength at high temperature with low creep and high resistance to corrosion and oxidation (particularly at the outer rim) are also high on the list of required properties.

Fonda[15] has reported the results of tests in which turbine disks were spun to destruction in a test pit. The tests were carried out at room temperature, utilizing a large number of disks of different materials and treated in different ways. He reported that center ductility seemed to be the common denominator in order that high and consistent bursting speeds be attained.

Hot-spin tests of rotors with blades attached were carried out by Saldin and DeHuff,[16] utilizing the materials 19-9DL, 16-25-6, and the British alloy G 18 B. They were studying the relative importance of ductility and strength, and they reported that, in addition to high strength, some minimum amount of ductility is required. One item of importance is that the ductile material gives warning of impending failure. The growth or gradual deformation causes rubbing of close-fitting parts, which can be heard. Immediate shutdown can prevent costly and disastrous failure. Also, local cracks may be detected at inspection periods. The strong but brittle disk gives little indication of trouble. It apparently operates satisfactorily, and then suddenly it bursts. The entire engine may be destroyed immediately.

It is a well-known fact that many materials are less ductile at high temperature than at room temperature, so that room-temperature ductility is not a reliable guide. Also, many materials can be processed to produce low strength and high ductility or high strength and lower ductility, so that a compromise must be adopted.

Practically all gas-turbine rotor disks are forgings, common materials being the iron-base alloys, such as 16-25-6, 19-9DL, and Discaloy. These alloys are rich in nickel and chromium (see Table 12-3), which are expensive and critical materials. At present, disk failures in operating engines are rare, and thought is being given to the possibility of reducing the amount of scarce, critical materials now used in disks. Careful attention to the production of ingots used for disk manufacture and careful control in the forging process itself have produced great improvements in the past few years with practically no change in the basic materials. The key seems to be greater consistency, reliability, and soundness in manufacture, giving higher strength and improved ductility.

A major development in disk design and manufacture is the composite disk. Where disk cooling is employed, the center portion is made of a different material from the rim, which must withstand higher temperatures. The two are welded together to form the completed disk. Operation in jet engines has shown this construction to be very satisfactory.

Steam-turbine experience has indicated that for large, long-life units the single-piece rotor forging is superior, but this conclusion is not necessarily true for gas-turbine rotors, at least at the present time. The production of a sound, large forging of shaft and disks is a sizable undertaking, particularly for a unit of several stages. It is done for steam turbines, but the higher temperature materials required for the gas turbine are another matter, and much difficulty has been experienced in attempts to produce sound forgings of large size. However, it has been done in some instances. Multistage rotors are more often built as an assembly of several disks connected by welding or bolting. Cracking of welds has given trouble in some instances, while bolting, too, has its possibilities of loosening over long periods of time at high temperatures. The composite-disk idea can be used for each disk of a multistage unit, regardless of how the disks are to be fastened together to form the completed rotor. Thus most efficient and economical use of materials may be realized.

A development which is receiving attention is the casting of the entire turbine wheel, including blades, in one piece. In some sizes, this method of manufacture is feasible. Large sizes present greater difficulties than small ones, and the presence of a flaw requiring rejection of the entire piece becomes a costly proposition in a large precision casting of expensive material.

12-10. Combustion Chambers. The parts which contact the hot combustion gases must be corrosion-resistant to a large degree as well as heat-resistant, although the stress is ordinarily low. A fuel of the nature of kerosene is perfectly satisfactory for gas turbines, but some turbines may operate on high-test gasoline to minimize fuel-supply problems at airports which must supply both jet engines and standard piston engines. This high-octane gasoline has a high tetraethyl lead content, and the decomposition products (chiefly lead bromide and lead oxide) can be very corrosive. Satisfactory combustion-chamber liners are made from materials such as 25-20 chromium-nickel steel and Inconel, a nickel-base alloy having high chromium content.

In looking to the future, large-size stationary gas turbines and those used in ships and railroads should be capable of burning low-grade inexpensive fuels if they are to compete economically with other forms of power equipment. Such fuels will probably contain contaminants, which can produce corrosive substances in the turbine in larger amounts than those produced by more highly refined and costly fuels. Therefore corrosion resistance will continue to be of importance regardless of whether tetraethyl lead is present in the fuel or not. In marine service, salt corrosion is a factor.

Another approach is the coating of the combustion-chamber inner surfaces with a refractory material. This practice could allow the use of less critical materials, but there are problems of differential expansion and possible cracking of the refractory.

The exhaust nozzle of the jet engine and the ductwork enclosing the after burner also require corrosion-resistant material, and alloys such as Inconel are used for these parts.

12-11. Bolting Materials. There are many places in the gas turbine where bolts are used in assembling the various parts, and the high-temperature regions require special consideration. A bolt must be capable of being tightened and should hold that tightness for a long period of engine operation. Cracking should not occur in the thread roots or under the head, and the bolts should allow relatively easy disassembly and reuse. With high-temperature long-time loading, creep relaxation causes a bolted joint to become loose. Also, the characteristics of the material may change with time, load, and temperature, allowing further loosening and possible brittleness.

In service, bolts may be tightened periodically in order to counteract the loosening resulting from creep relaxation. This brings up the problem of how often tightening must be done. In addition, the continued stretching from retightening brings the bolt closer and closer to actual failure. How tight a bolt should be, how many times and how often it should be retightened, and when it should be discarded because of imminent failure are questions which are not easy to answer, particularly if operating service is intermittent and variable. Relaxation tests are used to evaluate performance of bolting materials. Stress-time data are taken, with strain and temperature constant.

Hot bolting materials should have high strength at elevated temperatures with high resistance to creep relaxation and should not be brittle. They should exhibit good stability, with minimum loss of the desirable properties originally present. They should resist corrosion. All these attributes are essential in other parts of the turbine, too, of course. Samples of successful bolting alloys are Inconel X, K-42-B, N-155, S-590, and S-816.

12-12. Compressors. The high-temperature requirements of turbines are not present in compressors, and so the problem of materials is not so acute. Satisfactory materials are lower in cost and in greater supply. Corrosion problems are not so great, although salt brought in with sea air could be detrimental in some instances.

In aircraft units, where light weight is desirable, much use has been made of aluminum alloys and magnesium alloys. Centrifugal rotors machined from 25 ST and 14 ST aluminum forgings have given good service. Cast magnesium alloys as well as aluminum alloys have proved to be satisfactory for compressor casings and accessories. These alloys are given special heat-treatments to minimize distortion in service, such as a solution treatment followed by a stabilizing treatment. In some axial-flow compressors, aluminum alloys have been used for the wheels of the early stages, with the last stage made of stainless steel (such as 410) better to resist the higher temperatures present near the compressor discharge. The blades of the axial-flow compressor are often all made of stainless steel, but some compressor blades have been aluminum alloy.

Interest is growing in the use of titanium and its alloys for lightweight, strong components that must withstand temperatures higher than those which the usual aluminum and magnesium alloys can accommodate. Undoubtedly this metal will find increased usage where lightness, strength, and corrosion resistance are necessary. Another alloy gaining prominence is a magnesium–rare-earth–zirconium combination. This alloy is light in weight and can carry high stress in the 300 to 600 F temperature range, making it superior to aluminum alloys such as 25 ST aluminum for casings and other parts.

In gas turbines designed for use on ships or locomotives and for stationary power plants, light weight is not so essential. Iron and steel may be substituted for aluminum or magnesium where it is found advisable and economical to do so.

12-13. Materials Used in Representative Gas Turbines. In this article several gas turbines will be surveyed, with emphasis on the materials actually used in their construction. One of the earliest uses for gas turbines in this country was for supercharging diesel engines, and this use continues. The supercharging unit is a gas turbine, through which passes the hot diesel-engine exhaust, connected to a centrifugal blower. Typical materials are as follows:

Turbine wheel and buckets..........	19-9DL
Shaft..............................	SAE 4130 steel
Compressor impeller and diffusers....	Cast aluminum
Turbine nozzle blades...............	18-8 alloy
Turbine and blower casings..........	Meehanite (a high-strength cast iron which has been treated with calcium silicide)

Another early and important application of gas turbines is for supercharging aircraft engines. Typical materials that have been used in turbosuperchargers are as follows:

Turbine wheel...............	Timken alloy 16-25-6
Turbine buckets..............	Vitallium
Diaphragm..................	25-20 chrome-nickel steel
Nozzle box..................	nickel-chrome-molybdenum austenitic steel
Compressor impeller..........	25 ST aluminum alloy
Compressor casing............	Aluminum alloy

As an example of jet-engine construction, the well-known General Electric J-47 will be cited.

Turbine wheel-hub section	A1S1-4340 ferritic alloy
Turbine wheel-rim section	Timken alloy 16-25-6
Turbine buckets	S-816
Diaphragm	Vitallium
Combustion-chamber liners	310 stainless steel

Other alloys used in this engine include Hastelloy C, Hastelloy B, Inconel and Inconel X (a variation of Inconel), 6059, 25-12, 19-9DL, and several others. The Westinghouse J-34 jet engine has made use of K-42-B, Refractaloy, Discaloy, and, for turbine blades and nozzle vanes, Haynes Alloy 36 (see Art. 12-8).

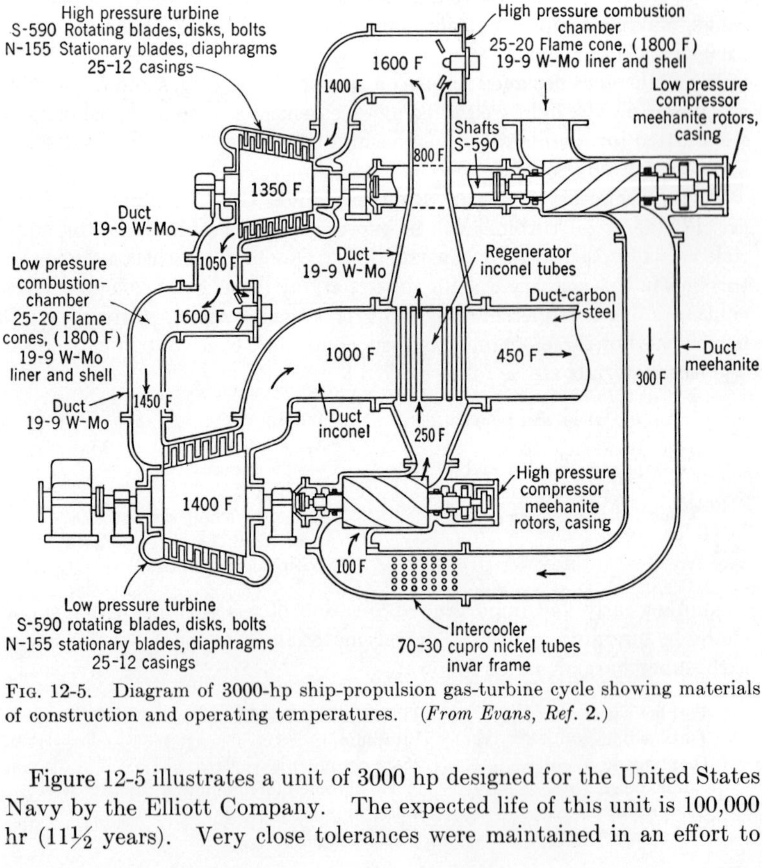

FIG. 12-5. Diagram of 3000-hp ship-propulsion gas-turbine cycle showing materials of construction and operating temperatures. (*From Evans, Ref.* 2.)

Figure 12-5 illustrates a unit of 3000 hp designed for the United States Navy by the Elliott Company. The expected life of this unit is 100,000 hr (11½ years). Very close tolerances were maintained in an effort to produce a machine of high thermal efficiency. Low stresses were used in order that creep rate be kept very low. The figure shows the gas temperatures present in various parts of the cycle and the materials of construction. A thermal efficiency of 35 per cent was anticipated, and this figure demonstrates that the gas turbine has the capability of competing with other established power plants in at least some types of service. It is probably not very important whether this particular turbine power plant "lives" its anticipated life of 11½ years. Its chief value probably lies in the experience gained during operation under various conditions. Many valuable guides uncovered can be useful in future designs.

Several small-size gas turbines have been built for specific services, only one of which will be mentioned here. The Solar Aircraft Company has built for the Navy a gas-turbine-driven pump. The turbine is rated at 45 hp at 40,300 rpm. Hastelloy B is used for the turbine impeller casting and type 321 stainless steel for shrouds, turbine nozzles, combustion chamber, tail pipe, and framework. The maximum gas temperature in this unit is 1250 F.

These few samples will serve to indicate the specific materials which have been used in specific gas turbines designed for several types of service.

12-14. Ceramics and Cermets. The demand for high-temperature materials has stimulated interest and study in the field of ceramics. Silicates (glasses) come to mind immediately, but there are many glass-free ceramics, among them carbon, graphites, and refractory oxides, carbides, borides, and nitrides. Some ceramics possess higher strength and greater creep resistance at temperatures above 1800 F than the best metallic alloys. The availability of most ceramics is adequate. Supply is not so critical a problem as for some of the heavy metals. Great corrosion and oxidation resistance is another desirable attribute. The single largest drawback to the use of ceramics in gas turbines is their lack of ductility—they are brittle. The situation is not much different in the heavy metals. High strength at high temperature can often be attained, but at the sacrifice of ductility.

Another interesting property of ceramics is their greater strength in compression than in tension, sometimes 10 times as great. This fact, along with the drawback of brittleness, suggests that much is to be gained by designing gas-turbine parts specifically for ceramics. Proper design could do much to improve performance of ceramic materials. As a specific example, turbine blades might be mounted on a rotating rim, with blades pointing radially inward instead of outward. Centrifugal forces would then load the blades in compression instead of in tension. There would still be problems, of course, of column action and flexure, but advantage would be taken of the good compressive strength in such a design. Another approach might be prestressing in compression by some means. Because of brittleness, local stress raisers must be avoided. Blade roots would have to be designed carefully, for example.

There is another aspect to the problem of using ceramics which could not be alleviated so readily by design, and that is their very poor resistance to shock, both thermal and mechanical. Rapid temperature changes and steep temperature gradients cause cracking, although not all ceramics have the same resistance to this type of failure. Those with high thermal conductivity are more resistant to thermal shock than those with low conductivity. A blade of high conductivity, however, will heat the wheel rim more, giving rise to a separate problem. Resistance to mechanical shock is important because this type of loading may occur in assembly of parts (for example, if they are dropped) as well as in operation of the turbine. The designer might do everything possible to eliminate steep temperature gradients and stress concentrations so that a satisfactorily operating engine is produced for steady-load operation or where load changes are very gradual. However, much operating service requires rapid load changes and rapid accelerations and decelerations. A jet pilot who needs a sudden burst of power may need it *immediately*, or he may be dead. Lack of resistance to thermal and mechanical shock is a serious fault of ceramic bodies which is not easily remedied by mechanical design.

Investigations regarding the use of ceramic materials in gas turbines have taken two directions. The first is the study of the use of ceramics in forming complete solid parts, such as turbine rotor blades and stator blades. The second is the study of the use of ceramics as coating materials to prolong the life of hot parts subject to corrosion, such as combustion chambers. Such coatings not only increase the service life of existing alloys in many instances but may also allow the use of less costly, less critical materials.

For solid bodies, in order to gain desired ductility, a whole new field of study has been opened, the field of *cermets*, or combinations of ceramics and metals. The ceramic gives resistance to deformation under stress at high temperatures (1800 to 2500 F) plus oxidation and corrosion resistance. The metallic constituent provides increased thermal conductivity and shock resistance. The number of possible combinations is tremendous, and although progress has been made, much study and evaluation remain to be done. Better, faster, and simpler laboratory tests are needed which will better correlate with performance in actual operating service, particularly for the rapidly moving parts. The establishment and use of such laboratory procedures would do much to speed the process of sorting and evaluating the large number of possible promising materials, not only for cermets and ceramics, but for the more usual materials as well. Needless to say, much effort is being expended in this direction.

Properties of one sintered carbide cermet which has been developed will be cited as an example of the promising potentialities of ceramals. The material consists chiefly of titanium carbide (80 per cent) plus other carbides, with nickel as auxiliary binding metal. The material weighs only two-thirds as much as steel and the usual heat-resisting alloys and has twice the resistance to deflection under load. Its stress-rupture strength at 1600 F is about three times that of many good alloys, and its superiority is even greater at higher temperatures. It resists oxidation for long periods at 1800 F and exhibits very good resistance to thermal shock. Experimental turbine blades have been run 110 hr alongside more conventional blades with turbine inlet-gas temperatures of 2200 F. A large proportion of the usual alloy blades failed during this time. In NACA tests, a turbine blade of this material was coated with a thin layer of 80 per cent chromium, 20 per cent glass. In a 200-hr test at 1800 F, no corrosion was evident, and the coating had sufficient ductility to creep with the base material without failure. This is very promising performance.

The cost of titanium has been high during the early history of its production and use ($5 per pound of sponge metal in 1951), and this fact has retarded rapid widespread use of the metal. Continued development of refining techniques will unquestionably continue to bring the cost down. One large difficulty is the affinity of atmospheric gases for titanium during the refining operation, and also for the pure metal when used for long periods above 1000 F, causing it to become brittle. Corrosion resistance, however, is excellent.

The manufacture of a ceramic body usually involves powdering of the constituents, mixing with plasticizing and bonding agents, and forming of the desired shape by pressing or extruding. Drying and curing occur next, and then the body is sintered to form the final product. Close control is required of raw materials, impurities, powder size, forming method, and the sintering atmosphere, time, and temperature if consistent properties in the finished product are to be obtained.

Ceramics and cermets have a promising future in gas turbines. Although their use has been mainly in experimental and developmental units, there is little doubt that the desirable characteristics of these materials will not be ignored in the design of the higher performance, higher-temperature gas turbines of the future.

12-15. A Look to the Future. The entire field of metallurgy is under intensive investigation at the present time, and improvements and changes can be expected to occur continuously. For obvious reasons concerning production, supply, and maintenance, materials and designs must be chosen and frozen for reasonable periods of time, of course, but this fact should not and will not preclude the use of superior materials and methods when they are found. The whole field of design and application of the gas turbine has developed rapidly.

The search continues for strong, lightweight materials which can perform satisfactorily at high temperatures. This search might be considered to consist in two general approaches. One is the improvement of existing types of alloys by changes in composition and treatment, and the other is the investigation of entirely new materials and combinations. Enough has been written in this chapter to give a broad view of present materials and methods, and the subject will not be pursued in this limited space.

As new materials are investigated and found promising, changes in the design of components for the gas turbine often become necessary in order to take maximum advantage of the newly discovered desirable properties. For example, a part to be made of a relatively brittle ceramic may have to be redesigned in order to perform satisfactorily when made of the new material. A turbine blade-root shape which gave satisfactory service when made of present metallic alloys had to be changed to reduce stress concentrations and prevent cracking when made of ceramic material in an experimental design. Another example might be cited involving strength-weight ratio. A lighter material for turbine or compressor blades produces lower centrifugal stress, and this may allow higher speeds or reduced weight and size of fastenings and possibly of the disks to which the blades are attached. Even for stationary parts, higher strength-weight ratio may allow reduction of size and weight. Still another example is the effect of thermal coefficients of expansion. Straight substitution, without redesign, of a material having a different coefficient from the present one can result in new problems of deformation and stress. Thus it is obvious that the discovery of new materials alone is not sufficient. Designers must learn how they can be used, and manufacturers must learn how they can be produced quickly and economically in the desired forms and shapes. A material which shows promise in the laboratory is still a long way from use in a production turbine.

Along with the development of new materials, another avenue of approach shows high promise—the cooling of the hot parts. This approach is already used in cooling the turbine disk, combustion chamber, and certain other parts by suitable flows of air, but more can be done. In experimental turbines, blades which were simply hollow shells were cooled by a flow of air bled from the compressor to the blades through the turbine wheel. For effective cooling, large amounts of air were necessary, and performance of the whole unit suffered. Increased heat-transfer surface was obtained by packing the shell with hollow tubes. This type of blade reduced metal temperatures several hundred degrees with a greatly reduced air flow, and the blade was found to be relatively easy to fabricate.

It seems inevitable that more widespread adoption of cooling schemes such as this will occur. Less critical and costly materials may be used for the hot parts, or higher temperatures and better performance can be expected from present types of materials. Care will be necessary in controlling temperature gradients and thermal expansion, but these problems can probably be solved satisfactorily.

Figure 12-6 shows some of the blade-cooling arrangements which have been tried. The hollow blade packed with tubes has given very effective cooling with a relatively small air flow and would appear to be relatively

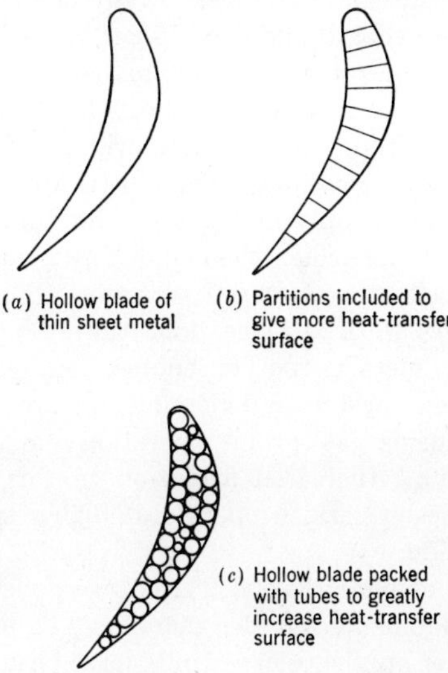

FIG. 12-6. Examples of blade-cooling schemes which utilize flow of air through hollow blades.

easy to fabricate. Cooling air can be bled to the blade through a hollow-shaft and split-disk arrangement which can serve to cool the disk as well as the blades. The air leaves the blade at the tip and mixes with the main gas stream.

In the German Jumo 004 and BMW003 jet engines, both turbine nozzles and turbine blades were hollow and cooled by air. Air was tapped from the compressor for cooling the nozzles, while a separate centrifugal fan made as part of the turbine disk furnished the air for the blades on the periphery of the disk. Performance suffered because of the excessive power requirements for the air flows used and because of the compromises in design shapes which were necessary.

It would seem logical to collect the heated air after it has passed over the hot surfaces of the turbine parts and feed it to the combustion chambers, and this has been done in some turbines. A thermodynamic gain results from the regenerative effect.

The use of water or fuel as liquid coolants has been proposed, because of the higher rates of heat transfer which might be attained. Of course, the equipment becomes additionally complex, and this fact must be considered. It is probably too early to say what cooling methods will ultimately prove to be best suited for use in gas turbines for various services. Many schemes are being investigated and evaluated.

In times of national emergency, increased emphasis is placed on the reduction in use of strategic materials. Many of the present high-temperature alloys require cobalt, and over 75 per cent of the world's cobalt comes from Africa. Nickel, chromium, manganese, and tungsten are other metals which must largely be imported. Reduced use of these materials is desirable, and future designs will call for less critical and strategic metals wherever possible. The North American continent evidently has an adequate supply of titanium, another factor in favor of increasing use of this particular element. The availability of ceramic materials, too, is excellent.

Much development work is being done toward utilizing more effectively the strategic metals, too, of course. Alloys containing large amounts of chromium, tungsten, and molybdenum are receiving attention.

Noise produced during gas-turbine operation remains a large problem for some applications. Increased attention to this problem is to be expected. Future designs which operate at higher speeds may further aggravate the situation.

The advent of the jet engine has been largely responsible for the impetus in gas-turbine development, but designers and builders have also put much effort into other applications. Units have been built for light aircraft, for automobiles and trucks, for helicopters, and for driving pumps and compressors. Others have driven boats and ships, locomotives, and electric generators for the production of power.

Attempts are being made to burn solid fuels in some experimental units, in order to reduce fuel costs. Longer life and reliability, increased power and efficiency, and lower cost are basic aims. In some applications the gas turbine competes successfully at the present time with other more conventional types of power-producing apparatus, and in some it does not. For high-speed aircraft, the jet engine reigns supreme. For locomotive, ship, and stationary power plants, prospects look good. For automobiles for the public use, prospects are poor. Cost is high (at present), and gas mileage is very low.

A whole new industry with a host of problems to be solved has been opened by the rapid development of the gas turbine. New materials, processes, designs, and applications appear almost daily. The entire field has developed rapidly and it is obvious that the gas turbine is here to stay. One more type of power-production equipment has taken a permanent place in the line-up of power machinery.

REFERENCES

1. Mochel, N. L.: Metallurgical Considerations of Gas Turbines, *Trans. ASME*, vol. 69, no. 6, p. 651, August, 1947.
2. Evans, C. T., Jr.: Materials for Power Gas Turbines, *Trans. ASME*, August, 1947, p. 601.
3. Scott, Howard, and R. B. Gordon: Precipitation-hardened Alloys for Gas Turbine Service-I and II, *Trans. ASME*, August, 1947, pp. 583, 593.
4. Evans, C. T., Jr.: Wrought Heat Resisting Alloys for Gas Turbine Service, *Metal Progress*, November, 1945.
5. Sweeny, W. O.: Haynes Alloys for High Temperature Service, *Trans. ASME*, August, 1947, p. 569.
6. Crawford, C. A.: Nickel-Chromium Alloys for Gas-turbine Service, *Trans. ASME*, August, 1947, p. 609.
7. Emmert, H. D.: Current Design Practices for Gas Turbine Power Elements, *Trans. ASME*, vol. 72, no. 2, February, 1950.
8. Mochel, N. L.: Metals for Gas Turbines, *Mech. Eng.*, June, 1950, p. 462.
9. Scott, Howard: Gas Turbine Alloys, 10 Years Later, *Metal Progress*, October, 1950, p. 503.
10. Colwell, A. T., and R. E. Cummings: Turbine Engine Blading: Manufacturing Technique and Fastening Methods, *Trans. SAE*, July, 1948, p. 419; Seven Ways to Produce Turbine Blades, *J. SAE*, June, 1950.
11. Tatnal, F. G. Testing Materials at High Temperature, *Mech. Eng.*, November, 1949, p. 906.
12. Duckworth, W. H., and I. E. Campbell: The Outlook for Ceramics in Gas Turbines, *Mech. Eng.*, February, 1950, p. 128.
13. Clark, F. H.: "Metals at High Temperatures," Reinhold Publishing Corporation, New York, 1950.
14. Grant, N. J., A. F. Frederickson, and M. E. Taylor: A Summary of Heat Resistant Alloys from 1200 F to 1800 F, *Iron Age*, Mar. 18, Apr. 8, 15, 1948.
15. Fonda, L. B.: High-temperature Disk-forging Developments for Aircraft Gas Turbines, *Trans. ASME*, vol. 70, 1948.
16. Saldin, H. B., and P. G. DeHuff: Hot-spin Tests of Bladed Jet-engine Rotors, *Trans. ASME*, July, 1949.
17. Ault, G. M., and G. C. Deutsch: NACA Studies Ceramals for Turbine Blading, *J. SAE*, May, 1950.

APPENDIX

Conversion Constants

Pressure in psf (lb per sq ft) = 144 × psi (lb per sq in.)
Pressure in psi = 0.491 × pressure in in. Hg (mercury) at 32 F
= 0.489 × in. Hg at 70 F
1 standard atmosphere = 29.921 in. Hg at 32 F = 760 mm Hg at 32 F
= 14.696 psi = 1.0332 kg per cm^2
1 metric (technical) atmosphere = 1.00 kg per cm^2 = 14.223 psi
Psia = barometric pressure in psi ± psig

Degrees Fahrenheit (F) = degrees centigrade (C) × 1.8 + 32

Deg C = $(F - 32) \times \frac{1}{1.8}$

Degrees Rankine (R) = Deg F + 459.7 ≈ F + 460
Degrees Kelvin (K) = Deg R × 5/9 = Deg C + 273.16

1 Btu = 778.26 ft-lb ≈ 778 ft-lb
= 251.996 gram-calories ≈ 0.252 kg-cal

1 Btu per lb = $\frac{1}{1.8}$ × kg-cal per kg = 2.326 joules per gram

1 Btu/(lb)(F) = 1 kg-cal/(kg)(C)
1 horsepower (hp) = 550 ft-lb per sec = 33,000 ft-lb per min
= 2544 Btu per hr = 0.7067 Btu per sec
= 0.7455 kilowatt (kw)
1 kw = 1.341 hp
1 metric hp (*cheval vapeur*) = 0.9863 hp
1 hp-hr = 2544 Btu = 0.7455 kw-hr
1 kw-hr = 3413 Btu = 1.341 hp-hr = 3600 joules

1 inch = 2.540 centimeters = $\frac{1}{39.37}$ meter

1 mile = 5280 ft = 1.609 kilometers
1 mile per hour (mph) = 88 fpm = 1.467 fps

1 lb (mass) = 453.6 grams = 0.4536 kg = 7000 grains = $\frac{1}{32.174}$ slug

1 gallon = 231 $in.^3$
1 ft^3 = 7.48 gal = 1728 $in.^3$
1 barrel (petroleum industry) = 42 gallons

INDEX

A

Admission, partial, 158, 170
Aerodynamic blading design, 204
Air, atmospheric, 29
  dry, analysis of, 31
  humidity in, 29
  properties table, 36–62
  specific heat, 26, 33, 36–62
  tables, 31, 36–62
  used in combustion, 331–335
Air rate, 109
Airfoil sections, 156
Airfoil theory, 188, 204, 209–210
Airplanes, Comet, 275-276
  compound engine, 284–286
  cruising range, 272
  cyclone engine, 284–285
  forces on, 255
  horsepower requirements, 266-270
  jet-operated, 13
  performance, 281-285
  propeller-type, 282-285
  reciprocating engine-turbine, 284–286
  take-off, 21,276
Alloys, heat-resistant, 451–462
Angle, of attack, 188, 203–204
  of incidence, 203, 210
Angular momentum, 407–409
API degrees, 343–344
  specific gravity table, 345
Approach velocity, 151
Aromatics in petroleum, 340–341
Aspect ratio, 204
Available energy, 101
Avogadro's law, 25
Axial flow process, 140, 186

B

Balancing, 415–438
  dynamic, 420–425
  machines for, 423–425
Balancing, rotors, flexible, 419–420
    rigid, 419–423
  static, 420, 423
Bearings, 21, 103, 163
  cooling, 303
  losses in, 107
Beau de Rochas, 2
Bending, centrifugal, 160
Blades and blading, airfoil theory, 210
  allowable speeds, 174
  angle, 136, 137
    turning, 156
  balance gas moment, 385
  cambered, 210
  centrifugal bending, 160, 385
  centrifugal stress, 139, 383–387
  compressor, 202
  constant reaction, 169
  construction, 159
  cooling, 473
  corrosion of, 163
  design, aerodynamic, 200
    angles, 140
  diagram, 136
    calculations, 149
    efficiency, 144, 146, 150, 156
    driving force, 137, 384
  edge-thickness factors, 172, 173
  equation, kinetic form, 141
  fastenings, 163, 405–407
  fir-tree root, 139, 407, 429
  fixed row, 146
  friction, 136, 142
    gas, 146
    profile, 156–157
  gaging, 162, 174, 175
  geometry, 204
  height, 161
  impulse, 159, 160, 172–174
  inlet angles, 173
  leakage, 156
  life, 162, 382, 387, 417, 447
  manufacture, 162, 453–457, 461, 463

Blades and blading, material, 302–303, 451–462
- mean diameter, 168
- mean radius, 138
- moments, 385–391
- moving, 146
- nomenclature, 203
- number of, 172
- outlet angles, 173
- overlap, 173
- pitch, 161, 173, 175
- profile, 141
- reaction, 156, 160, 174
- ring area, 200
- rotor, power, 142
- round-nose, 139
- secondary loss, 157
- shock loss, 148
- shrouded, 161, 391
- solidity, 187
- speed, 147, 168, 174
  - mean, 138
  - optimum, 132
  - root, 139
  - tangential, 136
  - tip, 139, 303
- speed ratio, 142, 153, 157–158
- stresses, 383–392
- taper, 384
- trailing edge, 140
- turbine, 167
  - temperatures, 173
- turbulence, 142
- twist, 134, 161, 387
- velocity, 132, 138
- velocity distribution, 205
- velocity ratio, 143
- vibration, 161, 426–427
- widths, 173, 174

Blowers, characteristics, 225–226
- displacement, 184

Boilers, gas-turbine systems, 314
- waste-heat, 313

Boundary gas, 186

Boundary layer, 74, 83, 186

Brayton cycle, 71, 92, 94, 95

Buckets,
- (*See also* Blades and blading), 132, 135

Bunker C fuel, 15, 356–359, 377

Burner, annular-type, 98, 371
- straight-flow, 98, 366-369

C

Camber, 187

Camber angle, 203

Camber line, 203, 204

Cambered blade, 210

Carry-over velocity, 151

Cascade spacing, 187

Cascade tests, 157, 205, 212

Centrifugal bending, 160

Centrifugal field, 205, 225–227

Centrifugal forces, 383–396

Ceramics, 469–470

Cetane number, 353

Channel passage, blade, 199, 202

Closed-cycle system, 115
- accumulator, 123
- advantages, 115, 117
- efficiency, 117, 120
- flow rate, 119
- gas-turbine, 116
- gases, 118
  - helium, 118
  - hydrogen, 118
- governing, 123–124
- maximum temperatures, 116
- pressure ratio, 115
- specific volume of gas, 115

Coal operation, 301, 379

Coefficient, drag, 157, 204, 208, 209
- lift, 157, 208, 209, 211

Coefficients of expansion, 462

Combustion, calculations, 99, 329–335, 359–366
- chambers, 295, 304, 366–375, 465
- chart for products, 363
- coal-fired, 328
- energy equation, 359–364
- fuel, 296–297, 353–359
- heat releases, 329
- ignition for, 305, 369–371, 375
- primary air, 368, 369

Combustion process, 328

Combustor, 12
- annular chamber, 368, 371
- General Electric, 367, 372–373
- Ghost engine, 374–375
- reverse-flow-type, 98
- Westinghouse, 369–372
- zones, 368

Compression, isentropic, 104

Compression, pressure, 16
  ratio, 4–6
Compressors, axial-flow, 20, 140, 184–222
    blade diagram, 189
    blading nomenclature, 203
    by-pass, 213
    design, 212, 320
    efficiency, 11, 66, 186, 196–197
    General Electric, 217-219, 276-279
    Lysholm, 184, 242–243
    moving row, 185, 189
    nonsymmetric stage, 191
    performance characteristics, 104, 187, 193
    power, 193
    pressure ratio, 188
    pressure rise, 185, 189, 192–194, 212, 223–228
    process, 186
    rotor, 217
    stability, 186
    stage, 100, 104, 111, 185
      efficiency, 196, 200
      performance, 196
      vortex, 191
    stalling, 187
    stress calculations, 381–405
    surge, 186–188
    symmetric, 190
    types, 184
    velocity diagram, 190, 192
  centrifugal, 10, 19–20, 223–243
    capacity, 237
    design, 235, 317-319
    diffusion vanes, 224, 228
    efficiency, 225–226, 238–239
    Elliott, 15, 229–230, 242–243
    eye, 228, 231–233, 235
    impeller, 9, 223–225, 228, 231–233
    Mach number, 237
    power required, 235, 239
    pressure-field derivation, 225–227
    pressure ratio, 236–240
    pressure rise, 223–224, 235
    radial vane, 224, 233
    secondary flow, 234
    slip, 240
    stress calculations, 381–405
    surge, 225
    swirl, 232
    velocity of whirl, 232
Compressors, lobe-type, 184
  Roots, 184
  scavenging air, 7
  Solar units, 316-321
    velocity triangles, 224
Constant reaction, 169
Continuity of mass, 191
Conversion constants, 477
Cooling, blade, 473
  turbine, 292–303
Creep, 382, 405, 417, 442, 444–445
Creep rupture, 444
Crude oil, 342
  distillation range, 343
Curtis, Charles G., 11
Curtis stage, 158
Curtis turbine, 132, 145, 147, 153
Cycle, 70–73
  Brayton, 71, 92, 94, 95
  closed, 115–123
  combination, 313
  Diesel, 71
  four-stroke, 2, 6
  gas-turbine (open), 96, 97, 111
  ideal-regenerative, 94
  Joule, 92
  open, 107, 109, 111
    diagrammatic layout, 112
    efficiency, 112
  Otto, 71
  performance, 106
  power, 70
  two-stroke, 6

D

Deflection angles, 204
Deformation, 382, 441
Degree of reaction defined, 149
Density of metals, 461
Design, constant-reaction, 168
  free vortex, 168
  vortex, 168
Design conditions, 155, 381–383
Design velocity ratio, variations in, 155
Dew point, combustion products, 329
Diagram power, 198
Diaphragms, 165
Diesel cycle, 71
Diesel plant, 299, 315

Diffusers, 12, 74, 76, 79–84
  design, 83
  efficiency, 80–83
  forms, 83
  pressure rise, 83
Diffusing exhaust, 135
Diffusion, effectiveness, 80
  process of, 12, 186, 225
Diffusion vanes, 224–228
Disk, centrifugal stress in, 394–396
  friction, 150
  thermal stress in, 402
  turbine, 463
  uniform strength, 397–401
Drag coefficient, 204, 208, 209, 255
Dynamic magnification factor, 415–417
Dynamometer, 201

E

Edge-thickness coefficient, 174
Efficiency, compressor, 11, 66, 93, 106, 186, 196–197, 225–226, 238–239
  cycle, 93
  diagram, 143, 146
  internal, 104, 105
  mechanical, 105, 106
  open-cycle, 107
  over-all, 105
  pressure ratio, 93
  ram, 80
  ram diffuser, 80
  regenerative cycle, 95, 96, 109
  speed ratio, 154
  stage, 100, 103, 104, 154
    reaction-turbine, 150–151
  thermal, 93, 109, 117
  turbine, 93, 102, 103
Energy, available, 101
Energy equation, steady-flow, 22
Engine thrust, 19, 254
Engines, diesel, 1, 5
  dual-fuel, 328
  external-combustion, 70
  internal-combustion, 1, 70
  jet, 19, 272-284
    high-velocity, 19
    JT-6B, 19
    tail-pipe, 19
    turbo-fan, 276, 278-281
  turbojet, 19-21, 276-282
    (*See also* Jet engines)
Engines, propeller-jet (*see* Propeller-jet engines)
  reciprocating, 1, 2, 70, 253, 284–286
    aircraft, 4
  spark-ignition, 1, 71
  Wright, Cyclone, 4
Enthalpy, definition, 23
Entropy, 27, 34
  pressure change for air, 63–65
Escher-Wyss Engineering Works, 117, 118, 124

F

Fanning loss, 150
Fatigue, 417, 446
Fir-tree root, 133, 407, 429
Fire point, 351
Flash point, 351
Floating-piston engine, 243–249
Flow coefficient, 199, 238
Flow continuity, 75
Flow process, axial, 140, 186
Flue-gas analyses, 330–333
Fluid angle, 203
Fluid deflection, 202
Fluid-flow angles, 202
Free-piston compressor, 243–249
  bounce cylinder, 245
  diesel engine with, 245–247
  Oppenheim and London, 246–247
  Pescara, 244, 249
  Sulzer engine, 247–248
Free-vortex distribution, 214
Frequency, blade, 425–428
  natural, 433–434
  rotor, 434
  vane, 429–434
Frictional reheat, 172
Fuel, consumption, 19, 21
  jet-engine, 269–274
Fuel-air ratio charts, 375–378
Fuel oils, 353–359
  Bunker C, 356, 377
  cetane number, 353
  grade requirements, 354–355
  heating value, 357–358
  preheat temperature, 359
  residual, 356
Full admission, 170

G

g_c, 22, **137**
Gaging, 162, 174, 175
Gaging correction, 162
Gas, perfect, 24
- real, 24
- thermodynamic relationships for, 24
- work, 27

Gas constant, 24, 30, 76
Gas density, 30
Gas pressure, radial variation in, 205
Gas turbines, for aircraft use, 270-284
- Armstrong-Siddeley Mamba, 282–283
- bearings and lubrication, 21, 276, 277
- blade speed, 321
- capacity, altitude variation, 298
- control, 123–125
- development, 11
- exhaust, 18
- fuel consumption, 269–274, 284, 318, 321
- General Electric turbojet, 21-22, 276-279
- governor, 319
- historical, 1–2, 11
- Houdry process unit, 290–291
- igniter, 279
- inlet temperature, effect on, 18
- internal-combustion, 70
- jet units, 276
- liquid fuels for, 328–359
- material, 321
  - used, 467–469
- Pratt and Whitney, 279-282
- pumping-service gas line, 321–323
- for reciprocating engine, 284–286
- smoke conditions, 16
- solar units, 315–321
- solid fuels for, 370
- starting, 277, 317, 320
- stationary plant, 17
- temperatures, 2, 21, 277, 279, 283

Gasoline, 376
General Electric Company, 21, 276-279
Ghost engine, 166, 276-277, 374-375
- De Havilland, 272-276

Glands, 103, 150
- losses in,

Goblin turbine, 166
Gyroscopic effects, 406–409

H

Heat, specific, air, 26, 36–62
- charts, 26, 28
- hydrogen, 28
- instantaneous, 25
- mean, 25
- table of gases, 30
- water-vapor, 28

Heat-exchange equipment, 309
Heat-treatment, metals, 450, 453
Heating value, fuels, 356–358, 360
Horsepower stage, 201
Hub stresses, 160
Huey Station, 17
Hydrocarbon, properties, 338–341
- structural formulas, 336
- tests, 351–353
- viscosity-temperature chart, 350

Hydrogen-carbon ratio, 328

I

Igniter, 277, 317, 369–371, 375
Impeller, 9, 223–225, 228, 272
- construction and design, 233
- double-inlet, 231
- Ghost, 228–229, 272
- power, 232

Impulse-blade profile construction, 179
Impulse-blading efficiency, 143
Impulse turbines, 132, 142
- velocity-compounded, 132, 133, 145, 154

Inducer, 228, 229, 231–232
Intercooler, 299, 309, 311–312
Intercooling, 111
Internal energy, 25
Interstage bleed, 188
Isentropic enthalpy drop, 149–151, 168
Isentropic expansion, 29
- change, 101

J

Jet engines, 19-21, 276-282
- afterburning, 277, 279
- Allison, 283
- axial versus centrifugal, 270–272
- axial-flow velocity, 261
- combustion chambers, 274, 276, 279, 366-375, 465

Jet engines, comparison of types, 263–264
  control system, 286–287
  De Havilland Ghost engine, 272-275
  double Mamba, 282–283
  economy of, 276, 281-283
  efficiency, over-all, 265–269
    propulsion, 257–258
    thermal, 265
  exit-loss horsepower, 258–259
  frontal area, 254, 272
  fuel-air ratio, 257
  fuel rate, 254
  General Electric, 20-21, 276-279
  igniters, 277, 317, 369–371, 375
  layout of plant, 254
  length, 21, 279
  nomenclature, 254
  power, 257, 264–265
  Pratt and Whitney, 279-282
  propeller- (*see* Propeller-jet engines)
  range of flight, 272
  specific fuel consumption, 21, 270, 276-277, 281, 282, 283
  starting, 277
  thrust, 254–256, 278
    temperature and, 270
  weight, 279, 283
  weight-thrust ratio, 21, 279
  (*See also* Engines)
Joule's cycle, 92

K

k, 25
k_b, values, 142
Keenan and Kaye, 60
  gas tables, 68–69
Kenthalpy, 29, 74
Kerosene, 377–378

L

Leaving velocity, 151
Lift coefficient, 208, 209, 211
Lift force, 204
Limit of stability, 188
Locomotives, Allis Chalmers, 300–303
  Brown-Boveri, 307
  cleaning, 306
  diesel, 16
Locomotives, Elliott, 14
  exhaust temperature, 296
  fuel rate, 297
  gas-turbine, 13, 15
  General Electric, 294–297
  governing, 302
  railway, 1, 133
  requirements, 293–295
  starting, 297, 303, 305
  Westinghouse, 303–306
Lost-wax process, 457

M

Mach number, 79, 156, 200, 204
  centrifugal compressor, 237
Marine power plant, 165
Marine units, 314–315
  Annapolis test turbine, 290–293
Mass, 23
  flow rate, 87, 168, 171
Materials, 382
  alloying elements, 449
  ceramics and ceramals, 469–470
  factors for selection, 441–446
  failure, 443–446
  fatigue strength, 446–448
  heat-resistant alloys, tables, 451–457
  heat-treatments for, 450
  high-temperature requirements, 442
  manufacture of blades, 453, 457, 461, 463
  metals in turbines, 451–462
  processing, 448–463
  yield strength, 446
Mechanical efficiency, compressor, 239
Metallurgy, 441–474
  alloying elements, 449
  heat-resistant alloys, 451–457
  heat-treating, 450, 453
  powder, 462–463
  table of metals used, 451–462
Momentum changes, 86–88, 407–409
Moving stream, pressure rise, 80
  temperature, 80, 84

N

Naphthenes, 339–341
Natural-gas pumping, 321–323
Newton's second law, 23, 86, 137

Newton's third law, 253
Nitargon, 30
Nitrogen, 26, 30
  atmospheric, 334
Noise, 437, 474
Nozzle relationships, 73–77, 170–171
  acoustic (sonic) velocity, 77
  converging passage, 76
  divergent, 76
  efficiencies, 74
  pressure ratio, 77
  supersonic, 74
  throat, 77
Nozzles, 132, 135
  angles, 173
  number of, 171
  partitions, 173

O

Olefins, 337–339, 341
Open cycle (*see* Cycle)
Otto cycle, 2, 71

P

Paraffins, 335–339
Parsons, John, 11
Petroleum, aromatics in, 340–341
  compounds, 335–342
  crude oil, 342–343
  fuels, 328, 353–359
Pilot fuel, 8
Piping, double-wall, 292
Pound mole, 25
Power equation, 25
Power input measurement, 201
Power plants, Allis Chalmers, 290–293
  arrangements, 299
  Boeing unit, 323–324
  Brown-Boveri, 307–308
  gas-pumping service, 321–323
  General Electric, 298–299, 321–322
  installation costs, 298
  Solar unit, 316-320
  units, 290–315
  waste-heat boiler, 299, 312
  Westinghouse, 307
Precombustion chamber, 6
Pressure coefficient, 200
  centrifugal compressor, 236–237
Pressure compounding, 132
Pressure-expansion ratio, 97
Pressure factor, 199
Pressure loss, 97, 98, 108, 169
  combustor, 97
  real system, 97
  regenerator, 98
Pressure ratio, 13, 106, 111, 115, 236–237
  compression, 15
  relative, 33
  stagnation, 78–79
  static, 78–79
  total, 78, 79
Profile friction, 156
Profile loss, 156, 157, 204
Propeller-jet engines, 253, 268, 282-285
  comparison of types, 263
  division of power, 286
  equivalent horsepower, 268–269, 271
  total thrust, 268
Propeller-tip speeds, 4
Propeller-type airplane, 253, 263, 282-285
Propulsion, aircraft, 253–287
    efficiency, 257–262
    rocket, 261
    systems, 262–265

R

°R, degrees Rankine, 24
Radial equilibrium, 169, 205, 206, 208
Railway locomotive, 1, 133
Ram effect, 4
  air, 9
  efficiency defined, 80
Rateau stage, 156, 158–159
Rateau turbine, 133, 134, 147, 170
Reaction-blade shapes, 159
Reaction blading, 148
  equations, 151
Reaction stage, 158
Reaction turbine (*see* Turbines)
Reciprocating engines, 1, 2, 70, 253, 284–286
Reduction gearing, 4
Regenerative heating, effectiveness and benefits of, 111
Regenerator, 14, 291, 300, 309–311
  effectiveness, 99, 100, 309
  heat transfer in, 310
Reheat, 151
Reheat factor, 103

Relative-pressure development, 66–67
Relative velocity, 136, 191
Residual velocity loss, 191
Resonance, 418–419, 425, 427–428
Reversible-adiabatic process, 27, 29
Reynolds number, 204
Rocket, 253, 261, 264–265
  efficiency of, 262
Rotor, energy transfer, 146
  power delivered to, 150, 198
  stresses, 392–405

S

St. Denis station, 119
Saybolt viscosimeter, 347
Secondary flow, 156, 161
Shaft, critical speed, 435
  power, 103
  seals, 163
Slug, 23, 137
Solidity, 175, 204, 215
Solidity values, 205
Sound wave, 76
Spark plugs, dual, 4
Specific gravity, °API table, 345
Specific heat, air, 26
  carbon dioxide, 26
  oxygen, 26
Specific power, 109, 111
Specific volume, 172
Stability limit, 188, 225
Stage, compressor (*see* Compressors)
  Curtis, 158
  horsepower, 201
  Rateau, 156, 158–159
  reaction, 158
  symmetric, 151
Stage efficiency, diagram, 150–151
  enthalpy, 101, 150
Stage pressure factor, 199
Stagger angle, 204
Stall, 187, 188
Standard atmosphere, 31, 32
  altitude, pressure, temperature, 32
Starting motor, 13
Steam power plant, 70
Stellite, 21
Stream tube, 86
Stress, 381–409
  blades, 383–392
  calculations, 381–405
Stress, centrifugal load, 384–387
  disk, 394–395
  drum rotor, 392–393
  elastic behavior, 381
  fatigue, 417
  plastic behavior, 381, 396, 405
  rotor, centrifugal-compressor, 401
    turbine, 392, 402
  steady load, 383
  thermal, 392, 403–404
Supercharger, 4, 9
  centrifugal-type, 5
  turbine-driven, 8
Supercharging, 2, 7
Surge, 204, 225
Surge line, 186
Symmetric stage, 151
Synchronous speed operation, 124

T

Temperature, moving stream, 84
  recovery factor, 85
  rise, 199
  stagnation, 84, 85
  static, 84, 85
  thermometer, 84
  total, 84
Tensile strength, metals, 454–460
Test-performance methods, 305
Thermal efficiency, 16
  jet engine, 265–268
Thermodynamic properties, 23
Thermodynamic sink, 13
Thermodynamics, first law of, 22
Throttling, 29
Thrust, 254
  equivalent, 268–269
  gross, 255–256
  horsepower, 256–258
  net, 256
Tip-clearance area, 156
Tip leakage, 142, 149
Ts diagram, areas on, 92
Turbines, 132–183
  aircraft, 170
  Curtis, 132, 145, 147, 153
  De Havilland, 166
  design procedure, 167
  design requirements, 169
  expansion ratio, 169

Turbines, gas (*see* Gas turbines)
gas-flow capacity, 141
governing, 123–124
horsepower, 138, 164
impulse (*see* Impulse turbines)
impulse-type, 132, 164, 170, 178
blade width, 178
material, 133
power, 138
pressure-compounded, 159
Rateau, 133, 134, 147, 170
reaction, 11, 134, 136, 137, 147, 150, 151
diagram stage efficiency, 150
enthalpy drop, 150
speed ratio, 147
velocity triangle, 147
rotor, 132, 134, 135
stage efficiency, 100–101, 149, 150
temperature, 170
type comparison, 153–154
velocity-compounded, 132, 133, 145, 154
blade-speed ratio, 145, 153
Turbosupercharger, 8
Elliott-Buchi, 10
Turning angles, 161, 203
Turning vanes, 135

V

Valve timing, 7
Vanadium, 163
Vanes, turning, 135
Velocity, axial, 172, 199
carry-over, 151
sonic, 76
supersonic, 76
of whirl, 137, 211, 213
Velocity loss, residual, 191
Velocity triangles, 178
Vibration, 415–438
blades, 425–428
combustion chambers, 434
compressor rotor, 430–434
frequency ratio, 415–417
impeller, 429–430
modes, 426, 428, 430–433
resonant frequency, 415
torsional, 436
Viscosity, 344, 346–351
Furol, 349
Saybolt, 347–349
Vitallium, 21
Volume-compression ratios, 6

W

Water injection, compressor, 249–250
Weight, 23
Whirl velocity, 137, 211, 213
Work-done factor, 199
Working medium, air, 13
Wright-Cyclone engine, 4, 284-285

SOME DOVER SCIENCE BOOKS

CHANCE, LUCK AND STATISTICS: THE SCIENCE OF CHANCE,
Horace C. Levinson

Theory of probability and science of statistics in simple, non-technical language. Part I deals with theory of probability, covering odd superstitions in regard to "luck," the meaning of betting odds, the law of mathematical expectation, gambling, and applications in poker, roulette, lotteries, dice, bridge, and other games of chance. Part II discusses the misuse of statistics, the concept of statistical probabilities, normal and skew frequency distributions, and statistics applied to various fields—birth rates, stock speculation, insurance rates, advertising, etc. "Presented in an easy humorous style which I consider the best kind of expository writing," Prof. A. C. Cohen, Industry Quality Control. Enlarged revised edition. Formerly titled *The Science of Chance*. Preface and two new appendices by the author. Index. xiv + 365pp. 5⅜ x 8. Paperbound $2.00

BASIC ELECTRONICS,
prepared by the U.S. Navy Training Publications Center

A thorough and comprehensive manual on the fundamentals of electronics. Written clearly, it is equally useful for self-study or course work for those with a knowledge of the principles of basic electricity. Partial contents: Operating Principles of the Electron Tube; Introduction to Transistors; Power Supplies for Electronic Equipment; Tuned Circuits; Electron-Tube Amplifiers; Audio Power Amplifiers; Oscillators; Transmitters; Transmission Lines; Antennas and Propagation; Introduction to Computers; and related topics. Appendix. Index. Hundreds of illustrations and diagrams. vi + 471pp. 6½ x 9¼. Paperbound $2.75

BASIC THEORY AND APPLICATION OF TRANSISTORS,
prepared by the U.S. Department of the Army

An introductory manual prepared for an army training program. One of the finest available surveys of theory and application of transistor design and operation. Minimal knowledge of physics and theory of electron tubes required. Suitable for textbook use, course supplement, or home study. Chapters: Introduction; fundamental theory of transistors; transistor amplifier fundamentals; parameters, equivalent circuits, and characteristic curves; bias stabilization; transistor analysis and comparison using characteristic curves and charts; audio amplifiers; tuned amplifiers; wide-band amplifiers; oscillators; pulse and switching circuits; modulation, mixing, and demodulation; and additional semiconductor devices. Unabridged, corrected edition. 240 schematic drawings, photographs, wiring diagrams, etc. 2 Appendices. Glossary. Index. 263pp. 6½ x 9¼. Paperbound $1.25

GUIDE TO THE LITERATURE OF MATHEMATICS AND PHYSICS,
N. G. Parke III

Over 5000 entries included under approximately 120 major subject headings of selected most important books, monographs, periodicals, articles in English, plus important works in German, French, Italian, Spanish, Russian (many recently available works). Covers every branch of physics, math, related engineering. Includes author, title, edition, publisher, place, date, number of volumes, number of pages. A 40-page introduction on the basic problems of research and study provides useful information on the organization and use of libraries, the psychology of learning, etc. This reference work will save you hours of time. 2nd revised edition. Indices of authors, subjects, 464pp. 5⅜ x 8. Paperbound $2.75

THE RISE OF THE NEW PHYSICS (formerly THE DECLINE OF MECHANISM), *A. d'Abro*

This authoritative and comprehensive 2-volume exposition is unique in scientific publishing. Written for intelligent readers not familiar with higher mathematics, it is the only thorough explanation in non-technical language of modern mathematical-physical theory. Combining both history and exposition, it ranges from classical Newtonian concepts up through the electronic theories of Dirac and Heisenberg, the statistical mechanics of Fermi, and Einstein's relativity theories. "A must for anyone doing serious study in the physical sciences," *J. of Franklin Inst.* 97 illustrations. 991pp. 2 volumes.

T3, T4 Two volume set, paperbound $5.50

THE STRANGE STORY OF THE QUANTUM, AN ACCOUNT FOR THE GENERAL READER OF THE GROWTH OF IDEAS UNDERLYING OUR PRESENT ATOMIC KNOWLEDGE, *B. Hoffmann*

Presents lucidly and expertly, with barest amount of mathematics, the problems and theories which led to modern quantum physics. Dr. Hoffmann begins with the closing years of the 19th century, when certain trifling discrepancies were noticed, and with illuminating analogies and examples takes you through the brilliant concepts of Planck, Einstein, Pauli, de Broglie, Bohr, Schroedinger, Heisenberg, Dirac, Sommerfeld, Feynman, etc. This edition includes a new, long postscript carrying the story through 1958. "Of the books attempting an account of the history and contents of our modern atomic physics which have come to my attention, this is the best," H. Margenau, Yale University, in *American Journal of Physics*. 32 tables and line illustrations. Index. 275pp. 5⅜ x 8.

T518 Paperbound $2.00

GREAT IDEAS AND THEORIES OF MODERN COSMOLOGY, *Jagjit Singh*

The theories of Jeans, Eddington, Milne, Kant, Bondi, Gold, Newton, Einstein, Gamow, Hoyle, Dirac, Kuiper, Hubble, Weizsäcker and many others on such cosmological questions as the origin of the universe, space and time, planet formation, "continuous creation," the birth, life, and death of the stars, the origin of the galaxies, etc. By the author of the popular *Great Ideas of Modern Mathematics*. A gifted popularizer of science, he makes the most difficult abstractions crystal-clear even to the most non-mathematical reader. Index. xii + 276pp. 5⅜ x 8½ T925 Paperbound $2.00

GREAT IDEAS OF MODERN MATHEMATICS: THEIR NATURE AND USE, *Jagjit Singh*

Reader with only high school math will understand main mathematical ideas of modern physics, astronomy, genetics, psychology, evolution, etc., better than many who use them as tools, but comprehend little of their basic structure. Author uses his wide knowledge of non-mathematical fields in brilliant exposition of differential equations, matrices, group theory, logic, statistics, problems of mathematical foundations, imaginary numbers, vectors, etc. Original publications, appendices. indexes. 65 illustr. 322pp. 5⅜ x 8. T587 Paperbound $2.00

THE MATHEMATICS OF GREAT AMATEURS, *Julian L. Coolidge*

Great discoveries made by poets, theologians, philosophers, artists and other non-mathematicians: Omar Khayyam, Leonardo da Vinci, Albrecht Dürer, John Napier, Pascal, Diderot, Bolzano, etc. Surprising accounts of what can result from a non-professional preoccupation with the oldest of sciences. 56 figures. viii + 211pp. 5⅜ x 8½. S1009 Paperbound $2.00

COLLEGE ALGEBRA, *H. B. Fine*
Standard college text that gives a systematic and deductive structure to algebra; comprehensive, connected, with emphasis on theory. Discusses the commutative, associative, and distributive laws of number in unusual detail, and goes on with undetermined coefficients, quadratic equations, progressions, logarithms, permutations, probability, power series, and much more. Still most valuable elementary-intermediate text on the science and structure of algebra. Index. 1560 problems, all with answers. x + 631pp. 5⅜ x 8. Paperbound $2.75

HIGHER MATHEMATICS FOR STUDENTS OF CHEMISTRY AND PHYSICS, *J. W. Mellor*
Not abstract, but practical, building its problems out of familiar laboratory material, this covers differential calculus, coordinate, analytical geometry, functions, integral calculus, infinite series, numerical equations, differential equations, Fourier's theorem, probability, theory of errors, calculus of variations, determinants. "If the reader is not familiar with this book, it will repay him to examine it," *Chem. & Engineering News.* 800 problems. 189 figures. Bibliography. xxi + 641pp. 5⅜ x 8. Paperbound $2.50

TRIGONOMETRY REFRESHER FOR TECHNICAL MEN, *A. A. Klaf*
A modern question and answer text on plane and spherical trigonometry. Part I covers plane trigonometry: angles, quadrants, trigonometrical functions, graphical representation, interpolation, equations, logarithms, solution of triangles, slide rules, etc. Part II discusses applications to navigation, surveying, elasticity, architecture, and engineering. Small angles, periodic functions, vectors, polar coordinates, De Moivre's theorem, fully covered. Part III is devoted to spherical trigonometry and the solution of spherical triangles, with applications to terrestrial and astronomical problems. Special time-savers for numerical calculation. 913 questions answered for you! 1738 problems; answers to odd numbers. 494 figures. 14 pages of functions, formulae. Index. x + 629pp. 5⅜ x 8. Paperbound $2.00

CALCULUS REFRESHER FOR TECHNICAL MEN, *A. A. Klaf*
Not an ordinary textbook but a unique refresher for engineers, technicians, and students. An examination of the most important aspects of differential and integral calculus by means of 756 key questions. Part I covers simple differential calculus: constants, variables, functions, increments, derivatives, logarithms, curvature, etc. Part II treats fundamental concepts of integration: inspection, substitution, transformation, reduction, areas and volumes, mean value, successive and partial integration, double and triple integration. Stresses practical aspects! A 50 page section gives applications to civil and nautical engineering, electricity, stress and strain, elasticity, industrial engineering, and similar fields. 756 questions answered. 556 problems; solutions to odd numbers. 36 pages of constants, formulae. Index. v + 431pp. 5⅜ x 8. Paperbound $2.00

INTRODUCTION TO THE THEORY OF GROUPS OF FINITE ORDER, *R. Carmichael*
Examines fundamental theorems and their application. Beginning with sets, systems, permutations, etc., it progresses in easy stages through important types of groups: Abelian, prime power, permutation, etc. Except 1 chapter where matrices are desirable, no higher math needed. 783 exercises, problems. Index. xvi + 447pp. 5⅜ x 8. Paperbound $3.00

FIVE VOLUME "THEORY OF FUNCTIONS" SET BY KONRAD KNOPP

This five-volume set, prepared by Konrad Knopp, provides a complete and readily followed account of theory of functions. Proofs are given concisely, yet without sacrifice of completeness or rigor. These volumes are used as texts by such universities as M.I.T., University of Chicago, N. Y. City College, and many others. "Excellent introduction . . . remarkably readable, concise, clear, rigorous," *Journal of the American Statistical Association.*

ELEMENTS OF THE THEORY OF FUNCTIONS,
Konrad Knopp
This book provides the student with background for further volumes in this set, or texts on a similar level. Partial contents: foundations, system of complex numbers and the Gaussian plane of numbers, Riemann sphere of numbers, mapping by linear functions, normal forms, the logarithm, the cyclometric functions and binomial series. "Not only for the young student, but also for the student who knows all about what is in it," *Mathematical Journal.* Bibliography. Index. 140pp. 5⅜ x 8. Paperbound $1.50

THEORY OF FUNCTIONS, PART I,
Konrad Knopp
With volume II, this book provides coverage of basic concepts and theorems. Partial contents: numbers and points, functions of a complex variable, integral of a continuous function, Cauchy's integral theorem, Cauchy's integral formulae, series with variable terms, expansion of analytic functions in power series, analytic continuation and complete definition of analytic functions, entire transcendental functions, Laurent expansion, types of singularities. Bibliography. Index. vii + 146pp. 5⅜ x 8. Paperbound $1.35

THEORY OF FUNCTIONS, PART II,
Konrad Knopp
Application and further development of general theory, special topics. Single valued functions. Entire, Weierstrass, Meromorphic functions. Riemann surfaces. Algebraic functions. Analytical configuration, Riemann surface. Bibliography. Index. x + 150pp. 5⅜ x 8. Paperbound $1.35

PROBLEM BOOK IN THE THEORY OF FUNCTIONS, VOLUME 1.
Konrad Knopp
Problems in elementary theory, for use with Knopp's *Theory of Functions,* or any other text, arranged according to increasing difficulty. Fundamental concepts, sequences of numbers and infinite series, complex variable, integral theorems, development in series, conformal mapping. 182 problems. Answers. viii + 126pp. 5⅜ x 8. Paperbound $1.35

PROBLEM BOOK IN THE THEORY OF FUNCTIONS, VOLUME 2,
Konrad Knopp
Advanced theory of functions, to be used either with Knopp's *Theory of Functions,* or any other comparable text. Singularities, entire & meromorphic functions, periodic, analytic, continuation, multiple-valued functions, Riemann surfaces, conformal mapping. Includes a section of additional elementary problems. "The difficult task of selecting from the immense material of the modern theory of functions the problems just within the reach of the beginner is here masterfully accomplished," *Am. Math. Soc.* Answers. 138pp. 5⅜ x 8.
Paperbound $1.50

NUMERICAL SOLUTIONS OF DIFFERENTIAL EQUATIONS,
H. Levy & E. A. Baggott

Comprehensive collection of methods for solving ordinary differential equations of first and higher order. All must pass 2 requirements: easy to grasp and practical, more rapid than school methods. Partial contents: graphical integration of differential equations, graphical methods for detailed solution. Numerical solution. Simultaneous equations and equations of 2nd and higher orders. "Should be in the hands of all in research in applied mathematics, teaching," *Nature*. 21 figures. viii + 238pp. 5⅜ x 8. Paperbound $1.85

ELEMENTARY STATISTICS, WITH APPLICATIONS IN MEDICINE AND THE BIOLOGICAL SCIENCES, *F. E. Croxton*

A sound introduction to statistics for anyone in the physical sciences, assuming no prior acquaintance and requiring only a modest knowledge of math. All basic formulas carefully explained and illustrated; all necessary reference tables included. From basic terms and concepts, the study proceeds to frequency distribution, linear, non-linear, and multiple correlation, skewness, kurtosis, etc. A large section deals with reliability and significance of statistical methods. Containing concrete examples from medicine and biology, this book will prove unusually helpful to workers in those fields who increasingly must evaluate, check, and interpret statistics. Formerly titled "Elementary Statistics with Applications in Medicine." 101 charts. 57 tables. 14 appendices. Index. vi + 376pp. 5⅜ x 8. Paperbound $2.00

INTRODUCTION TO SYMBOLIC LOGIC,
S. Langer

No special knowledge of math required—probably the clearest book ever written on symbolic logic, suitable for the layman, general scientist, and philosopher. You start with simple symbols and advance to a knowledge of the Boole-Schroeder and Russell-Whitehead systems. Forms, logical structure, classes, the calculus of propositions, logic of the syllogism, etc. are all covered. "One of the clearest and simplest introductions," *Mathematics Gazette*. Second enlarged, revised edition. 368pp. 5⅜ x 8. Paperbound $2.00

A SHORT ACCOUNT OF THE HISTORY OF MATHEMATICS,
W. W. R. Ball

Most readable non-technical history of mathematics treats lives, discoveries of every important figure from Egyptian, Phoenician, mathematicians to late 19th century. Discusses schools of Ionia, Pythagoras, Athens, Cyzicus, Alexandria, Byzantium, systems of numeration; primitive arithmetic; Middle Ages, Renaissance, including Arabs, Bacon, Regiomontanus, Tartaglia, Cardan, Stevinus, Galileo, Kepler; modern mathematics of Descartes, Pascal, Wallis, Huygens, Newton, Leibnitz, d'Alembert, Euler, Lambert, Laplace, Legendre, Gauss, Hermite, Weierstrass, scores more. Index. 25 figures. 546pp. 5⅜ x 8. Paperbound $2.25

INTRODUCTION TO NONLINEAR DIFFERENTIAL AND INTEGRAL EQUATIONS,
Harold T. Davis

Aspects of the problem of nonlinear equations, transformations that lead to equations solvable by classical means, results in special cases, and useful generalizations. Thorough, but easily followed by mathematically sophisticated reader who knows little about non-linear equations. 137 problems for student to solve. xv + 566pp. 5⅜ x 8½. Paperbound $2.00

AN INTRODUCTION TO THE GEOMETRY OF N DIMENSIONS,
D. H. Y. Sommerville

An introduction presupposing no prior knowledge of the field, the only book in English devoted exclusively to higher dimensional geometry. Discusses fundamental ideas of incidence, parallelism, perpendicularity, angles between linear space; enumerative geometry; analytical geometry from projective and metric points of view; polytopes; elementary ideas in analysis situs; content of hyper-spacial figures. Bibliography. Index. 60 diagrams. 196pp. 5⅜ x 8.
Paperbound $1.50

ELEMENTARY CONCEPTS OF TOPOLOGY, *P. Alexandroff*

First English translation of the famous brief introduction to topology for the beginner or for the mathematician not undertaking extensive study. This unusually useful intuitive approach deals primarily with the concepts of complex, cycle, and homology, and is wholly consistent with current investigations. Ranges from basic concepts of set-theoretic topology to the concept of Betti groups. "Glowing example of harmony between intuition and thought," David Hilbert. Translated by A. E. Farley. Introduction by D. Hilbert. Index. 25 figures. 73pp. 5⅜ x 8.
Paperbound $1.00

ELEMENTS OF NON-EUCLIDEAN GEOMETRY,
D. M. Y. Sommerville

Unique in proceeding step-by-step, in the manner of traditional geometry. Enables the student with only a good knowledge of high school algebra and geometry to grasp elementary hyperbolic, elliptic, analytic non-Euclidean geometries; space curvature and its philosophical implications; theory of radical axes; homothetic centres and systems of circles; parataxy and parallelism; absolute measure; Gauss' proof of the defect area theorem; geodesic representation; much more, all with exceptional clarity. 126 problems at chapter endings provide progressive practice and familiarity. 133 figures. Index. xvi + 274pp. 5⅜ x 8.
Paperbound $2.00

INTRODUCTION TO THE THEORY OF NUMBERS, *L. E. Dickson*

Thorough, comprehensive approach with adequate coverage of classical literature, an introductory volume beginners can follow. Chapters on divisibility, congruences, quadratic residues & reciprocity. Diophantine equations, etc. Full treatment of binary quadratic forms without usual restriction to integral coefficients. Covers infinitude of primes, least residues. Fermat's theorem. Euler's phi function, Legendre's symbol, Gauss's lemma, automorphs, reduced forms, recent theorems of Thue & Siegel, many more. Much material not readily available elsewhere. 239 problems. Index. I figure. viii + 183pp. 5⅜ x 8.
Paperbound $1.75

MATHEMATICAL TABLES AND FORMULAS,
compiled by Robert D. Carmichael and Edwin R. Smith

Valuable collection for students, etc. Contains all tables necessary in college algebra and trigonometry, such as five-place common logarithms, logarithmic sines and tangents of small angles, logarithmic trigonometric functions, natural trigonometric functions, four-place antilogarithms, tables for changing from sexagesimal to circular and from circular to sexagesimal measure of angles, etc. Also many tables and formulas not ordinarily accessible, including powers, roots, and reciprocals, exponential and hyperbolic functions, ten-place logarithms of prime numbers, and formulas and theorems from analytical and elementary geometry and from calculus. Explanatory introduction. viii + 269pp. 5⅜ x 8½.
Paperbound $1.25

A SOURCE BOOK IN MATHEMATICS,
D. E. Smith

Great discoveries in math, from Renaissance to end of 19th century, in English translation. Read announcements by Dedekind, Gauss, Delamain, Pascal, Fermat, Newton, Abel, Lobachevsky, Bolyai, Riemann, De Moivre, Legendre, Laplace, others of discoveries about imaginary numbers, number congruence, slide rule, equations, symbolism, cubic algebraic equations, non-Euclidean forms of geometry, calculus, function theory, quaternions, etc. Succinct selections from 125 different treatises, articles, most unavailable elsewhere in English. Each article preceded by biographical introduction. Vol. I: Fields of Number, Algebra. Index. 32 illus. 338pp. 5⅜ x 8. Vol. II: Fields of Geometry, Probability, Calculus, Functions, Quaternions. 83 illus. 432pp. 5⅜ x 8.

Vol. 1 Paperbound $2.00, Vol. 2 Paperbound $2.00,
The set $4.00

FOUNDATIONS OF PHYSICS,
R. B. Lindsay & H. Margenau

Excellent bridge between semi-popular works & technical treatises. A discussion of methods of physical description, construction of theory; valuable for physicist with elementary calculus who is interested in ideas that give meaning to data, tools of modern physics. Contents include symbolism; mathematical equations; space & time foundations of mechanics; probability; physics & continua; electron theory; special & general relativity; quantum mechanics; causality. "Thorough and yet not overdetailed. Unreservedly recommended," *Nature* (London). Unabridged, corrected edition. List of recommended readings. 35 illustrations. xi + 537pp. 5⅜ x 8. Paperbound $3.00

FUNDAMENTAL FORMULAS OF PHYSICS,
ed. by D. H. Menzel

High useful, full, inexpensive reference and study text, ranging from simple to highly sophisticated operations. Mathematics integrated into text—each chapter stands as short textbook of field represented. Vol. 1: Statistics, Physical Constants, Special Theory of Relativity, Hydrodynamics, Aerodynamics, Boundary Value Problems in Math, Physics, Viscosity, Electromagnetic Theory, etc. Vol. 2: Sound, Acoustics, Geometrical Optics, Electron Optics, High-Energy Phenomena, Magnetism, Biophysics, much more. Index. Total of 800pp. 5⅜ x 8.

Vol. 1 Paperbound $2.25, Vol. 2 Paperbound $2.25,
The set $4.50

THEORETICAL PHYSICS,
A. S. Kompaneyets

One of the very few thorough studies of the subject in this price range. Provides advanced students with a comprehensive theoretical background. Especially strong on recent experimentation and developments in quantum theory. Contents: Mechanics (Generalized Coordinates, Lagrange's Equation, Collision of Particles, etc.), Electrodynamics (Vector Analysis, Maxwell's equations, Transmission of Signals, Theory of Relativity, etc.), Quantum Mechanics (the Inadequacy of Classical Mechanics, the Wave Equation, Motion in a Central Field, Quantum Theory of Radiation, Quantum Theories of Dispersion and Scattering, etc.), and Statistical Physics (Equilibrium Distribution of Molecules in an Ideal Gas, Boltzmann Statistics, Bose and Fermi Distribution. Thermodynamic Quantities, etc.). Revised to 1961. Translated by George Yankovsky, authorized by Kompaneyets. 137 exercises. 56 figures. 529pp. 5⅜ x 8½.

Paperbound $2.50

MATHEMATICAL PHYSICS, *D. H. Menzel*

Thorough one-volume treatment of the mathematical techniques vital for classical mechanics, electromagnetic theory, quantum theory, and relativity. Written by the Harvard Professor of Astrophysics for junior, senior, and graduate courses, it gives clear explanations of all those aspects of function theory, vectors, matrices, dyadics, tensors, partial differential equations, etc., necessary for the understanding of the various physical theories. Electron theory, relativity, and other topics seldom presented appear here in considerable detail. Scores of definition, conversion factors, dimensional constants, etc. "More detailed than normal for an advanced text . . . excellent set of sections on Dyadics, Matrices, and Tensors," *Journal of the Franklin Institute*. Index. 193 problems, with answers. x + 412pp. 5⅜ x 8. Paperbound $2.50

THE THEORY OF SOUND, *Lord Rayleigh*

Most vibrating systems likely to be encountered in practice can be tackled successfully by the methods set forth by the great Nobel laureate, Lord Rayleigh. Complete coverage of experimental, mathematical aspects of sound theory. Partial contents: Harmonic motions, vibrating systems in general, lateral vibrations of bars, curved plates or shells, applications of Laplace's functions to acoustical problems, fluid friction, plane vortex-sheet, vibrations of solid bodies, etc. This is the first inexpensive edition of this great reference and study work. Bibliography, Historical introduction by R. B. Lindsay. Total of 1040pp. 97 figures. 5⅜ x 8. Vol. 1 Paperbound $2.50, Vol. 2 Paperbound $2.50, The set $5.00

HYDRODYNAMICS, *Horace Lamb*

Internationally famous complete coverage of standard reference work on dynamics of liquids & gases. Fundamental theorems, equations, methods, solutions, background, for classical hydrodynamics. Chapters include Equations of Motion, Integration of Equations in Special Gases, Irrotational Motion, Motion of Liquid in 2 Dimensions, Motion of Solids through Liquid-Dynamical Theory, Vortex Motion, Tidal Waves, Surface Waves, Waves of Expansion, Viscosity, Rotating Masses of Liquids. Excellently planned, arranged; clear, lucid presentation. 6th enlarged, revised edition. Index. Over 900 footnotes, mostly bibliographical. 119 figures. xv + 738pp. 6⅛ x 9¼. Paperbound $4.00

DYNAMICAL THEORY OF GASES, *James Jeans*

Divided into mathematical and physical chapters for the convenience of those not expert in mathematics, this volume discusses the mathematical theory of gas in a steady state, thermodynamics, Boltzmann and Maxwell, kinetic theory, quantum theory, exponentials, etc. 4th enlarged edition, with new material on quantum theory, quantum dynamics, etc. Indexes. 28 figures. 444pp. 6⅛ x 9¼. Paperbound $2.75

THERMODYNAMICS, *Enrico Fermi*

Unabridged reproduction of 1937 edition. Elementary in treatment; remarkable for clarity, organization. Requires no knowledge of advanced math beyond calculus, only familiarity with fundamentals of thermometry, calorimetry. Partial Contents: Thermodynamic systems; First & Second laws of thermodynamics; Entropy; Thermodynamic potentials: phase rule, reversible electric cell; Gaseous reactions: van't Hoff reaction box, principle of LeChatelier; Thermodynamics of dilute solutions: osmotic & vapor pressures, boiling & freezing points; Entropy constant. Index. 25 problems. 24 illustrations. x + 160pp. 5⅜ x 8. Paperbound $1.75

APPLIED OPTICS AND OPTICAL DESIGN,
A. E. Conrady

With publication of vol. 2, standard work for designers in optics is now complete for first time. Only work of its kind in English; only detailed work for practical designer and self-taught. Requires, for bulk of work, no math above trig. Step-by-step exposition, from fundamental concepts of geometrical, physical optics, to systematic study, design, of almost all types of optical systems. Vol. 1: all ordinary ray-tracing methods; primary aberrations; necessary higher aberration for design of telescopes, low-power microscopes, photographic equipment. Vol. 2: (Completed from author's notes by R. Kingslake, Dir. Optical Design, Eastman Kodak.) Special attention to high-power microscope, anastigmatic photographic objectives. "An indispensable work," *J., Optical Soc. of Amer.* Index. Bibliography. 193 diagrams. 852pp. 6⅛ x 9¼.
Two volume set, paperbound $7.00

MECHANICS OF THE GYROSCOPE, THE DYNAMICS OF ROTATION,
R. F. Deimel, Professor of Mechanical Engineering at Stevens Institute of Technology

Elementary general treatment of dynamics of rotation, with special application of gyroscopic phenomena. No knowledge of vectors needed. Velocity of a moving curve, acceleration to a point, general equations of motion, gyroscopic horizon, free gyro, motion of discs, the damped gyro, 103 similar topics. Exercises. 75 figures. 208pp. 5⅜ x 8. Paperbound $1.75

STRENGTH OF MATERIALS,
J. P. Den Hartog

Full, clear treatment of elementary material (tension, torsion, bending, compound stresses, deflection of beams, etc.), plus much advanced material on engineering methods of great practical value: full treatment of the Mohr circle, lucid elementary discussions of the theory of the center of shear and the "Myosotis" method of calculating beam deflections, reinforced concrete, plastic deformations, photoelasticity, etc. In all sections, both general principles and concrete applications are given. Index. 186 figures (160 others in problem section). 350 problems, all with answers. List of formulas. viii + 323pp. 5⅜ x 8.
Paperbound $2.00

HYDRAULIC TRANSIENTS,
G. R. Rich

The best text in hydraulics ever printed in English . . . by former Chief Design Engineer for T.V.A. Provides a transition from the basic differential equations of hydraulic transient theory to the arithmetic integration computation required by practicing engineers. Sections cover Water Hammer, Turbine Speed Regulation, Stability of Governing, Water-Hammer Pressures in Pump Discharge Lines, The Differential and Restricted Orifice Surge Tanks, The Normalized Surge Tank Charts of Calame and Gaden, Navigation Locks, Surges in Power Canals—Tidal Harmonics, etc. Revised and enlarged. Author's prefaces. Index. xiv + 409pp. 5⅜ x 8½. Paperbound $2.50

Prices subject to change without notice.

Available at your book dealer or write for free catalogue to Dept. Adsci, Dover Publications, Inc., 180 Varick St., N.Y., N.Y. 10014. Dover publishes more than 150 books each year on science, elementary and advanced mathematics, biology, music, art, literary history, social sciences and other areas.